Student's Solutions Manual

for use with

Beginning and Intermediate Algebra

Julie Miller
Daytona Beach Community College

Molly O'Neill
Daytona Beach Community College

Prepared by
David French
Tidewater Community College
with contributions by Tammy Fisher-Vasta

Boston Burr Ridge, IL Dubuque, IA Madison, WI New York San Francisco St. Louis
Bangkok Bogotá Caracas Kuala Lumpur Lisbon London Madrid Mexico City
Milan Montreal New Delhi Santiago Seoul Singapore Sydney Taipei Toronto

The McGraw-Hill Companies

Student's Solutions Manual for use with
BEGINNING AND INTERMEDIATE ALGEBRA
Julie Miller and Molly O'Neill

Published by McGraw-Hill Higher Education, an imprint of The McGraw-Hill Companies, Inc., 1221 Avenue of the Americas, New York, NY 10020.

This book is printed on recycled, acid-free paper containing 10% postconsumer waste.

2 3 4 5 6 7 8 9 0 QSR/QSR 0 9 8 7 6 5

ISBN 0-07-298492-9

www.mhhe.com

Contents

Chapter R	2
Chapter 1	7
Chapter 2	21
Chapter 3	48
Chapter 4	66
Chapter 5	86
Chapter 6	105
Chapter 7	132
Chapter 8	155
Chapter 9	211
Chapter 10	256
Chapter 11	288
Chapter 12	315
Chapter 13	342
Chapter 14	378
Beginning Algebra Review	399

Chapter R

Section R.1 Practice Exercises

1. Numerator 7; denominator 8; proper

3. Numerator 9; denominator 5; improper

5. Numerator 6; denominator 6; improper

7. Numerator 12; denominator 1; improper

9. $\frac{3}{4}$

11. $\frac{4}{3}$

13. $\frac{1}{6}$

15. $\frac{2}{2}$

17. $\frac{5}{2}$ or $2\frac{1}{2}$

19. $\frac{6}{2}$ or 3

21. The set of whole numbers includes the number 0 and the set of natural numbers does not.

23. Answers may vary. One example would be $\frac{3}{6}$.

25. Prime

27. Composite

29. Composite

31. Prime

33. $2 \times 2 \times 3 \times 3$

35. $2 \times 3 \times 7$

37. $2 \times 5 \times 11$

39. $3 \times 3 \times 3 \times 5$

41. $\frac{3}{15} = \frac{\cancel{3}}{\cancel{3} \times 5} = \frac{1}{5}$

43. $\frac{6}{16} = \frac{\cancel{2} \times 3}{\cancel{2} \times 2 \times 2 \times 2} = \frac{3}{8}$

45. $\frac{42}{48} = \frac{\cancel{2} \times \cancel{3} \times 7}{\cancel{2} \times 2 \times 2 \times 2 \times \cancel{3}} = \frac{7}{8}$

47. $\frac{48}{64} = \frac{\cancel{2} \times \cancel{2} \times \cancel{2} \times \cancel{2} \times 3}{\cancel{2} \times \cancel{2} \times \cancel{2} \times \cancel{2} \times 2 \times 2} = \frac{3}{4}$

49. $\frac{110}{176} = \frac{\cancel{2} \times 5 \times \cancel{11}}{\cancel{2} \times 2 \times 2 \times 2 \times \cancel{11}} = \frac{5}{8}$

51. $\frac{150}{200} = \frac{\cancel{2} \times 3 \times \cancel{5} \times \cancel{5}}{\cancel{2} \times 2 \times 2 \times \cancel{5} \times \cancel{5}} = \frac{3}{4}$

53. False: Only when adding/subtracting fractions is it necessary to have a common denominator.

55. $\frac{11}{9} = \frac{9}{9} + \frac{2}{9} = 1\frac{2}{9}$

57. $\frac{7}{2} = \frac{6}{2} + \frac{1}{2} = 3\frac{1}{2}$

59. $5\frac{2}{5} = \frac{25}{5} + \frac{2}{5} = \frac{27}{5}$

61. $1\frac{7}{8} = \frac{8}{8} + \frac{7}{8} = \frac{15}{8}$

63. $\frac{10}{13} \times \frac{26}{15} = \frac{2 \times 2 \times \cancel{5} \times \cancel{13}}{3 \times \cancel{5} \times \cancel{13}} = \frac{4}{3} = 1\frac{1}{3}$

65. $\frac{3}{7} \div \frac{9}{14} = \frac{3}{7} \times \frac{14}{9} = \frac{2 \times \cancel{3} \times \cancel{7}}{3 \times \cancel{3} \times \cancel{7}} = \frac{2}{3}$

67. $\frac{9}{10} \times 5 = \frac{9}{10} \times \frac{5}{1} = \frac{3 \times 3 \times \cancel{5}}{2 \times \cancel{5}} = \frac{9}{2} = 4\frac{1}{2}$

69. $4\frac{3}{5}\div\frac{1}{10}=\frac{23}{5}\times\frac{10}{1}=\frac{2\times\cancel{5}\times23}{\cancel{5}}=\frac{46}{1}=46$

71. $3\frac{1}{5}\times\frac{7}{8}=\frac{16}{5}\times\frac{7}{8}$
$=\frac{\cancel{2}\times\cancel{2}\times\cancel{2}\times2\times7}{\cancel{2}\times\cancel{2}\times\cancel{2}\times5}$
$=\frac{14}{5}$
$=2\frac{4}{5}$

73. $1\frac{2}{9}\div6=\frac{11}{9}\times\frac{1}{6}=\frac{11}{9\times6}=\frac{11}{54}$

75. $\frac{1}{4}$ of $\$1200=\frac{1}{4}\times\frac{1200}{1}=\frac{1200}{4}=\$300$

77. $\frac{1}{3}$ of $12=\frac{1}{3}\times\frac{12}{1}=\frac{12}{3}=4$, 4 eggs

79. $6\text{ lb}\div\frac{3}{4}\text{ lb}=\frac{6}{1}\times\frac{4}{3}=\frac{24}{3}=8$, 8 jars

81. $4\text{ servings}\div\frac{1}{3}\text{ serving}=\frac{4}{1}\times\frac{3}{1}$
$=\frac{12}{1}$
$=12$ servings

83. $21\text{ tsp}\div1\frac{3}{4}\text{ tsp}=\frac{21}{1}\times\frac{4}{7}$
$=\frac{3\times4\times\cancel{7}}{\cancel{7}}$
$=12$ batches

85. $1\times24=24$

87. $2^3\times5=40$

89. $2\times3^2\times5=90$

91. $\frac{5}{14}+\frac{1}{14}=\frac{6}{14}=\frac{\cancel{2}\times3}{\cancel{2}\times7}=\frac{3}{7}$

93. $\frac{17}{24}-\frac{5}{24}=\frac{12}{24}=\frac{1}{2}$

95. $\frac{1}{8}+\frac{3}{4}=\frac{1}{8}+\frac{6}{8}=\frac{7}{8}$

97. $\frac{3}{8}-\frac{3}{10}=\frac{15}{40}-\frac{12}{40}=\frac{3}{40}$

99. $\frac{7}{26}-\frac{2}{13}=\frac{7}{26}-\frac{4}{26}=\frac{3}{26}$

101. $\frac{7}{18}+\frac{5}{12}-\frac{1}{6}=\frac{14}{36}+\frac{15}{36}-\frac{6}{36}=\frac{23}{36}$

103. $\frac{3}{4}-\frac{1}{20}+\frac{2}{5}=\frac{15}{20}-\frac{1}{20}+\frac{8}{20}$
$=\frac{22}{20}$
$=\frac{11}{10}$
$=1\frac{1}{10}$

105. $\frac{5}{12}+\frac{5}{16}-\frac{1}{3}=\frac{20}{48}+\frac{15}{48}-\frac{16}{48}=\frac{19}{48}$

107. $2\frac{1}{8}+1\frac{3}{8}=\frac{17}{8}+\frac{11}{8}$
$=\frac{28}{8}$
$=\frac{\cancel{2}\times\cancel{2}\times7}{\cancel{2}\times\cancel{2}\times2}$
$=\frac{7}{2}$
$=3\frac{1}{2}$

109. $1\frac{5}{6}-\frac{7}{8}=\frac{11}{6}-\frac{7}{8}=\frac{44}{24}-\frac{21}{24}=\frac{23}{24}$

111. $1\frac{1}{6}+3\frac{3}{4}=\frac{7}{6}+\frac{15}{4}$
$=\frac{14}{12}+\frac{45}{12}$
$=\frac{59}{12}$
$=4\frac{11}{12}$

113. $1-\frac{7}{8}=\frac{8}{8}-\frac{7}{8}=\frac{1}{8}$

115. $3\frac{5}{6}\text{ ft}+1\frac{3}{4}\text{ ft}=\frac{23}{6}+\frac{7}{4}$
$=\frac{46}{12}+\frac{21}{12}$
$=\frac{67}{12}$
$=5\frac{7}{12}\text{ ft}$

117. $\frac{1}{2}+\frac{1}{3}+\frac{3}{4}=\frac{6}{12}+\frac{4}{12}+\frac{9}{12}=\frac{19}{12}=1\frac{7}{12}$ cups

119. $5\frac{1}{2}-4\frac{5}{6}=\frac{11}{2}-\frac{29}{6}=\frac{33}{6}-\frac{29}{6}=\frac{4}{6}=\frac{2}{3}$ miles

121. $2\frac{3}{4}-1\frac{1}{2}=\frac{11}{4}-\frac{3}{2}=\frac{11}{4}-\frac{6}{4}=\frac{5}{4}=1\frac{1}{4}$ hours

123. $48\frac{3}{4}+40\frac{1}{4}+50\frac{1}{2}+70+35$
$=\frac{195}{4}+\frac{161}{4}+\frac{202}{4}+105$
$=\frac{558}{4}+\frac{420}{4}$
$=\frac{978}{4}$
$=244\frac{1}{2}$ yards

125. $\frac{7}{8}+\frac{16}{15}\approx 1+1=2$

127. $\frac{21}{4}+\frac{98}{100}+\frac{80}{41}\approx 5+1+2=8$

Section R.2 Practice Exercises

1. Only b, e, and i.

3. $P=2w+2l$
$=2(32)+2(22)$
$=64\text{ cm}+44\text{ cm}$
$=108\text{ cm}$

$A=lw$
$=(32)(22)$
$=704\text{ cm}^2$

5. $P=4s$
$=4(0.25)$
$=1\text{ mile}$

$A=s^2$
$=(0.25)^2$
$=0.0625\text{ ft}^2$

7. $P=2\frac{1}{3}+5\frac{1}{6}+4$
$=\frac{14}{6}+\frac{31}{6}+\frac{24}{6}$
$=\frac{69}{6}$
$=\frac{23}{2}$
$=11\frac{1}{2}$ in.

9. $C=2\pi r=2(3.14)(5)=31.4$ ft

11. a, f, g

13. $A=bh=(0.04)(0.01)=0.0004\text{ m}^2$

15. $A=\frac{1}{2}bh=0.5(16)(5)=40$ square miles

17. $A=\pi r^2=(3.14)(6.5)^2=132.665\text{ cm}^2$

19. $A=\frac{1}{2}(b_1+b_2)h=0.5(14+8)6=66\text{ in}^2$

21. $A=\frac{1}{2}bh=0.5(4)(3)=6\text{ km}^2$

23. $V=\pi r^2h=(3.14)(2)^2(6)=75.36\text{ ft}^3$

25. $V=lwh=(6.5)(1.5)(4)=39\text{ in}^3$

27. $V=\frac{4}{3}\pi r^3=\frac{4}{3}(3.14)(9)^3=3052.08\text{ in}^3$

29. $V=\frac{1}{3}\pi r^2h=\frac{1}{3}(3.14)(3)^2(12)=113.04\text{ cm}^3$

31. $V=s^3=(3.2)^3=32.768\text{ ft}^3$

33. $P=20+2(16)+2(5)+20=82$ ft

35. $\begin{pmatrix}\text{Area of}\\ \text{outer square}\end{pmatrix} - \begin{pmatrix}\text{Area of}\\ \text{inner square}\end{pmatrix} = 10^2 - 8^2$
$= 100 - 64$
$= 36 \text{ in}^2$

37. $\begin{pmatrix}\text{Area of outer}\\ \text{rectangle}\end{pmatrix} - \begin{pmatrix}\text{Area of inner}\\ \text{circle}\end{pmatrix}$
$= (6.2)(4.1) - (3.14)(1.8)^2$
$= 15.2464 \text{ cm}^2$

39. The wall is $(20)(80) = 1600 \text{ ft}^2$.

(a) $1600 \text{ ft}^2 \div \$50 = \$0.31/\text{ft}^2$

(b) The areas are $(20)(8) = 160 \text{ ft}^2$, $(16)(8) = 128 \text{ ft}^2$, and $(16)(8) = 128 \text{ ft}^2$. The total area is $160 + 128 + 128 = 416 \text{ ft}^2$. The total price is 416(0.31) = $129.

41. Perimeter

43. (a) $A = \pi r^2 = (3.14)(4) = 50.24 \text{ in}^2$.

(b) $A = \pi r^2 = (3.14)(6) = 113.04 \text{ in}^2$.

(c) Two 8-inch pizzas have $2(50.24) = 100.48 \text{ in}^2$. One 12-inch pizza has 113.04 in^2. Therefore one 12-inch pizza has more.

45. $V = \pi r^2 h = (3.14)(3.2)^2(9) = 289.3824 \text{ cm}^3$

47. True

49. True

51. True

53. True

55. 45°; 45° + 45° = 90°

57. Not possible; acute angles are less than 90°.

59. Answers vary. One example: 100°, 80°

61. (a) $\angle 1, \angle 3$ and $\angle 2, \angle 4$

(b) $\angle 1, \angle 2$; $\angle 2, \angle 3$; $\angle 3, \angle 4$; $\angle 1, \angle 4$

(c) Since $\angle 1$ and $\angle 4$ are supplementary, and $\angle 1 + \angle 4 = 180°$, $\angle 1$ must equal 100°. It follows that $\angle 2 = 80°$ and $\angle 3 = 100°$.

63. 90° – 33° = 57

65. 90° – 12° = 78°

67. 90° – 30° = 60°

69. 90° – 70° = 20°

71. 180° – 33° = 147°

73. 180° – 122° = 58°

75. 180° – 45° = 135°

77. 180° – 135° = 45°

79. $\angle 7$

81. $\angle 1$

83. $\angle 1$

85. $\angle 5$

87. $\angle a = 45°$ (supplementary angles); $\angle b = 135°$ (vertical angles are equal); $\angle c = 45°$; $\angle d = 135°$ (Corresponding angles are equal); $\angle e = 45°$; $\angle f = 135°$; $\angle g = 45°$

89. Scalene (no equal sides

91. Isosceles (Two sides equal)

93. No; a 90° angle plus an angle greater than 90° would make the sum of the angles greater than 180°.

95. 40°; the sum of angles of a triangle is 180°.

97. 37°; the angles of a right triangle are complementary.

99. $\angle a = 80°$ (The sum of angles of a triangle is 180°); $\angle b = 80°$ (Vertical angles are equal); $\angle c = 100°$ (Supplementary angles); $\angle d = 100°$ $\angle e = 65°$; $\angle f = 115°$; $\angle g = 115°$; $\angle h = 35°$; $\angle i = 145°$; $\angle j = 145°$

101. $\angle a = 70°$; $\angle b = 65°$; $\angle c = 65°$; $\angle d = 110°$; $\angle e = 70°$; $\angle f = 110°$; $\angle g = 115°$; $\angle h = 115°$; $\angle i = 65°$; $\angle j = 70°$; $\angle k = 65°$

Chapter 1

Section 1.1 Practice Exercises

1. 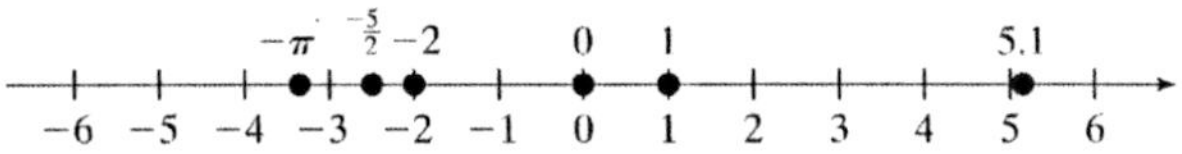

3. Terminating decimal

5. Repeating decimal

7. Terminating decimal

9. Terminating decimal

11. Non-terminating, non-repeating decimal

13. Terminating decimal

15. Terminating decimal

17. Terminating decimal

19. Repeating decimal

21. Non-terminating, non-repeating decimal

23. $0.29, 3.8, \frac{1}{9}, \frac{1}{3}, \frac{1}{8}, \frac{1}{5}, 5, 2, -0.125, -3.24, -3, -6, \frac{7}{20}, \frac{5}{8}, 0.\overline{2}, 0.\overline{6}$

25. The counting numbers, $\{1, 2, 3, \ldots\}$

27. The set of all real numbers that are not rational.

29. The set of all rational and irrational numbers.

31. Answers vary; π, $\sqrt{2}$, 3π

33. Answers vary; $-2, 0, -5$

35. Answers vary; $-1, \frac{1}{2}, 0$

37. Yes

39. $\frac{0}{5}, 1$

41. $\sqrt{11}, \sqrt{7}$

43. 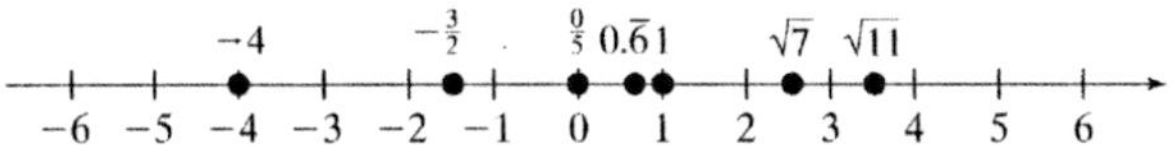

45.

Long Beac, Philadelp, Dallas, Kansas Ci, Denver

−7 0 390 720 5130

(a) >

(b) <

(c) <

(d) <

47.

Webb, Fukushim, Mallon, Kane, Inkster

−7 −6 −3 0 2

(a) <

(b) >

(c) >

(d) >

49. Opposite –2, absolute value 2

51. Opposite 2.5, absolute value 2.5

53. Opposite $\frac{1}{3}$, absolute value $\frac{1}{3}$

55. Opposite $-\frac{1}{9}$, absolute value $\frac{1}{9}$

57. 5.1

59. 7

61. False; $|m|$ is never negative.

63. True; 8 is to the left of 10.

65. False; 19 is equal to 19.

67. True; −1 is equal to −1.

69. False; $-\frac{1}{5}$ is to the left of 0.

71. False; 6 is to the right of −10.

73. False; 10 is equal to 10.

75. False; 3 is to the right of 1.

77. True; $\frac{1}{3}$ is equal to $\frac{1}{3}$.

79. False; 13 is equal to 13.

81. True; −6 is to the left of 6.

83. True; 11 is equal to 11.

85. True; 21 is equal to 21.

87. For all $a \geq 0$, all positive real numbers.

Section 1.2 Practice Exercises

1. $-\sqrt{9}$ -2 0 7

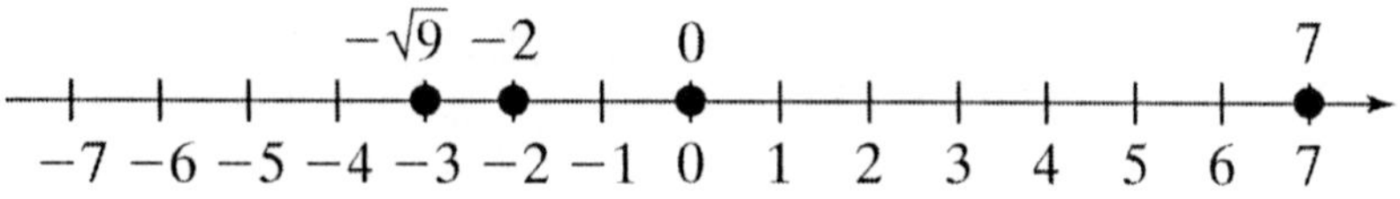

−7 −6 −5 −4 −3 −2 −1 0 1 2 3 4 5 6 7

3. $\sqrt{5}$ π

−7 −6 −5 −4 −3 −2 −1 0 1 2 3 4 5 6 7

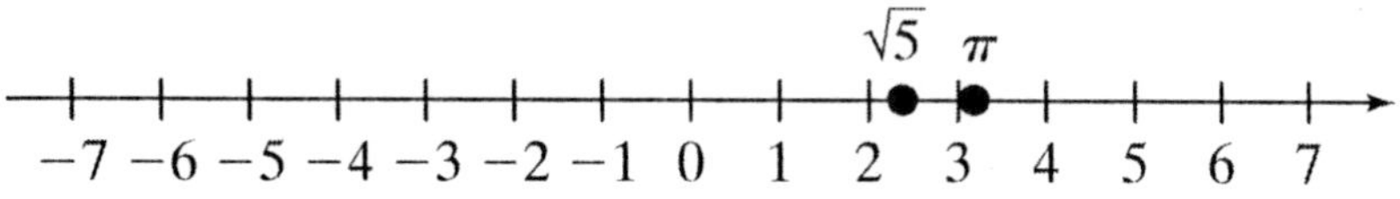

5. **(a)** <

(b) >

(c) >

(d) <

7. $\frac{1}{6}\cdot\frac{1}{6}\cdot\frac{1}{6}\cdot\frac{1}{6}=\left(\frac{1}{6}\right)^4$

9. $a\cdot a\cdot a\cdot b\cdot b=a^3b^2$

11. $(5c)^5$

13. $8yx^6$

15. $x^3 = x \cdot x \cdot x$

17. $(2b^3) = 2b \cdot 2b \cdot 2b$

19. $10y^5 = 10 \cdot y \cdot y \cdot y \cdot y \cdot y$

21. $2wz^2 = 2 \cdot w \cdot z \cdot z$

23. $5^2 = 5 \cdot 5 = 25$

25. $\left(\frac{1}{7}\right)^2 = \frac{1}{7} \cdot \frac{1}{7} = \frac{1}{49}$

27. $(0.25)^3 = 0.25 \cdot 0.25 \cdot 0.25 = 0.015625$

29. $2^6 = 2 \cdot 2 \cdot 2 \cdot 2 \cdot 2 \cdot 2 = 64$

31. $\sqrt{81} = 9$

33. $\sqrt{4} = 2$

35. $\sqrt{100} = 10$

37. $\sqrt{16} = 4$

39. $8 + 2 \cdot 6 = 8 + 12 = 20$

41. $(8 + 2)6 = 10 \cdot 6 = 60$

43. $4 + 2 \div 2 \cdot 3 + 1 = 4 + 3 + 1 = 8$

45. $\frac{1}{4} \cdot \frac{2}{3} - \frac{1}{6} = \frac{1}{6} - \frac{1}{6} = 0$

47. $\frac{9}{8} - \frac{1}{3} \cdot \frac{3}{4} = \frac{9}{8} - \frac{1}{4} = \frac{9}{8} - \frac{2}{8} = \frac{7}{8}$

49. $3[5 + 2(8 - 3)] = 3[5 + 2(5)] = 3[15] = 45$

51. $10 + |-6| = 10 + 6 = 16$

53. $21 - |8 - 2| = 21 - 6 = 15$

55. $2^2 + \sqrt{9} \cdot 5 = 4 + 15 = 19$

57. $\sqrt{9 + 16} - 2 = \sqrt{25} - 2 = 5 - 2 = 3$

59. $\frac{7 + 3(8 - 2)}{(7 + 3)(8 - 2)} = \frac{7 + 18}{(10)(6)} = \frac{25}{60} = \frac{5}{12}$

61. $\frac{15 - 5(3 \cdot 2 - 4)}{10 - 2(4 \cdot 5 - 16)} = \frac{15 - 5(2)}{10 - 2(4)} = \frac{5}{2}$

63. $[4^2 \cdot (6 - 4) \div 8] + [7 \cdot (8 - 3)]$
$= [16 \cdot 2 \div 8] + [7 \cdot 5]$
$= 4 + 35$
$= 39$

65. $48 - 13 \cdot 3 + [(50 - 7 \cdot 5) + 2]$
$= 48 - 39 + [15 + 2]$
$= 26$

67. $y - 3 = 18 - 3 = 15$

69. $\frac{15}{t} = \frac{15}{5} = 3$

71. $2(c + 1) - 5 = 2(4 + 1) - 5 = 10 - 5 = 2$

73. $5 + 6d = 5 + 6 \cdot \frac{2}{3} = 5 + 4 = 9$

75. $p^2 + \frac{2}{9} = \left(\frac{2}{3}\right)^2 + \frac{2}{9} = \frac{4}{9} + \frac{2}{9} = \frac{6}{9} = \frac{2}{3}$

77. $5(x + 2.3) = 5(1.1 + 2.3) = 5 \cdot 3.4 = 17$

79. $A = lw = (160 \text{ ft})(360 \text{ ft}) = 57{,}600 \text{ ft}^2$

81. $A = \frac{1}{2}(b_1 + b_2)h = \frac{1}{2}(8 + 6)3 = \frac{1}{2} \cdot 42 = 21 \text{ ft}^2$

83. **(a)** x

(b) Yes, 5 has an exponent of 1 $(5^1 = 5)$

85. $3x$

87. $\frac{x}{7}$ or $x \div 7$

89. $2 - a$

91. $2y + x$

93. $4(x + 12)$

95. $21 - 2x$

97. $t - 14$

99. The sum of 5 and r

101. The difference of s and 14

103. The quotient of 5 and the product of 2 and p

105. One more than the product of 7 and x

107. 5, squared

109. The square root of 5

111. 7, cubed

113. The sum of 2 and the square of x

115. The sum of 3 and the square root of r

117. **(a)** Always perform multiplication and/or division from left to right.

(b) Always perform addition and/or subtraction from left to right.

119. With multiplication and/or division, always work from left to right. The same is true for addition and/or subtraction.

121. $\dfrac{5-\sqrt{9}}{\sqrt{\frac{4}{9}}+\frac{1}{3}} = \dfrac{5-3}{\frac{2}{3}+\frac{1}{3}} = \dfrac{2}{1} = 2$

123. $\dfrac{|-4|^2}{2^2+\sqrt{144}} = \dfrac{16}{4+12} = \dfrac{16}{16} = 1$

125.

```
(4+6)/(8-3)
                 2
110-5*(2+1)-4
                91
100-2*(5-3)^3
                84
```

127–129.

```
3+(4-1)²
                12
(12-6+1)²
                49
3*8-√(32+2²)
                18
```

131.

```
√(18-2)
                 4
(4*3-3*3)^3
                27
(20-3²)/(26-2²)
                .5
```

Section 1.3 Practice Exercises

1. Rational

3. Rational

5. Irrational

7. Rational

9. >

11. >

13. >

15. $6 + (-3) = 3$

17. $2 + (-5) = -3$

19. $-19 + 2 = -17$

21. $-4 + 11 = 7$

23. $-16 + (-3) = -19$

25. $-2 + (-21) = -23$

27. $0 + (-5) = -5$

29. $-3 + 0 = -3$

31. $-16 + 16 = 0$

33. $41 + (-41) = 0$

35. $4 + (-9) = -5$

37. $7 + (-2) + (-8) = -3$

39. $-17 + (-3) + 20 = -20 + 20 = 0$

41. $-3 + (-8) + (-12) = -11 + (-12) = -23$

43. $-42 + (-3) + 45 + (-6) = -45 + 39 = -6$

45. $-5 + (-3) + (-7) + 4 + 8 = -8 + (-3) + 8 = -3$

47. $-5 + 13 + (-11)$, $-3°$

49. $3 + (-5) + 14$, 12-yd gain

51. Subtract the smaller absolute value from the larger, then keep the sign of the number with the larger absolute value.

53. 21.3

55. $-\frac{2}{7}+\frac{1}{14}=-\frac{4}{14}+\frac{1}{14}=-\frac{3}{14}$

57. $-2.1+\left(-\frac{3}{10}\right)=-2.1+-0.3=-2.4$ or $-\frac{12}{5}$

59. $\frac{3}{4}+(-0.5)=0.75+(-0.5)=0.25$ or $\frac{1}{4}$

61. $8.23 + (-8.23) = 0$

63. $-\frac{7}{8}+0=-\frac{7}{8}$

65. $-\frac{2}{3}+\left(-\frac{1}{9}\right)+2=-\frac{6}{9}+\left(-\frac{1}{9}\right)+\frac{18}{9}=\frac{11}{9}$

67. –23.08

69. 494.686

71. –0.002117

73. **(a)** $52.23 + (-52.95) = \$0.72$

(b) Yes

75. **(a)** $100 + 200 + (-500) + 300 + 100 + (-200)$

(b) \$0

77. $x+y+\sqrt{z}=-3+(-2)+\sqrt{16}=-5+4=-1$

79. $y+3\sqrt{z}=-2+3\sqrt{16}$
$=-2+3\cdot 4$
$=-2+12$
$=10$

81. $|x|+|y|=|-3|+|-2|=3+2=5$

83. $-6 + (-10); -16$

85. $-3 + 8; 5$

87. $-21 + 17; -4$

89. $3(-14 + 20); 18$

91. $(-7 + (-2)) + 5; -4$

Section 1.4 Practice Exercises

1. Answers will vary, 0, 1, 2, 3, 4

3. Answers will vary, $-\sqrt{2}, \sqrt{3}, \pi, \sqrt{5}, 2\pi$

5. Answers will vary, 1, 2, 3, 4, 5

7. $\sqrt{6}$

9. $-7 + 10$

11. –3

13. –12

15. 4

17. 3

19. $3 - 5 = 3 + (-5) = -2$

21. $3 - (-5) = 3 + 5 = 8$

23. $-3 - 5 = -3 + (-5) = -8$

25. $-3 - (-5) = -3 + 5 = 2$

27. $23 - 17 = 6$

29. $23 - (-17) = 23 + 17 = 40$

31. $-23 - 17 = -23 + (-17) = -40$

33. $-23 - (-17) = -23 + 17 = -6$

35. $-6 - 14 = -6 + (-14) = -20$

37. $-7 - 17 = -7 + (-17) = -24$

39. $13 - (-12) = 13 + 12 = 25$

41. $-14 - (-9) = -14 + 9 = -5$

43. $-\frac{6}{5}-\frac{3}{10}=-\frac{12}{10}+\left(-\frac{3}{10}\right)=-\frac{15}{10}=-\frac{3}{2}$

45. $\frac{3}{8}-\left(-\frac{4}{3}\right)=\frac{9}{24}+\frac{32}{24}=\frac{41}{24}$

47. $\frac{1}{2}-\frac{1}{10}=\frac{5}{10}-\frac{1}{10}=\frac{4}{10}=\frac{2}{5}$

49. $-\frac{11}{12}-\left(-\frac{1}{4}\right)=-\frac{11}{12}+\frac{3}{12}=-\frac{8}{12}=-\frac{2}{3}$

51. $6.8-(-2.4)=6.8+2.4=9.2$

53. $3.1-8.82=3.10+(-8.82)=-5.72$

55. $-4-3-2-1=-4+(-3)+(-2)+(-1)=-10$

57. $6-8-2-10=6+(-8)+(-2)+(-10)=-14$

59. $-36.75-14.25=-51$

61. $-112.846+(-13.03)-47.312=-173.188$

63. $0.085-(-3.14)+(0.018)=3.243$

65. $6-(-7)$; 13

67. $3-18$; -15

69. $-5-(-11)$; 6

71. $-1-(-13)$; 12

73. $-32-20$; -52

75. $6+8-(-2)-4+1=14+2-3=13$

77. $-1-7+(-3)-8+10=-8+(-11)+10=-9$

79. $-\frac{13}{10}+\frac{8}{15}-\left(-\frac{2}{5}\right)=-\frac{39}{30}+\frac{16}{30}+\frac{12}{30}=-\frac{11}{30}$

81. $\frac{2}{3}+\frac{5}{9}-\frac{4}{3}-\left(-\frac{1}{6}\right)=\frac{12}{18}+\frac{10}{18}-\frac{24}{18}+\frac{3}{18}=\frac{1}{18}$

83. $2-(-8)+7+3-15=2+8+10-15=5$

85. $-6+(-1)+(-8)+(-10)=-6+(-19)=-25$

87.
$$\begin{aligned}-6-1-8-10&=-6+(-1)+(-8)+(-10)\\&=-7+(-18)\\&=-25\end{aligned}$$

89. $8848-(-11033\text{ m})=19{,}881\text{ m}$

91. $200+400+600+800-1000$; \$1000

93. $113°-(-39°)=152°$

95. $(a+b)-c=(-2+(-6))-(-1)=-8+7=-7$

97.
$$\begin{aligned}a-(b+c)&=-2-(-6+(-1))\\&=-2-(-7)\\&=-2+7\\&=5\end{aligned}$$

99. $(a-b)-c=(-2-(-6))-(-1)=(4)+1=5$

101.
$$\begin{aligned}a-(b-c)&=-2-(-6-(-1))\\&=-2-(-5)\\&=-2+5\\&=3\end{aligned}$$

103. $\sqrt{29+(-4)}-7=\sqrt{25}-7=5-7=-2$

105.
$$\begin{aligned}|10+(-3)|-|-12+(-6)|&=|7|-|-18|\\&=7-18\\&=-11\end{aligned}$$

107. $\frac{3-4+5}{4+(-2)}=\frac{4}{2}=2$

109.
```
-8+(-5)
                -13
4+(-5)+(-1)
                 -2
627-(-84)
                711
```

111–113.
```
-0.06-0.12
                -.18
-3.2+(-14.5)
               -17.7
-472+(-518)
                -990
```

115.
```
-12-9+4
                -17
209-108+(-63)
                 38
```

Section 1.5 Practice Exercises

1. True; $4 > 1$

3. False; $-25 \not\geq -14$

5. True; $20 \leq 20$

7. False; $0 \not> 0$

9. $4 + 4 + 4 + 4 + 4$

11. $(-2) + (-2) + (-2)$

13. $(-2)(-7) = 14$

15. $-5 \cdot 0 = 0$

17. No number multiplied by 0 equals 6.

19. $2 \cdot 3 = 6$

21. $2(-3) = -6$

23. $(-2)3 = -6$

25. $(-2)(-3) = 6$

27. $24 \div 3 = 8$

29. $24 \div (-3) = -8$

31. $(-24) \div 3 = -8$

33. $(-24) \div (-3) = 8$

35. $-6 \cdot 0 = 0$

37. Undefined

39. $0\left(\frac{2}{5}\right) = 0$

41. $0 \div \left(-\frac{1}{10}\right) = 0$

43. $\frac{-14}{-7} = 2$

45. $\frac{13}{-65} = \frac{\cancel{13}}{-5 \cdot \cancel{13}} = -\frac{1}{5}$

47. $\frac{-9}{6} = \frac{-3 \cdot \cancel{3}}{2 \cdot \cancel{3}} = -\frac{3}{2}$

49. $\frac{-30}{-100} = \frac{-3 \cdot 10}{-10 \cdot 10} = \frac{3}{10}$

51. $\frac{26}{-13} = -2$

53. $(1.72)(-4.6) = -7.912$

55. $-0.02(-4.6) = 0.092$

57. $\frac{14.4}{-2.4} = -6$

59. $\frac{-5.25}{-2.5} = 2.1$

61. $(-3)^2 = 9$

63. $-3^2 = -9$

65. $\left(-\frac{2}{3}\right)^3 = \left(-\frac{2}{3}\right)\left(-\frac{2}{3}\right)\left(-\frac{2}{3}\right) = -\frac{8}{27}$

67. $(-0.2)^4 = 0.0016$

69. $-0.2^4 = -0.0016$

71. $-|-3| = -3$

73. $-(-3) = 3$

75. $-|7| = -7$

77. $|-7| = 7$

79. $(-2)(-5)(-3) = (10)(-3) = -30$

81. $(-8)(-4)(-1)(-3) = (32)(3) = 96$

83. $100 \div (-10) \div (-5) = (-10) \div (-5) = 2$

85. $\frac{2}{5} \cdot \frac{1}{3} \cdot \left(-\frac{10}{11}\right) = \frac{2}{15} \cdot \left(-\frac{10}{11}\right) = -\frac{20}{165} = -\frac{4}{33}$

87. $\left(1\frac{1}{3}\right)\div 3\div\left(-\frac{7}{9}\right)=\frac{4}{3}\cdot\frac{1}{3}\div\left(-\frac{7}{9}\right)$
$=\frac{4}{9}\cdot\left(-\frac{9}{7}\right)$
$=-\frac{4}{7}$

89. $-12\div(-6)\div(-2)=2\div(-2)=-1$

91. $-2(3)+3$; a loss of \$3

93. $87\div(-3)=-29$

95. $-4(-12)=48$

97. $2.8(-5.1)=-14.28$

99. $(-6.8)\div(-0.02)=340$

101. $\left(-\frac{2}{15}\right)\left(\frac{25}{3}\right)=-\frac{50}{45}=-\frac{5\cdot 10}{5\cdot 9}=-\frac{10}{9}$

103. $\left(-\frac{7}{8}\right)\div\left(-\frac{9}{16}\right)=\left(-\frac{7}{8}\right)\cdot\left(-\frac{16}{9}\right)$
$=\frac{112}{72}$
$=\frac{\cancel{8}\cdot 14}{\cancel{8}\cdot 9}$
$=\frac{14}{9}$

105. $12\div(-2)(4)=(-6)(4)=-24$

107. $\left(-\frac{12}{5}\right)\div(-6)\cdot\left(\frac{1}{8}\right)=\left(-\frac{12}{5}\right)\cdot\left(-\frac{1}{6}\right)\cdot\left(\frac{1}{8}\right)$
$=\frac{12}{30}\cdot\frac{1}{8}$
$=\frac{2}{5}\cdot\frac{1}{8}$
$=\frac{2}{40}$
$=\frac{1}{20}$

109. No, parentheses are needed around the quantity $5x$.

111. $-3.75(0.3)=-1.125$

113. $\left(\frac{16}{5}\right)\div\left(-\frac{8}{9}\right)=\frac{16}{5}\cdot\left(-\frac{9}{8}\right)=-\frac{144}{40}=-\frac{18}{5}$

115. $-0.4+6(-0.42)=-2.92$

117. $-\frac{1}{4}-6\left(-\frac{1}{3}\right)=-\frac{1}{4}+2=-\frac{1}{4}+\frac{8}{4}=\frac{7}{4}$

119. **(a)** $-4-3-2-1=-4+(-3)+(-2)+(-1)$
$=-10$

(b) $-4(-3)(-2)(-1)=12(2)=24$

(c) Part (a) is subtraction; part (b) is multiplication.

121. $8-2^3\cdot 5+3-(-6)=8-8\cdot 5+3+6$
$=8-40+9$
$=-23$

123. $-(2-8)^2\div(-6)\cdot 2=-36\div(-6)\cdot 2$
$=6\cdot 2$
$=12$

125. $\frac{6(-4)-2(5-8)}{-6-3-5}=\frac{-24+6}{-14}=\frac{-18}{-14}=\frac{9}{7}$

127. $\frac{-4+5}{(-2)\cdot 5+10}=\frac{1}{-10+10}=\frac{1}{0}=$ undefined

129. $|-5|-|-7|=5-7=-2$

131. $-|-1|-|5|=-1-5=-6$

133. $\frac{|2-9|-|5-7|}{10-15}=\frac{|-7|-|-2|}{-5}=\frac{7-2}{-5}=\frac{5}{-5}=-1$

135. $x^2-2y=(-2)^2-2(-4)=4+8=12$

137. $4(2x-z)=4(2(-2)-6)$
$=4(-4-6)$
$=4(-10)$
$=-40$

139. $\dfrac{3x+2y}{y}=\dfrac{3(-2)+2(-4)}{-4}$
$=\dfrac{-6+(-8)}{-4}$
$=\dfrac{-14}{-4}$
$=\dfrac{7}{2}$

141. $\dfrac{x+2y}{x-2y}=\dfrac{-2+2(-4)}{-2-2(-4)}$
$=\dfrac{-2+(-8)}{-2+8}$
$=\dfrac{-10}{6}$
$=-\dfrac{5}{3}$

143. For $x = 2$, $x^2+6=2^2+6=4+6=10$
For $x = -2$, $x^2+6=(-2)^2+6=4+6=10$

145–147.

```
-6(5)
                -30
-5.2/2.6
                 -2
(-5)(-5)(-5)(-5)
                625
```

149.

```
(-5)^4
                625
-5^4
               -625
-2.4²
              -5.76
```

151.

```
(-2.4)²
               5.76
(-1)(-1)(-1)
                 -1
```

153.

```
-8.4/-2.1
                  4
90/(-5)(2)
                -36
```

Section 1.6 Practice Exercises

1. $-13-(-5)=-13+5=-8$

3. $18\div(-4)=-\dfrac{18}{4}=-\dfrac{9}{2}=-4.5$

5. $\dfrac{25}{21}-\dfrac{6}{7}=\dfrac{25}{21}-\dfrac{18}{21}=\dfrac{7}{21}=\dfrac{1}{3}$

7. $\left(-\dfrac{3}{5}\right)\left(\dfrac{4}{27}\right)=-\dfrac{12}{135}=-\dfrac{4}{45}$

9. Reciprocal

11. Zero

13. b

15. i

17. g

19. d

21. h

23. $6(5x+1)=6(5x)+6(1)=30x+6$

25. $-2(a+8)=-2a+(-2)(8)=-2a-16$

27. $3(5c-d)=3(5c)-3d=15c-3d$

29. $-7(y-2)=-7y-(-7)(2)=-7y+14$

31. $\dfrac{1}{3}(m-3)=\dfrac{1}{3}m-\dfrac{1}{3}\cdot 3=\dfrac{1}{3}m-1$

33. $\dfrac{3}{8}(4+8s)=\dfrac{3}{8}(4)+\dfrac{3}{8}(8s)$
$=\dfrac{12}{8}+\dfrac{24}{8}s$
$=\dfrac{3}{2}+3s$

35. $-\dfrac{2}{3}(x-6)=-\dfrac{2}{3}x-\left(-\dfrac{2}{3}\right)(6)$
$=-\dfrac{2}{3}x+\dfrac{12}{3}$
$=-\dfrac{2}{3}x+4$

37. $-(2p+10)=-2p-10$

39. $-(-3w-5z)=3w+5z$

41. $4(x+2y-z)=4(x)+4(2y)-4(z)$
$=4x+8y-4z$

43. $-(-6w+x-3y)=6w-x+3y$

45. $4(92) = 4(90 + 2);\ 360 + 8 = 368$

47. $4(902) = 4(900 + 2);\ 3600 + 8 = 3608$

49. Term: $3xy$, coefficient 3; Term: $-6x^2$, coefficient -6; Term: y, coefficient 1; Term: -17, coefficient -17

51. Term: x^4, coefficient 1; Term: $-10xy$, coefficient -10; Term: 12, coefficient 12; Term: $-y$, coefficient -1.

53. The variables have different exponents.

55. The variables are the same *and* raised to the same power.

57. Answers vary: $2x, -5x, x$

59. $5k - 10k - 12k + 16 + 7 = -17k + 23$

61. $$\begin{aligned} 9x - 7y + 12x + 14y &= 9x + 12x - 7y + 14y \\ &= 21x + 7y \end{aligned}$$

63. $$\begin{aligned} \frac{1}{4}a + b - \frac{3}{4}a - 5b &= \frac{1}{4}a - \frac{3}{4}a + b - 5b \\ &= -\frac{2}{4}a - 4b \\ &= -\frac{1}{2}a - 4b \end{aligned}$$

65. $2.8z - 8.1z + 6 - 15.2 = -5.3z - 9.2$

67. $-3(2x - 4) + 10 = -6x + 12 + 10 = -6x + 22$

69. $4(w + 3) - 12 = 4w + 12 - 12 = 4w$

71. $5 - 3(x + 4) = 5 - 3x + 12 = 17 - 3x$

73. $$\begin{aligned} -3(2t + 4) + 8(2t - 4) &= -6t - 12 + 16t - 32 \\ &= -6t + 16t - 12 - 32 \\ &= 10t - 44 \end{aligned}$$

75. $2(w - 5) - (2w + 8) = 2w - 10 - 2w - 8 = -18$

77. $-\frac{1}{3}(6t + 9) + 10 = -2t - 3 + 10 = -2t + 7$

79. $10(5.1a - 3.1) + 4 = 51a - 31 + 4 = 51a - 27$

81. $$\begin{aligned} -4m + 2(m - 3) + 2m &= -4m + 2m - 6 + 2m \\ &= -6 \end{aligned}$$

83. $$\begin{aligned} \frac{1}{2}(10q - 2) + \frac{1}{3}(2 - 3q) &= 5q - 1 + \frac{2}{3} - q \\ &= 4q - \frac{1}{3} \end{aligned}$$

85. $$\begin{aligned} 7n - 2(n - 3) - 6 + n &= 7n - 2n + 6 - 6 + n \\ &= 6n \end{aligned}$$

87. $$\begin{aligned} &6(x + 3) - 12 - 4(x - 3) \\ &= 6x + 18 - 12 - 4x + 12 \\ &= 2x + 18 \end{aligned}$$

89. $$\begin{aligned} 6.1(5.3z - 4.1) - 5.8 &= 32.33z - 25.01 - 5.8 \\ &= 32.33z - 30.81 \end{aligned}$$

91. Associative property of multiplication

93. Associative property of addition

95. Distributive property of multiplication over addition

97. Identity property of addition

99. Identity property of multiplication

101. Inverse property of addition

103. Equivalent

105. Not equivalent; not like terms.

107. Not equivalent; subtraction is not commutative

109. Equivalent

111. $14\frac{2}{7} + \left(2\frac{1}{3} + \frac{2}{3}\right)$; fractions have common denominators.

113. **(a)** $$\begin{aligned} &10 + (1 + 9) + (2 + 8) + (3 + 7) \\ &\quad + (4 + 6) + 5 \\ &= 55 \end{aligned}$$

(b) $$\begin{aligned} &(1 + 19) + (2 + 18) + (3 + 17) + (4 + 16) \\ &\quad + (5 + 15) + (6 + 14) + (7 + 13) \\ &\quad + (8 + 12) + (9 + 11) + 10 = 210 \end{aligned}$$

Chapter 1 Review Exercises

1. **(a)** 7, 1

(b) 7, –4, 0, 1

(c) 7, 0, 1

(d) $7, \frac{1}{3}, -4, 0, -0.\overline{2}, 1$

(e) $-\sqrt{3}, \pi$

(f) $7, \frac{1}{3}, -\sqrt{3}, -0.\overline{2}, \pi, 1, -4, 0$

3. $|-6| = 6$

5. $|0| = 0$

7. False

9. True

11. True

13. True

15. $\frac{7}{y}$ or $7 \div y$

17. $a - 5$

19. $13z - 7$

21. $3y + 12$

23. $2p - 5$

25. $15^2 = 225$

27. $\left(\frac{1}{4}\right)^2 = \frac{1}{16}$

29. $\left(\frac{3}{2}\right)^3 = \frac{27}{8}$

31. $|-11| + |5| - (7 - 2) = 11 + 5 - 5 = 11$

33. $22 - 3(8 \div 4)^2 = 22 - 3(2)^2 = 22 - 12 = 10$

35. $14 + (-10) = 4$

37. $-12 + (-5) = -17$

39. $-\frac{8}{11} + \frac{1}{2} = -\frac{16}{22} + \frac{11}{22} = -\frac{5}{22}$

41. $\left(-\frac{5}{2}\right) + \left(-\frac{1}{5}\right) = -\frac{25}{10} + \left(-\frac{2}{10}\right) = -\frac{27}{10}$

43. $2.9 + (-7.18) = -4.28$

45. $-5 + (-7) + 20 = -12 + 20 = 8$

47. If a and b are both negative or if a is negative with a greater absolute value than b or if b is negative with a greater absolute value than a.

49. $13 - 25 = -12$

51. $-8 - (-7) = -8 + 7 = -1$

53. $-\frac{7}{9} - \frac{5}{6} = -\frac{14}{18} - \frac{15}{18} = -\frac{29}{18}$

55. $7 - 8.2 = -1.2$

57. $-16.1 - (-5.9) = -16.1 + 5.9 = -102$

59. $\frac{11}{2} - \left(-\frac{1}{6}\right) - \frac{7}{3} = \frac{33}{6} + \frac{1}{6} - \frac{14}{6} = \frac{20}{6} = \frac{10}{3}$

61. $6 - 14 - (-1) - 10 - (-21) - 5$
$= 6 - 14 + 1 - 10 + 21 - 5$
$= -8 - 9 + 16$
$= -17 + 16$
$= -1$

63. $-7 - (-18)$
$-7 - (-18) = 11$

65. $7 - 13$
$7 - 13 = -6$

67. $(6 + (-12)) - 21$
$(6 + (-12)) - 21 = -6 - 21 = -27$

69. $10(-17) = -170$

71. $(-52) \div 26 = -2$

73. $\frac{7}{4}\div\left(-\frac{21}{2}\right)=\frac{7}{4}\cdot\left(-\frac{2}{21}\right)=-\frac{14}{84}=-\frac{1}{6}$

75. $-\frac{21}{5}\cdot 0=0$

77. $0\div(-14)=0$

79. $(-0.45)(-5)=2.25$

81. $-\frac{21}{14}=-\frac{3\cdot 7}{2\cdot 7}=-\frac{3}{2}$

83. $(5)(-2)(3)=(-10)(3)=-30$

85. $\left(-\frac{1}{2}\right)\left(\frac{7}{8}\right)\left(-\frac{4}{7}\right)=\left(-\frac{7}{16}\right)\left(-\frac{4}{7}\right)=\frac{7\cdot 4}{16\cdot 7}=\frac{1}{4}$

87. $40\div 4\div(-5)=10\div(-5)=-2$

89. $9-4[-2(4-8)-5(3-1)]$
$=9-4[-2(-4)-5(2)]$
$=9-4[8-10]$
$=9-4[-2]$
$=9+8$
$=17$

91. $\frac{2}{3}-\left(\frac{3}{8}+\frac{5}{6}\right)\div\frac{5}{3}=\frac{2}{3}-\left(\frac{9}{24}+\frac{20}{24}\right)\cdot\frac{3}{5}$
$=\frac{16}{24}-\frac{29}{24}\cdot\frac{3}{5}$
$=\frac{16}{24}-\frac{29}{40}$
$=\frac{80}{120}-\frac{87}{120}$
$=-\frac{7}{120}$

93. $3(x+2)\div y=3(4+2)\div(-9)$
$=18\div(-9)$
$=-2$

95. $w+xy-\sqrt{z}=12+(6)(-5)-\sqrt{25}$
$=12+(-30)-5$
$=-23$

97. $x=\mu+z\sigma$
$x=(100)+(-1.96)(15)$
$x=70.6$

99. True

101. True

103. True

105. $2+3=3+2$

107. $5+(-5)=0$

109. $5\cdot 2=2\cdot 5$

111. $3\cdot\frac{1}{3}=1$

113. $5x-2y=5x+(-2y)$; changing subtraction to addition of the opposite. Then, use commutative property of addition.

115. $3y$, $10x$, 12, xy

117. **(a)** $3a+3b-4b+5a-10$
$=3a+5a+3b-4b-10$
$=8a-b-10$

(b) $-6p+2q+9-13q-p+7$
$=-6p-p+2q-13q+9+7$
$=-7p-11q+16$

119. $2p-(p+5)+3=2p-p-5+3=p-2$

121. $\frac{1}{2}(-6z)+q-4\left(3q+\frac{1}{4}\right)=-3q+q-12q-1$
$=-14q-1$

123. $-4[2(x+1)-(3x+8)]=-4[2x+2-3x-8]$
$=-4[-x-6]$
$=4x+24$

Chapter 1 Test

1. Rational; all repeating decimals are rational numbers.

3. **(a)** False

(b) True

(c) True

(d) True

5. **(a)** Twice the difference of a and b

(b) The difference twice a and b

7. $18 + (-12) = 6$

9. $-\frac{1}{8} + \left(-\frac{3}{4}\right) = -\frac{1}{8} + \left(-\frac{6}{8}\right) = -\frac{7}{8}$

11. $-14 + (-2) - 16 = -14 + (-18) = -32$

13. $38 \div 0 = \text{undefined}$

15. $-22 \cdot 0 = 0$

17. $$\begin{aligned}\frac{2}{5} \div \left(-\frac{7}{10}\right) \cdot \left(-\frac{7}{6}\right) &= \frac{2}{5} \cdot \left(-\frac{10}{7}\right) \cdot \left(-\frac{7}{6}\right)\\ &= -\frac{4}{7} \cdot \left(-\frac{7}{6}\right)\\ &= \frac{28}{42}\\ &= \frac{2}{3}\end{aligned}$$

19. $$\begin{aligned}8 - [(2-4) - (8-9)] &= 8 - [(-2) - (-1)]\\ &= 8 - [-1]\\ &= 8 + 1\\ &= 9\end{aligned}$$

21. $\frac{|4-10|}{2-3(5-1)} = \frac{|-6|}{2-3(4)} = \frac{6}{-10} = -\frac{3}{5}$

23. $3k - 20 + (-9k) + 12 = -6k - 8$

25. $$\begin{aligned}\frac{1}{2}(12p-4) + \frac{1}{3}(2-6p) &= 6p - 2 + \frac{2}{3} - 2p\\ &= 4p - \frac{4}{3}\end{aligned}$$

27. $$\begin{aligned}-x^2 - 4y + z &= -(-6)^2 - 4(2) + (-7)\\ &= -36 - 8 + (-7)\\ &= -51\end{aligned}$$

29. $(-8)(-6)(-4)(-3) = (48)(12) = 576$

31. $12 - (-4)$

$12 - (-4) = 12 + 4 = 16$

Chapter 2

Section 2.1 Practice Exercises

1. Expression

3. Equation

5. Equation

7. Expression

9. Substitute the value into the variable in the equation, simplify, and determine if the left-hand side is equal to the right-hand side.

11. No; $3^2 - 2 \neq 0$

$$7 \neq 0$$

13. Yes; $(-4) + 5 = 1$

$$1 = 1$$

15. **(a)** No

(b) Yes

(c) No

(d) Yes

17.
$$\begin{aligned} x + 6 &= 5 \\ x + 6 + (-6) &= 5 + (-6) \\ x &= -1 \end{aligned}$$

19.
$$\begin{aligned} q - 14 &= 6 \\ q - 14 + 14 &= 6 + 14 \\ q &= 20 \end{aligned}$$

21.
$$\begin{aligned} 2 + m &= -15 \\ -2 + 2 + m &= -15 - 2 \\ m &= -17 \end{aligned}$$

23.
$$\begin{aligned} -23 &= y - 7 \\ -23 + 7 &= y - 7 + 7 \\ -16 &= y \text{ or } y = 16 \end{aligned}$$

25.
$$\begin{aligned} 5 &= z - \frac{1}{2} \\ 5 + \frac{1}{2} &= z - \frac{1}{2} + \frac{1}{2} \\ \frac{11}{2} &= z \text{ or } z = \frac{11}{2} \end{aligned}$$

27.
$$\begin{aligned} x + \frac{5}{2} &= \frac{1}{2} \\ x + \frac{5}{2} - \frac{5}{2} &= \frac{1}{2} - \frac{5}{2} \\ x &= -\frac{4}{2} = -2 \end{aligned}$$

29.
$$\begin{aligned} 4.1 &= 2.8 + a \\ 4.1 - 2.8 &= -2.8 + 2.8 + a \\ a &= 1.3 \end{aligned}$$

31.
$$\begin{aligned} 4 + c &= 4 \\ -4 + 4 + c &= 4 - 4 \\ c &= 0 \end{aligned}$$

33.
$$\begin{aligned} -6.02 + c &= -8.15 \\ 6.02 - 6.02 + c &= -8.15 + 6.02 \\ c &= -2.13 \end{aligned}$$

35.
$$\begin{aligned} 3.245 + t &= -0.0225 \\ 3.245 - 3.245 + t &= -0.0225 + 3.245 \\ t &= -3.2675 \end{aligned}$$

37. Let x = the number. $-8 + x = 42$;
$$\begin{aligned} -8 + x &= 42 \\ 8 - 8 + x &= 42 + 8 \\ x &= 50 \end{aligned}$$

39. Let x = the number. $x - (-6) = 18$;
$$\begin{aligned} x - (-6) &= 18 \\ x + 6 - 6 &= 18 - 6 \\ x &= 12 \end{aligned}$$

41. Let x = the number. $x+\frac{5}{8}=\frac{13}{8}$;

$$x+\frac{5}{8}=\frac{13}{8}$$
$$x+\frac{5}{8}-\frac{5}{8}=\frac{13}{8}-\frac{5}{8}$$
$$x=\frac{8}{8}=1$$

43. $6x=54$
$$\frac{6x}{6}=\frac{54}{6}$$
$$x=9$$

45. $12=-3p$
$$\frac{12}{-3}=\frac{-3p}{-3}$$
$-4=p$ or $p=-4$

47. $-5y=0$
$$\frac{-5y}{-5}=\frac{0}{5}$$
$$y=0$$

49.
$$-\frac{y}{5}=3$$
$$-\frac{y}{5}\cdot(-5)=3(-5)$$
$$y=-15$$

51.
$$\frac{4}{5}=-t$$
$$\frac{4}{5}(-1)=-t(-1)$$
$$t=-\frac{4}{5}$$

53.
$$\frac{2}{5}a=-4$$
$$\frac{5}{2}\cdot\frac{2}{5}a=\frac{5}{2}(-4)=-\frac{20}{2}$$
$$a=-10$$

55.
$$-\frac{1}{5}b=-\frac{4}{5}$$
$$(-5)\left(-\frac{1}{5}b\right)=(-5)\left(-\frac{4}{5}\right)$$
$$b=4$$

57.
$$-41=-x$$
$$(-1)(-41)=(-1)(-x)$$
$$x=41$$

59. $3.81=-0.03p$
$$\frac{3.81}{-0.03}=\frac{-0.03p}{-0.03}$$
$$p=-127$$

61. $5.82y=-15.132$
$$\frac{5.82y}{5.82}=\frac{-15.132}{5.82}$$
$$y=-2.6$$

63. $x\cdot 7=-63$ or $7x=-63$
$$\frac{7x}{7}=\frac{-63}{7}$$
$$x=-9$$

65.
$$\frac{x}{12}=\frac{1}{3}$$
$$12\cdot\frac{x}{12}=12\cdot\frac{1}{3}$$
$$x=4$$

67.
$$a-9=1$$
$$a-9+9=1+9$$
$$a=10$$

69. $1=-9x$
$$\frac{1}{-9}=\frac{-9x}{-9}$$
$$x=-\frac{1}{9}$$

71.
$$-\frac{2}{3}h=8$$
$$\left(-\frac{3}{2}\right)\left(-\frac{2}{3}h\right)=\left(-\frac{3}{2}\right)\cdot 8$$
$$h=-\frac{24}{2}=-12$$

73.
$$\frac{2}{3}+t=8$$
$$-\frac{2}{3}+\frac{2}{3}+t=-\frac{2}{3}+8$$
$$t=\frac{22}{3}$$

75.
$$\frac{r}{3}=-12$$
$$3\cdot\frac{r}{3}=3(-12)$$
$$r=-36$$

77.
$$k+16=32$$
$$k+16-16=32-16$$
$$k=16$$

79.
$$16k=32$$
$$\frac{16k}{16}=\frac{32}{16}$$
$$k=2$$

81.
$$7=-4q$$
$$\frac{7}{-4}=\frac{-4q}{-4}$$
$$q=-\frac{7}{4}$$

83.
$$-4+q=7$$
$$-4+4+q=4+7$$
$$q=11$$

85.
$$-\frac{1}{3}d=12$$
$$(-3)\left(-\frac{1}{3}d\right)=(-3)(12)$$
$$d=-36$$

87.
$$4=\frac{1}{2}+z$$
$$4-\frac{1}{2}=\frac{1}{2}-\frac{1}{2}+z$$
$$z=\frac{7}{2}$$

89.
$$1.2y=4.8$$
$$\frac{1.2y}{1.2}=\frac{4.8}{1.2}$$
$$y=4$$

91.
$$4.8=1.2+y$$
$$4.8-1.2=1.2-1.2+y$$
$$y=3.6$$

93.
$$0.0034=y-0.405$$
$$0.0034+0.405=y-0.405+0.405$$
$$y=0.4084$$

95.
$$A=lw$$
$$322=l(23)$$
$$\frac{322}{23}=\frac{l(23)}{23}$$
$$l=14\text{ cm}$$

97.
$$A=lw$$
$$18.18=(4.5)w$$
$$\frac{18.18}{4.5}=\frac{4.5w}{4.5}$$
$$w=4.04\text{ ft}$$

99.
$$d=rt$$
$$d=(4.5)(t)$$
$$\frac{6}{4.5}=\frac{4.5t}{4.5}$$
$$t=1\frac{1}{3}\text{ hours}$$

101. $t=45\text{ minutes}=\frac{3}{4}\text{ hours}$
$$d=rt$$
$$6=r\left(\frac{3}{4}\right)$$
$$6\cdot\frac{4}{3}=r\left(\frac{3}{4}\right)\cdot\frac{4}{3}$$
$$r=8\text{ mph}$$

Section 2.2 Practice Exercises

1. $-3(4t)+5t-6=-12t+5t-6=-7t-6$

3. $5z+2-7z-3z=5z-10z+2=-5z+2$

5. $-(-7+9)+(3p-1)=7p-9+3p-1$
$=10p-10$

7. $5(3a)+5(3+a)=15a+15+5a=20a+15$

9. To simplify an expression, clear parentheses and combine like terms. To solve an equation, isolate the variable.

11.
$$-7y=21$$
$$\frac{-7y}{-7}=\frac{21}{-7}$$
$$y=-3$$

13.
$$z-23=-28$$
$$z-23+23=-28+23$$
$$z=-5$$

15.
$$12=b+\frac{1}{5}$$
$$12-\frac{1}{5}=b+\frac{1}{5}-\frac{1}{5}$$
$$b=\frac{59}{5}=11\frac{4}{5}$$

17.
$$-\frac{3}{10}=-6h$$
$$\left(-\frac{1}{6}\right)\left(-\frac{3}{10}\right)=\left(-\frac{1}{6}\right)(-6h)$$
$$h=\frac{3}{60}=\frac{1}{20}$$

19. First use the addition property, then the division property.

21.
$$5x+2=-13$$
$$5x+2-2=-13-2$$
$$\frac{5x}{5}=\frac{-15}{5}$$
$$x=-3$$

23.
$$-7w-5=-19$$
$$-7w-5+5=-19+5$$
$$\frac{-7w}{-7}=\frac{-14}{-7}$$
$$w=2$$

25.
$$4q+5=2$$
$$4q+5-5=2-5$$
$$\frac{4q}{4}=\frac{-3}{4}$$
$$q=-\frac{3}{4}$$

27.
$$-9=4n-1$$
$$-9+1=4n-1+1$$
$$\frac{-8}{4}=\frac{4n}{4}$$
$$n=-2$$

29.
$$2b-\frac{1}{4}=5$$
$$2-\frac{1}{4}+\frac{1}{4}=5+\frac{1}{4}$$
$$\frac{1}{2}\cdot 2b=\frac{21}{4}\cdot\frac{1}{2}$$
$$b=\frac{21}{8}$$

31.
$$-1.8+2.4a=-6.6$$
$$1.8-1.8+2.4a=-6.6+1.8$$
$$\frac{2.4a}{2.4}=\frac{-4.4}{2.4}$$
$$a=-2$$

33.
$$\frac{6}{7}=\frac{1}{7}+\frac{5}{3}r$$
$$\frac{6}{7}-\frac{1}{7}=-\frac{1}{7}+\frac{1}{7}+\frac{5}{3}r$$
$$\left(\frac{3}{5}\right)\left(\frac{5}{7}\right)=\left(\frac{3}{5}\right)\left(\frac{5}{3}r\right)$$
$$r=\frac{3}{7}$$

35.
$$5v-3-4v=13$$
$$v-3+3=13+3$$
$$v=16$$

37. $6u-5-8u=-7$
$-2u-5+5=-7+5$
$\frac{-2u}{-2}=\frac{-2}{-2}$
$u=1$

39. $6b-20=14+5b$
$6b-5b-20+20=14+20+5b-5b$
$b=34$

41. $-6x-7=-3-8x$
$-6x+8x-7+7=-3+7-8x+8x$
$\frac{2x}{2}=\frac{4}{2}$
$x=2$

43. $-7t+4=-6t$
$7t-7t+4=-6t+7t$
$t=4$

45. $\frac{3}{7}x-\frac{1}{4}=-\frac{4}{7}x-\frac{5}{4}$
$\frac{3}{7}x+\frac{4}{7}x-\frac{1}{4}+\frac{1}{4}=-\frac{4}{7}x+\frac{4}{7}x-\frac{5}{4}+\frac{1}{4}$
$x=-1$

47. $4(t+15)=20$
$4t+60=20$
$4t+60-60=20-60$
$\frac{4t}{4}=\frac{-40}{4}$
$t=-10$

49. $4(2k+1)-1=5$
$8k+4-1=5$
$8k+3-3=5-3$
$\frac{8k}{8}=\frac{2}{8}$
$k=\frac{1}{4}$

51. $4(w-5)-3w=2$
$4w-20-3w=2$
$w-20+20=2+20$
$w=22$

53. $5(4+p)=3(3p-1)-9$
$20+5p=9p-3-9$
$20+12=9p-5p$
$\frac{32}{4}=\frac{4p}{4}$
$p=8$

55. $-5y+2(2y+1)=2(5y-1)-7$
$-5y+4y+2=10y-2-7$
$-y+2=10y-9$
$-11y=-11$
$y=1$

57. $5-(6k+1)=2[(5k-3)-(k-2)]$
$4-6k=2[4k-1]$
$4-6k=8k-2$
$-14k=-6$
$k=\frac{-6}{-14}=\frac{6}{14}=\frac{3}{7}$

59. $0.4z-0.15=0.65-0.3(6-2z)$
$0.4z-0.15=0.65-1.8+0.6z$
$0.4z-0.15=-1.15+0.6z$
$-0.2z=-1$
$z=5$

61. $7(0.4m-0.1)=5.2m+0.86$
$2.8m-0.7=5.2m+0.86$
$-2.4m=1.56$
$m=-0.65$

63. No solution

65. Contradiction; no solution

67. Conditional equation; $7x+3=6x-12$
$x=-15$

69. Identity; all real numbers

71. Identity; all real numbers

73. Conditional equation; $8x+14=14-2x$
$x=0$

75. $x+a=10$
$-5+a=10$
$a=15$

77.
$$\begin{aligned} ax &= 12 \\ a(3) &= 12 \\ a &= 4 \end{aligned}$$

79.
$$\begin{aligned} 2(q+1)-1.5 &= 5(0.4q+0.16) \\ 2q+0.5 &= 2q+8 \\ 0.5 &\neq 8 \end{aligned}$$
Contradiction

Section 2.3 Practice Exercises

1.
$$\begin{aligned} 25x &= -15 \\ \frac{25x}{25} &= \frac{-15}{25} \\ x &= -\frac{3}{5} \end{aligned}$$

3.
$$\begin{aligned} 34 &= m-12 \\ 34+12 &= m-12+12 \\ m &= 46 \end{aligned}$$

5.
$$\begin{aligned} 2+3(n-6)-2b &= 6 \\ 2+3b-18-2b &= 6 \\ b-16+16 &= 6+16 \\ b &= 22 \end{aligned}$$

7.
$$\begin{aligned} 7x+2 &= 7(x-12) \\ 7x+2 &= 7x-84 \\ 2 &\neq -84 \end{aligned}$$
Contradiction

9. 18, 36

11. 100; 1000; 10,000

13.
$$\begin{aligned} \frac{1}{2}x+3 &= 5 \\ 2\left(\frac{1}{2}x+3\right) &= 2(5) \\ x+6 &= 10 \\ x &= 4 \end{aligned}$$

15.
$$\begin{aligned} \frac{1}{6}y+2 &= \frac{5}{12} \\ 12\left(\frac{1}{6}y+2\right) &= 12\left(\frac{5}{12}\right) \\ 2y+24 &= 5 \\ \frac{2y}{2} &= -\frac{19}{2} \\ y &= -\frac{19}{2} \end{aligned}$$

17.
$$\begin{aligned} \frac{1}{3}q+\frac{3}{5} &= \frac{1}{15}q-\frac{2}{5} \\ 15\left(\frac{1}{3}q+\frac{3}{5}\right) &= 15\left(\frac{1}{15}q-\frac{2}{5}\right) \\ 5q+9 &= q-6 \\ 4q &= -15 \\ q &= -\frac{15}{4} \end{aligned}$$

19.
$$\begin{aligned} \frac{12}{5}w+7 &= 31-\frac{3}{5}w \\ 5\left(\frac{12}{5}w+7\right) &= 5\left(31-\frac{3}{5}w\right) \\ 12w+35 &= 155-3w \\ 15w &= 120 \\ w &= 8 \end{aligned}$$

21.
$$\begin{aligned} \frac{1}{4}(3m-4)-\frac{1}{5} &= \frac{1}{4}m+\frac{3}{10} \\ 20\left[\frac{1}{4}(3m-4)-\frac{1}{5}\right] &= 20\left(\frac{1}{4}m+\frac{3}{10}\right) \\ 5(3m-4)-4 &= 5m+6 \\ 15m-20-4 &= 5m+6 \\ 15m-24 &= 5m+6 \\ 10m &= 30 \\ m &= 3 \end{aligned}$$

23.
$$\begin{aligned} \frac{1}{6}(5s+3) &= \frac{1}{2}(s+11) \\ 6\left[\frac{1}{6}(5s+3)\right] &= 6\left[\frac{1}{2}(s+11)\right] \\ 5s+3 &= 3(s+11) \\ 5s+3 &= 3s+33 \\ 2s &= 30 \\ s &= 15 \end{aligned}$$

25. $\frac{2}{3}x+4=\frac{2}{3}x-6$
$4 \neq -6$
No solution

27. $\frac{1}{6}(2c-1)=\frac{1}{3}c-\frac{1}{6}$
$6\left(\frac{1}{6}(2c-1)\right)=6\left(\frac{1}{3}c-\frac{1}{6}\right)$
$2c-1=2x-1$
$-1=-1$
All real numbers

29. $\frac{2x+1}{3}+\frac{x-1}{3}=5$
$3\left(\frac{2x+1}{3}+\frac{x-1}{3}\right)=3(5)$
$2x+1+x-1=15$
$3x=15$
$x=5$

31. $\frac{3w-2}{6}=1-\frac{w-1}{3}$
$6\left(\frac{3w-2}{6}\right)=6\left(1-\frac{w-1}{3}\right)$
$3w-2=6-2(w-1)$
$3w-2=8-2w$
$5w=10$
$w=2$

33. Let x = the number. Then, $\frac{1}{2}x=-8$
$2\left(\frac{1}{2}x\right)=2(-8)$
$x=-16$

35. Let x = the number. $\frac{2}{5}+2x=\frac{11}{5}+x$
$5\left(\frac{2}{5}+2x\right)=5\left(\frac{11}{5}+x\right)$
$2+10x=11+5x$
$5x=9$
$x=\frac{9}{5}$

37. Let x = the number. $2x+\frac{3}{4}=4x-\frac{1}{8}$
$8\left(2x+\frac{3}{4}\right)=8\left(4x-\frac{1}{8}\right)$
$16x+6=32x-1$
$16x=7$
$x=\frac{7}{16}$

39. $9.2y-4.3=50.9$
$9.2y=55.2$
$y=\frac{55.2}{9.2}=6$

41. $21.1w+4.6=10.9w+35.2$
$10.2w=30.6$
$w=\frac{30.6}{10.2}=3$

43. $0.2p-1.4=0.2(p-7)$
$0.2p-1.4=0.2p-1.4$
$0=0$
All real numbers

45. $0.20x+53.60=x$
$0.80x=53.60$
$x=67$

47. $0.15(90)+0.05p=0.10(90+p)$
$13.5+0.05p=9+0.10p$
$-0.05p=-4.5$
$p=90$

49. $0.40(y+10)+0.60y=2$
$0.40y+4+0.60y=2$
$y=-2$

51. $2b+23=6b-5$
$-4b=-28$
$\frac{-4b}{-4}=\frac{-28}{-4}$
$b=7$

53. $\frac{y}{4}=-2$
$4\cdot\frac{y}{4}=4(-2)$
$y=-8$

55. $$\begin{aligned} 0.5(2a-3)-0.1&=0.4(6+2a)\\ a-1.5-0.1&=2.4+0.8a\\ 0.2a&=4\\ a&=20 \end{aligned}$$

57. $$\begin{aligned} -6x&=0\\ x&=\frac{0}{-6}=0 \end{aligned}$$

59. $$\begin{aligned} 9.8h+2&=3.8h+20\\ 6h&=18\\ h&=3 \end{aligned}$$

61. $$\begin{aligned} \frac{1}{4}(x+4)&=\frac{1}{5}(2x+3)\\ 20\left(\frac{1}{4}(x+4)\right)&=20\left(\frac{1}{5}(2x+3)\right)\\ 5(x+4)&=4(2x+3)\\ 5x+20&=8x+12\\ -3x&=-8\\ x&=\frac{-8}{-3}=\frac{8}{3} \end{aligned}$$

63. $$\begin{aligned} 2z-7&=2(z-13)\\ 2z-7&=2z-26\\ -7&\neq-26 \end{aligned}$$
No solution

65. $$\begin{aligned} \frac{4}{5}w&=10\\ \frac{5}{4}\cdot\frac{4}{5}w&=\frac{5}{4}\cdot10\\ w&=\frac{50}{4}=\frac{25}{2} \end{aligned}$$

67. $$\begin{aligned} 4b-8-b&=-3b+2(3b-4)\\ 3b-8&=-3b+6b-8\\ 3b-8&=3b-8\\ 0&=0 \end{aligned}$$
All real numbers

69. $$\begin{aligned} -3a+1&=19\\ -3a&=18\\ a&=\frac{18}{-3}=-6 \end{aligned}$$

71. $$\begin{aligned} 3(4h-2)-(5h-8)&=8-(2h+3)\\ 12h-6-5h+8&=8-2h-3\\ 7h+2&=5-2h\\ 9h&=3\\ h&=\frac{3}{9}=\frac{1}{3} \end{aligned}$$

73. $$\begin{aligned} \frac{3}{8}t-\frac{5}{8}&=\frac{1}{2}t+\frac{1}{8}\\ 8\left(\frac{3}{8}t-\frac{5}{8}\right)&=8\left(\frac{1}{2}t+\frac{1}{8}\right)\\ 3t-5&=4t+1\\ -t&=6\\ t&=-6 \end{aligned}$$

75. $$\begin{aligned} \frac{1}{2}a+0.4&=-0.7-\frac{3}{5}a\\ 10\left(\frac{1}{2}a+0.4\right)&=10\left(-0.7-\frac{3}{5}a\right)\\ 5a+4&=-7-6a\\ 11a&=-11\\ a&=-1 \end{aligned}$$

77. $$\begin{aligned} 0.8+\frac{7}{10}b&=\frac{3}{2}b-0.8\\ 10\left(0.8+\frac{7}{10}b\right)&=10\left(\frac{3}{2}b-0.8\right)\\ 8+7b&=15b-8\\ -8b&=-16\\ b&=\frac{-16}{-8}=2 \end{aligned}$$

Section 2.4 Practice Exercises

1. $$\begin{aligned} x+16&=-31\\ x+16-16&=-31-16\\ x&=-47 \end{aligned}$$

3. $$\begin{aligned} x-6&=-3\\ x-6+6&=-3+6\\ x&=3 \end{aligned}$$

5. $$\begin{aligned} x-16&=-1\\ x-16+16&=-1+16\\ x&=15 \end{aligned}$$

7. Contradiction; $4t-8=1+4t$

$$-8\neq 1$$

No solution

9. Conditional; $-5y-9=15$

$$y=-7$$

11. Identity; $14m-14-5m=9m-14$

$$9m-14=9m-14$$
$$0=0$$

All real numbers

13. Let x represent the unknown number.

Twice(sum of x and 7) = 8

$$2(x+7)=8$$
$$2x+14=8$$
$$2x=-6$$
$$x=-3$$

The number is –3.

15. Let x represent the unknown number.

5(the difference of x and 3) = 4 less than $4\cdot x$

$$5(x-3)=4x-4$$
$$5x-15=4x-4$$
$$x=11$$

The number is 11.

17. Let x represent the unknown number.

(x added to 5) = Twice x

$$x+5=2x$$
$$x=5$$

The number is 5.

19. Let x represent the unknown number.

(Sum of $6x$ and 10) = (difference of x and 15)

$$6x+10=x-15$$
$$5x=-25$$
$$x=-5$$

The number is –5.

21. Let x represent the unknown number.

(3 added to $5x$) = (43 more than x)

$$3+5x=43+x$$
$$4x=40$$
$$x=10$$

The number is 10.

23. Let x represent the unknown number.

Triple(difference of x and 4) = 6 more than x

$$3(x-4)=6+x$$
$$3x-12=6x+x$$
$$2x=18$$
$$x=9$$

The number is 9.

25. Let x represent the number of Republicans. Then the number of Democrats is $104+x$.

$$\begin{pmatrix}\text{number of}\\\text{Republicans}\end{pmatrix}+\begin{pmatrix}\text{number of}\\\text{Democrats}\end{pmatrix}=434$$
$$x+(104+x)=434$$
$$2x=330$$
$$x=165$$

There are 165 Republicans and 269 Democrats.

27. Let x = the length of the first piece. Then, the length of the second piece is $x+20$.

$$\begin{pmatrix}\text{length of}\\\text{the 1st piece}\end{pmatrix}+\begin{pmatrix}\text{length of}\\\text{2nd piece}\end{pmatrix}=86\text{ cm}$$
$$x+(x+20)=86$$
$$2x=66$$
$$x=33$$

The length of the pieces are 33 cm and 53 cm.

29. Let x = length of the Congo River. Then, the length of the Nile is $x+2455$.

$$\begin{pmatrix}\text{length of}\\\text{Congo}\end{pmatrix}+\begin{pmatrix}\text{length of}\\\text{Nile}\end{pmatrix}=11{,}195\text{ km}$$
$$x+(x+2455)=11195$$
$$2x=8740$$
$$x=4370$$

The length of the Congo is 4370 km; the length of the Nile is 6825 km.

31. **(a)** Let x = smallest integer. Then, $x + 1$ is the next integer and $x + 2$ is the largest.

(b) Let x = largest integer. Then, $x - 1$ is the next and $x - 2$ is the smallest.

33. **(a)** Let x = smallest odd integer. Then, $x + 2$ is the next, and $x + 4$ is the largest.

(b) Let x = largest odd integer. then, $x - 2$ is the next, and $x - 4$ is the smallest.

35. Let x = the first page number and $x + 1$ be the next page number.

$$\begin{pmatrix}\text{1st page}\\\text{number}\end{pmatrix}+\begin{pmatrix}\text{2nd page}\\\text{number}\end{pmatrix}=941$$
$$x+x+1=941$$
$$2x=940$$
$$x=470$$

The page numbers are 470 and 471.

37. Let x = first odd integer. Then $x + 2$ is the next odd integer and $x + 4$ is the third.

$$\begin{pmatrix}\text{3 times the}\\\text{smallest}\end{pmatrix}=\begin{pmatrix}\text{9 more than}\\\text{twice the largest}\end{pmatrix}$$
$$3x=2(x+4)+9$$
$$3x=2x+8+9$$
$$x=17$$

The numbers are 17, 19, and 21.

39. Let x = smallest integer. Then, $x + 1$ is the next integer and $x + 2$ is the largest.

$$(\text{3 times the largest})=\begin{pmatrix}\text{47 + the sum}\\\text{of the two smaller}\end{pmatrix}$$
$$3(x+2)=47+x+x+1$$
$$3x+6=2x+48$$
$$x=42$$

The numbers are 42, 43, and 44.

41. Let x, $x + 1$, and $x + 2$ be the lengths of the sides of the triangle.

$$\text{Perimeter}=\begin{pmatrix}\text{sum of the}\\\text{lengths of the}\\\text{three sides}\end{pmatrix}$$
$$42=x+x+1+x+2$$
$$42=3x+3$$
$$3x=39$$
$$x=13$$

The sides are 13 in., 14 in., and 15 in.

43. Let x, $x + 1$, $x + 2$, $x + 3$, $x + 4$ be the lengths of the sides of the pentagon.

$$\text{Perimeter}=\begin{pmatrix}\text{sum of the}\\\text{lengths of the}\\\text{five sides}\end{pmatrix}$$
$$80=x+x+1+x+2+x+3+x+4$$
$$80=5x+10$$
$$5x=70$$
$$x=14$$

The sides are 14 in., 15 in., 16 in., 17 in., and 18 in.

45. Let x = area of New Guinea. Then, the area of Greenland is $3x - 201{,}900$.

$$\begin{pmatrix}\text{Area of}\\\text{Greenland}\end{pmatrix}=2{,}175{,}600\text{ km}^2$$
$$3x-201{,}900=2{,}175{,}600$$
$$3x=2{,}377{,}500$$
$$x=792{,}500$$

The area of New Guinea is $792{,}500\text{ km}^2$.

47. Let x = the land area of Africa. Then, the land area of Asia is $x + 14{,}514{,}000$.

$$\begin{pmatrix}\text{area of}\\\text{Africa}\end{pmatrix}+\begin{pmatrix}\text{area of}\\\text{Asia}\end{pmatrix}=74{,}644{,}000\text{ km}^2$$
$$x+x+14{,}514{,}000=74{,}644{,}000$$
$$2x=60{,}130{,}000$$
$$x=30{,}065{,}000$$

The area of Africa is $30{,}065{,}000\text{ km}^2$. The area of Asia is $44{,}579{,}000\text{ km}^2$.

Section 2.5 Practice Exercises

1. Let x = the first integer. Then $x + 2$ is the next consecutive odd integer.
$$x+(x+2)=-280$$
$$2x=-282$$
$$x=-71$$
The numbers are –69 and –71.

3. $$15\left(\frac{a}{15}+6\right)=15\left(\frac{a}{5}+\frac{2}{3}\right)$$
$$a+90=3a+10$$
$$2a=80$$
$$a=40$$

5. $57\% = 0.57$

7. $135\% = 1.35$

9. $0.69 = 69\%$

11. $0.006 = 0.6\%$

13. Let x = the percent. $45=x(360)$
$$360x=45$$
$$x=0.125$$
$$x=12.5\%$$

15. Let x = the percent. $544=x(640)$
$$640x=544$$
$$x=0.85$$
$$x=85\%$$

17. Let x = the number. $0.5\% \text{ of } 150=x$
$$0.005(150)=x$$
$$x=0.75$$

19. Let x = the number. $x=42\% \text{ of } 740$
$$x=0.42(740)$$
$$x=310.8$$

21. Let x = the number. $177=20\% \text{ of } x$
$$177=0.20x$$
$$x=\frac{177}{0.20}=885$$

23. Let x = the number. $275=12.5\% \text{ of } x$
$$2200=0.125x$$
$$x=\frac{275}{0.125}=2200$$

25. $$\frac{(\text{number of African American})}{(\text{Total number})}\cdot 100$$
$$=\frac{21,748}{42,700}\cdot 100$$
$$=0.509\cdot 100$$
$$=50.9\%$$

27. $$\frac{(\text{number of Caucasian})}{(\text{Total number})}\cdot 100=\frac{13,174}{42,700}\cdot 100$$
$$=0.309\cdot 100$$
$$=30.9\%$$

29. Let x = total price.
$$(\text{price})+(7\% \text{ of the price})=x$$
$$74.95+(74.95)(0.07)=x$$
$$74.95+5.2465=x$$
$$x=80.20$$
The price is \$80.20.

31. Let x = price.
$$(6.5\% \text{ of the price})=\$1.04$$
$$0.065x=1.04$$
$$x=\frac{1.04}{0.065}=16.00$$
The price of the screwdrivers is \$16.00.

33. Let x = tax rate.
$$(\text{price})+(x\% \text{ tax rate of price})=\$1890$$
$$1800+x(1800)=1890$$
$$1800x=90$$
$$x=\frac{90}{1800}$$
$$=0.05$$
The tax is 5%.

35. Let x = original price.
$$(\text{price})-(30\% \text{ off price})=\$20.97$$
$$x-0.30x=20.97$$
$$0.70x=20.97$$
$$x=\frac{20.97}{0.70}=29.96$$
The original price is \$29.96.

37. Let x = commission.
$$(25\% \text{ of profit}) = \text{commission}$$
$$0.25(18,250) = x$$
$$x = 4562.50$$
The commission is $4562.50.

39. Let x = amount sold.
$$(3\% \text{ of amount sold}) = \$116.37$$
$$0.03x = 116.37$$
$$x = \frac{116.37}{0.03} = 3879$$
The amount sold is $3,879.

41. Let x = rate.
$$\begin{pmatrix} x\% \text{ of} \\ \text{amount sold} \end{pmatrix} = \$2300 - \$500 \begin{cases} \text{sales have} \\ \text{to exceed} \\ \$500 \end{cases}$$
$$x(180) = 1800$$
$$x = 10$$
The rate is 10%.

43. $I = Prt$
$I = (\$3000)(3.5\%)(4 \text{ years})$
$I = (105)(4) = \$420$

45. Let P = amount borrowed.
$$\text{Total paid} = P + Prt$$
$$\$1260 = P + P(12\%)(1)$$
$$1260 = P + 0.12P$$
$$1.12P = 1260$$
$$P = \$1200$$

47. Interest = \$1950 − \$1500 = \$450
$$I = Prt$$
$$450 = 1500(r)(5)$$
$$450 = 7500r$$
$$r = 0.06 = 6\%$$

49. **(a)** $I = Prt$
$$I = (\$2000)(11\%)\left(\frac{3}{2}\right)$$
$$I = (2000)(0.165)$$
$$I = \$330$$

(b) $(\text{principal} + \text{interest}) = P + I$
$$= \$2000 + \$330$$
$$= \$2330$$

51. Let x = amount sold over $200.
$$\begin{pmatrix} 4\% \text{ on amount} \\ \text{sold over } \$200 \end{pmatrix} = \$25.80$$
$$0.04x = 25.80$$
$$x = \frac{25.80}{0.04} = 645$$
Diane sold $645 over $200.

53. Let x = original price.
$$(\text{sale price}) = x - 0.15x$$
$$5\% \text{ of sale price} = \$2.55$$
$$0.05(x - 0.15x) = \$2.55$$
$$0.05(0.85x) = \$2.55$$
$$0.0425x = \$2.55$$
$$x = 60$$
The original price is $60.

Section 2.6 Practice Exercises

1.
$$3(2y+3) - 4(-y+1) = 7y - 10$$
$$6y + 9 + 4y - 4 = 7y - 10$$
$$10y + 5 = 7y - 10$$
$$3y = -15$$
$$y = -5$$

3.
$$\frac{1}{2}(x-3) + \frac{3}{4} = 3x - \frac{3}{4}$$
$$(4)\left[\frac{1}{2}(x-3) + \frac{3}{4}\right] = (4)\left[3x - \frac{3}{4}\right]$$
$$2(x-3) + 3 = 12x - 3$$
$$2x - 3 = 12x - 3$$
$$-10x = 0$$
$$x = 0$$

5.
$$0.5(y+2) - 0.3 = 0.4y + 0.5$$
$$0.5y + 1 - 0.3 = 0.4y + 0.5$$
$$0.1y = -0.2$$
$$y = -2$$

7.
$$P = a + b + c$$
$$P - b - c = a + b + c - b - c$$
$$a = P - b - c$$

9.
$$x = y - z$$
$$y = x + z$$

11.
$$p = 250 + q$$
$$p - 250 = 250 - 250 + q$$
$$q = p - 250$$

13. $d = rt$
$$\frac{d}{r} = \frac{rt}{r}$$
$$t = \frac{d}{r}$$

15. $PV = nrt$
$$\frac{PV}{nr} = t$$

17. $x - y = 5$
$$x = 5 + y$$

19. $3x + y = -19$
$$y = -19 - 3x$$

21. $2x + 3y = 6$
$$3y = 6 - 2x$$
$$y = \frac{6 - 2x}{3} = -\frac{2}{3}x + 2$$

23. $-2x - y = 9$
$$-2x = 9 + y$$
$$x = \frac{9 + y}{-2} = -\frac{9 + y}{2}$$

25. $4x - 3y = 12$
$$-3y = 12 - 4x$$
$$y = \frac{12 - 4x}{-3} = \frac{4}{3}x - 4$$

27. $ax + by = c$
$$by = c - ax$$
$$y = \frac{c - ax}{b}$$

29.
$$A = P(1 + rt)$$
$$A = P + Prt$$
$$A - P = Prt$$
$$t = \frac{A - P}{Pr} = \frac{A}{Pr} - \frac{1}{r}$$

31.
$$a = 2(b + c)$$
$$a = 2b + 2c$$
$$a - 2b = 2c$$
$$c = \frac{a - 2b}{2}$$

33. $Q = \frac{x + y}{2}$
$$2Q = x + y$$
$$y = 2Q - x$$

35. $M = \frac{a}{S}$
$$a = MS$$

37. $P = I^2R$
$$R = \frac{P}{I^2}$$

39. Let x = width. Then, length equals $x + 2$.
$$P = 2w + 2l$$
$$24 = 2x + 2(x + 2)$$
$$24 = 2x + 2x + 4$$
$$4x = 20$$
$$x = 5$$
The width is 5 feet; the length is 7 feet.

41. Let x = width. Then, the length is $2x - 5$.
$$P = 2w + 2l$$
$$590 = 2x + 2(2x - 5)$$
$$590 = 2x + 4x - 10$$
$$6x = 600$$
$$x = 100$$
the width is 100 m; the length is 195 m.

43. Let x = smallest angle. Then the middle angle is $2x$ and the largest angle is $3x$.
(sum of angles of a triangle) $= 180°$
$$x + 2x + 3x = 180$$
$$6x = 180$$
$$x = 30°$$
The angles are 30°, 60°, 90°.

45. Let x = largest angle. Then, the middle angle is $x - 30°$, and the smallest angle is $\frac{1}{2}$ of $x = \frac{1}{2}x$.

(sum of angles of a triangle) $= 180°$

$$x + x - 30 + \frac{1}{2}x = 180$$
$$\frac{5}{2}x = 210$$
$$x = 210 \cdot \frac{2}{5} = 84°$$

The angles are 84°,
84° – 30° = 54°
$\frac{1}{2}(84°) = 42°$

47. The sum of complementary angles is 90°.

$$(3x + 5) + (2x) = 90$$
$$5x = 85$$
$$x = 17$$

The angles are 3(17) + 5 = 56°, 2(17) = 34°.

49. Adjacent Supplementary angles form a Straight angle. The words *supplementary* and *straight* both begin with the same letter.

51. Let x = one angle. Then, $2x$ is the other angle.

(sum of the two angles) $= 90°$

$$x + 2x = 90$$
$$3x = 90$$
$$x = 30$$

The angles are 30° and 60°.

53. Let x = one angle. Then, $3x$ is the other angle.

(sum of the angles) $= 180°$

$$x + 3x = 180$$
$$4x = 180$$
$$x = 45°$$

The angles are 45° and 135°.

55. Let x = one angle. Then $4x + 6$ is the other angle.

(sum of the angles) $= 180°$

$$x + 4x + 6 = 180$$
$$5x = 174$$
$$x = 34.8$$

The angles are 34.8° and 145.2°.

57. Vertical angles are equal.

$$3y + 26 = 5y - 54$$
$$-2y = -80$$
$$y = 40$$

The angles are 3(40) + 26 = 146° and 5(40) – 54 = 146°.

59. **(a)** $A = bh$

(b) $A = bh$

$$\frac{A}{h} = \frac{bh}{h}$$
$$b = \frac{A}{h}$$

(c) $b = \frac{A}{h}$

$$b = \frac{40}{5} = 8 \text{ m}$$

61. **(a)** $P = s + s + s + s = 4s$

(b) $P = 4s$

$$\frac{P}{4} = \frac{4s}{4}$$
$$s = \frac{P}{4}$$

(c) $s = \frac{P}{4}$

$$s = \frac{921.6}{4} = 230.4 \text{ m}$$

63. **(a)** $V = lwh$

(b) $V = lwh$

$$\frac{V}{lw} = \frac{lwh}{lw}$$
$$h = \frac{V}{lw}$$

(c) $h = \frac{V}{lw}$

$h = \frac{45}{(4.5)(5.0)} = \frac{45}{22.5}$

$h = 2 \text{ feet}$

65. $V = \frac{4}{3}\pi r^3$

$\frac{3}{4\pi} \cdot V = \frac{\not{3}}{\not{4}\pi} \cdot \frac{\not{4}}{\not{3}}\pi r^3$

$r^3 = \frac{3V}{4\pi}$

67. $V = \frac{1}{3}lwh$

$3V = lwh$

$\frac{3V}{lw} = \frac{lwh}{lw}$

$h = \frac{3V}{lw}$

69. (a) $A = \pi r^2$

$A = (3.14)(20)^2 = 3.14(400)$

$A = 1256 \text{ ft}^2$

(b) $V = \pi r^2 h$

$V = (3.14)(20)^2(8) = 3.14(400)(8)$

$V = 10{,}048 \text{ ft}^3$

71. (a) $A = bh$

$A = (7)(4) = 28 \text{ m}^2$

(b) $A = \frac{1}{2}bh$

$A = 0.5(7)(4) = 14 \text{ m}^2$

(c) The area of the triangle is exactly one-half the area of the parallelogram.

73.

```
880/(2π)
          140.0563499
1600/(π(4)²)
          31.83098862
```

75.

```
20/((-0.05)*(5))
                  -80
10/(0.5(6+4))
                    2
```

Section 2.7 Practice Exercises

1. $3x + 7 = 13$

$3x = 6$

$x = 2$

3. $4x + 5y = 20$

$5y = 20 - 4x$

$y = \frac{20 - 4x}{5} = -\frac{4}{5}x + 4$

5. $0.3(a + 4) = 7.1a - 2.2$

$10(0.3)(a + 4) = (10)(7.1a - 2.2)$

$3(a + 4) = 71a - 22$

$3a + 12 = 71a - 22$

$-68a = -34$

$a = \frac{-34}{-68} = 0.5$

7. $-7 = -4r - 6$

$-1 = -4r$

$r = \frac{-1}{-4} = \frac{1}{4}$

9. (a) $\text{Revenue} = (\text{price})\left(\begin{matrix}\text{number of} \\ \text{tickets sold}\end{matrix}\right)$

$R = (6)(200) = \$1200$

(b) $R = 6x$

(c) $R = 6(750 - x) = 4500 - 6x$

11. Let x = number of \$3 tickets

	\$3 tickets	\$2 tickets	Total
Number of tickets	x	$81 - x$	81
Value of tickets	$3x$	$2(81 - x)$	215

$$\begin{pmatrix}\text{Value of}\\ \text{\$3 tickets}\end{pmatrix}+\begin{pmatrix}\text{Value of}\\ \text{\$2 tickets}\end{pmatrix}=\$215$$

$$\begin{aligned}3x+2(81-x)&=215\\ 3x+162-2x&=215\\ x&=53\end{aligned}$$

53 tickets were sold at \$3; 28 sold at \$2.

13. Let x = amount of \$2 tickets. then $208-x$ is the amount of \$1 tickets.

$$\begin{pmatrix}\text{value of}\\ \text{\$2 tickets}\end{pmatrix}+\begin{pmatrix}\text{value of}\\ \text{\$1 tickets}\end{pmatrix}=\$320$$

$$\begin{aligned}2x+1(208-x)&=320\\ 2x+208-x&=320\\ x&=112\end{aligned}$$

112 \$2 tickets sold; 96 \$1 tickets

15. **(a)** 10% of 20 pounds, $0.10(20)=2$ lb

(b) 10% of x pounds, $0.10x$

(c) 10% of $(x+3)$ pounds, $0.10(x+3)=0.10x+0.30$

17. Let x = pounds of \$12 coffee.

	\$12 coffee	\$8 coffee	Total
Number of pounds	x	$50-x$	50
Value of coffee	$12x$	$8(50-x)$	50(8.8) = \$440

$$\begin{pmatrix}\text{value of}\\ \text{\$12 coffee}\end{pmatrix}+\begin{pmatrix}\text{value of}\\ \text{\$8 coffee}\end{pmatrix}=\$440$$

$$\begin{aligned}12x+8(50-x)&=440\\ 12x+400-8x&=440\\ 4x&=40\\ x&=10\end{aligned}$$

10 lb of \$12 coffee; 40 lb of \$8 coffee

19. Let x = pounds of raisins. Then, $(6-x)$ is the pounds of granola.

$$\begin{pmatrix}\text{value of}\\ \text{raisins}\end{pmatrix}+\begin{pmatrix}\text{value of}\\ \text{granola}\end{pmatrix}=6\text{ lb}\cdot\$2.29$$

$$\begin{aligned}1.69x+2.59(6-x)&=13.74\\ 1.69x+15.54-2.59x&=13.74\\ -0.90x&=-1.80\\ x&=2\end{aligned}$$

2 lb of raisins; 4 lb of granola

21. Let x = pounds of black tea. Then, $(4-x)$ is the pounds of orange pekoe tea.

$$\begin{pmatrix}\text{cost of}\\ \text{black tea}\end{pmatrix}+\begin{pmatrix}\text{cost of}\\ \text{orange}\\ \text{pekoe tea}\end{pmatrix}=\begin{pmatrix}\text{\$2.50 per}\\ \text{pound for}\\ \text{4 lb}\end{pmatrix}$$

$$\begin{aligned}\$2.20x+\$3.00(4-x)&=\$2.50(4)\\ 2.2x+12-3x&=10\\ -0.8x&=-2\\ x&=2.5\end{aligned}$$

2.5 lb of black tea; 1.5 lb of orange pekoe tea

23. **(a)** Interest = Principal(rate)(time)

$$\begin{aligned}I&=Prt\\ I&=(5000)(0.06)(1)\\ I&=\$300\end{aligned}$$

(b) Prt

$x(0.06)(1)$

$0.06x$

(c) Prt

$(20000-x)(0.06)(1)$

$0.06(20{,}000-x)$ or $1200-0.06x$

25. Let x = amount invested at 8%.

	6% account	8% account	Total
Principal	$800+x$	x	$800+2x$
Interest	$0.06(800+x)$	$0.08x$	\$104

$$\begin{pmatrix}\text{Interest}\\\text{at } 6\%\end{pmatrix}+\begin{pmatrix}\text{Interest}\\\text{at } 8\%\end{pmatrix}=\$104$$
$$0.06(800+x)+0.08x=104$$
$$48+0.06x+0.08x=104$$
$$0.14x=56$$
$$x=400$$
There is \$400 at 8%; \$1200 at 6%.

27. Let x = amount invested at 8%

	8% account	12% account	Total
Principal	x	$12500-x$	\$12500
Interest	$0.08x$	$0.12(12500-x)$	\$1160

$$\begin{pmatrix}\text{Interest}\\\text{at } 8\%\end{pmatrix}+\begin{pmatrix}\text{Interest}\\\text{at } 12\%\end{pmatrix}=\$1160$$
$$0.08x+0.12(12500-x)=1160$$
$$0.08x+1500-0.12x=1160$$
$$-0.04x=-340$$
$$x=8500$$
\$8500 at 8%; \$4000 at 12%

29. Let x = amount at 5%. Then, $2x$ is amount at 6%.
$$\begin{pmatrix}\text{Interest}\\\text{at } 5\%\end{pmatrix}+\begin{pmatrix}\text{Interest}\\\text{at } 6\%\end{pmatrix}=\begin{pmatrix}\text{total}\\\text{interest}\end{pmatrix}$$
$$0.05x+0.06(2x)=\$765$$
$$0.05x+0.12x=765$$
$$0.17x=765$$
$$x=4500$$
\$4500 at 5%; \$9000 at 6%

31. Let x = amount at 6%. Then, $(8750-x)$ is the amount at 8%.
$$\begin{pmatrix}\text{Interest}\\\text{at } 6\%\end{pmatrix}=\begin{pmatrix}\text{Interest}\\\text{at } 8\%\end{pmatrix}$$
$$0.06x=0.08(8750-x)$$
$$0.06x=700-0.08x$$
$$0.14x=700$$
$$x=5000$$
\$5000 at 6%; \$3750 invested at 8%

33. **(a)** distance = (rate)(time)
distance = (60)(5) = 300 miles

(b) (rate)(time) = $(x)(5) = 5x$

(c) (rate)(time) = $(x+12)(5) = 5(x+12)$

35. Let x = time traveled by the 2nd car.
$$\begin{pmatrix}\text{distance of}\\\text{1st car}\end{pmatrix}=\begin{pmatrix}\text{distance of}\\\text{2nd car}\end{pmatrix}$$
$$(\text{rate})(\text{time})=(\text{rate})(\text{time})$$
$$55(4)=44(x)$$
$$44x=220$$
$$x=5$$
The car travels 5 hours.

37. Let x = rate of the slower car.

	Distance	Rate	Time
Faster car	$2(x+4)$	$x+4$	2
Slower car	$2(x)$	x	2

$$\begin{pmatrix}\text{distance of}\\\text{1st car}\end{pmatrix}+\begin{pmatrix}\text{distance of}\\\text{2nd car}\end{pmatrix}=192\text{ miles}$$
$$2(x+4)+2x=192$$
$$4x+8=192$$
$$4x=184$$
$$x=46$$
the slower car is traveling at 46 mph. The faster car is traveling at 50 mph.

39. Let x = rate of Cessna. Then, $(x+10)$ is the rate of the Piper.
$$\begin{pmatrix}\text{distance}\\\text{of Cessna}\end{pmatrix}+\begin{pmatrix}\text{distance}\\\text{of Piper}\end{pmatrix}=(\text{total distance})$$
$$(\text{rate})(\text{time})+(\text{rate})(\text{time})=690$$
$$x(3)+(x+10)(3)=690$$
$$3x+3x+30=690$$
$$6x=660$$
$$x=110$$
The speed of the Cessna is 110 mph; the speed of the Piper is 120 mph.

41. Let x = rate hiking uphill. Then, $(x + 1)$ is the rate hiking downhill.

$$\begin{pmatrix}\text{distance}\\ \text{hiking uphill}\end{pmatrix}=\begin{pmatrix}\text{distance}\\ \text{hiking downhill}\end{pmatrix}$$
$$\begin{aligned}(\text{rate})(\text{time}) &= (\text{rate})(\text{time})\\ x(6) &= (x+1)(3)\\ 6x &= 3x+3\\ 3x &= 3\\ x &= 1\end{aligned}$$

The rate hiking down to the lake is 2 mph.

43. Let x = speed of slower boat. Then, $2x$ is the speed of the faster boat.

$$\begin{pmatrix}\text{distance of}\\ \text{slower boat}\end{pmatrix}+60\text{ miles}=\begin{pmatrix}\text{distance of}\\ \text{faster boat}\end{pmatrix}$$
$$\begin{aligned}(\text{rate})(\text{time})+60 &= (\text{rate})(\text{time})\\ (x)(3)+60 &= (2x)(3)\\ 3x+60 &= 6x\\ 3x &= 60\\ x &= 20\end{aligned}$$

The speeds are 20 mph and 40 mph.

45. **(a)** 70 nickels $\cdot$ \$0.05 = \$3.50

(b) $(x \text{ nickels})(0.05) = 0.05x$

(c) $(30 + x)(0.05) = 0.05(30 + x)$

47. Let x = number of quarters. Then, $(12 - x)$ is the number of dimes.

	Quarters	Dimes	Total
Number	x	$12 - x$	12
Value	$0.25x$	$0.10(12 - x)$	\$1.65

$$\begin{pmatrix}\text{value of}\\ \text{quarters}\end{pmatrix}+\begin{pmatrix}\text{value of}\\ \text{dimes}\end{pmatrix}=\$1.65$$
$$\begin{aligned}0.25x+0.10(12-x) &= 1.65\\ 0.25x+1.2-0.10x &= 1.65\\ 0.15x &= 0.45\\ x &= 3\end{aligned}$$

There are 3 quarters and 9 dimes.

49. Let x = number of \$10 bills. Then, $2x$ is the number of \$20 bills.

$$\begin{pmatrix}\text{value of}\\ \$10\text{ bills}\end{pmatrix}+\begin{pmatrix}\text{value of}\\ \$20\text{ bills}\end{pmatrix}=\$200$$
$$\begin{aligned}10x+20(2x) &= 200\\ 50x &= 200\\ x &= 4\end{aligned}$$

She has four \$10 bills and eight \$20 bills.

Section 2.8 Practice Exercises

1. **(a)** $3(x + 2) - (2x - 7)$
$3x + 6 - 2x + 7$
$x + 13$

(b) $-(5x - 1) - 2(x + 6)$
$-5x + 1 - 2x - 12$
$-7x - 11$

(c)
$$\begin{aligned}3(x+2)-(2x-7) &= -(5x-1)-2(x+6)\\ x+13 &= -7x-11\\ 8x &= -24\\ x &= -3\end{aligned}$$

3. Let x = points of U.S. Then, $(x + 5)$ is the points of Europe.

$$\begin{pmatrix}\text{points of}\\ \text{U.S.}\end{pmatrix}+\begin{pmatrix}\text{points of}\\ \text{Europe}\end{pmatrix}=28$$
$$\begin{aligned}x+x+5 &= 28\\ 2x &= 23\\ x &= 11\frac{1}{2}\end{aligned}$$

The U.S. scored 11.5 points; Europe scored 16.5 points.

5. $[6, \infty)$

7. $(-\infty, 2.1]$

9. $(-2, 7]$

11. $\left(\frac{3}{4}, \infty\right)$

13. $(-1, 8)$

15. $(-\infty, -14)$

17. $[18, \infty)$

19. $(-\infty, -0.6)$

21. $[-3.5, -7.1]$

23. (a) $x+3-3=6-3$
$x=3$

(b) $x+3-3>6-3$
$x>3$

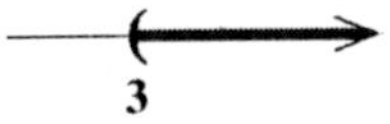

3

25. (a) $p-4+4=9+4$
$p=13$

(b) $p-4+4\le 9+4$
$p\le 13$

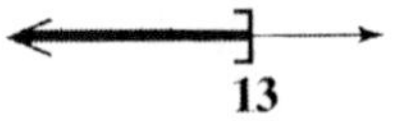

13

27. (a) $\frac{4c}{4}=\frac{-12}{4}$
$c=-3$

(b) $\frac{4c}{4}<\frac{-12}{4}$
$c<-3$

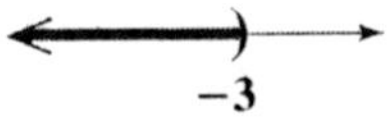

−3

29. (a) $\frac{-10z}{-10}=\frac{15}{-10}$
$z=-\frac{15}{10}=-\frac{3}{2}$

(b) $-10z\le 15$
$\frac{-10z}{-10}\le\frac{15}{-10}$
$z\ge -\frac{3}{2}$

$-\frac{3}{2}$

31. $-2(-2)+5<4$
$9 \not< 4$; No

33. $4(1+7)-1>2+1$
$10>3$; Yes

35. $x+5-5\le 6-5$
$x\le 1$
$(-\infty, 1]$

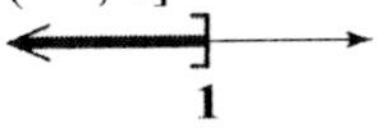

1

37. $q-7+7>3+7$
$q>10$
$(10, \infty)$

10

39. $4<1+z$
$1-1+z>4-1$
$z>3$
$(3, \infty)$

3

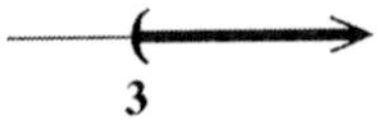

41. $2\ge a-6$
$2+6\ge a-6+6$
$8\ge a$ or $a\le 8$
$(-\infty, 8]$

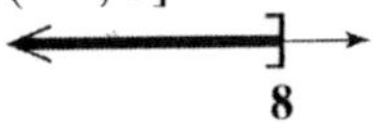

8

43. $3c>6$
$\frac{3c}{3}>\frac{6}{3}$
$c>2$
$(2, \infty)$

2

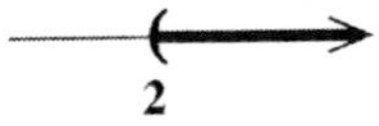

45. $-3c>6$
$\frac{-3c}{-3}>\frac{6}{-3}$
$c<-2$
$(-\infty, -2)$

−2

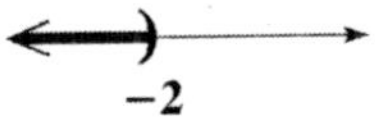

47. $-h\le -14$
$(-1)(-h)\le(-1)(-14)$
$h\ge 14$
$[14, \infty)$

14

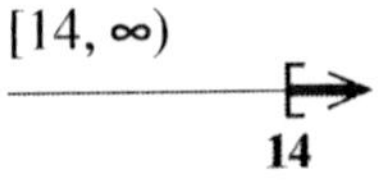

49. $12 \ge -\frac{x}{2}$

$(-2)(12) \ge (-2)\left(-\frac{x}{2}\right)$

$-24 \le x$ or $x \ge -24$

$[-24, \infty)$

−24

51. $-2 \le p+1 < 4$

$-2-1 \le p+1-1 < 4-1$

$-3 \le p < 3$

$[-3, 3)$

−3 3

53. $-3 < 6h-3 < 12$

$0 < 6h < 15$

$0 < h < \frac{15}{6}$

$0 < h < \frac{5}{2}$

$\left(0, \frac{5}{2}\right)$

$0 \quad \frac{5}{2}$

55. $5 < \frac{1}{2}x < 6$

$2(5) < 2\left(\frac{1}{2}x\right) < 2(6)$

$10 < x < 12$

$(10, 12)$

10 12

57. $-5 \le 4x-1 < 15$

$-4 \le 4x < 16$

$-1 \le x < 4$

$[-1, 4)$

−1 4

59. $0.6z \ge 54$

$\frac{0.6z}{0.6} \ge \frac{54}{0.6}$

$z \ge 90$

$[90, \infty)$

90

61. $-\frac{2}{3}y < 6$

$\left(-\frac{3}{2}\right)\left(-\frac{2}{3}y\right) < \left(-\frac{3}{2}\right)(6)$

$y > -9$

$(-9, \infty)$

−9

63. $-2x-4 \le 11$

$-2x \le 15$

$\frac{-2x}{-2} \le \frac{15}{-2}$

$x \ge -\frac{15}{2}$

$\left[-\frac{15}{2}, \infty\right)$

$-\frac{15}{2}$

65. $-7b-3 \le 2b$

$-9b \le 3$

$b \ge -\frac{1}{3}$

$\left[-\frac{1}{3}, \infty\right)$

$-\frac{1}{3}$

67. $4n+2 < 6n+8$

$-2n < 6$

$n > -3$

$(-3, \infty)$

−3

69. $8-6(x-3)>-4x+12$
$8-6x+18>-4x+12$
$-2x>-14$
$x<7$

$(-\infty, 7)$

7

71. $\frac{7}{6}p+\frac{4}{3}\geq\frac{11}{6}p-\frac{7}{6}$
$6\left(\frac{7}{6}p+\frac{4}{3}\right)\geq6\left(\frac{11}{6}p-\frac{7}{6}\right)$
$7p+8\geq11p-7$
$-4p\geq-15$
$p\leq\frac{15}{4}$

$\left(-\infty, \frac{15}{4}\right]$

$\frac{15}{4}$

73. $-1.2a-0.4<-0.4a+2$
$-0.8a<2.4$
$a>-3$

$(-3, \infty)$

-3

75. **(a)** A [93, 100]
B+ [89, 93)
B [84, 89)
C+ [80, 84)
C [75, 80)
F [0, 75)

(b) B

(c) C

77. $s\geq110$

79. $t>90$

81. $h\leq2$

83. $t\geq100°$

85. $d\leq10$

87. $2<h<5$

89. Let x = August rainfall.
$\left(\text{Average summer rainfall}\right)\geq(7.4\text{ inches})$
$\left(\frac{5.9+6.1+x}{3}\right)\geq7.4$
$12+x\geq22.2$
$x\geq10.2$
More than 10.2 inches of rain is needed.

91. **(a)** Let x = number of birdhouses ordered.
$(\text{original price})-\left(\text{percent discount}\right)=\left(\text{total cost}\right)$
$\left(\text{\$9 of number ordered}\right)-\left(\text{10\% of original cost}\right)=\left(\text{total cost}\right)$
$9x-0.10(9x)=\left(\text{total cost}\right)$
$9(190)-0.10(9)(190)=1539$
The total cost is \$1539.

(b) Total cost for 200 birdhouses is
9(200) – 0.20(9)(200) = 1440, or \$1440.
The cost of 190 is \$1539. Therefore, it costs more to purchase 190 birdhouses.

93. **(a)** Revenue > Cost
$2.00x > 75 + 0.17x$

(b) $2.00x>75+0.17x$
$1.83x>75$
$x>41$
Profit occurs when more than 41 lemonades are sold.

95. $3(x+2)-(2x-7)\leq(5x-1)-2(x+6)$
$3x+6-2x+7\leq5x-1-2x-12$
$x+13\leq3x-13$
$-2x\leq-26$
$x\geq13$

$[13, \infty)$

13

97.
$$-2-\frac{w}{4}\le\frac{1+w}{3}$$
$$12\left(-2-\frac{w}{4}\right)\le12\left(\frac{1+w}{3}\right)$$
$$-24-3w\le4(1+w)$$
$$-7w\le28$$
$$w\ge-4$$
$[-4, \infty)$

−4

99. $-0.703<0.122p-2.472$
$$1.769<0.122p$$
$$14.5<p \text{ or } p>14.5$$
$(14.5, \infty)$

14.5

Chapter 2 Review Exercises

1. **(a)** Equation

(b) Expression

(c) Equation

(d) Equation

3. b and d

5.
$$a+6=-2$$
$$a+6-6=-2-6$$
$$a=-8$$

7. $-\frac{3}{4}+k=\frac{9}{2}$
$$k=\frac{9}{2}+\frac{3}{4}=\frac{18}{4}+\frac{3}{4}=\frac{21}{4}=5\frac{1}{4}$$

9. $-5x=21$
$$x=\frac{21}{-5}=-\frac{21}{5}=-4\frac{1}{5}$$

11.
$$-\frac{2}{5}k=\frac{4}{7}$$
$$-\frac{5}{2}\left(-\frac{2}{5}k\right)=-\frac{5}{2}\cdot\frac{4}{7}$$
$$k=-\frac{20}{14}=-\frac{10}{7}$$

13. Let x = the number.
$$\begin{pmatrix}\text{quotient of}\\ x \text{ and } -6\end{pmatrix}=-10$$
$$\frac{x}{-6}=-10$$
$$x=(-10)(-6)=60$$
The number is 60.

15. Let x = the number.
$$x-4=-12$$
$$x=-8$$
The number is −8.

17. $4d+2=6$
$$4d=4$$
$$d=1$$

19. $-7c=-3c-9$
$$-4c=-9$$
$$c=\frac{-9}{-4}=\frac{9}{4}$$

21. $\frac{b}{3}+1=0$
$$\frac{b}{3}=-1$$
$$b=(-1)(3)=-3$$

23. $-3p+7=5p+1$
$$-8p=-6$$
$$p=\frac{-6}{-8}=\frac{3}{4}$$

25. $4a-9=3(a-3)$
$$4a-9=3a-9$$
$$a=0$$

27. $7b+3(b-1)+16=2(b+8)$
$$7b+3b-3+16=2b+16$$
$$10b+13=2b+16$$
$$8b=3$$
$$b=\frac{3}{8}$$

29. A contradiction has no solution and an identity is true for all real numbers.

31.
$$\frac{x}{8}-\frac{1}{4}=\frac{1}{2}$$
$$8\left(\frac{x}{8}-\frac{1}{4}\right)=8\left(\frac{1}{2}\right)$$
$$x-2=4$$
$$x=6$$

33.
$$\frac{4z+7}{5}=z+2$$
$$5\left(\frac{4z+7}{5}\right)=5(z+2)$$
$$4z+7=5z+10$$
$$-z=3$$
$$z=-3$$

35.
$$\frac{1}{10}p-3=\frac{2}{5}p$$
$$10\left(\frac{1}{10}p-3\right)=10\left(\frac{2}{5}p\right)$$
$$p-30=4p$$
$$-3p=3p$$
$$p=-10$$

37.
$$-\frac{1}{4}(2-3t)=\frac{3}{4}$$
$$4\left(-\frac{1}{4}(2-3t)\right)=4\cdot\frac{3}{4}$$
$$-2+3t=3$$
$$3t=5$$
$$t=\frac{5}{3}$$

39. $17.3-2.7q=10.55$
$$-2.7q=-6.75$$
$$q=2.5$$

41. $5.74a+9.28=2.24a-5.42$
$$3.5a=-14.7$$
$$a=-4.2$$

43. $0.05x+0.10(24-x)=0.75(24)$
$$0.05x+2.4-0.10x=18$$
$$-0.05x=15.6$$
$$x=-312$$

45. $100-(t-6)=-(t-1)$
$$100-t+6=-t+1$$
$$106\neq 1$$
No solution

47. $5t-(2t+14)=3t-14$
$$3t-14=3t-14$$
$$0=0$$
All real numbers

49. $4w+5[2(w+4)-3]=2(w-3)-5$
$$4w+5[2w+8-3]=2w-6-5$$
$$4w+10w+25=2w-11$$
$$12w=-36$$
$$w=-3$$

51. Let x = the number.
$$\left(\text{twice the number}\right)+(4)=40$$
$$2x+4=40$$
$$2x=36$$
$$x=18$$
The number is 18.

53. Let x = the number.
$$12+(x+2)=44$$
$$14+x=44$$
$$x=30$$
The number is 30.

55. Let x = the number.
$$3x=2x-7$$
$$x=-7$$
The number is −7.

57. Let x = the 1st integer. Then, $x+2$ is the next even consecutive integer and $x+4$ is the largest.
$$\left(\text{3 times the largest}\right)=76+\left(\text{sum of other two integers}\right)$$
$$3(x+4)=76+x+(x+2)$$
$$3x+12=78+2x$$
$$x=66$$
The numbers are 66, 68, 70.

59. Let $x = 1^{st}$ integer. Then, $x + 1$ is the next and $x + 2$ is the largest.

$$P = (\text{sum of the sides})$$
$$78 = x + (x+1) + (x+2)$$
$$78 = 3x + 3$$
$$3x = 75$$
$$x = 25$$

The sides are 25 in., 26 in., and 27 in.

61. Let x = minimum salary of 1980.

$$(\text{salary of } 1985) = (\text{twice salary of } 1980)$$
$$60000 = 2x$$
$$x = 30000$$

The minimum salary was $30,000 in 1980.

63. Let x = the number

$$x = (35\% \text{ of } 68)$$
$$x = 0.35(68)$$
$$x = 23.8$$

65. Let x = the percent.

$$53.5 = x\% \text{ of } 428$$
$$53.5 = x(428)$$
$$x = \frac{53.5}{428} = 0.125 = 12.5\%$$

67. Let x = a number.

$$24 = 15\% \text{ of } x$$
$$24 = 0.15x$$
$$x = \frac{24}{0.15} = 160$$

69. Let x = sale price.

$$\left(\begin{matrix}\text{original}\\ \text{price}\end{matrix}\right) - \left(\begin{matrix}12\%\\ \text{discount}\end{matrix}\right) = (\text{sale price})$$
$$\$29.99 - 12\%(29.99) = x$$
$$29.99 - 0.12(29.99) = x$$
$$x = 26.39$$

The sale price is $26.39.

71. Let x = amount sold.

$$(14\% \text{ of } x) = \$238$$
$$0.14x = 238$$
$$x = 1700$$

She sold $1700.

73. **(a)** $I = Prt$

$$I = (3000)(0.08)(3.5)$$
$$I = 840$$

$840 is earned.

(b) Total amount = $P + I$

Total amount = 3000 + 840 = 3840

$3840 is total amount.

75. $C = K - 273$

$C + 273 = K$ or $K = C + 273$

77. $P = 4s$

$$\frac{P}{4} = s \text{ or } s = \frac{P}{4}$$

79. $2x + 5y = -2$

$$5y = -2 - 2x$$
$$y = \frac{-2 - 2x}{5}$$

81. Let x = width of rectangular area.

$$\left(\begin{matrix}\text{Area of}\\ \text{semicircle}\end{matrix}\right) + \left(\begin{matrix}\text{Area of}\\ \text{rectangular}\end{matrix}\right) = 85 \text{ ft}^2$$
$$\frac{(4)^2(3.14)}{2} + (12)(x) = 85$$
$$25.12 + 12x = 85$$
$$12x = 59.88$$
$$x = 4.99$$

The width is 5 ft.

83. Let x = second angle. Then, $(2x - 6)$ is the first angle.

$$(\text{complementary angles sum}) = 90°$$
$$x + (2x - 6) = 90$$
$$3x = 96$$
$$x = 32$$

The angles are 32°, 58°.

85. Let x = amount invested in savings. Then, $0.5x$ is the amount invested in stock.

$$\left(\begin{matrix}\text{Interest on}\\ \text{savings}\end{matrix}\right) + \left(\begin{matrix}\text{Interest}\\ \text{on stock}\end{matrix}\right) = \$285$$
$$(3.5\% \text{ of } x) + (12\% \text{ of } 0.5x) = \$285$$
$$0.035x + 0.12(0.5x) = 285$$
$$0.095x = 285$$
$$x = 3000$$

$3000 invested in savings, $1500 in stocks.

87. Let x = number of ice creams purchased. Then, $24 - x$ is the number of popsicles purchased.

	Popsicles	Ice cream	Total
No.	$24 - x$	x	24
Cost	$\$1(24 - x)$	$\$1.50(x)$	\$29

$$\begin{pmatrix}\text{Cost of}\\ \text{ice cream}\end{pmatrix} + \begin{pmatrix}\text{Cost of}\\ \text{popsicles}\end{pmatrix} = \$29$$
$$1(24 - x) + 1.50(x) = 29$$
$$24 - x + 1.5x = 29$$
$$0.5x = 5$$
$$x = 10$$

She buys 10 ice creams on a stick and 14 popsicles.

89. Let x = speed of Miller family. Then, $x - 5$ is the speed of the O'Neill family.

$$\begin{pmatrix}\text{distance}\\ \text{traveled by}\\ \text{Millers}\end{pmatrix} + \begin{pmatrix}\text{distance}\\ \text{traveled by}\\ \text{ONiells}\end{pmatrix} = 210 \text{ miles}$$
$$x(2) + (x - 5)(2) = 210$$
$$4x - 10 = 210$$
$$4x = 220$$
$$x = 55$$

The Millers travel 55 mph; the O'Neills travel 50 mph.

91. **(a)** $5 \cdot \begin{pmatrix}\text{number}\\ \text{of plants}\end{pmatrix} - \begin{pmatrix}\text{2\% of}\\ \text{price}\end{pmatrix}$

$= 5(130) - 0.02(5 \cdot 130)$
$= 650 - 13$
$= 637$
\$637

(b) Three hundred plants cost \$1410 $(5 \cdot 300 - 0.06(1500)) = 1410$; 295 plants cost \$1416 $(5 \cdot 295 - 0.04 \cdot 1475) = 1416$. Therefore it costs more to purchase 295 plants.

93. $3w - 4 > -5$

$3w > -1$

$w > -\frac{1}{3}$

$\left(-\frac{1}{3}, \infty\right)$

$-\frac{1}{3}$

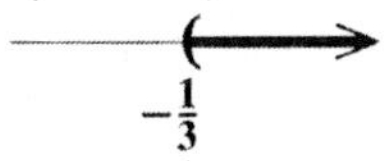

95. $5(y + 2) \le -4$

$5y + 10 \le -4$

$5y \le -14$

$y \le -\frac{14}{5}$

$\left(-\infty, -\frac{14}{5}\right]$

$-\frac{14}{5}$

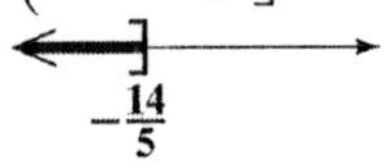

97. $1.3 > 0.4t - 12.5$

$13.8 > 0.4t$

$34.5 > 5$ or $t < 34.5$

$(-\infty, 34.5)$

34.5

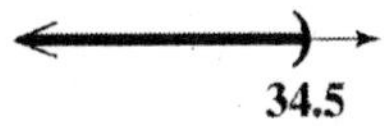

99. $\frac{6}{5}h - \frac{1}{5} \le \frac{3}{10} + h$

$10\left(\frac{6}{5}h - \frac{1}{5}\right) \le 10\left(\frac{3}{10} + h\right)$

$12h - 2 \le 3 + 10h$

$2h \le 5$

$h \le \frac{5}{2}$

$\left(-\infty, \frac{5}{2}\right]$

$\frac{5}{2}$

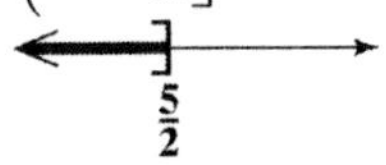

101. $-2 \le z + 4 \le 9$

$-2 - 4 \le z + 4 - 4 \le 9 - 4$

$-6 \le z \le 5$

$[-6, 5]$

-6 $\quad$ 5

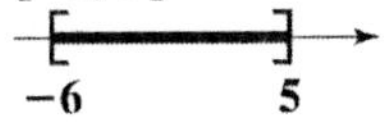

103. **(a)** Let x = number of hot dogs.

$$(\text{Revenue}) > (\text{Cost})$$
$$1.50x > 33 + 0.4x$$

(b)
$$1.50x > 33 + 0.4x$$
$$1.1x > 33$$
$$x > 30$$

A profit is realized if more than 30 hot dogs are sold.

Chapter 2 Test

1.
$$t + 3 = -13$$
$$t + 3 - 3 = -13 - 3$$
$$t = -16$$

3.
$$\frac{t}{8} = -\frac{2}{9}$$
$$(8)\frac{t}{8} = (8)\left(-\frac{2}{9}\right)$$
$$t = -\frac{16}{9}$$

5.
$$2(p - 4) = p + 7$$
$$2p - 8 = p + 7$$
$$p = 15$$

7.
$$\frac{3}{7} + \frac{2}{5}x = -\frac{1}{5}x + 1$$
$$35\left(\frac{3}{7} + \frac{2}{5}x\right) = 35\left(-\frac{1}{5}x + 1\right)$$
$$15 + 14x = -7x + 35$$
$$21x = 20$$
$$x = \frac{20}{21}$$

9.
$$\frac{4y - 3}{5} = y + 2$$
$$5\left(\frac{4y - 3}{5}\right) = 5(y + 2)$$
$$4y - 3 = 5y + 10$$
$$-y = 13$$
$$y = -13$$

11.
$$-5(x + 2) + 8x = -2 + 3x - 8$$
$$-5x - 10 + 8x = -10 + 3x$$
$$3x - 10 = 3x - 10$$
$$0 = 0$$

All real numbers

13.
$$C = 2\pi r$$
$$\frac{C}{2\pi} = \frac{2\pi r}{2\pi}$$
$$r = \frac{C}{2\pi}$$

15. Let x = the first integer. Then, the next four consecutive integers are $x + 1$, $x + 2$, $x + 3$, $x + 4$. the perimeter is the sum of the sides.

$$(\text{sum of the sides}) = 315 \text{ in.}$$
$$x + x + 1 + x + 2 + x + 3 + x + 4 = 315$$
$$5x + 10 = 315$$
$$5x = 305$$
$$x = 61$$

The sides are 61 in., 62 in., 63 in., 64 in., and 65 in.

17. Let x = cost of shoes before tax.

$$(\text{cost of shoes}) + (\text{sales tax}) = \$87.74$$
$$x + (0.07x) = 87.74$$
$$1.07 = 87.74$$
$$x = 82$$

The shoes cost \$82.00.

19. Let x = one angle. Then, $x + 26°$ is the other angle.

$$(\text{sum of complementary angles}) = 90°$$
$$x + x + 26 = 90$$
$$2x = 64$$
$$x = 32$$

The angles are 32° and 58°.

21. Let x = the number of deaths from the Florida hurricane. Then, $x + 6164$ is the number of deaths from the Texas hurricane.

$$\begin{pmatrix}\text{no. of deaths}\\ \text{from Florida}\\ \text{hurrican}\end{pmatrix} + \begin{pmatrix}\text{no. of deaths}\\ \text{from Texas}\\ \text{hurricane}\end{pmatrix} = 9836$$
$$x + x + 6164 = 9836$$
$$2x = 3672$$
$$x = 1836$$

The number of deaths from Texas hurricane is 8000.

23. $5x+14>-2x$

$7x>-14$

$x>-2$

$-(2, \infty)$

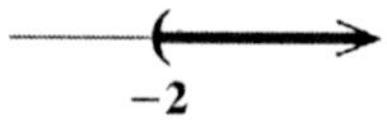

25. $-13\le 3p+2\le 5$

$-15\le 3p\le 3$

$\dfrac{-15}{3}\le\dfrac{3p}{3}\le\dfrac{3}{3}$

$-5\le p\le 1$

$[-5, 1]$

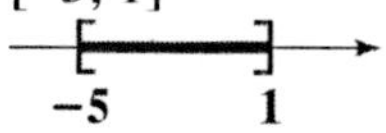

Cumulative Review Exercises Chapters 1–2

1. $-12 + 5 - (-3) = -7 + 3 = -4$

3. $19.8 \div (-7.2) = -2.75$

5. $-\dfrac{2}{3}+\left(\dfrac{1}{2}\right)^2=-\dfrac{2}{3}+\dfrac{1}{4}=-\dfrac{8}{12}+\dfrac{3}{12}=-\dfrac{5}{12}$

7. $\sqrt{5-(-20)-3^2}=\sqrt{5+20-9}=\sqrt{16}=4$

9.

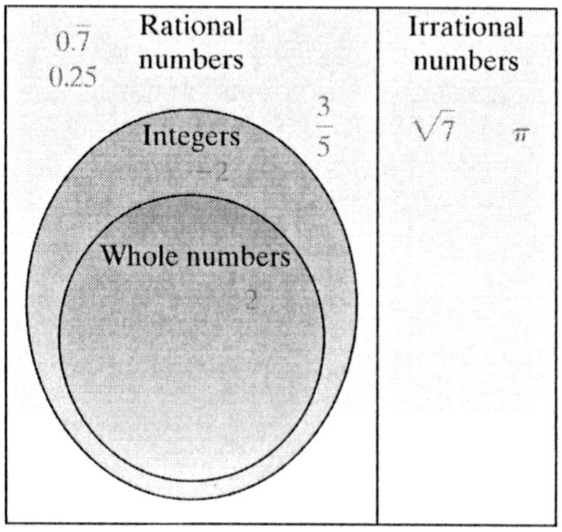

11. (a) $-\dfrac{5}{2}$

(b) $\dfrac{1}{3}$

(c) Not possible.

13. Since $ab = 1ab$, the coefficient is 1.

15. $-2.5x-5.2=12.8$

$-2.5x=18$

$x=-7.2$

17. Let x = the 1st odd integer. Then, $x + 2$ is the next odd integer.

$x+(x+2)=156$

$2x=154$

$x=77$

The numbers are 77 and 79.

19. $A=\dfrac{1}{2}bh$

$41=0.5(12)(h)$

$41=6h$

$h=6.83$

The height is 6.83 cm.

21. Let x = speed at which she runs. Then, $x + 12$ is the speed at which she bicycles.

distance = (rate)(time)

$\begin{pmatrix}\text{distance}\\ \text{ran}\end{pmatrix}+\begin{pmatrix}\text{distance}\\ \text{bicycled}\end{pmatrix}=30.4 \text{ miles}$

$(x(0.8))+(x+12)(1.2))=30.4$

$0.8x+1.2x+14.4=30.4$

$2x=16$

$x=8$

Her running speed is 8 mph and bicycling speed is 20 mph.

23. $-|-10|=-(10)=-10$

Chapter 3

Section 3.1 Practice Exercises

1.

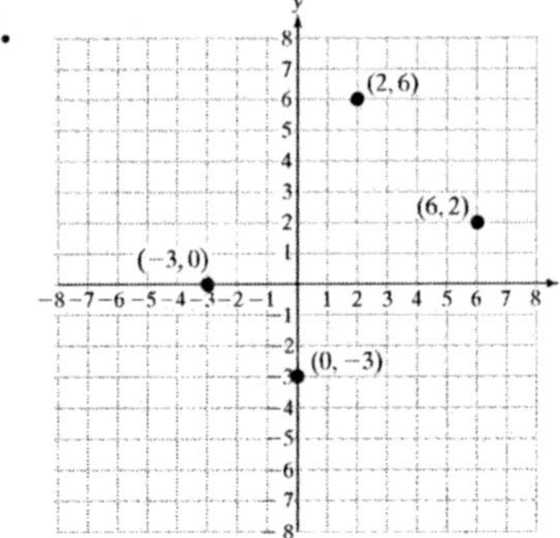

3.

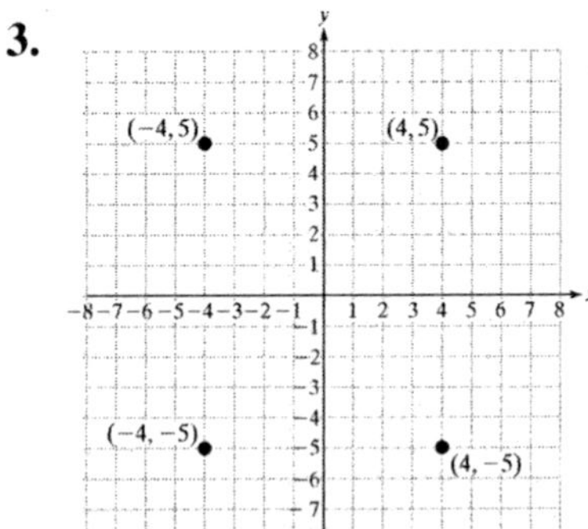

5.

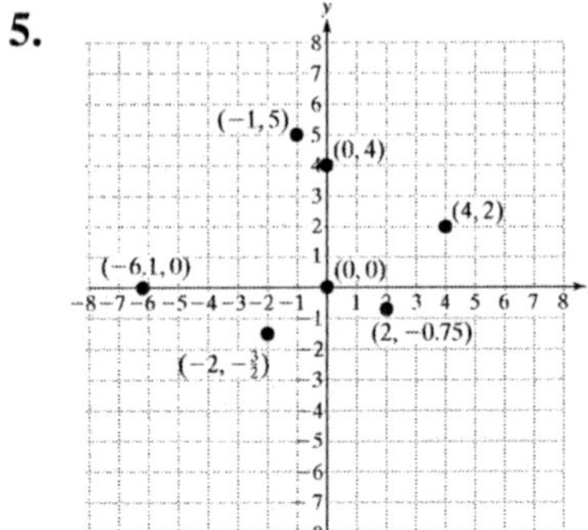

7. Quadrant IV

9. Quadrant II

11. Quadrant III

13. Quadrant I

15. (0, −5) lies on the y-axis.

17. $\left(\frac{7}{8}, 0\right)$ is located on the x-axis.

19. $A(-4, 2)$, $B\left(\frac{1}{2}, 4\right)$, $C(3, -4)$, $D(-3, -4)$, $E(0, -3)$, $F(5, 0)$

21. **(a)** (250, 225), (175, 193), (315, 330), (220, 209), (450, 570), (400, 480), (190, 185); the first ordered pair represents 250 people in attendance who spent $225 on popcorn.

(b)

23. **(a)** In the year 1710, the population of the U.S. colonies was 332,000.

(b)

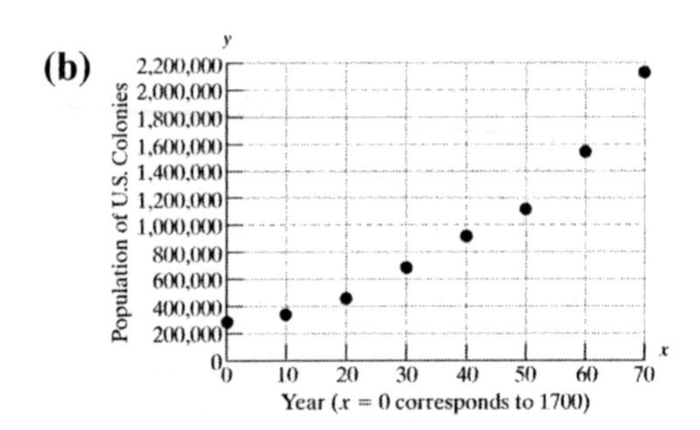

25. **(a)** (1, −10.2), (2, −9.0), (3, −2.5), (4, 5.7), (5, 13.0), (6, 18.3), (7, 20.9), (8, 19.6), (9, 14.8), (10, 8.7), (11, 2.0), (12, −6.9)

(b)

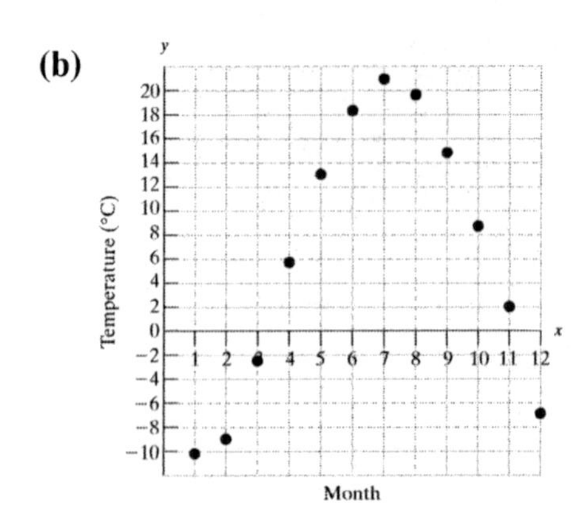

27. **(a)** Month 10

(b) 30

(c) Between months 3 and 5; also between months 10 and 12

(d) Between months 8 and 9

(e) Month 3

(f) 80 patients

29. **(a)** (1, 89.25), (2, 92.50), (3, 91.25), (4, 93.00), (5, 90.25); on day 1 the price per share of the stock was \$89.25.

(b) \$1.75

(c) –\$2.75

31. **(a)** $A(400, 200)$, $B(200, -150)$, $C(-300, -200)$, $D(-300, 250)$, $E(0, 450)$

(b) 450 m

Section 3.2 Practice Exercises

1. Yes; $(-2) + (-3) = -5$

3. Yes; $1 = 3(1) - 2$

5. No; $0 \neq -\frac{5}{2}(-2) + 5$

7.

x	y
2	1
0	3
–1	4
3	0

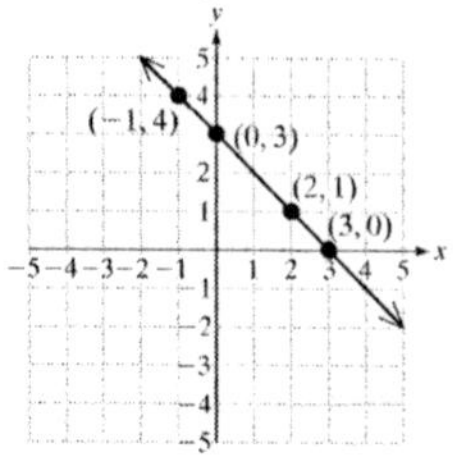

9.

x	y
3	–3
0	–6
6	0
7	1

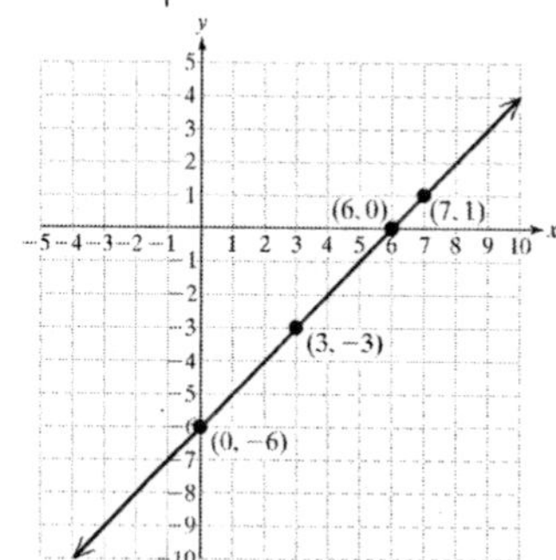

11.

x	y
0	–2
3	0
2	$-\frac{2}{3}$

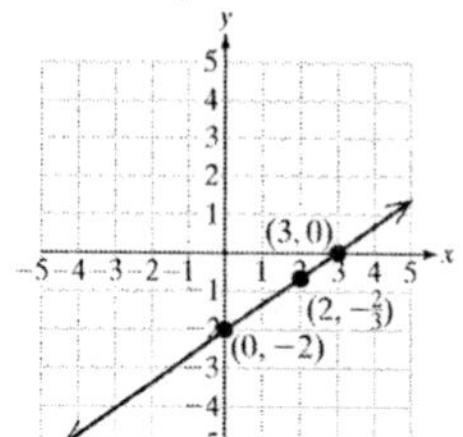

13.

x	y
1	6
2	11
–1	–4

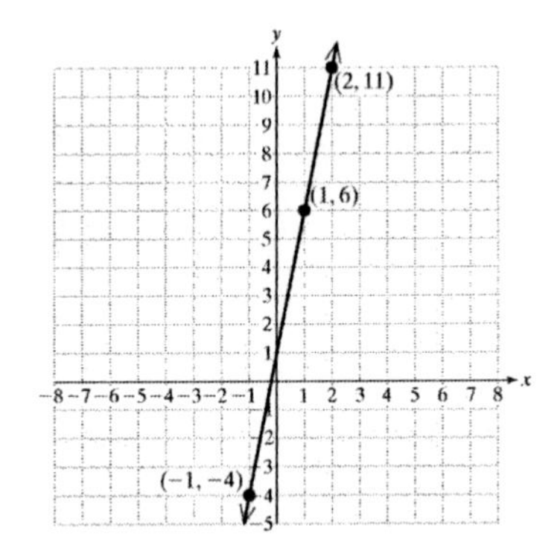

15.

x	y
7	−3
−14	−9
0	−5

y
x
−6 −4 −2 2 4 6 8 10
(7, −3)
(0, −5)

17.

x	y
1	$\frac{7}{3}$
0	4
−1	$\frac{17}{3}$

y
x
$(-1, \frac{17}{3})$
(0, 4)
$(1, \frac{7}{3})$
−7 −6 −5 −4 −3 −2 −1 1 2 3 4 5 6 7

19.

x	y
0	5.8
1	2.4
2	−1

y
x
(1, 2.4)
(2, −1)
−5 −4 −3 −2 −1 1 2 3 4 5

21.

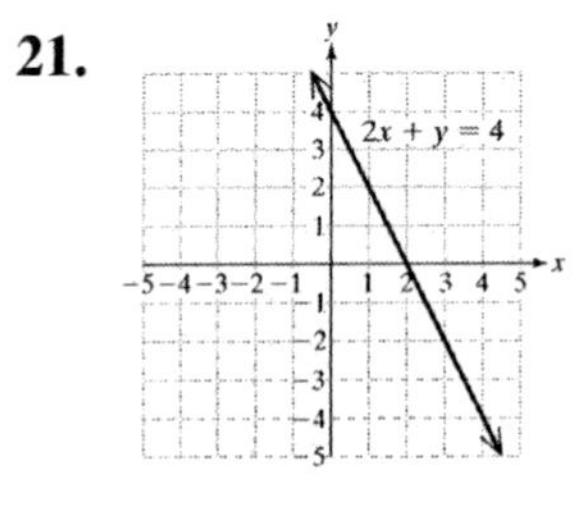

23.

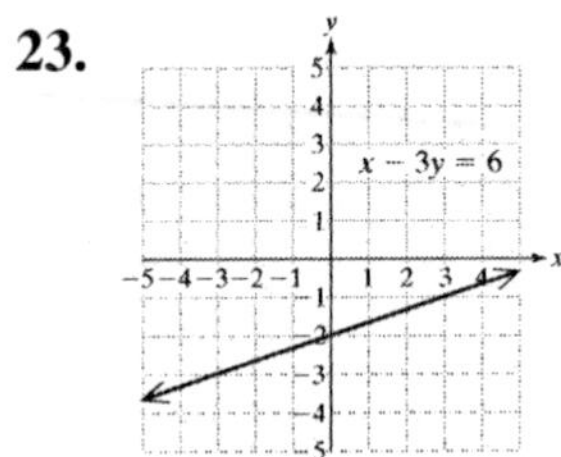

25.

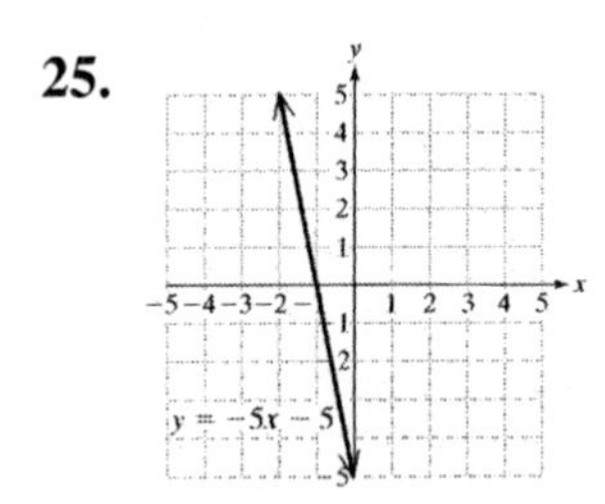

27.

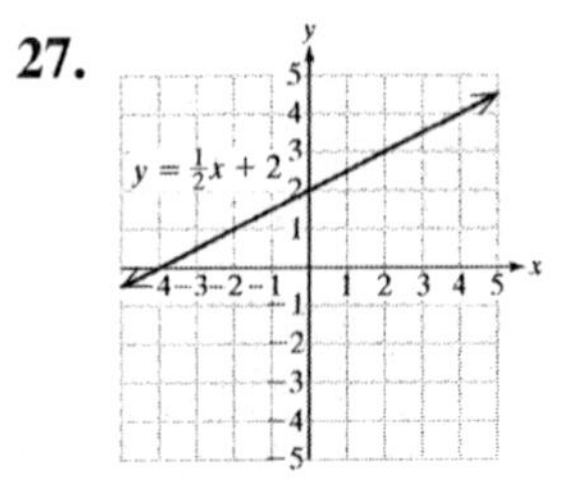

29.

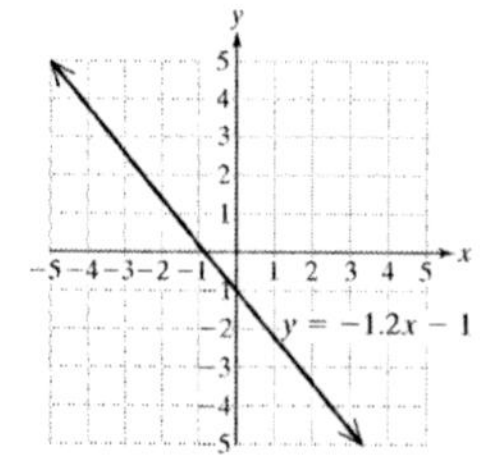

31. (a) $y = 0.69(55) - 20 = 17.95$

(b)
$$80.05 = 0.69x - 20$$
$$100.05 = 0.69x$$
$$x = 145$$

(c) (55, 17.95) For 55 pounds of aluminum, the students will be paid $17.95.
(145, 80.05) To collect $80.05, the students need 145 pounds of aluminum.

(d)

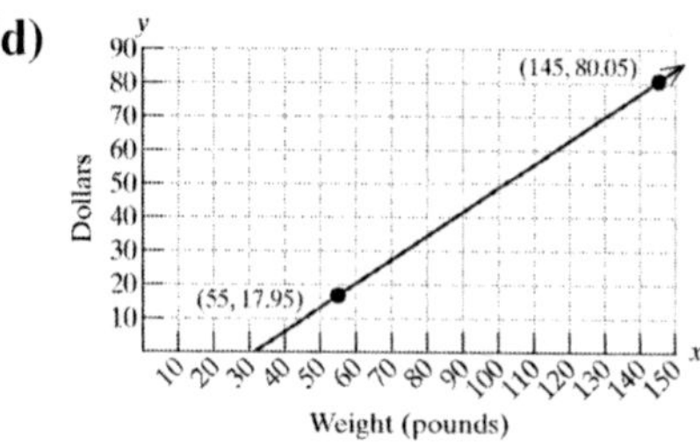

33. (a) $y = -1531(1) + 11{,}599 = 10{,}068$

(b)
$$7006 = -1531x + 11{,}599$$
$$-4593 = -1531x$$
$$\frac{-4593}{-1531} = x$$
$$x = 3$$

(c) (1, 10,068) After 1 year, the value of the Accent is $10,068.
(3, 7006) After 3 years, the value of the Accent is $7,006.

35. (a) $y = 1136(1) + 8790$
$y = \$9926$ million

(b) $y = 1136(3) + 8790$
$y = \$13{,}334$ million

(c)
$$12{,}198 = 1136x + 8790$$
$$3408 = 1136x$$
$$x = 3$$
When $x = 3$, the corresponding year is 1997.

(d)
$$14{,}470 = 1136x + 8790$$
$$5680 = 1136x$$
$$x = 5$$
When $x = 5$, the corresponding year is 1999.

37.

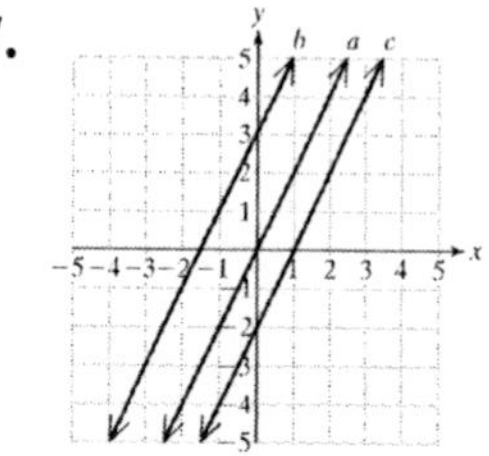

The lines are all parallel (they do not intersect).

39. (a)

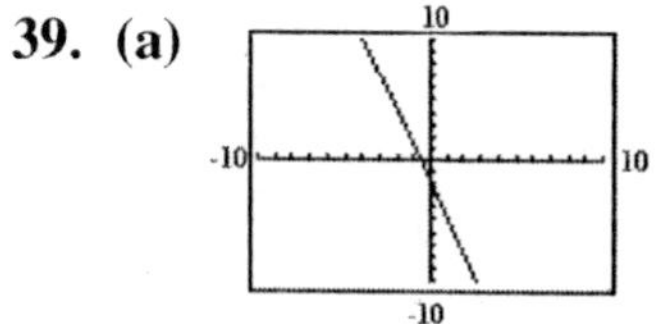

(b)

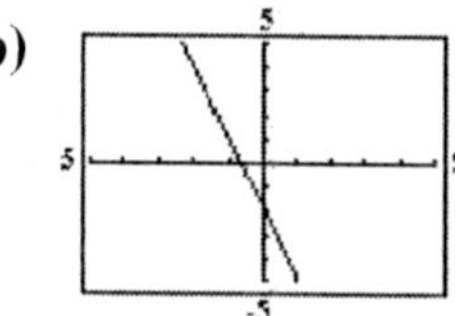

41.

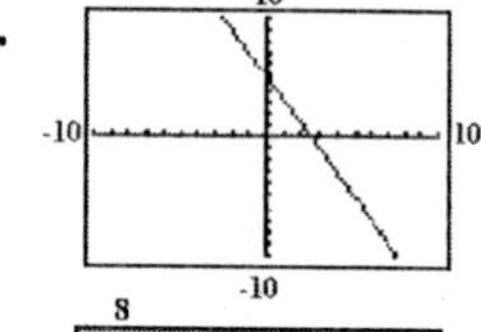

43.

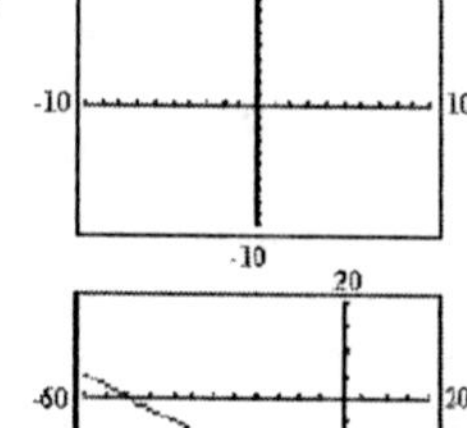

Section 3.3 Practice Exercises

1. II

3. III

5. **(a)** Yes; $0+2(3)=6$
$6=6$

(b) No; $1+2(2)=6$
$5\neq 6$

(c) Yes; $-4+2(-5)=6$
$6=6$

(d) Yes; $8+2(-1)=6$
$6=6$

7. An x-intercept is a point $(a, 0)$ where a graph intersects the x-axis.

9. Substitute $x = 0$, and solve for y.

11. x-intercept $(0, 0)$; y-intercepts $(0, 0)$, $(0, -3)$

13. x-intercept $(-2, 0)$;
y-intercepts $(0, 4)$, $(0, -3)$

15. x-intercepts $(2, 0)$, $(-2, 0)$; y-intercept $(0, 2)$

17. x-intercept
$2x-4(0)=8$
$2x=8$
$x=4$
$(4, 0)$

y-intercept
$2(0)-4y=8$
$-4y=8$
$y=-2$
$(0, -2)$

19. x-intercept
$0=5x+10$
$x=-2$
$(-2, 0)$

y-intercept
$y=5(0)+10$
$y=10$
$(0, 10)$

21. x-intercept
$x=0+2$
$x=2$
$(2, 0)$

y-intercept
$0=y+2$
$y=-2$
$(0, -2)$

23. x-intercept
$0=4x$
$x=0$
$(0, 0)$

y-intercept
$y=4(0)$
$y=0$
$(0, 0)$

25. x-intercept
$0=-\frac{1}{2}x+3$
$\frac{1}{2}x=3$
$x=6$
$(6, 0)$

y-intercept
$y=-\frac{1}{2}(0)+3$
$y=0+3$
$y=3$
$(0, 3)$

27. x-intercept
$4x-7(0)=9$
$4x=9$
$x=\frac{9}{4}$
$\left(\frac{9}{4}, 0\right)$

y-intercept
$4(0)-7y=9$
$-7y=9$
$y=-\frac{9}{7}$
$\left(0, -\frac{9}{7}\right)$

29. **(a)** False; $x = 3$ is vertical.

(b) True

31. Vertical

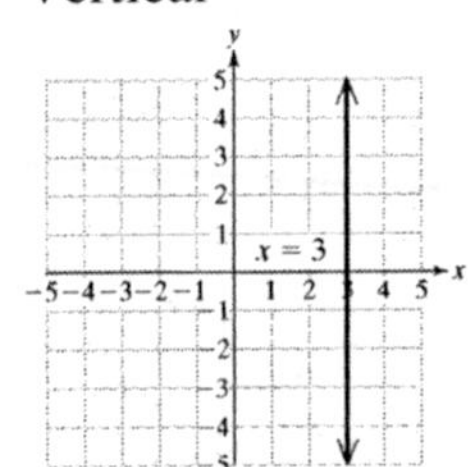

33. Horizontal $y = -4$

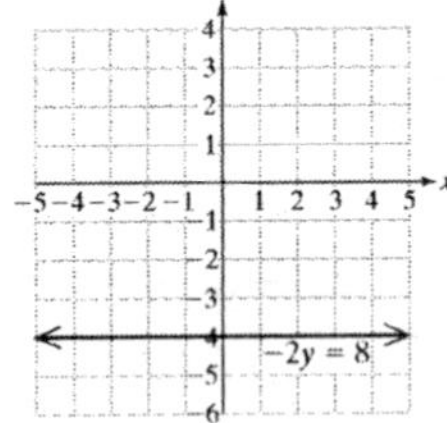

35. Vertical $x = 4$

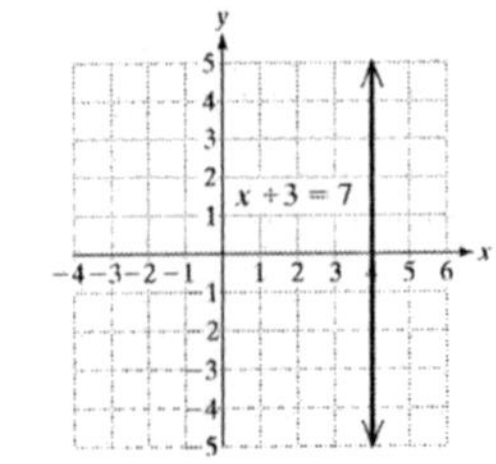

37. Horizontal $y = 0$

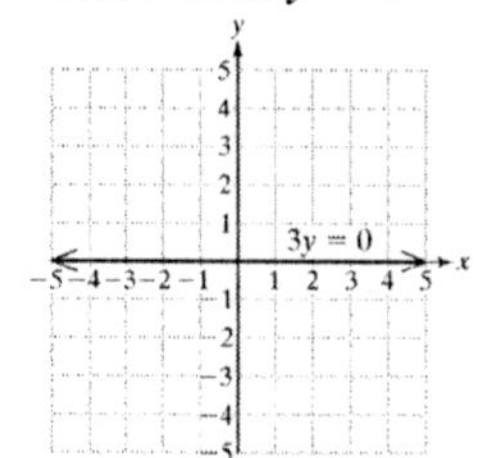

39. Vertical $x = \dfrac{3}{2}$

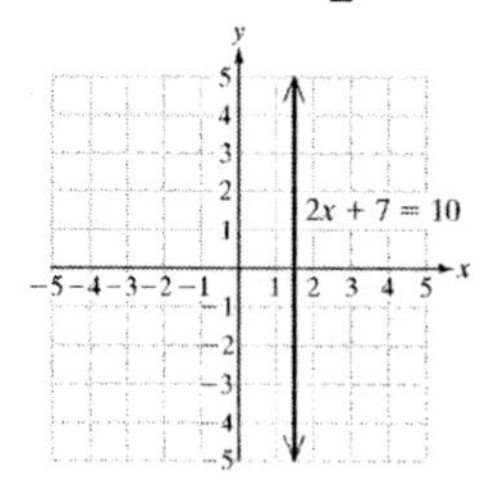

41. Horizontal $y = \dfrac{6}{4} = \dfrac{3}{2}$

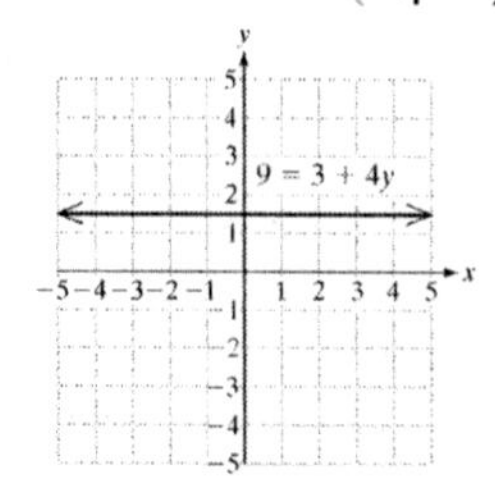

43. A horizontal line may not have an x-intercept. A vertical line may not have a y-intercept.

45. y-axis

47. a (horizontal line); b (intersects origin only); c (vertical line)

49. x-intercept

$x - 3(0) = -9$

$x = -9$

$(-9, 0)$

y-intercept

$0 - 3y = -9$

$y = 3$

$(0, 3)$

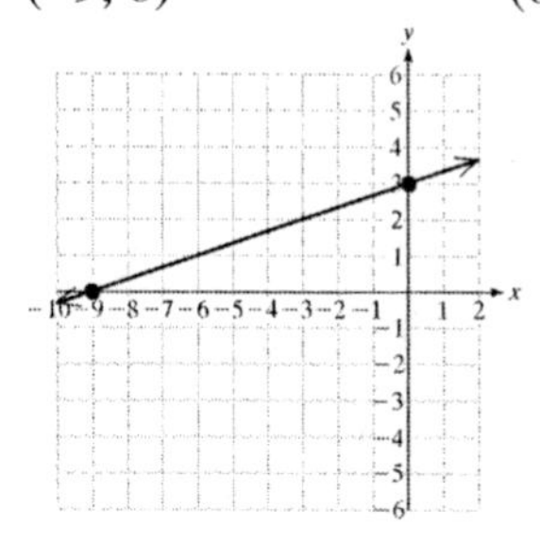

51. x-intercept

$0 = -\dfrac{3}{4}x + 2$

$x = \dfrac{8}{3}$

$\left(\dfrac{8}{3}, 0\right)$

y-intercept

$y = -\dfrac{3}{4}(0) + 2$

$y = 2$

$(0, 2)$

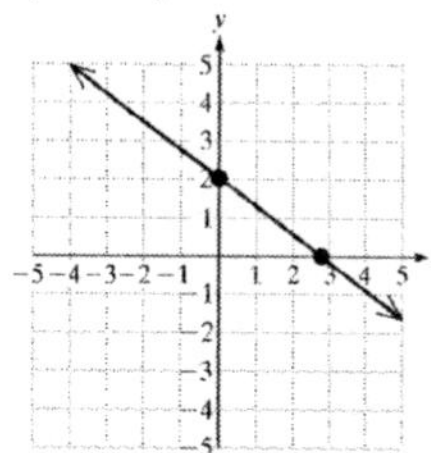

53. x-intercept

$2x + 8 = 0$

$x = -4$

$(-4, 0)$

y-intercept

$2(0) + 8 = y$

$y = 8$

$(0, 8)$

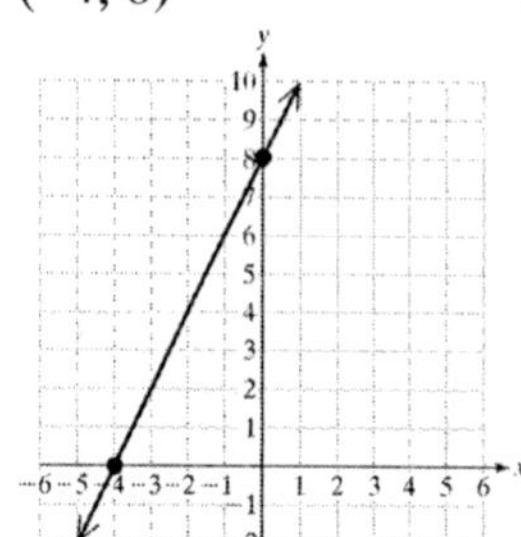

55. x-intercept

$2x - 2 = 0$

$x = 1$

$(1, 0)$

y-intercept

none

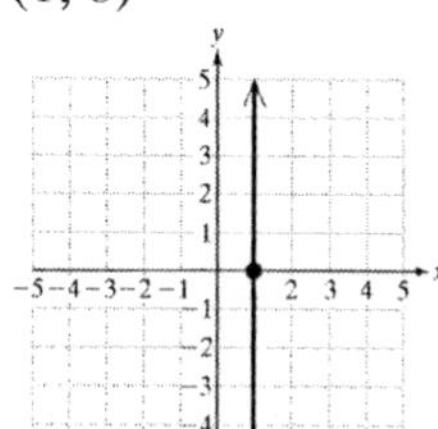

57. x-intercept: none

y-intercept: $y = -2$

$(0, -2)$

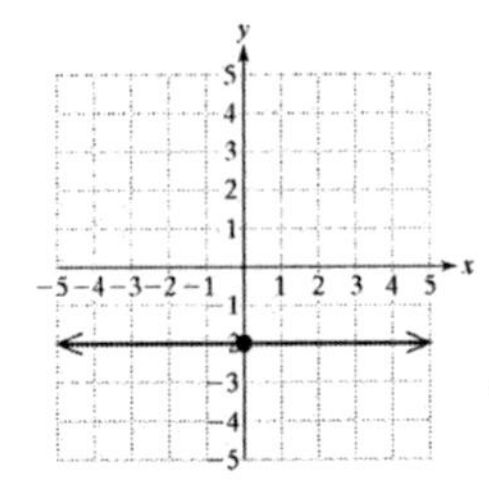

59. x-intercept

$x = \dfrac{5}{4}$

$\left(\dfrac{5}{4}, 0\right)$

y-intercept: none

61. x-intercept

$20x = -40(0) + 200$

$x = 10$

$(10, 0)$

y-intercept

$20(0) = -40y + 200$

$y = 5$

$(0, 5)$

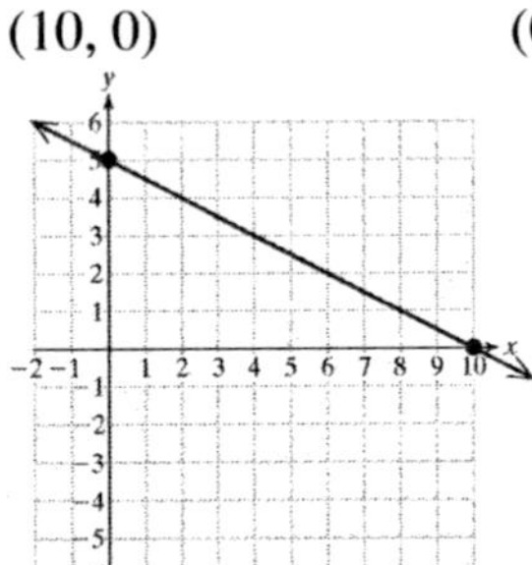

63. x-intercept

$-8.1x - 10.8(0) = 16.2$

$x = -2$

$(-2, 0)$

y-intercept

$-8.1(0) - 10.8y = 16.2$

$y = -1.5$

$(0, -1.5)$

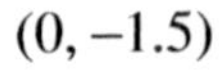

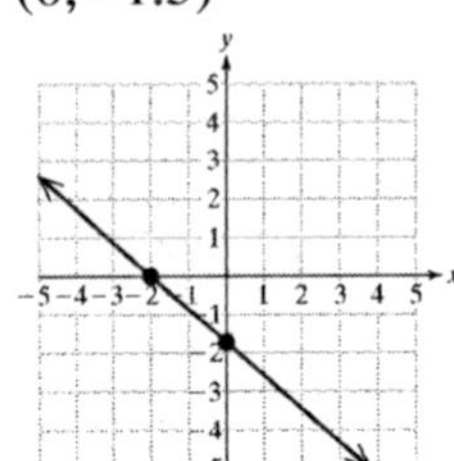

65. x-intercept

$x = -5(0)$

$x = 0$

$(0, 0)$

y-intercept

$0 = -5y$

$y = 0$

$(0, 0)$

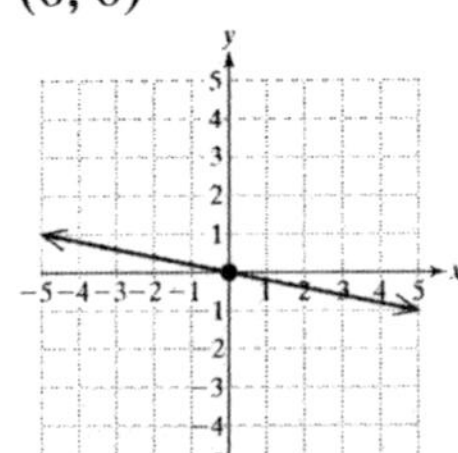

67. x-intercept: none

y-intercept: $y = -4$

$(0, -4)$

69. x-intercept

$x = 1$

$(1, 0)$

y-intercept: none

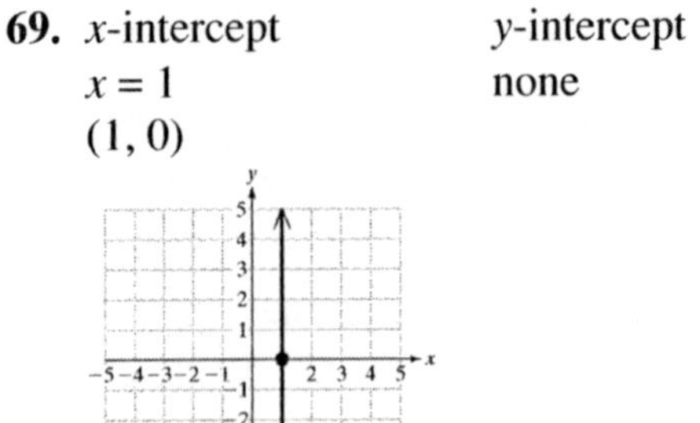

71. **(a)** $y = -400(1) + 800 = 400$; \$400 refund

(b) $y = -400(1.5) + 800 = 200$; \$200 refund

(c) $y = -400(0) + 800 = 800$; if the stereo does not work (0 years), the refund is \$800.

(d) $0 = -400x + 800$

$x = 2$

After 2 years, there will be no refund (\$0).

73. **(a)** $N = 20000 - 400(10) = 16000$; 16,000 tickets

(b) $N = 20000 - 400(40) = 4000$; 4000 tickets

(c) $N = 20000 - 400(0) = 20000$; if tickets cost \$0 (free) there will be 20,000 sold.

(d) $0 = 20000 - 400x$

$x = 50$

If tickets cost \$50, there will be 0 sold.

75. $x = 0$

77. $x = 4$

79. $y = 5$

81. x-intercept

$y = 2x - 4$

$0 = 2x - 4$

$4 = 2x$

$2 = x$

$(2, 0)$

y-intercept

$y = 2x - 4$

$y = 2(0) - 4$

$y = -4$

$(0, -4)$

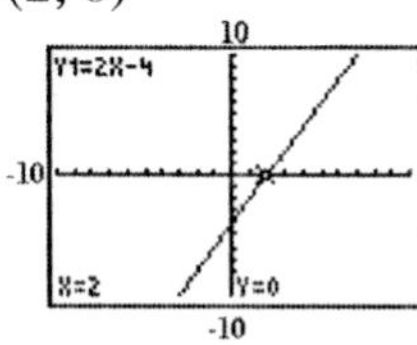

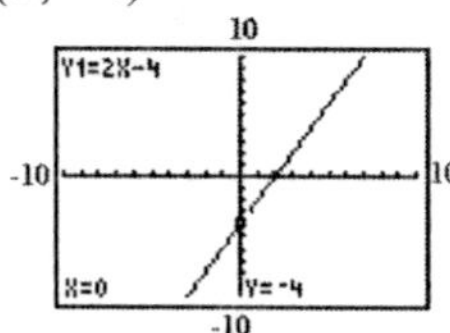

83. x-intercept

$3x + 4(0) = 6$

$x = 2$

$(2, 0)$

y-intercept

$3(0) + 4y = 6$

$y = 1.5$

$(0, 1.5)$

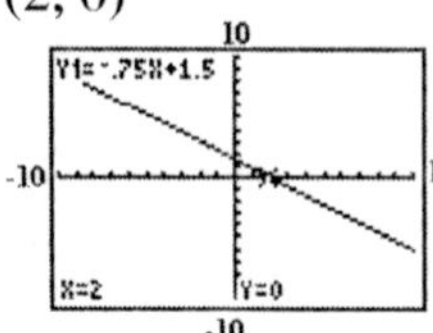

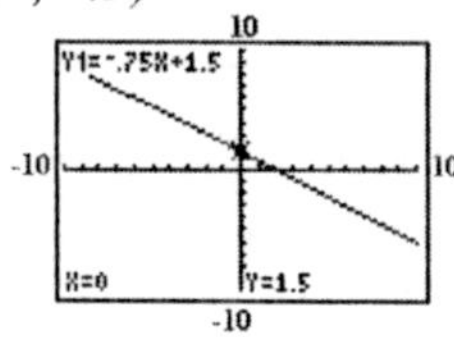

85. x-intercept

$0 = x - 15$

$x = 15$

$(15, 0)$

y-intercept

$y = 0 - 15$

$y = -15$

$(0, -15)$

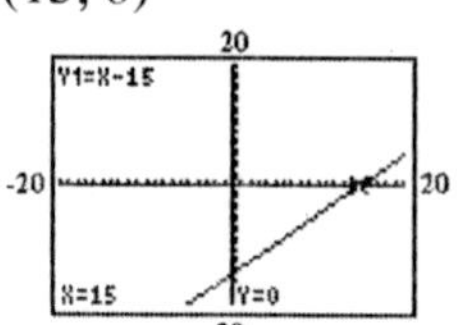

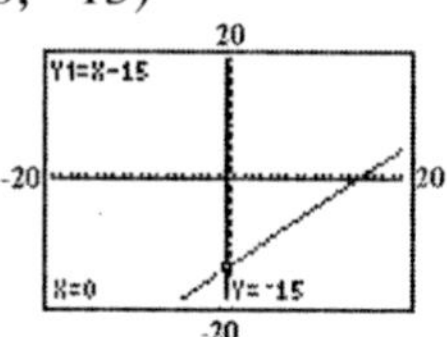

Section 3.4 Practice Exercises

1. x-intercept

$x - 3(0) = 6$

$x = 6$

$(6, 0)$

y-intercept

$0 - 3y = 6$

$y = -2$

$(0, -2)$

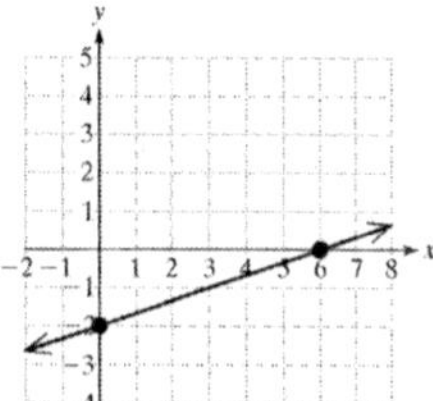

3. x-intercept

$x - 5 = 2$

$x = 7$

$(7, 0)$

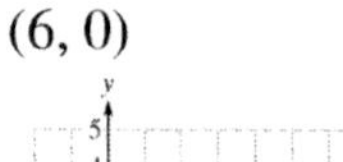

y-intercept

none

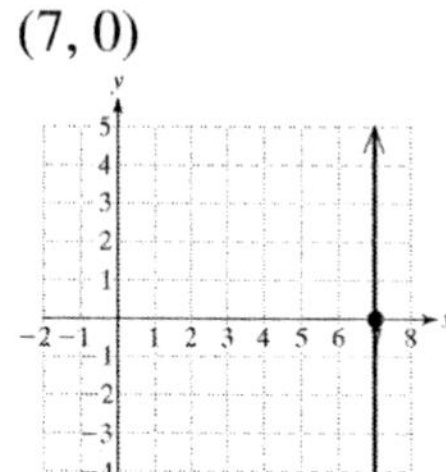

5. x-intercept

none

y-intercept

$2y - 3 = 0$

$y = \frac{3}{2}$

$\left(0, \frac{3}{2}\right)$

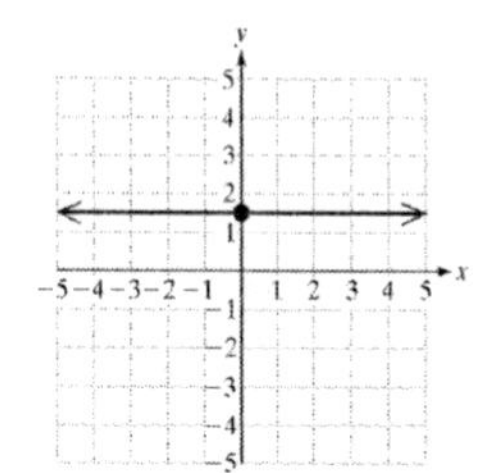

7. x-intercept
$2x = 4(0)$
$x = 0$
$(0, 0)$

y-intercept
$2(0) = 4y$
$y = 0$
$(0, 0)$

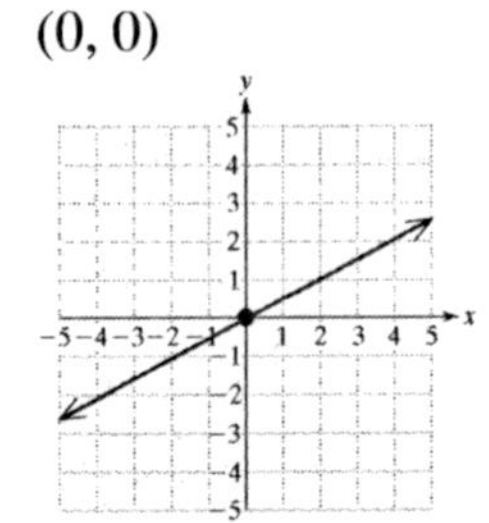

9. undefined

11. positive

13. Negative

15. Zero

17. Undefined

19. Positive

21. $m = \dfrac{-1-4}{-1-2} = \dfrac{-5}{-3} = \dfrac{5}{3}$

23. $m = \dfrac{0-3}{-1-(-2)} = \dfrac{-3}{1} = -3$

25. $m = \dfrac{3-3}{-2-5} = \dfrac{0}{-7} = 0$

27. $m = \dfrac{5-(-7)}{2-2} = \dfrac{12}{0} = \text{undefined}$

29. $m = \dfrac{-\frac{4}{5} - \frac{3}{5}}{\frac{1}{4} - \frac{1}{2}} = \dfrac{-\frac{7}{5}}{-\frac{1}{4}} = \dfrac{28}{5}$

31. $m = \dfrac{6\frac{1}{2} - \left(-1\frac{1}{4}\right)}{-5 - 3\frac{3}{4}} = \dfrac{7\frac{3}{4}}{-8\frac{3}{4}} = \dfrac{\frac{31}{4}}{-\frac{35}{4}} = -\dfrac{31}{35}$

33. $m = \dfrac{1.1-(-3.4)}{-3.2-6.8} = \dfrac{4.5}{-10} = -0.45$

35. $m = \dfrac{-4.80-1.75}{-1.50-(-5.50)} = \dfrac{-6.55}{4} = -1.6375$

37. $m = \dfrac{24000-35000}{2000-1994} = \dfrac{-11000}{6} = -1833.\overline{3}$

39. $\dfrac{3 \text{ units up}}{4 \text{ units right}} = \dfrac{3}{4}$

41. $\dfrac{3 \text{ units up}}{3 \text{ units left}} = \dfrac{3}{-3} = -1$

43. $\dfrac{0 \text{ units up}}{5 \text{ units right}} = \dfrac{0}{5} = 0$

45. (a) $m = \dfrac{1069-304}{1996-1980} = \dfrac{765}{16} \approx 47.8$

(b) The number of male inmates increased by 765 thousand in 16 years, or approximately 47.8 thousand inmates per year.

47. (a) $m = \dfrac{12800-10100}{1996-1990} = 450$
The median income for women in the United States increased by \$450/year.

(b) No, the rate of increase in the women's median income is less than the rate of increase in the men's median income.

49.

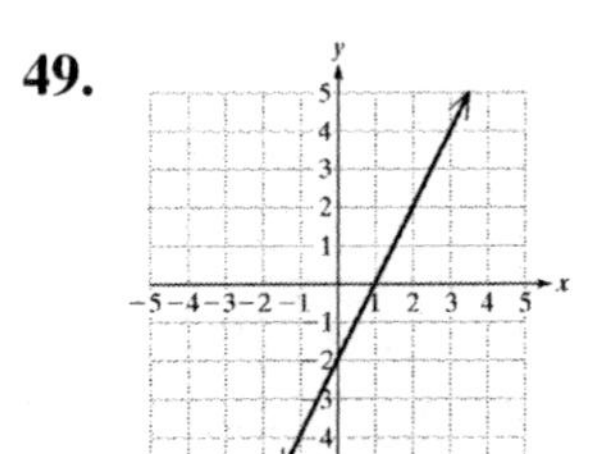

51.

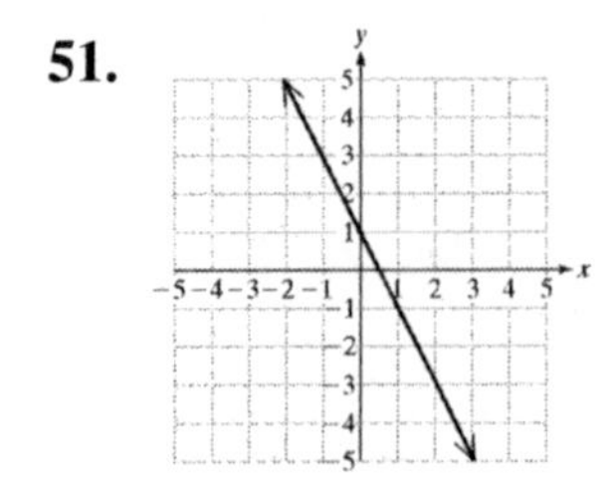

53.

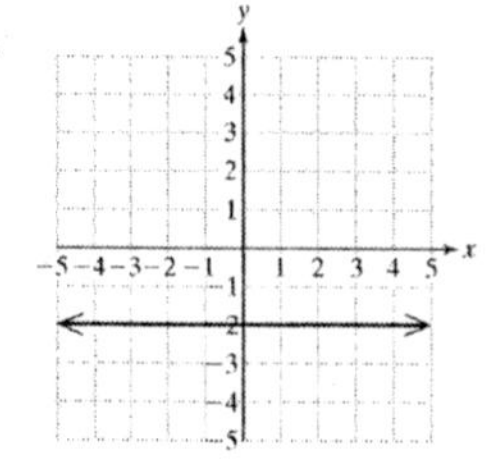

55.

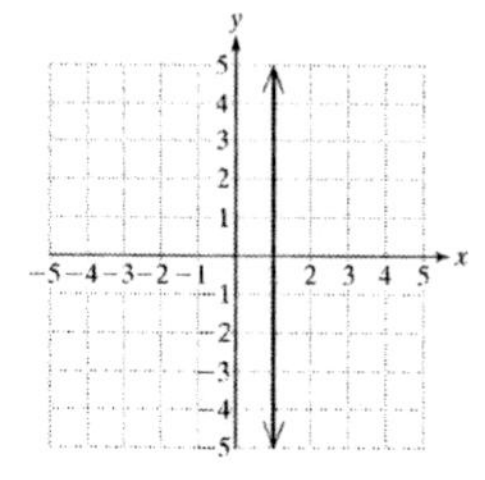

57.

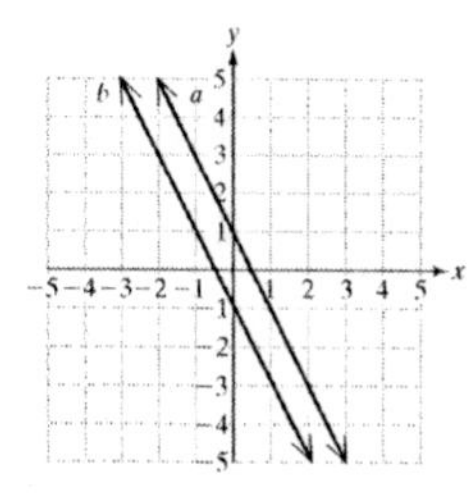

59.

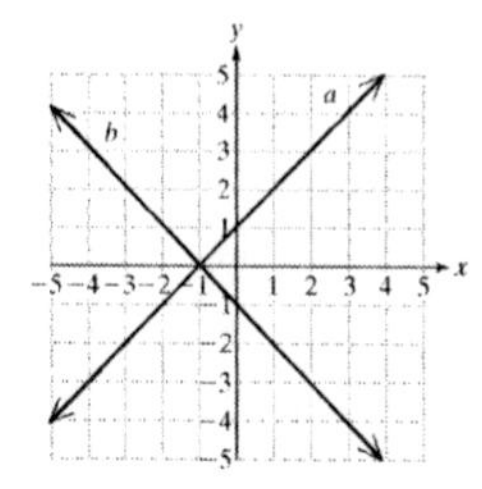

61.

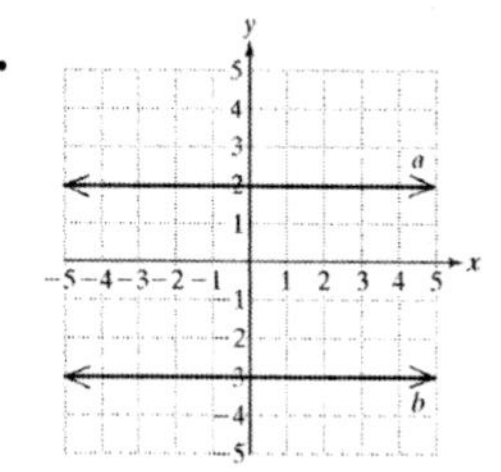

63. **(a)** $\frac{2}{3}$

(b) $-\frac{3}{2}$

65. **(a)** Undefined

(b) 0

67. $m_1 = \frac{4-0}{-2-0} = \frac{4}{-2} = -2$

$m_2 = \frac{-1-(-5)}{-1-1} = \frac{4}{-2} = -2$

Parallel

69. $m_1 = \frac{-8-(-4)}{-1-3} = \frac{-4}{-4} = 1$

$m_2 = \frac{2-(-5)}{-2-5} = \frac{7}{-7} = -1$

Perpendicular

71. $m_1 = \frac{-5-5}{-2-3} = \frac{-10}{-5} = 2$

$m_2 = \frac{-3-0}{-4-2} = \frac{-3}{-6} = \frac{1}{2}$

Neither

73. $m_1 = \frac{-6.7-(-6.7)}{-2.3-4.5} = \frac{0}{-6.8} = 0$

$m_2 = \frac{-6.7-(-6.7)}{-1.4-(-2.2)} = \frac{0}{0.8} = 0$

Parallel

75. $\frac{y \text{ feet}}{18 \text{ feet}} = \frac{1}{4}$

$y = 18 \cdot \frac{1}{4} = \frac{18}{4} = \frac{9}{2} = 4\frac{1}{2}$ feet

77. **(a)** $P = 11.50x = 11.50(20 \text{ hr}) = \$230.00$

(b) $P = 11.50x = 11.50(21 \text{ hr}) = \$241.50$

(c) $P = 11.50x = 11.50(22 \text{ hr}) = \$253.00$

(d) $m = \frac{\$241.50 - \$230.00}{21 \text{ hr} - 20 \text{ hr}} = 11.50$

Jorge's pay increases \$11.50 for each additional hour worked.

79. $m=\dfrac{(s-t)-(s+t)}{(c-2d)-(3c-d)}$
$=\dfrac{s-s-t-t}{c-3c-2d+d}$
$=\dfrac{-2t}{-2c-d}$
$=\dfrac{2t}{2c+d}$

81. $a(0)+by=c$
$by=c$
$y=\dfrac{c}{b}$
$\left(0,\dfrac{c}{b}\right)$

83. From point (–3, 4) move up 1 unit and move right 4 units to the point (1, 5).

Section 3.5 Practice Exercises

1. x-intercept
$x-5(0)=10$
$x=10$
(10, 0)

y-intercept
$0-5y=10$
$y=-2$
(0, –2)

3. x-intercept
none

y-intercept
$3y=-9$
$y=-3$
(0, –3)

5. x-intercept
$-4x=6(0)$
$x=0$
(0, 0)

y-intercept
$-4(0)=6y$
$y=0$
(0, 0)

7. x-intercept
$-x+3=8$
$x=-5$
(–5, 0)

y-intercept
none

9. $2x-5y=4$
$-5y=-2x+4$
$y=\dfrac{-2x+4}{-5}=\dfrac{2}{5}x-\dfrac{4}{5}$
$m=\dfrac{2}{5}$; y-intercept $\left(0,-\dfrac{4}{5}\right)$

11. $3x-y=5$
$y=3x-5$
$m=3$; y-intercept (0, –5)

13. $x+y=6$
$y=-x+6$
$m=-1$; y-intercept (0, 6)

15. Not possible; slope is undefined and no y-intercept.

17. $-8y=2$
$y=-\dfrac{1}{4}$
$m=0$, y-intercept $\left(0,-\dfrac{1}{4}\right)$

19. $3y=2x$
$y=\dfrac{2}{3}x$
$m=\dfrac{2}{3}$; y-intercept (0, 0)

21.

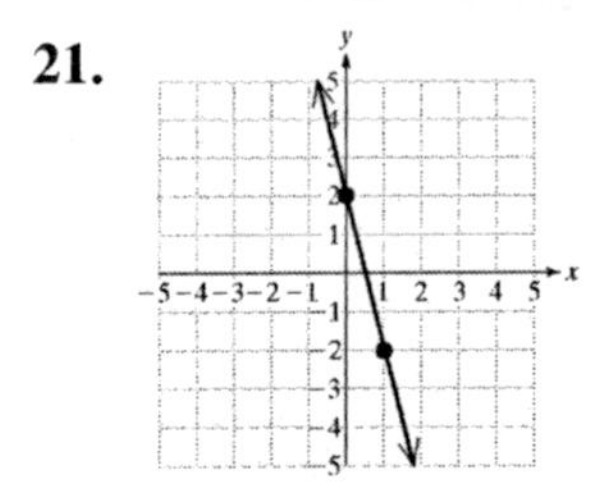

23.

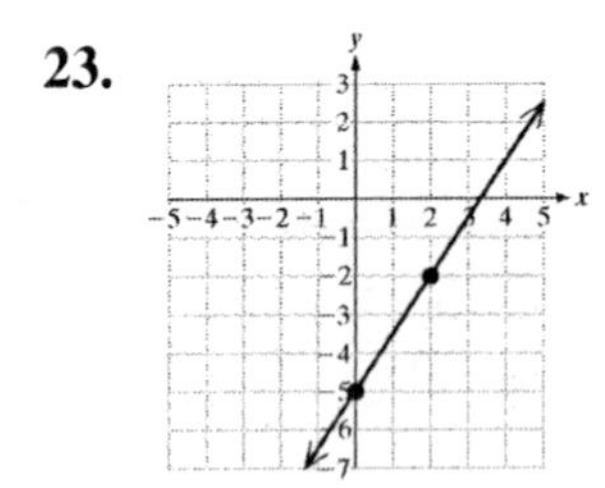

25. $x-2y=6$
$$-2y=-x+6$$
$$y=\frac{1}{2}x-3$$

27. $2x+y=9$
$$y=-2x+9$$

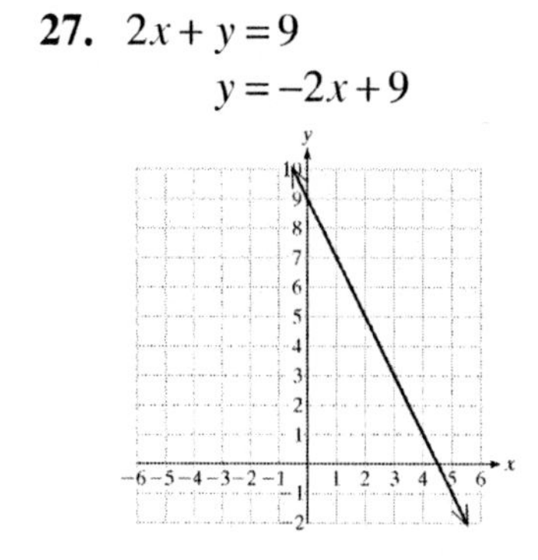

29. $3y+6x=-1$
$$3y=-6x-1$$
$$y=-2x-\frac{1}{3}$$

31. $2x=-4y+6$
$$-4y=2x-6$$
$$y=-\frac{1}{2}x+\frac{3}{2}$$

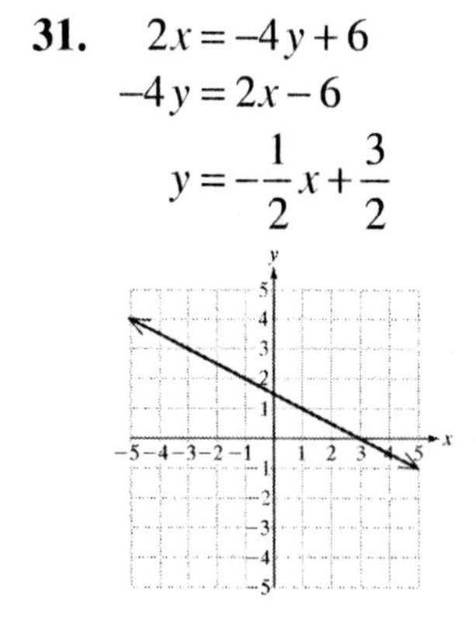

33. $x+y=0$
$$y=-x$$

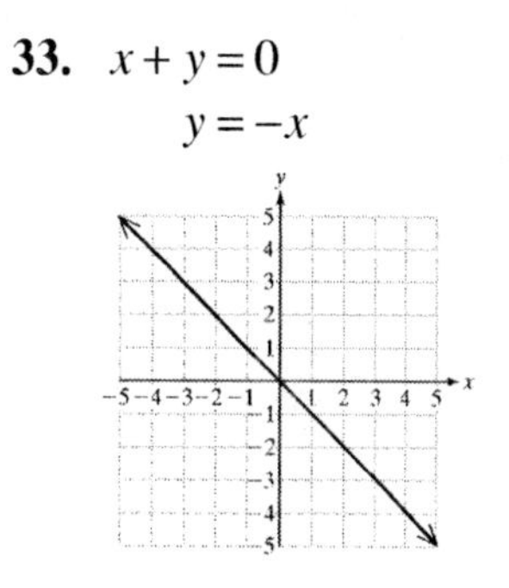

35. $0.2x-0.5y=0.1$
$$-0.5y=-0.2x+0.1$$
$$y=\frac{2}{5}x-\frac{1}{5}$$

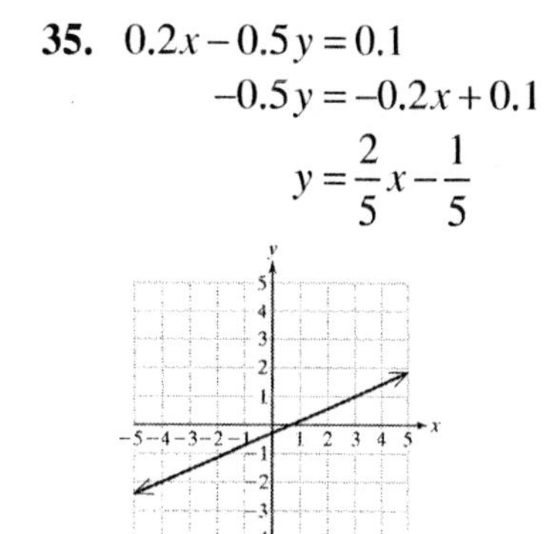

37. $5y=9x$
$$y=\frac{9}{5}x$$

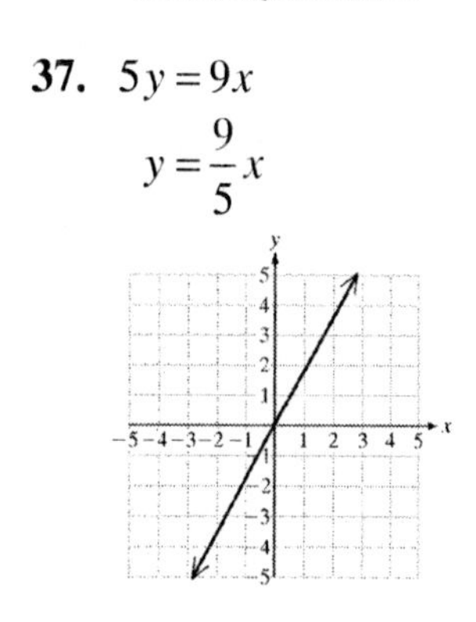

39. $3y+2=0$
$$y=-\frac{2}{3}$$

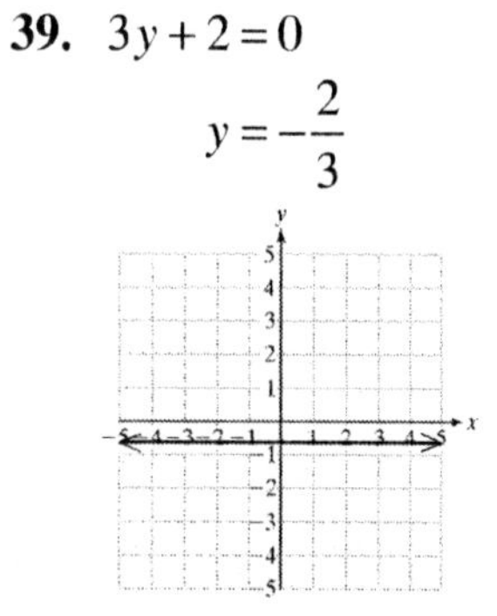

41. $x = 2$

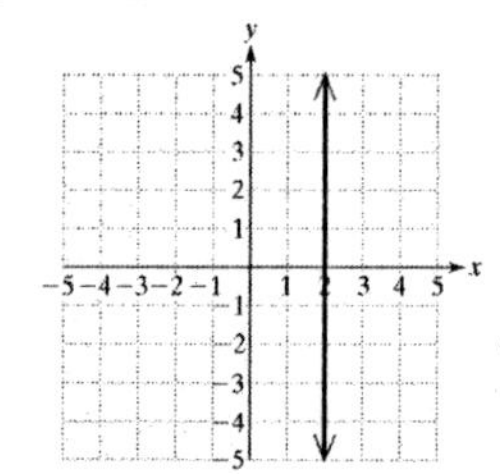

43. $\frac{1}{2}x + \frac{1}{4}y = \frac{1}{2}$

$2x + y = 2$

$y = 2 - 2x$

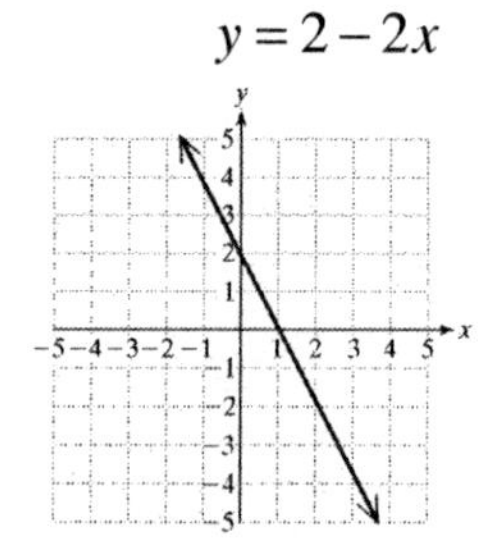

45. Perpendicular

47. Parallel

49. Neither

51. l_1: $m = -2$

l_2: $m = \frac{1}{2}$

perpendicular

53. l_1: $m = \frac{4}{5}$

l_2: $m = \frac{5}{4}$

neither

55. l_1: $m = -9$

l_2: $m = -9$

parallel

57. Vertical and horizontal lines; perpendicular

59. Vertical lines; parallel

61. l_1: $m = -\frac{2}{3}$

l_2: $m = \frac{3}{2}$

perpendicular

63. l_1: $m = -2$

l_2: $m = -\frac{1}{2}$

neither

65. l_1: $m = \frac{1}{5}$

l_2: $m = \frac{1}{5}$

parallel

67. $y = mx + b$

$y = -\frac{1}{3}x + 2$

69. $y = 5x$

71. $y = 6x - 2$

73. **(a)** $m = 49.95$; the cost increases \$49.95 per day.

(b) (0, \$31.95); the cost to rent the car for 0 days is \$31.95.

(c) $C = 49.95(7) + 31.95 = 381.6$; \$381.60

75. **(a)** $y = 2.5(12) + 45 = 75$; \$75

(b) $m = 2.5$; it costs \$2.50 per mile to tow the car.

(c) (0, 45); towing a car 0 miles will cost \$45. (There must be a \$45 fee before mileage is charged.)

77. **(a–c)**

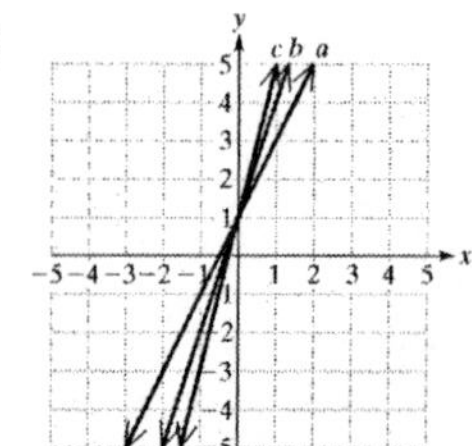

79. **(a–c)**

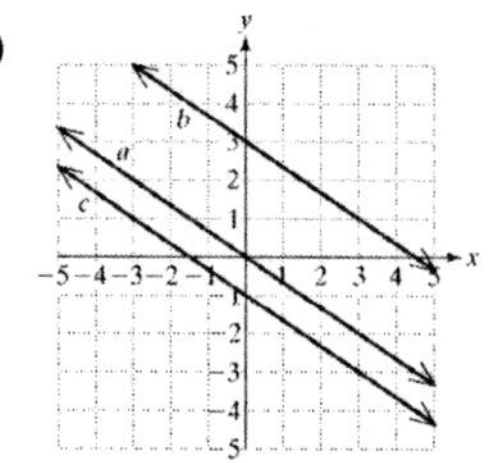

81. $ax + by = c$

$$by = -ax + c$$

$$y = \frac{-a}{b}x + \frac{c}{b}$$

$$m = \frac{-a}{b}$$

83. $m = -\frac{6}{7}$

85. $m = \frac{11}{8}$

87. Parallel

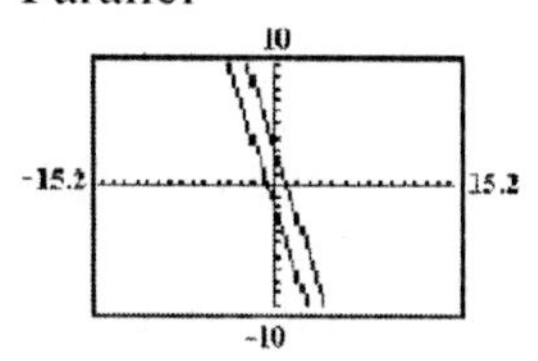

89. They are not parallel because their slopes are not equal.

Chapter 3 Review Exercises

1.

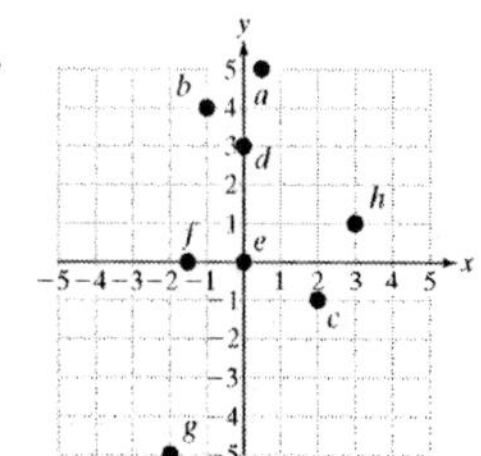

3. III

5. IV

7. IV

9. x-axis

11. **(a)** (1, 26.25), (2, 28.50), (3, 28.00), (4, 27.00), (5, 24.75); on day 1, the price was $26.25.

(b) Day 2

(c) $28.50 – $26.25 = $2.25

13. $5(0) - 3(4) = 12$

$$-12 \neq 12$$

No

15. $1 = \frac{1}{3}(9) - 2$

$$1 = 3 - 2 = 1$$

Yes

17.

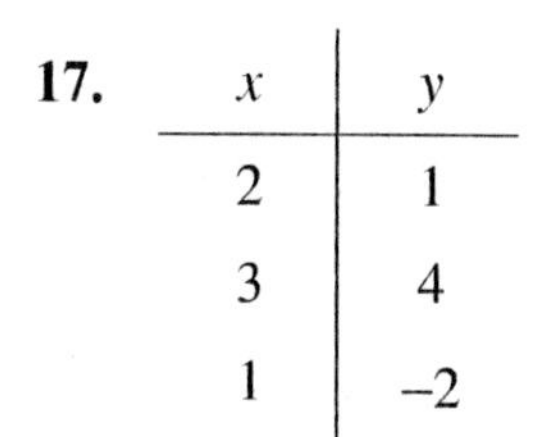

x	y
2	1
3	4
1	–2

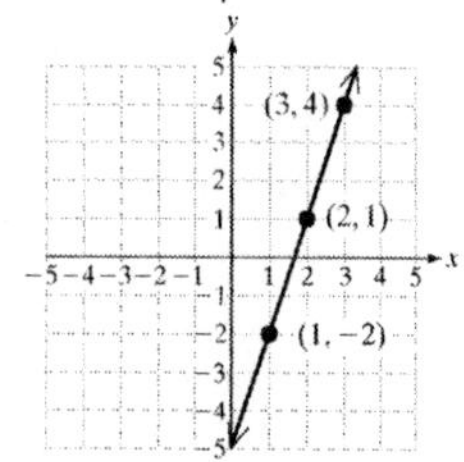

19.

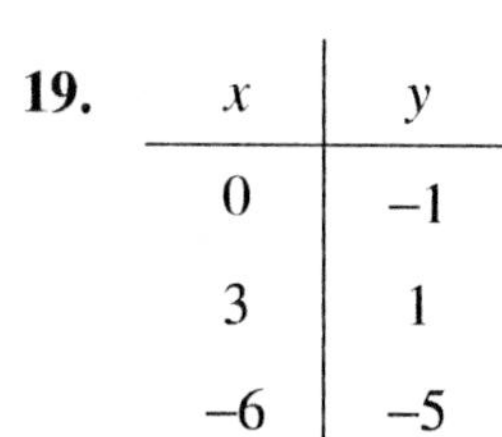

x	y
0	–1
3	1
–6	–5

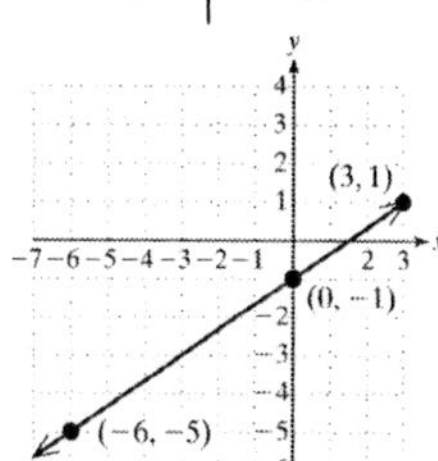

21.

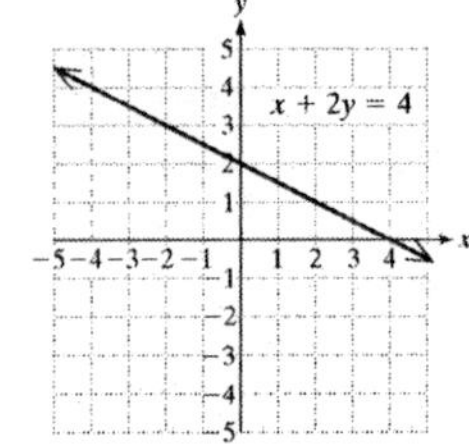

23.

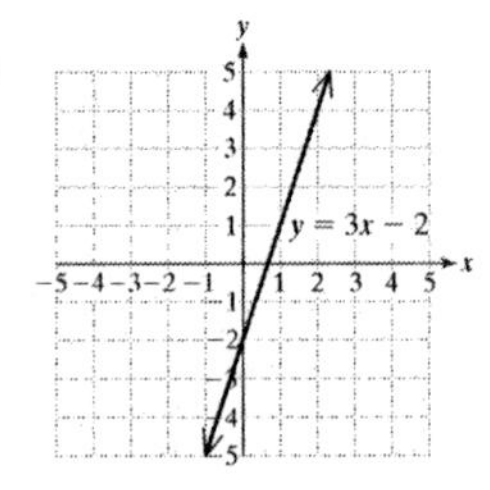

25. **(a)** $y = 1.54(12.4) = \$19.10$

(b)

gallons, x	Cost, y
5.0	7.70
7.5	11.55
10.0	15.40
12.5	19.25
15.0	23.10

(c)

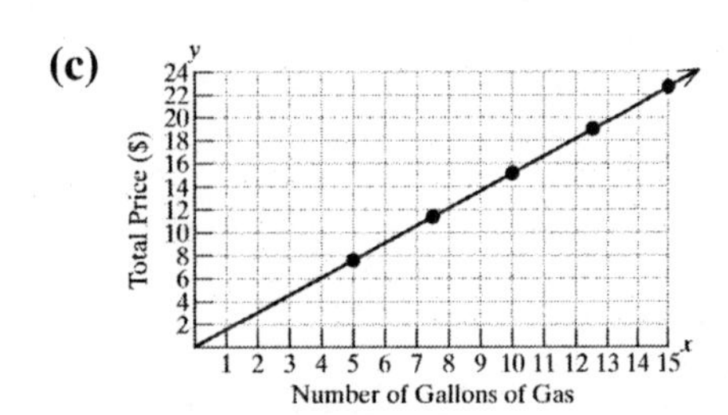

(d) $\$10.01 = 1.54x$

$$x = \frac{10.01}{1.54} = 6.5 \text{ gallons}$$

(e) $\$26.18 = 1.54x$

$$x = \frac{26.18}{1.54} = 17 \text{ gallons}$$

27. Vertical

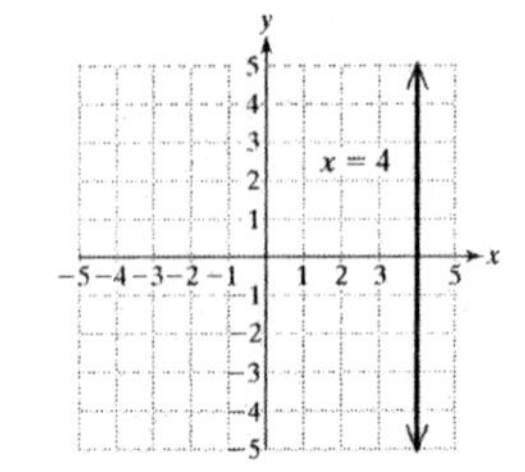

29. Horizontal

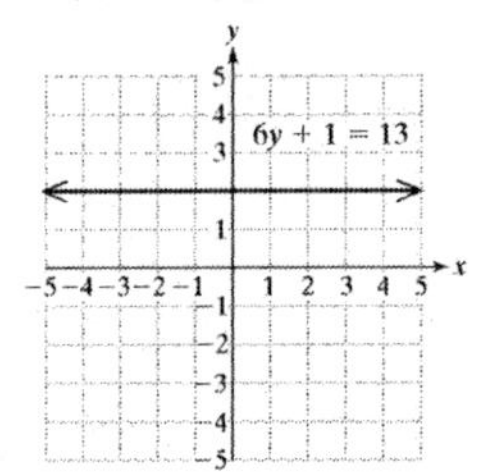

31. x-intercept

$$-4x + 8(0) = 12$$
$$-4x = 12$$
$$x = -3$$

$(-3, 0)$

y-intercept

$$-4(0) + 8y = 12$$
$$8y = 12$$
$$y = \frac{12}{8} = \frac{3}{2}$$

$\left(0, \frac{3}{2}\right)$

33. x-intercept

$0 = 8x$

$x = 0$

$(0, 0)$

y-intercept

$y = 8(0)$

$y = 0$

$(0, 0)$

35. x-intercept

none

y-intercept

$6y = -24$

$y = -4$

$(0, -4)$

37. x-intercept

$2x + 5 = 0$

$$x = -\frac{5}{2}$$

$\left(-\frac{5}{2}, 0\right)$

y-intercept

none

39. **(a)** $V = -5000(0) + 30000 = 30000$

$(0, 30{,}000)$

(b) At the time of purchase ($n = 0$), the value of the car is $30,000.

(c) $V = -5000(3) + 30000 = \$15{,}000$

41. 

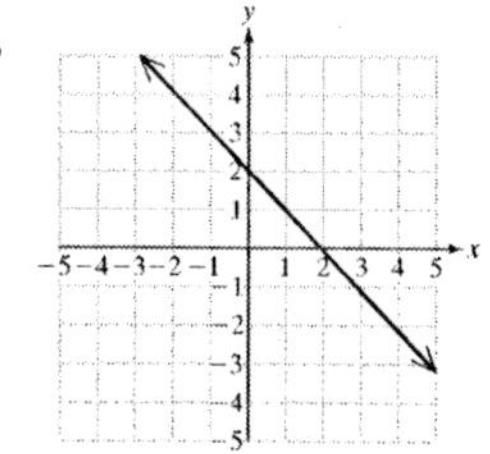

43.

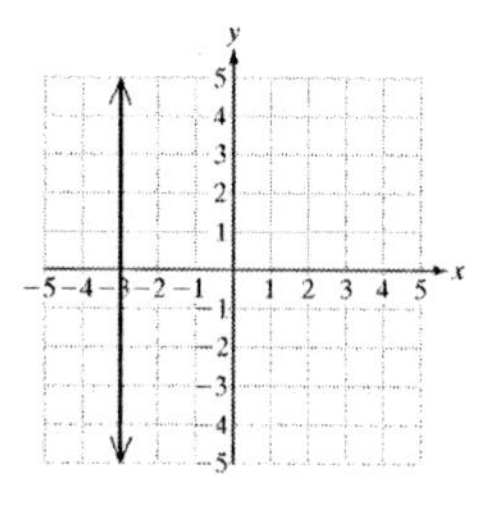

45. $m = \dfrac{-4}{2} = -2$

47. $m = \dfrac{8-0}{0-(-1)} = \dfrac{8}{1} = 8$

49. $m = 0$

51. **(a)** 0

(b) Undefined

53. $m_1 = \dfrac{9-1}{-1-(-2)} = \dfrac{8}{1} = 8$

$m_2 = \dfrac{10-(-6)}{2-0} = \dfrac{16}{2} = 8$

Parallel

55. $m_1 = \dfrac{-8-1}{1-1} = \dfrac{-9}{10} = \text{undefined}$

$m_2 = \dfrac{-5-(-5)}{3-0} = \dfrac{0}{3} = 0$

Perpendicular

57. $3x + 4y = 12$

$$4y = -3x + 12$$

$$y = -\frac{3}{4}x + 3$$

$m = -\dfrac{3}{4}$ y-intercept $(0, 3)$

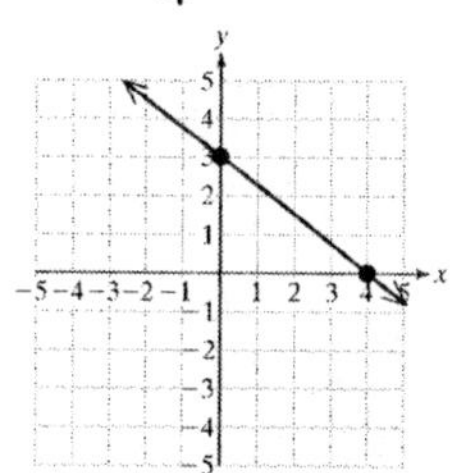

59. $3x - y = 4$

$$-y = -3x + 4$$

$$y = 3x - 4$$

$m = 3$ y-intercept $(0, -4)$

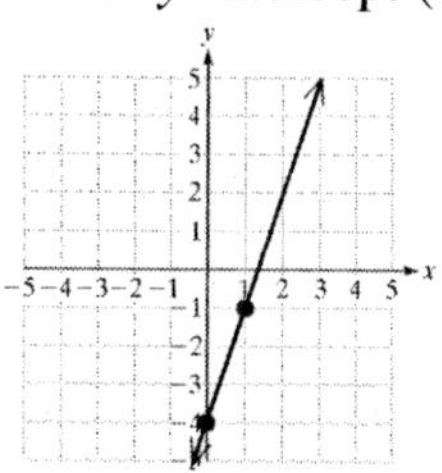

61. $5y - 8 = 4$

$$5y = 12$$

$$y = \frac{12}{5}$$

$m = 0$ y-intercept $\left(0, \dfrac{12}{5}\right)$

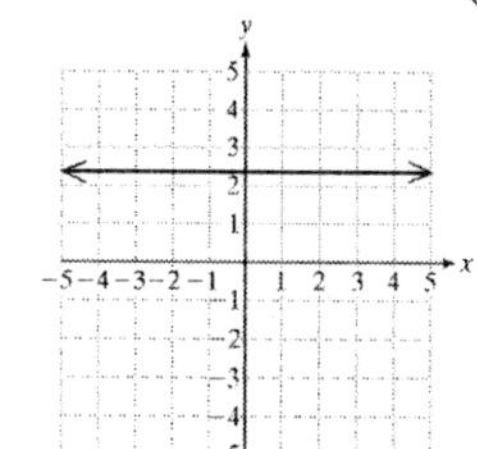

63. $y - x = 0$

$y = x$

$m = 1$ y-intercept $(0, 0)$

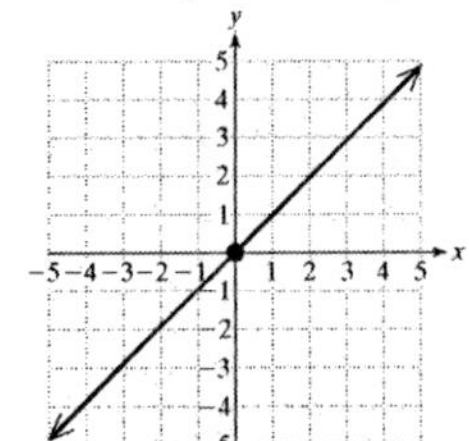

65. $m_1 = \frac{2}{5}$ $m_2 = -\frac{5}{2}$ Perpendicular

67. $m_1 = \frac{1}{3}$ $m_2 = \frac{1}{3}$ Parallel

69. $y = 2$

Chapter 3 Test

1. **(a)** II

(b) IV

(c) III

3. 0

5. **(a)** $2(0) - 6 = 6$

$-6 \neq 6$

No

(b) $2(4) - 2 = 6$

$6 = 6$

Yes

(c) $2(3) - 0 = 6$

$6 = 6$

Yes

(d) $2\left(\frac{9}{2}\right) - 3 = 6$

$6 = 6$

Yes

7. **(a)** $y = 220 - 18 = 202$ beats per minute

(b) (20, 200), (30, 190), (40, 180) (50, 170), (60, 160)

9. $x = \frac{7}{5}$; vertical

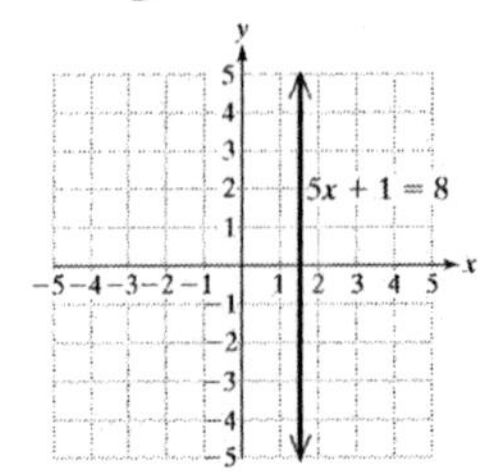

11. $\frac{4 \text{ ft}}{10 \text{ ft}} = \frac{2}{5}$

13. **(a)** $x + 4y = -16$

$y = -\frac{1}{4}x - 4$

$m = -\frac{1}{4}$

(b) $m = 4$

15. x-intercept

$0 = 8x + 2$

$x = -\frac{1}{4}$

$\left(-\frac{1}{4}, 0\right)$

y-intercept

$y = 8(0) + 2$

$y = 2$

$(0, 2)$

17. x-intercept

$x - 3 = 0$

$x = 3$

$(3, 0)$

y-intercept

none

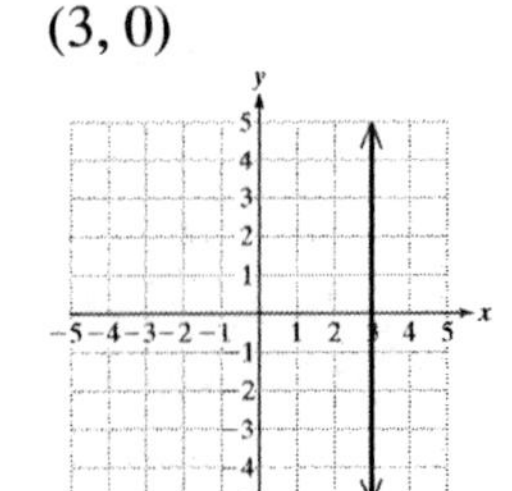

19. $m_1 = \frac{3}{2}$ $m_2 = -\frac{2}{3}$; perpendicular

21. $2x + 3y = -9$

$$3y = -2x - 9$$

$$y = -\frac{2}{3}x - 2$$

$m = -\frac{2}{3}$ $(0, -3)$

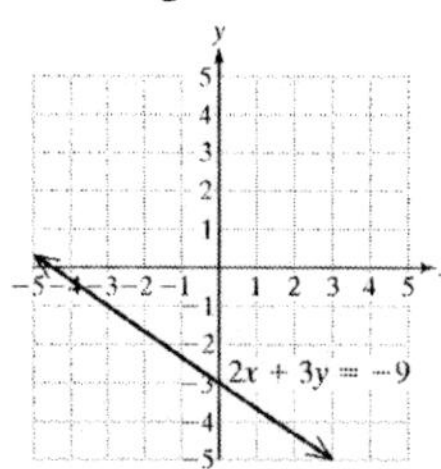

23. **(a)** $y = 0.15(1400) + 400 = \$610$

(b)

$$730 = 0.15x + 400$$

$$0.15x = 330$$

$$x = \$2200$$

(c) $y = 0.15(0) + 400 = 400$ $(0, 400)$; if Gerard has \$0 in sales, his salary will be \$400.

(d) $m = 0.15$; the slope represents the commission rate. That is, he makes \$0.15 in income for every \$1 sold.

Cumulative Review Exercises Chapters 1–3

1. **(a)** Rational

(b) Rational

(c) Irrational

(d) Rational

3. $32 \div 2 \cdot 4 + 5 = 16 \cdot 4 + 5 = 69$

5. $16 - 5 - (-7) = 16 - 5 + 7 = 18$

7. $(-2.1)(-6)$; $(-2.1)(-6) = 12.6$

9. $6x - 10 = 14$

$$6x = 24$$

$$x = 4$$

11.

$$\frac{2}{3}y - \frac{1}{6} = y + \frac{4}{3}$$

$$6\left(\frac{2}{3}y - \frac{1}{6}\right) = 6\left(y + \frac{4}{3}\right)$$

$$4y - 1 = 6y + 8$$

$$-2y = 9$$

$$y = -\frac{9}{2}$$

13. Let x = the area of Maine.

712 less than $29x = 267277$

$$29x - 712 = 267277$$

$$29x = 267989$$

$$x = 9241$$

The area of Maine is 9241 mi^2.

15.

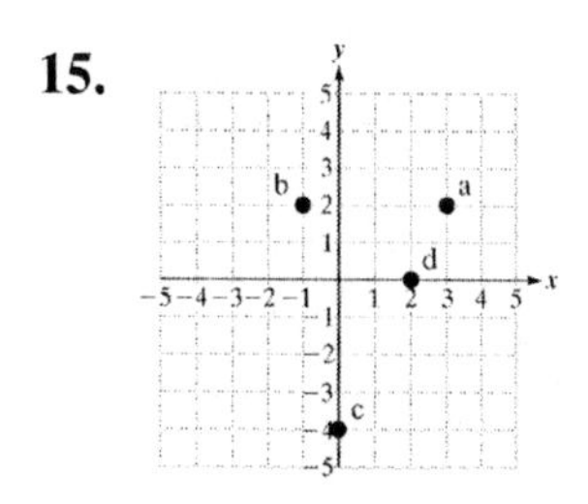

17. x-intercept

$$-2x + 4(0) = 4$$

$$-2x = 4$$

$$x = -2$$

$(-2, 0)$

y-intercept

$$-2(0) + 4y = 4$$

$$4y = 4$$

$$y = 1$$

$(0, 1)$

19. $3x + 2y = -12$

$$2y = -3x - 12$$

$$y = -\frac{3}{2}x - 6$$

$m = -\frac{3}{2}$ y-intercept $(0, -6)$

21. Since the equation can be simplified to $x = 1$, the equation is a vertical line. It does not cross the y-axis, only the x-axis.

Chapter 4

Section 4.1 Practice Exercises

1. Base: r; exponent: 4

3. Base: 5; exponent: 2

5. Base: –4; exponent: 8

7. Base: x; exponent: 1

9. y

11. x

13. No; $(-5)^2 = 25$ however, $-5^2 = -25$.

15. Yes; $(-2)^5 = -32$ and $-2^5 = -32$

17. $\left(\frac{1}{2}\right)^3 = \frac{1}{8}$ and $\frac{1}{2^3} = \frac{1}{8}$

19. $\left(\frac{3}{10}\right)^2 = \frac{9}{100}$ and $(0.3)^2 = 0.09$

21. **(a)** $x^4 \cdot x^3 = (x \cdot x \cdot x \cdot x)(x \cdot x \cdot x) = x^7$

(b) $5^4 \cdot 5^3 = (5 \cdot 5 \cdot 5 \cdot 5)(5 \cdot 5 \cdot 5) = 5^7$

23. $z^5 z^3 = z^{5+3} = z^8$

25. $a \cdot a^8 = a^{1+8} = a^9$

27. $4^5 \cdot 4^9 = 4^{5+9} = 4^{14}$

29. $9^4 \cdot 9 = 9^{4+1} = 9^5$

31. $c^5 c^2 c^7 = c^{5+2+7} = c^{14}$

33. **(a)** $\frac{p^8}{p^3} = \frac{p \cdot p \cdot p \cdot p \cdot p \cdot \not{p} \cdot \not{p} \cdot \not{p}}{\not{p} \cdot \not{p} \cdot \not{p}} = p^5$

(b) $\frac{8^8}{8^3} = \frac{8 \cdot 8 \cdot 8 \cdot 8 \cdot 8 \cdot \not{8} \cdot \not{8} \cdot \not{8}}{\not{8} \cdot \not{8} \cdot \not{8}} = 8^5$

35. $\frac{x^8}{x^6} = x^{8-6} = x^2$

37. $\frac{a^{10}}{a} = a^{10-1} = a^9$

39. $\frac{7^{13}}{7^6} = 7^{13-6} = 7^7$

41. $\frac{5^8}{5} = 5^{8-1} = 5^7$

43. $\frac{y^{13}}{y^{12}} = y^{13-12} = y$

45. $\frac{h^3 h^8}{h^7} = \frac{h^{3+8}}{h^7} = h^{11-7} = h^4$

47. $\frac{x^9 x}{x^5} = \frac{x^{9+1}}{x^5} = x^{10-5} = x^5$

49. $\frac{7^2 \cdot 7^6}{7} = \frac{7^{2+6}}{7} = 7^{8-1} = 7^7$

51. $\frac{x^{13}}{x^3 x^4} = \frac{x^{13}}{x^{3+4}} = x^{13-7} = x^6$

53. $\frac{10^{20}}{10^3 \cdot 10^8} = \frac{10^{20}}{10^{3+8}} = 10^{20-11} = 10^9$

55. $\frac{6^8 \cdot 6^5}{6^2 \cdot 6} = \frac{6^{13}}{6^3} = 6^{13-3} = 6^{10}$

57. $\frac{z^3 z^{11}}{z^4 z^6} = \frac{z^{14}}{z^{10}} = z^4$

59. $(5a^2b)(8a^3b^4) = 8 \cdot 5 \cdot a^2 a^3 b b^4 = 40a^5b^5$

61. $(r^6s^4)(13r^2s) = 13r^6r^2s^4s = 13r^8s^5$

63. $\left(\frac{2}{3}m^{13}n^{8}\right)(24m^{7}n^{2})=\frac{2}{3}\cdot 24\cdot m^{13}m^{7}n^{8}n^{2}$
$=16m^{20}n^{10}$

65. $\frac{14c^4d^5}{7c^3d}=2c^{4-3}d^{5-1}=2cd^4$

67. $\frac{2x^3y^5}{8xy^3}=\frac{x^{3-1}y^{5-3}}{4}=\frac{x^2y^2}{4}$

69. $\frac{25h^3jk^5}{12h^2k}=\frac{25h^{3-2}jk^{5-1}}{12}=\frac{25hjk^4}{12}$

71. $A=\pi r^2$
$=(3.14)(8\text{ in.})^2$
$=3.14(64)$
$=201\text{ in}^2$

73. $V=\frac{4}{3}\pi r^3=\frac{4}{3}(3.14)(3)^3=113\text{ cm}^3$

75. The length of the napkin is 24 in. = 2 ft. The area of the napkin is $A=s^2=2^2=4\text{ ft}^2$. The total footage for 60 is $60(4)=240\text{ ft}^2$.

77. $A=\$5000(1+0.07)^2$
$=5000(1.07)^2$
$=\$5724.50$

79. $A=\$4000(1+0.06)^3$
$=4000(1.06)^3$
$=\$4764.06$

81. $x^n x^{n+1}=x^{n+n+1}=x^{2n+1}$

83. $p^{3m+5}p^{-m-2}=p^{3m+5-m-2}=p^{2m+3}$

85. $\frac{z^{b+1}}{z^b}=z^{b+1-b}=z$

87. $\frac{r^{3a+3}}{r^{3a}}=r^{3a+3-3a}=r^3$

89–91.
```
(1.06)^5
          1.338225578
(1.02)^40
          2.208039664
5000(1.06)^5
          6691.127888
```

93.
```
2000(1.02)^40
          4416.079327
3000(1+.06)^2
               3370.8
```

Section 4.2 Practice Exercises

1. $4^2\cdot 4^7=4^{2+7}=4^9$

3. $a^{13}\cdot a\cdot a^6=a^{13+1+6}=a^{20}$

5. $\frac{d^{13}d}{d^5}=d^{14-5}=d^9$

7. $\frac{7^{11}}{7^5}=7^{11-5}=7^6$

9. If multiplying two factors with the same base, add the exponents. If a quantity is raised to a power, multiply the exponents.

11. $(5^3)^4=5^{3\cdot 4}=5^{12}$

13. $(12^3)^2=12^{3\cdot 2}=12^6$

15. $(y^7)^2=y^{7\cdot 2}=y^{14}$

17. $(w^5)^5=w^{5\cdot 5}=w^{25}$

19. $(a^2a^4)^6=(a^6)^6=a^{6\cdot 6}=a^{36}$

21. $(y^3y^4)^2=(y^7)^2=y^{7\cdot 2}=y^{14}$

23. **(a)** $\frac{x^7}{x^5}=x^{7-5}=x^2$

(b) $x^7\cdot x^5=x^{7+5}=x^{12}$

(c) $(x^7)^5 = x^{7\cdot 5} = x^{35}$

25. **(a)** $\left(\frac{y^2}{z^3}\right)^4 = \frac{(y^2)^4}{(z^3)^4} = \frac{y^8}{z^{12}}$

(b) $(y^2z^3)^4 = (y^2)^4(z^3)^4 = y^8z^{12}$

27. $\left(\frac{2}{3}\right)^3 = \frac{2^3}{3^3} = \frac{8}{27}$

29. $\left(\frac{1}{4}\right)^2 = \frac{1^2}{4^2} = \frac{1}{16}$

31. $\left(\frac{x}{y}\right)^5 = \frac{x^5}{y^5}$

33. $\left(\frac{1}{t}\right)^4 = \frac{1^4}{t^4} = \frac{1}{t^4}$

35. $(-3a)^4 = (-3)^4a^4 = 81a^4$

37. $(-3abc)^3 = (-3)^3a^3b^3c^3 = -27a^3b^3c^3$

39. $\frac{9u^5}{3u^2} = \frac{9}{3}u^{5-2} = 3u^3$

41. $(2xy^3)(3x^2y) = 2\cdot 3\cdot xx^2y^2y = 6x^3y^3$

43. $\frac{a^4b^7}{a^2b} = a^{4-2}b^{7-1} = a^2b^6$

45. $\frac{(4m)^5}{(4m)^4} = (4m)^{5-4} = (4m)$

47. $\frac{(2a+b)^7}{(2a+b)^5} = (2a+b)^{7-5} = (2a+b)^2$

49. $(6u^2v^4)^3 = 6^3(u^2)^3(v^4)^3 = 216u^6v^{12}$

51. $5(x^2y)^4 = 5(x^2)^4(y)^4 = 5x^8y^4$

53. $\left(\frac{4}{rs^4}\right)^5 = \frac{4^5}{(rs^4)^5} = \frac{4^5}{(r)^5(s^4)^5} = \frac{1024}{r^5s^{20}}$

55. $\left(\frac{3p}{q^3}\right)^5 = \frac{(3p)^5}{(q^3)^5} = \frac{3^5p^5}{q^{3\cdot 5}} = \frac{243p^5}{q^{15}}$

57. $\frac{y^8(y^3)^4}{(y^2)^3} = \frac{y^8y^{12}}{y^6} = \frac{y^{20}}{y^6} = y^{14}$

59. $(x^2)^5(x^3)^7 = x^{10}x^{21} = x^{31}$

61. $(a^2b)^3(a^4b^3)^5 = a^6b^3a^{20}b^{15} = a^{26}b^{18}$

63. $\frac{(5a^3b)^4(a^2b)^4}{(5ab)^2} = \frac{5^4a^{12}b^4a^8b^4}{5^2a^2b^2} = 25a^{18}b^6$

65. $\frac{(21x^5y)(2x^8y^4)}{14xy} = \frac{42x^{13}y^5}{14xy} = 3x^{12}y^4$

67. $\left(\frac{2c^3d^4}{3c^2d}\right)^2 = \frac{4c^6d^8}{9c^4d^2} = \frac{4}{9}c^2d^6$

69. $(2c^3d^2)^5\left(\frac{c^6d^8}{4c^2d}\right)^3 = (32c^{15}d^{10})\left(\frac{c^{18}d^{24}}{64c^6d^3}\right)$

$= \frac{c^{27}d^{31}}{2}$

71. $(x^m)^2 = x^{2m}$

73. $(5a^{2n})^3 = 5^3a^{6n} = 125a^{6n}$

75. $\left(\frac{m^2}{n^3}\right)^b = \frac{(m^2)^b}{(n^3)^b} = \frac{m^{2b}}{n^{3b}}$

77. $\left(\frac{3a^3}{5b^4}\right)^n = \frac{3^na^{3n}}{5^nb^{4n}}$

79. $(2^2)^3 = 2^6$; $(2^3)^2 = 2^6$, they are the same.

81. $2^{(2^4)} = 2^{16}$ and $(2^2)^4 = 2^8 2^{(2^4)}$ is greater.

Section 4.3 Practice Exercises

1. $b^3b^8 = b^{3+8} = b^{11}$

3. $\dfrac{x^6}{x^2} = x^{6-2} = x^4$

5. $\dfrac{9^4 . 9^8}{9} = \dfrac{9^{12}}{9} = 9^{11}$

7. $(6ab^3c^2)^5 = 6^5a^5(b^3)^5(c^2)^5$
$= 7776a^5b^{15}c^{10}$

9. $\left(\dfrac{s^2t^5}{4}\right)^3 = \dfrac{(s^2t^5)^3}{4^3} = \dfrac{s^6t^{15}}{64}$

11. **(a)** $8^0 = 1$

(b) $\dfrac{8^4}{8^4} = 8^{4-4} = 8^0 = 1$

13. $p^0 = 1$

15. $5^0 = 1$

17. $-4^0 = -(4^0) = -1$

19. $(-6)^0 = 1$

21. $(8x)^0 = 1$

23. $-7x^0 = -7 \cdot 1 = -7$

25. $ab^0 = a \cdot 1 = a$

27. **(a)** $t^{-5} = \dfrac{1}{t^5}$

(b) $\dfrac{t^3}{t^8} = t^{3-8} = t^{-5} = \dfrac{1}{t^5}$

29. Subtract (−6). $\dfrac{x^4}{x^{-6}} = x^{4-(-6)} = x^{10}$

31. The exponent is only on the variable a.
$2a^{-3} = a \cdot \dfrac{1}{a^3} = \dfrac{2}{a^3}$

33. $\left(\dfrac{2}{7}\right)^{-3} = \left(\dfrac{7}{2}\right)^3 = \dfrac{7^3}{8^3} = \dfrac{343}{8}$

35. $\left(-\dfrac{1}{5}\right)^{-2} = (-5)^2 = 25$

37. $a^{-3} = \dfrac{1}{a^3}$

39. $12^{-1} = \dfrac{1}{12}$

41. $(4b)^{-2} = \dfrac{1}{(4b)^2} = \dfrac{1}{16b^2}$

43. $6x^{-2} = 6 \cdot \dfrac{1}{x^2} = \dfrac{6}{x^2}$

45. $w^{-4}w^{-2} = w^{-4+(-2)} = w^{-6} = \dfrac{1}{w^6}$

47. $x^{-8}x^4 = x^{-8+4} = x^{-4} = \dfrac{1}{x^4}$

49. $a^{-8}a^8 = a^{-8+8} = a^0 = 1$

51. $y^{17}y^{-13} = y^{17+(-13)} = y^4$

53. $(m^{-6}n^9)^3 = (m^{-6})^3(n^9)^3 = m^{-18}n^{27} = \dfrac{n^{27}}{m^{18}}$

55. $(-3j^{-5}k^6)^4 = (-3)^4(j^{-5})^4(k^6)^4$
$= 81j^{-20}k^{24}$
$= \dfrac{81k^{24}}{j^{20}}$

57. $\dfrac{p^3}{p^9} = p^{3-9} = p^{-6} = \dfrac{1}{p^6}$

59. $\dfrac{r^{-5}}{r^{-2}} = r^{-5-(-2)} = r^{-3} = \dfrac{1}{r^3}$

61. $\dfrac{7^3}{7^2 \cdot 7^8} = \dfrac{7^3}{7^{10}} = 7^{3-10} = 7^{-7} = \dfrac{1}{7^7}$

63. $\dfrac{a^{-1}b^2}{a^3b^8} = a^{-1-3}b^{2-8} = a^{-4}b^{-6} = \dfrac{1}{a^4b^6}$

65. $\dfrac{w^{-8}(w^2)^{-5}}{w^3} = \dfrac{w^{-8}w^{-10}}{w^3}$
$= \dfrac{w^{-18}}{w^3}$
$= w^{-21}$
$= \dfrac{1}{w^{21}}$

67. $(-8y^{-12})(2y^{16}z^{-2}) = (-8)(2)y^{-12+16}z^{-2}$
$= \dfrac{-16y^4}{z^2}$

69. $\dfrac{-18a^{10}b^6}{108a^{-2}b^6} = -\dfrac{1}{6}a^{10-(-2)}b^0 = -\dfrac{a^{12}}{6}$

71. $\dfrac{(-4c^{12}d^7)^2}{(5c^{-3}d^{10})^{-1}} = \dfrac{(-4)^2c^{12\cdot 2}d^{7\cdot 2}}{(5)^{-1}c^{(-3)(-1)}d^{10(-1)}}$
$= \dfrac{16c^{24}d^{14}}{5^{-1}c^3d^{-10}}$
$= 16(5)c^{24-3}d^{14-(-10)}$
$= 80c^{21}d^{24}$

73. $\left(\dfrac{2}{p^6p^3}\right)^{-3} = \left(\dfrac{2}{p^9}\right)^{-3}$
$= \left(\dfrac{p^9}{2}\right)^3$
$= \dfrac{p^{9\cdot 3}}{2^3}$
$= \dfrac{p^{27}}{8}$

75. $\left(\dfrac{5cd^{-3}}{10d^5}\right)^{-1} = \left(\dfrac{cd^{-3-5}}{2}\right)^{-1}$
$= \left(\dfrac{c}{2d^8}\right)^{-1}$
$= \dfrac{2d^8}{c}$

77. $\left(\dfrac{1}{2}\right)^{-1} + \left(\dfrac{1}{3}\right)^0 = \dfrac{2}{1} + 1 = 3$

79. $(2^5b^{-3})^{-3} = 2^{5(-3)}b^{(-3)(-3)} = 2^{-15}b^9 = \dfrac{b^9}{2^{15}}$

81. $\left(\dfrac{3x}{2y}\right)^{-4} = \left(\dfrac{2y}{3x}\right)^4 = \dfrac{(2y)^4}{(3x)^4} = \dfrac{16y^4}{81x^4}$

83. $(3ab^2)(a^2b)^3 = (3ab^2)(a^6b^3) = 3a^7b^5$

85. $\left(\dfrac{xy^2}{x^3y}\right)^4 = \dfrac{(xy^2)^4}{(x^3y)^4} = \dfrac{x^4y^8}{x^{12}y^4} = x^{-8}y^4 = \dfrac{y^4}{x^8}$

87. $\dfrac{(t^{-2})^3}{t^{-4}} = \dfrac{t^{-6}}{t^{-4}} = t^{-6-(-4)} = t^{-2} = \dfrac{1}{t^2}$

89. $\left(\dfrac{2w^2x^3}{3y^0}\right)^3 = \dfrac{2^3w^{2(3)}x^{3(3)}}{3} = \dfrac{8w^6x^9}{3}$

91. $\dfrac{q^3r^{-2}}{s^{-1}t^5} = \dfrac{q^3s}{r^2t^5}$

93. $\frac{(y^{-3})^2(y^5)}{(y^{-3})^{-4}} = \frac{y^{-6}y^5}{y^{12}} = \frac{y^{-1}}{y^{12}} = y^{-13} = \frac{1}{y^{13}}$

95. $\left(\frac{-2a^2b^{-3}}{a^{-4}b^{-5}}\right)^{-3} = (-2a^{2-(-4)}b^{-3-(-5)})^{-3}$
$= (-2a^6b^2)^{-3}$
$= \frac{1}{(-2a^6b^2)^3}$
$= -\frac{1}{8a^{18}b^6}$

97. $(5h^{-2}k^0)^3(5k^{-2})^{-4} = 5^3h^{-6}k^0 \cdot 5^{-4}k^8$
$= 5^{-1}h^{-6}k^8$
$= \frac{k^8}{5h^6}$

99. $5^{-1} + 2^{-2} = \frac{1}{5} + \frac{1}{4} = \frac{4}{20} + \frac{5}{20} = \frac{9}{20}$

101. $10^0 - 10^{-1} = 1 - \frac{1}{10} = \frac{10}{10} - \frac{1}{10} = \frac{9}{10}$

103. $\frac{4^{-1} + 3^{-2}}{1 + 2^{-3}} = \frac{\frac{1}{4} + \frac{1}{9}}{1 + \frac{1}{8}}$
$= \frac{\frac{9}{36} + \frac{4}{36}}{\frac{8}{8} + \frac{1}{8}}$
$= \frac{13}{36} \div \frac{9}{8}$
$= \frac{13}{36} \cdot \frac{8}{9}$
$= \frac{26}{81}$

Section 4.4 Practice Exercises

1. $a^3a^{-4} = a^{3+(-4)} = a^{-1} = \frac{1}{a}$

3. $10^3 \cdot 10^{-4} = 10^{3+(-4)} = 10^{-1} = \frac{1}{10}$

5. $\frac{x^3}{x^6} = x^{3-6} = x^{-3} = \frac{1}{x^3}$

7. $\frac{10^3}{10^6} = 10^{3-6} = 10^{-3} = \frac{1}{10^3}$

9. $\frac{z^9z^4}{z^3} = \frac{z^{13}}{z^3} = z^{13-3} = z^{10}$

11. $\frac{10^9 \cdot 10^4}{10^3} = \frac{10^{13}}{10^3} = 10^{13-3} = 10^{10}$

13. Move the decimal point between 2 and 3 and multiply by 10^{-10}; 2.3×10^{-10}

15. 6.8×10^7 gallons; 1.0×10^2 miles

17. 4.2×10^8

19. 8×10^{-6}

21. 1.7×10^{-24} g

23. 1.4115999×10^8 shares

25. Move the decimal point nine places to the left; 0.0000000031

27. 0.00005

29. 2800

31. 0.000000000001 g

33. 1600 and 2800 calories

35. $(2.5 \times 10^6)(2.0 \times 10^{-2}) = 2.5(2.0) \times 10^6 \cdot 10^{-2}$
$= 5.0 \times 10^4$

37. $(1.2 \times 10^4)(3 \times 10^7) = 1.2 \cdot 3 \times 10^4 \cdot 10^7$
$= 3.6 \times 10^{11}$

39. $\frac{7.7 \times 10^6}{3.5 \times 10^2} = \frac{7.7}{3.5} \times 10^{6-2} = 2.2 \times 10^4$

41. $\dfrac{9.0\times10^{-6}}{4.0\times10^{7}}=\dfrac{9.0}{4.0}\times10^{-6-7}=2.25\times10^{-13}$

43. $(8.0\times10^{10})(4.0\times10^{3})=8\cdot4\times10^{10}\cdot10^{3}$
$=32\times10^{13}$
$=3.2\times10^{14}$

45. $(3.2\times10^{-4})(7.6\times10^{-7})$
$=3.2\cdot7.6\times10^{-4}\cdot10^{-7}$
$=24.32\times10^{-11}$
$=2.432\times10^{-10}$

47. $\dfrac{2.1\times10^{11}}{7.0\times10^{-3}}=\dfrac{2.1}{7}\times10^{11-(-3)}$
$=0.3\times10^{14}$
$=3.0\times10^{13}$

49. $\dfrac{5.7\times10^{-2}}{9.5\times10^{-8}}=\dfrac{5.7}{9.5}\times10^{-2-(-8)}$
$=0.6\times10^{6}$
$=6.0\times10^{5}$

51. $6{,}000{,}000{,}000\times0.0000000023$
$=(6\times10^{9})(2.3\times10^{-9})$
$=13.8\times10^{0}$
$=1.38\times10^{1}$

53. $\dfrac{0.0000000003}{6000}=0.00000000000005$
$=5\times10^{-14}$

55. (thickness of paper)(no. of pieces)
$=(3\times10^{-3})(1.25\times10^{3})$
$=3(1.25)\times10^{0}$
$=3.75$ inches

57. $\dfrac{\text{amount spent}}{\text{no. of commercials}}$
$=\dfrac{\$6\times10^{8}}{3.5\times10^{5}}$
$=1.714\times10^{3}$
$=\$1714$ per commercial

59. **(a)** 65 million $=65{,}000{,}000=6.5\times10^{7}$

(b) $(6.5\times10^{7}$ years)(365 days)
$=2.3725\times10^{10}$ days

(c) $(6.5\times10^{7}$ years)(365 days)
$=2.3725\times10^{10}$ days;
$(2.3725\times10^{10}$ days)(24 hours)
$=5.694\times10^{11}$ hours

(d) $(6.5\times10^{7}$ years)
$=(5.694\times10^{11}$ hours)(3600 seconds)
$=2.04984\times10^{15}$ seconds

61. **(a)** 45% of $(1.5\times10^{6}$ workers)
$=0.45(1.5\times10^{6})$
$=6.75\times10^{5}$ workers

(b) 3% or $(1.5\times10^{6}$ workers)
$=0.03(1.5\times10^{6})$
$=4.5\times10^{4}$ workers

63. Let x = revenue 9 months earlier

$$\begin{pmatrix}\text{revenue}\\\text{9 months}\\\text{ago}\end{pmatrix}-\begin{pmatrix}\text{12\% of that}\\\text{revenue}\end{pmatrix}=\$6.337\times10^{10}$$

$x+0.12x=6.337\times10^{10}$
$1.12x=6.337\times10^{10}$
$x=\dfrac{6.337\times10^{10}}{1.12}$
$\approx\$5.66\times10^{10}$

65.

```
(5.2E6)*(4.6E-3)
            2.392E4
(2.19E-8)*(7.84E
-4)
        1.71696E-11
```

67.

```
(4.76E-5)/(2.38E
9)
             2E-14
(8.5E4)/(4.0E-1)
           2.125E5
```

69.

```
((5.0E-12)*(6.4E
-5))/((1.6E-8)*(
4.0E2))
             5E-11
```

Section 4.5 Practice Exercises

1. $\dfrac{p^3 \cdot 4p}{p^2} = \dfrac{4p^4}{p^2} = 4p^2$

3. $(6y^{-3})(2y^9) = 6 \cdot 2y^{-3}y^9 = 12y^6$

5. $\dfrac{8^3 \cdot 8^{-4}}{8^{-2} \cdot 8^6} = \dfrac{8^{-1}}{8^4} = 8^{-1-4} = 8^{-5} = \dfrac{1}{8^5}$

7. 3.0×10^7 is the number 3 multiplied by 10 to the 7th power; 3^7 is just 3 raised to the 7th power.

9. $-7x^4 + 7x^2 + 9x + 6$

11. Binomial; leading coefficient 10; degree 2

13. Monomial; leading coefficient 6; degree 2

15. Trinomial; leading coefficient −1; degree 4

17. Trinomial; leading coefficient 12; degree 4

19. Monomial; leading coefficient 5; degree 3

21. Binomial; leading coefficient 1; degree 4

23. The exponents on the variable x are different.

25. $23x^2y + 12x^2y = 35x^2y$

27. $(6y + 3x) + (4y - 3x) = 6y + 4y + 3x - 3x$
$= 10y$

29. $3b^2 + (5b^2 - 9) = 3b^2 + 5b^2 - 9 = 8b^2 - 9$

31. $(7y^2 + 2y - 9) + (-3y^2 - y)$
$= 7y^2 - 3y^2 - y + 2y - 9$
$= 4y^2 + y - 9$

33. $(6a + 2b - 5c) + (-2a - 2b - 3c)$
$= 6a - 2a + 2b - 2b - 5c - 3c$
$= 4a - 8c$

35. $\left(\dfrac{2}{5}a + \dfrac{1}{4}b - \dfrac{5}{6}\right) + \left(\dfrac{3}{5}a - \dfrac{3}{4}b - \dfrac{7}{6}\right)$
$= \dfrac{2}{5}a + \dfrac{3}{5}a + \dfrac{1}{4}b - \dfrac{3}{4}b - \dfrac{5}{6} - \dfrac{7}{6}$
$= a - \dfrac{1}{2}b - 2$

37. $\left(z - \dfrac{8}{3}\right) + \left(\dfrac{4}{3}z^2 - z + 1\right) = \dfrac{4}{3}z^2 + z - z - \dfrac{8}{3} + 1$
$= \dfrac{4}{3}z^2 - \dfrac{5}{3}$

39. $(7.9t^3 + 2.6t - 1.1) + (-3.4t^2 + 3.4t - 3.1)$
$= 7.9t^3 - 3.4t^2 + 2.6t + 3.4t - 1.1 - 3.1$
$= 7.9t^3 - 3.4t^2 + 6t - 4.2$

41. $(y^2 + 3) + (3y^3 - y^2 - 1) + (y^3 + 2y^2)$
$= 3y^3 + y^3 + y^2 - y^2 + 2y^2 + 3 - 1$
$= 4y^3 + 2y^2 + 2$

43. **(a)** after 1 sec: $h = -16(1)^2 + 150 = 134$ ft
after 1.5 sec:
$h = -16(1.5)^2 + 150 = 114$ ft
after 2 sec: $h = -16(2)^2 + 150 = 86$ ft

(b) when $t = 0$: $h = -16(0)^2 + 150 = 150$ ft

45. $-(4h - 5) = -4h + 5$

47. $-(-2m^2 + 3m - 15) = 2m^2 - 3m + 15$

49. $-(3v^3+5v^2+10v+22)$
$=-3v^3-5v^2-10v-22$

51. $-(-9t^4-8t-39)=9t^4+8t+39$

53. $4a^3b^2-12a^3b^2=-8a^3b^2$

55. $-32x^3-21x^3=-53x^3$

57. $(7a-7)-(12a-4)=7a-7-12a+4=-5a-3$

59. $(4k+3)-(-12k-6)=4k+3+12k+6=16k+9$

61. $25s-(23s-14)=25s-23s+14=2s+14$

63. $(5t^2-3t-2)-(2t^2+t+1)=5t^2-2t^2-3t-t-2-1=3t^3-4t-3$

65. $(10r-6s+2t)-(12r-3s-t)=10r-12r-6s+3s+2t+t=-2r-3s+3t$

67. $\left(\frac{7}{8}x+\frac{2}{3}y-\frac{3}{10}\right)-\left(\frac{1}{8}x+\frac{1}{3}y\right)=\frac{7}{8}x-\frac{1}{8}x+\frac{2}{3}y-\frac{1}{3}y-\frac{3}{10}=\frac{3}{4}x+\frac{1}{3}y-\frac{3}{10}$

69. $\left(\frac{2}{3}h^2-\frac{1}{5}h-\frac{3}{4}\right)-\left(\frac{4}{3}h^2-\frac{4}{5}h+\frac{7}{4}\right)=\frac{2}{3}h^2-\frac{4}{3}h^2-\frac{1}{5}h+\frac{4}{5}h-\frac{7}{4}-\frac{3}{4}=-\frac{2}{3}h^2+\frac{3}{5}h-\frac{5}{2}$

71. $(4.5x^4-3.1x^2-6.7)-(2.1x^4+4.4x)=2.4x^4-3.1x^2-4.4x-6.7$

73. $P=5a^2-2a+1$
$\begin{pmatrix}\text{missing}\\ \text{side}\end{pmatrix}+(a-3)+(2a^2-1)=P$
$\begin{pmatrix}\text{missing}\\ \text{side}\end{pmatrix}=5a^2-2a+1-(a-3)-(2a^2-1)$
$\begin{pmatrix}\text{missing}\\ \text{side}\end{pmatrix}=3a^2-a+5$

75. $(-2x^2+6x-21)-(3x^3-5x+10)=-3x^3-2x^2+6x+5x-21-10=-3x^3-2x^2+11x-31$

77. $(4b^3+6b-7)-(-12b^2+11b+5)=4b^3+12b^2+6b-11b-7-5=4b^3+12b^2-5b-12$

79. $(2ab^2+9a^2b)+(7ab^2-3ab+7a^2b)=2ab^2+7ab^2+9a^2b+7a^2b-3ab=9ab^2+16a^2b-3ab$

81. $(4z^5+z^3-3z+13)-(-z^4-8z^3+15)=4z^5+z^4+z^3-3z+13-15+8z^3=4z^5+z^4+9z^3-3z-2$

83. $(9x^4+2x^3-x+5)+(9x^3-3x^2+8x+3)-(7x^4-x+12)$
$=9x^4-7x^4+2x^3+9x^3-3x^2-x+x+8x+5+3-12$
$=2x^4+11x^3-3x^2+8x-4$

85. $(5w^2-3w+2)+(-4w+6)-(7w^2-10)$
$=5w^2-3w+2-4w+6-7w^2+10$
$=-2w^2-7w+18$

87. $(7p^2q-3pq^2)-(8p^2q+pq)+(4pq-pq^2)$
$=7p^2q-3pq^2-8p^2q-pq+4pq-pq^2$
$=-p^2q-4pq^2+3pq$

89. Answers will vary; x^3+3

91. Answers will vary; $8x^5$

93. Answers will vary; $-6x^2+x-2$

Section 4.6 Practice Exercises

1. $4x+5x=9x$

3. $(4x)(5x)=20x^2$

5. $-5a^3b-2a^3b=-7a^3b$

7. $(-5a^3b)(-2a^3b)=10a^6b^2$

9. $-c+4c^2=4c^2-c$

11. $(-c)(4c^2)=-4c^3$

13. $(4.3\times10^6)(2.3\times10^6)=4.3\cdot2.3\times10^{6+6}$
$=9.89\times10^{12}$

15. $(-2.1\times10^{-12})(9.3\times10^{-12})$
$=-2.1(9.3)\times10^{-12+(-12)}$
$=-19.53\times10^{-24}$
$=-1.953\times10^{-23}$

17. $8(4x)=32x$

19. $-10(5z)=-50z$

21. $(x^{10})(4x^3)=4x^{10+3}=4x^{13}$

23. $(4m^3n^7)(-3m^6n)=-12m^{3+6}n^{7+1}$
$=-12m^9n^8$

25. $8pq(2pq-3p+5q)$
$=8pq(2pq)-8pq(3p)+8pq(5q)$
$=16p^2q^2-24p^2q+40pq^2$

27. $(k^2-13k-6)(-4k)$
$=k^2(-4k)-13k(-4k)-6(-4k)$
$=-4k^3+52k^2+24k$

29. $-15pq(3p^2+p^3q^2-2q)$
$=-15pq(3p^2)-15pq(p^3q^2)-15pq(-2q)$
$=-45p^3q-15p^4q^3+30pq^2$

31. $(y-10)(y+9)=y^2+9y-10y-90$
$=y^2-y-90$

33. $(m-12)(m-2)=m^2-2m-12m+24$
$=m^2-14m+24$

35. $(p-2)(p+1)=p^2+p-2p-2$
$=p^2-p-2$

37. $(w+8)(w+3)=w^2+3w+8w+24$
$=w^2+11w+24$

39. $(p-3)(p-11)=p^2-11p-3p+33$
$=p^2-14p+33$

41. $(6x-1)(2x+5)=6x(2x)+5(6x)-2x-5$
$=12x^2+28x-5$

43. $(4a-9)(2a-1)=4a(2a)-4a-18a+9$
$=8a^2-22a-9$

45. $(3t-7)(3t+1)=3t(3t)+3t-21t-7$
$=9t^2-18t-7$

47. $(3x+4)(x+8)=3x^2+24x+4x+32$
$=3x^2+28x+32$

49. $(5s+3)(s^2+s-2)$
$=5s(s^2)+5s(s)-5s(2)+3s^2+3s-6$
$=5s^3+5s^2-10s+3s^2+3s-6$
$=5s^3+8s^2-7s-6$

51. $(3w-2)(9w^2+6w+4)$
$=27w^3+18w^2+12w-18w^2-12w-8$
$=27w^3-8$

53. $(3a-4b)(3a+4b)=(3a)^2-(4b)^2$
$=9a^2-16b^2$

55. $(9k+6)(9k-6)=(9k)^2-6^2=81k^2-36$

57. $\left(\frac{1}{2}-t\right)\left(\frac{1}{2}+t\right)=\left(\frac{1}{2}\right)^2-t^2=\frac{1}{4}-t^2$

59. $(u^3+5v)(u^3-5v)=(u^3)^2-(5v)^2$
$=u^6-25v^2$

61. $(a+b)^2=a^2+2ab+b^2$

63. $(x-y)^2=x^2-2xy+y^2$

65. $(2c+5)^2=(2c)^2+2(2c)(5)+5^2$
$=4c^2+20c+25$

67. $(3t^2-4s)^2=(3t^2)^2-2(3t^2)(4s)+(4s)^2$
$=9t^4-24st^2+16s^2$

69. **(a)** $(2+4)^2=(6)^2=36$

(b) $2^2+4^2=4+16=20$

(c) $36\neq 20;\ (a+b)^2\neq a^2+b^2$

71. $A=(2x+5)(2x-5)=(2x)^2-5^2=4x^2-25$

73. $A=(4p+5)^2$
$=(4p)^2+2(4p)(5)+5^2$
$=16p^2+40p+25$

75. $(7x+y)(7x-y)=(7x)^2-y^2=49x^2-y^2$

77. $(5s+3t)^2=(5s)^2+2(5s)(3t)+(3t)^2$
$=25s^2+30st+9t^2$

79. $(7x-3y)(3x-8y)$
$=7x(3x)-7x(8y)-3y(3x)+3y(8y)$
$=21x^2-65xy+24y^2$

81. $\left(\frac{2}{3}t+2\right)(3t+4)$
$=\frac{2}{3}t(3t)+\frac{2}{3}t(4)+2(3t)+2(4)$
$=2t^2+\frac{8}{3}t+6t+8$
$=2t^2+\frac{26}{3}t+8$

83. $(5z+3)(z^2+4z-1)$
$=5z(z^2)+5z(4z)-1(5z)+3z^2+3(4z)-3$
$=5z^3+23z^2+7z-3$

85. $\left(\frac{1}{3}m-n\right)^2=\left(\frac{1}{3}m\right)^2-2\left(\frac{1}{3}m\right)(n)+n^2$
$=\frac{1}{9}m^2-\frac{2}{3}mn+n^2$

87. $6w^2(7w-14)=6w^2(7w)-6w^2(14)$
$=42w^3-84w^2$

89. $(4y-8.1)(4y+8.1)$
$(4y)^2-(8.1)^2=16y^2-65.61$

91. $(3c^2+4)(7c^2-8)$
$=3c^2(7c^2)-(3c^2)(8)+4(7c^2)-4(8)$
$=21c^4+4c^2-32$

93. $(3.1x+4.5)^2$
$=(3.1x)^2+2(3.1x)(4.5)+(4.5)^2$
$=9.61x^2+27.9x+20.25$

95. $(k-4)^3=(k-4)(k-4)^2$
$=(k-4)(k^2-8k+16)$
$=k^3-8k^2+16k-4k^2+32k-64$
$=k^3-12k^2+48k-64$

97. $A=\frac{1}{2}bh$
$=\frac{1}{2}(5a^3-2)(6a^2)$
$=\frac{1}{2}(30a^5-12a^2)$
$=15a^5-6a^2$

99. V
$=s^3$
$=(3p-5)^3$
$=(3p-5)(3p-5)^2$
$=(3p-5)(9p^2-30p+25)$
$=27p^3-90p^2+75p-45p^2+150p-125$
$=27p^3-135p^2+225p-125$

101. $2a(3a-4)(a+5)=2a(3a^2+11a-20)$
$=6a^3+22a^2-40a$

103. $(x-3)(2x+1)(x-4)$
$=(x-3)(2x^2-7x-4)$
$=2x^3-7x^2-4x-6x^2+21x+12$
$=2x^3-13x^2+17x+12$

105. $(3x+5)(a+b)=6x^2-11x-35$
$3ax=6x^2$ and $5b=-35$
$a=2x$ and $b=-7$
$(2x-7)$

Section 4.7 Practice Exercises

1. $(6z^5-2z^3+z-6)-(10z^4+2z^3+z^2+z)$
$=6z^5-10z^4-4z^3-z^2-6$

3. $(10x+y)(x-3y)=10x^2-30xy+xy-3y^2$
$=10x^2-29xy-3y^2$

5. $(2w^3+5)^2=(2w^3)^2+2(2w^3)(5)+5^2$
$=4w^6+20w^3+25$

7. $\left(\frac{7}{8}w-1\right)\left(\frac{7}{8}w+1\right)=\left(\frac{7}{8}w\right)^2-1^2$
$=\frac{49}{64}w^2-1$

9. Use long division when the divisor is a polynomial with two or more terms.

11. (a) $\frac{15t^3+18t^2}{3t}=\frac{15t^3}{3t}+\frac{18t^2}{3t}=5t^2+6t$

(b) $3t(5t^2+6t)=15t^3+18t^2$

13. $(6a^2+4a-14)\div 2=\frac{6a^2}{2}+\frac{4a}{2}-\frac{14}{2}$
$=3a^2+2a-7$

15. $\frac{-5x^2-20x+5}{-5}=\frac{-5x^2}{-5}-\frac{20x}{-5}+\frac{5}{-5}$
$=x^2+4x-1$

17. $\frac{3p^3-p^2}{p}=\frac{3p^3}{p}-\frac{p^2}{p}=3p^2-p$

19. $(4m^2+8m)\div 4m^2=\frac{4m^2}{4m^2}+\frac{8m}{4m^2}=1+\frac{2}{m}$

21. $\frac{14y^4-7y^3+21y^2}{-7y^2}=\frac{14y^4}{-7y^2}-\frac{7y^3}{-7y^2}+\frac{21y^2}{-7y^2}$
$=-2y^2+y-3$

23. $(4x^3-24x^2-x+8)\div(4x)$
$=\frac{4x^3}{4x}-\frac{24x^2}{4x}-\frac{x}{4x}+\frac{8}{4x}$
$=x^2-6x-\frac{1}{4}+\frac{2}{x}$

25. $\dfrac{-a^3b^2+a^2b^2-ab^3}{-a^2b^2}$

$=\dfrac{-a^3b^2}{-a^2b^2}+\dfrac{a^2b^2}{-a^2b^2}-\dfrac{ab^3}{-a^2b^2}$

$=a-1+\dfrac{b}{a}$

27. $(6t^4-2t^3+3t^2-t+4)\div(2t^3)$

$=\dfrac{6t^4}{2t^3}-\dfrac{2t^3}{2t^3}+\dfrac{3t^2}{2t^3}-\dfrac{t}{2t^3}+\dfrac{4}{2t^3}$

$=3t-1+\dfrac{3}{2t}-\dfrac{1}{2t^2}+\dfrac{2}{t^3}$

29. (a) $z+2+\dfrac{1}{z+5}$

$$\begin{array}{r|l} & z+2 \\ z+5 & z^2+7z+11 \\ & \underline{-(z^2+5z)} \\ & 2z+11 \\ & \underline{-(2z+10)} \\ & 1 \end{array}$$

(b) $(z+5)(z+2)+1=z^2+7z+11$

31. $t+3$

$$\begin{array}{r|l} & t+3 \\ t+1 & t^2+4t+3 \\ & \underline{-(t^2+t)} \\ & 3t+3 \\ & \underline{-(3t+3)} \end{array}$$

33. $7b+4$

$$\begin{array}{r|l} & 7b+4 \\ b-1 & 7b^2-3b-4 \\ & \underline{-(7b^2-7b)} \\ & 4b-4 \\ & \underline{-(4b-4)} \end{array}$$

35. $k-6$

$$\begin{array}{r|l} & k-6 \\ 5k+1 & 5k^2-29k-6 \\ & \underline{-(5k^2+k)} \\ & -30k-6 \\ & \underline{-(-30k-6)} \end{array}$$

37. $2p^2+3p-4$

$$\begin{array}{r|l} & 2p^2+3p-4 \\ 2p+3 & 4p^3+12p^2+p-12 \\ & \underline{-(4p^3+6p^2)} \\ & 6p^2+p \\ & \underline{-(6p^2+9p)} \\ & -8p-12 \\ & \underline{-(-8p-12)} \end{array}$$

39. $k-2+\dfrac{-4}{k+1}$

$$\begin{array}{r|l} & k-2 \\ k+1 & k^2-k-6 \\ & \underline{-(k^2+k)} \\ & -2k-6 \\ & \underline{-(-2k-2)} \\ & -4 \end{array}$$

41. $2x^2-x+6+\dfrac{2}{2x-3}$

$$\begin{array}{r|l} & 2x^2-x+6 \\ 2x-3 & 4x^3-8x^2+15x-16 \\ & \underline{-(4x^3-6x^2)} \\ & -2x^2+15x \\ & \underline{-(-2x^2+3x)} \\ & 12x-16 \\ & \underline{-(12x-18)} \\ & 2 \end{array}$$

93. $(3.1x+4.5)^2$
$=(3.1x)^2+2(3.1x)(4.5)+(4.5)^2$
$=9.61x^2+27.9x+20.25$

95. $(k-4)^3=(k-4)(k-4)^2$
$=(k-4)(k^2-8k+16)$
$=k^3-8k^2+16k-4k^2+32k-64$
$=k^3-12k^2+48k-64$

97. $A=\frac{1}{2}bh$
$=\frac{1}{2}(5a^3-2)(6a^2)$
$=\frac{1}{2}(30a^5-12a^2)$
$=15a^5-6a^2$

99. V
$=s^3$
$=(3p-5)^3$
$=(3p-5)(3p-5)^2$
$=(3p-5)(9p^2-30p+25)$
$=27p^3-90p^2+75p-45p^2+150p-125$
$=27p^3-135p^2+225p-125$

101. $2a(3a-4)(a+5)=2a(3a^2+11a-20)$
$=6a^3+22a^2-40a$

103. $(x-3)(2x+1)(x-4)$
$=(x-3)(2x^2-7x-4)$
$=2x^3-7x^2-4x-6x^2+21x+12$
$=2x^3-13x^2+17x+12$

105. $(3x+5)(a+b)=6x^2-11x-35$
$3ax=6x^2$ and $5b=-35$
$a=2x$ and $b=-7$
$(2x-7)$

Section 4.7 Practice Exercises

1. $(6z^5-2z^3+z-6)-(10z^4+2z^3+z^2+z)$
$=6z^5-10z^4-4z^3-z^2-6$

3. $(10x+y)(x-3y)=10x^2-30xy+xy-3y^2$
$=10x^2-29xy-3y^2$

5. $(2w^3+5)^2=(2w^3)^2+2(2w^3)(5)+5^2$
$=4w^6+20w^3+25$

7. $\left(\frac{7}{8}w-1\right)\left(\frac{7}{8}w+1\right)=\left(\frac{7}{8}w\right)^2-1^2$
$=\frac{49}{64}w^2-1$

9. Use long division when the divisor is a polynomial with two or more terms.

11. (a) $\frac{15t^3+18t^2}{3t}=\frac{15t^3}{3t}+\frac{18t^2}{3t}=5t^2+6t$

(b) $3t(5t^2+6t)=15t^3+18t^2$

13. $(6a^2+4a-14)\div 2=\frac{6a^2}{2}+\frac{4a}{2}-\frac{14}{2}$
$=3a^2+2a-7$

15. $\frac{-5x^2-20x+5}{-5}=\frac{-5x^2}{-5}-\frac{20x}{-5}+\frac{5}{-5}$
$=x^2+4x-1$

17. $\frac{3p^3-p^2}{p}=\frac{3p^3}{p}-\frac{p^2}{p}=3p^2-p$

19. $(4m^2+8m)\div 4m^2=\frac{4m^2}{4m^2}+\frac{8m}{4m^2}=1+\frac{2}{m}$

21. $\frac{14y^4-7y^3+21y^2}{-7y^2}=\frac{14y^4}{-7y^2}-\frac{7y^3}{-7y^2}+\frac{21y^2}{-7y^2}$
$=-2y^2+y-3$

23. $(4x^3-24x^2-x+8)\div(4x)$
$=\frac{4x^3}{4x}-\frac{24x^2}{4x}-\frac{x}{4x}+\frac{8}{4x}$
$=x^2-6x-\frac{1}{4}+\frac{2}{x}$

25. $$\frac{-a^3b^2+a^2b^2-ab^3}{-a^2b^2}$$
$$=\frac{-a^3b^2}{-a^2b^2}+\frac{a^2b^2}{-a^2b^2}-\frac{ab^3}{-a^2b^2}$$
$$=a-1+\frac{b}{a}$$

27. $(6t^4-2t^3+3t^2-t+4)\div(2t^3)$
$$=\frac{6t^4}{2t^3}-\frac{2t^3}{2t^3}+\frac{3t^2}{2t^3}-\frac{t}{2t^3}+\frac{4}{2t^3}$$
$$=3t-1+\frac{3}{2t}-\frac{1}{2t^2}+\frac{2}{t^3}$$

29. (a) $z+2+\dfrac{1}{z+5}$

$$\begin{array}{r} z+2 \\ z+5\overline{)\,z^2+7z+11} \\ \underline{-(z^2+5z)} \\ 2z+11 \\ \underline{-(2z+10)} \\ 1 \end{array}$$

(b) $(z+5)(z+2)+1=z^2+7z+11$

31. $t+3$

$$\begin{array}{r} t+3 \\ t+1\overline{)\,t^2+4t+3} \\ \underline{-(t^2+t)} \\ 3t+3 \\ \underline{-(3t+3)} \end{array}$$

33. $7b+4$

$$\begin{array}{r} 7b+4 \\ b-1\overline{)\,7b^2-3b-4} \\ \underline{-(7b^2-7b)} \\ 4b-4 \\ \underline{-(4b-4)} \end{array}$$

35. $k-6$

$$\begin{array}{r} k-6 \\ 5k+1\overline{)\,5k^2-29k-6} \\ \underline{-(5k^2+k)} \\ -30k-6 \\ \underline{-(-30k-6)} \end{array}$$

37. $2p^2+3p-4$

$$\begin{array}{r} 2p^2+3p-4 \\ 2p+3\overline{)\,4p^3+12p^2+p-12} \\ \underline{-(4p^3+6p^2)} \\ 6p^2+p \\ \underline{-(6p^2+9p)} \\ -8p-12 \\ \underline{-(-8p-12)} \end{array}$$

39. $k-2+\dfrac{-4}{k+1}$

$$\begin{array}{r} k-2 \\ k+1\overline{)\,k^2-k-6} \\ \underline{-(k^2+k)} \\ -2k-6 \\ \underline{-(-2k-2)} \\ -4 \end{array}$$

41. $2x^2-x+6+\dfrac{2}{2x-3}$

$$\begin{array}{r} 2x^2-x+6 \\ 2x-3\overline{)\,4x^3-8x^2+15x-16} \\ \underline{-(4x^3-6x^2)} \\ -2x^2+15x \\ \underline{-(-2x^2+3x)} \\ 12x-16 \\ \underline{-(12x-18)} \\ 2 \end{array}$$

43. $a-3+\dfrac{18}{a+3}$

$$\begin{array}{r} a-3 \\ a+3\overline{)\ a^2+0a+9} \\ \underline{-(a^2+3a)} \\ -3a+9 \\ \underline{-(-3a-9)} \\ 18 \end{array}$$

45. $w^2+5w-2+\dfrac{1}{w^2-3}$

$$\begin{array}{r} w^2+5w-2 \\ w^2+0w-3\overline{)\ w^4+5w^3-5w^2-15w+7} \\ \underline{-(w^4+0w^3-3w^2)} \\ 5w^3-2w^2-15w \\ \underline{-(5w^3+0w^2-15w)} \\ -2w^2+0w+7 \\ \underline{-(-2w^2+0w+6)} \\ 1 \end{array}$$

47. n^2+n-6

$$\begin{array}{r} n^2+n-6 \\ 2n^2+3n-2\overline{)\ 2n^4+5n^3-11n^2-20n+12} \\ \underline{-(2n^4+3n^3-2n^2)} \\ 2n^3-9n^2-20n \\ \underline{-(2n^3+3n^2-2n)} \\ -12n^2-18n+12 \\ \underline{-(-12n^2-18n+12)} \end{array}$$

49. $x-1+\dfrac{8}{5x^2+5x+1}$

$$\begin{array}{r} x-1 \\ 5x^2+5x+1\overline{)\ 5x^3+0x^2-4x-9} \\ \underline{-(5x^3+5x^2+x)} \\ -5x^2-5x-9 \\ \underline{-(-5x^2-5x-1)} \\ -8 \end{array}$$

51. To check, multiply the divisor $(x-2)$ by the quotient (x^2+4).

$(x-2)(x^2+4)=x^3-2x^2+4x-8$ which does not equal x^3-8.

53. Monomial division;

$$\frac{9a^3}{3a}+\frac{12a^2}{3a}=3a^2+4a$$

55. Long division;

$p+2$

$$\begin{array}{r} p+2 \\ p^2-p-2\overline{)\ p^3+p^2-4p-4} \\ \underline{-(p^3-p^2-2p)} \\ 2p^2-2p-4 \\ \underline{-(2p^2-2p-4)} \end{array}$$

57. Long division;

$t^3-2t^2+5t-10+\dfrac{4}{t+2}$

$$\begin{array}{r} t^3-2t^2+5t-10 \\ t+2\overline{)\ t^4+0t^3+t^2+0t-16} \\ \underline{-(t^4+2t^3)} \\ -2t^3+t^2 \\ \underline{-(-2t^3-4t^2)} \\ 5t^2+0t \\ \underline{-(5t^2+10t)} \\ -10t-16 \\ \underline{-(-10t-20)} \\ 4 \end{array}$$

59. Long division;

$$w^2+3+\frac{1}{w^2-2}$$

$$\begin{array}{r} w^2+3 \\ w^2+0w-2\overline{)\ w^4+0w^3+w^2+0w-5} \\ \underline{-(w^4+0w^3-2w^2)} \\ 3w^2+0w-5 \\ \underline{-(3w^2+0w-6)} \\ 1 \end{array}$$

61. Long division;

$$n^2+4n+16$$

$$\begin{array}{r} n^2+4n+16 \\ n-4\overline{)\ n^3+0n^2+0n-64} \\ \underline{-(n^3-4n^2)} \\ 4n^2+0n \\ \underline{-(4n^2-16n)} \\ 16n-64 \\ \underline{-(16n-64)} \end{array}$$

63. Monomial division;

$$\frac{9r^3}{-3r^2}+\frac{-12r^2}{-3r^2}+\frac{9}{-3r^2}=-3r+4-\frac{3}{r^2}$$

65. $\dfrac{x^2-1}{x-1}=x+1$

67. $\dfrac{x^4-1}{x-1}=x^3+x^2+x+1$

69. $\dfrac{x^2}{x-1}=x+1+\dfrac{1}{x-1}$

71. $\dfrac{x^4}{x-1}=x^3+x^2+x+1+\dfrac{1}{x-1}$

Chapter 4 Review Exercises

1. Base 5; exponent 3

3. Base (–2); exponent 0

5. **(a)** $6^2=6\cdot 6=36$

(b) $(-6)^2=(-6)(-6)=36$

(c) $-6^2=-(6\cdot 6)=-36$

7. $5^3\cdot 5^{10}=5^{3+10}=5^{13}$

9. $x\cdot x^6\cdot x^2=x^{1+6+2}=x^9$

11. $\dfrac{10^7}{10^4}=10^{7-4}=10^3$

13. $\dfrac{b^9}{b}=b^{9-1}=b^8$

15. $\dfrac{k^2k^3}{k^4}=k^{5-4}=k^1=k$

17. $\dfrac{2^8\cdot 2^{10}}{2^3\cdot 2^7}=\dfrac{2^{18}}{2^{10}}=2^{18-10}=2^8$

19. You can only add exponents when the bases are the same.

21. $A=P(1+r)^t=6000(1+0.06)^3=\$7146.10$

23. $(7^3)^4=7^{3\cdot 4}=7^{12}$

25. $(p^4p^2)^3=(p^6)^3=p^{6\cdot 3}=p^{18}$

27. $\left(\dfrac{a}{b}\right)^2=\dfrac{a^2}{b^2}$

29. $\left(\dfrac{5}{c^2d^5}\right)^2=\dfrac{5^2}{(c^2d^5)^2}=\dfrac{25}{c^4d^{10}}$

31. $(2ab^2)^4=2^4a^4(b^2)^4=2^4a^4b^8$

33. $\left(\dfrac{-3x^3}{5y^2z}\right)^3 = \dfrac{(-3x^3)^3}{(5y^2z)^3} = \dfrac{-3^3(x^3)^3}{5^3(y^2)^3z^3} = -\dfrac{3^3x^9}{5^3y^6z^3}$

35. $\dfrac{a^4(a^2)^8}{(a^3)^3} = \dfrac{a^4a^{16}}{a^9} = \dfrac{a^{20}}{a^9} = a^{11}$

37. $\dfrac{(4h^2k)^2(h^3k)^4}{(2hk^3)^2} = \dfrac{4^2h^4k^2h^{12}k^4}{2^2h^2k^6} = \dfrac{16h^{16}k^6}{4h^2k^6} = 4h^{14}$

39. $\left(\dfrac{2x^4y^3}{4xy^2}\right)^2 = \dfrac{(2x^4y^3)^2}{(4xy^2)^2} = \dfrac{2^2x^8y^6}{4^2x^2y^4} = \dfrac{x^6y^2}{4}$

41. $8^0 = 1$

43. $1^0 = 1$

45. $2y^0 = 2(1) = 2$

47. $z^{-5} = \dfrac{1}{z^5}$

49. $(6a)^{-2} = \dfrac{1}{(6a)^2} = \dfrac{1}{36a^2}$

51. $4^0 + 4^{-2} = 1 + \dfrac{1}{4^2} = \dfrac{16}{16} + \dfrac{1}{16} = \dfrac{17}{16}$

53. $t^{-6}t^{-2} = t^{-8} = \dfrac{1}{t^8}$

55. $\dfrac{12x^{-2}y^3}{6x^4y^{-4}} = 2x^{-2-4}y^{3-(-4)} = 2x^{-6}y^7 = \dfrac{2y^7}{x^6}$

57. $(-2m^2n^{-4})^{-4} = (-2)^{-4}m^{-8}n^{16} = \dfrac{n^{16}}{2^4m^8} = \dfrac{n^{16}}{16m^8}$

59. $\dfrac{(k^{-6})^{-2}(k^3)}{5k^{-6}k^0} = \dfrac{k^{12}k^3}{5k^{-6}} = \dfrac{k^{15-(-6)}}{5} = \dfrac{k^{21}}{5}$

61. $\dfrac{7^0}{3^{-1} - 6^{-1}} = \dfrac{1}{\frac{1}{3} - \frac{1}{6}} = \dfrac{1}{\frac{1}{6}} = 6$

63. **(a)** $\$5.9148 \times 10^{12}$

(b) 4.2×10^{-3} in.

(c) 1.66241×10^8 km^2

65. $(4.1 \times 10^{-6})(2.3 \times 10^{11}) = (4.1)(2.3) \times 10^{-6+11} = 9.43 \times 10^5$

67. $\dfrac{2000}{0.000008} = 2.5 \times 10^8$

69. $5^{20} \approx 9.5367 \times 10^{13}$
This number has too many digits to fit on most calculator displays.

71. **(a)** $C = 2\pi r$
$= 2(3.14)(9.3 \times 10^7)$
$\approx 58.4 \times 10^7$
$= 5.84 \times 10^8$ miles

(b) $\dfrac{5.84 \times 10^8 \text{ miles}}{8.76 \times 10^3 \text{ hours}} \approx 0.667 \times 10^5$
$= 6.67 \times 10^4$ mph

73. **(a)** Trinomial

(b) degree 4

(c) leading coefficient 7

75. $(4x + 2) + (3x - 5) = 7x - 3$

77. $(9a^2-6)-(-5a^2+2a)=9a^2+5a^2-2a-6=14a^2-2a-6$

79. $$\left(5x^3-\frac{1}{4}x^2+\frac{5}{8}x+2\right)+\left(\frac{5}{2}x^3+\frac{1}{2}x^2-\frac{1}{8}x\right)=5x^3+\frac{5}{2}x^3-\frac{1}{4}x^2+\frac{1}{2}x^2+\frac{5}{8}x-\frac{1}{8}x+2$$
$$=\frac{15}{2}x^3+\frac{1}{4}x^2+\frac{1}{2}x+2$$

81. $(7x^2-5x)-(9x^2+4x+6)=7x^2-5x-9x^2-4x-6=-2x^2-9x-6$

83. Answers will vary; $-5x^2+x+1$

85. $P=2w+2l=2(w)+2(2w+3)=2w+4w+6=6w+6$

87. $(9a^6)(2a^2b^4)=18a^{6+2}b^4=18a^8b^4$

89. $(x^2+5x-3)(-2x)=x^2(-2x)+5x(-2x)-3(-2x)=-2x^3-10x^2+6x$

91. $(4t-1)(5t+2)=4t(5t)+4t(2)-1(5t)-1(2)=20t^2+8t-5t-2=20t^2+3t-2$

93. $(2a-6)(a+5)=2a(a)+2a(5)-6a-6(5)=2a^2+4a-30$

95. $(b-4)^2=b^2-2(b)(4)+4^2=b^2-8b+16$

97. $$(2w-1)(-w^2-3w-4)=-2w(w^2)-2w(3w)-2w(4)+w^2+3w+4$$
$$=-2w^3-6w^2-8w+w^2+3w+4$$
$$=-2w^3-5w^2-5w+4$$

99. $$\left(\frac{1}{3}r^4-s^2\right)\left(\frac{1}{3}r^4+s^2\right)=\left(\frac{1}{3}r^4\right)^2-(s^2)^2=\frac{1}{9}r^8-s^4$$

101. $$(2h+3)(h^4-h^3+h^2-h+1)=2h^5-2h^4+2h^3-2h^2+2h+3h^4-3h^3+3h^2-3h+3$$
$$=2h^5+h^4-h^3+h^2-h+3$$

103. $$\frac{20y^3-10y^2}{5y}=\frac{20y^3}{5y}-\frac{10y^2}{5y}=4y^2-2y$$

105. $$(12x^4-8x^3+4x^2)\div(-4x^2)=\frac{12x^4}{-4x^2}+\frac{-8x^3}{-4x^2}+\frac{4x^2}{-4x^2}=-3x^2+2x-1$$

107. $x+2$

$$\begin{array}{r} x+2 \\ x+5\overline{)\,x^2+7x+10} \\ -(x^2+5x) \\ \hline 2x+10 \\ -(2x+10) \\ \hline \end{array}$$

109. $p-3+\dfrac{5}{2p+7}$

$$\begin{array}{r} p-3 \\ 2p+7\overline{)\,2p^2\ +p-16} \\ -(2p^2+7p) \\ \hline -6p-16 \\ -(-6p-21) \\ \hline 5 \end{array}$$

111. $b^2+5b+25$

$$\begin{array}{r} b^2+5b+25 \\ b-5\overline{)\,b^3+0b^2+0b-125} \\ -(b^3-5b^2) \\ \hline 5b^2\ +0b \\ -(5b^2-25b) \\ \hline 25b-125 \\ -(25b-125) \\ \hline \end{array}$$

113. $y^2-4y+2+\dfrac{9y-4}{y^2+3}$

$$\begin{array}{r} y^2-4y+2 \\ y^2+0y+3\overline{)\,y^4-4y^3+5y^2\ -3y+2} \\ -(y^4+0y^3+3y^2) \\ \hline -4y^3+2y^2\ -3y \\ -(-4y^3+0y^2-12y) \\ \hline 2y^2\ +9y+2 \\ -(2y^2\ +0y+6) \\ \hline 9y-4 \end{array}$$

115. $2x^2-3x+2+\dfrac{1}{3x+2}$

$$\begin{array}{r} 2x^2-3x+2 \\ 3x+2\overline{)\,6x^3-5x^2+0x+5} \\ -(6x^3+4x^2) \\ \hline -9x^2+0x \\ -(-9x^2-6x) \\ \hline 6x+5 \\ -(6x+4) \\ \hline 1 \end{array}$$

117. $t^2-3t+1+\dfrac{-2t-6}{3t^2+t+1}$

$$\begin{array}{r} t^2-3t+1 \\ 3t^2+t+1\overline{)\,3t^4-8t^3+t^2-4t-5} \\ -(3t^4+t^3+t^2) \\ \hline -9t^3+0t^2-4t \\ -(-9t^3-3t^2-3t) \\ \hline 3t^2\ -t-5 \\ -(3t^2\ +t+1) \\ \hline -2t-6 \end{array}$$

Chapter 4 Test

1. $\dfrac{3^4\cdot 3^3}{3^6}=\dfrac{(3\cdot 3\cdot 3\cdot 3)(3\cdot 3\cdot 3)}{3\cdot 3\cdot 3\cdot 3\cdot 3\cdot 3}=3$

3. $\dfrac{q^{10}}{q^2}=q^{10-2}=q^8$

5. $\left(\dfrac{2x}{y^3}\right)^4=\dfrac{(2x)^4}{(y^3)^4}=\dfrac{2^4x^4}{y^{12}}=\dfrac{16x^4}{y^{12}}$

7. $c^{-3}=\dfrac{1}{c^3}$

9. $\dfrac{(s^2t)^3(7s^4t)^4}{(7s^2t^3)^2} = \dfrac{s^6t^3 7^4 s^{16}t^4}{7^2 s^4 t^6}$
$= \dfrac{7^4 s^{22} t^7}{7^2 s^4 t^6}$
$= 7^2 s^{18} t$
$= 49 s^{18} t$

11. $\left(\dfrac{6a^{-5}b}{8ab^{-2}}\right)^{-2} = \left(\dfrac{6a^{-5-1}b^{1-(-2)}}{8}\right)^{-2}$
$= \left(\dfrac{6a^{-6}b^3}{8}\right)^{-2}$
$= \left(\dfrac{3b^3}{4a^6}\right)^{-2}$
$= \left(\dfrac{4a^6}{3b^3}\right)^{2}$
$= \dfrac{16a^{12}}{9b^6}$

13. (a) 1440 minutes in one day;
1.68×10^5 m^3/min(1440 min/day)
$= 2.4192\times10^8$ m^3 in one day

(b) 2.4192×10^8 m^3 in one day
2.4192×10^8 m^3/day(365 days/year)
$= 8.83008\times10^{10}$ m^3/year

15. $(7w^2 - 11w - 6) + (8w^2 + 3w + 4) - (-9w^2 - 5w + 2)$
$= 7w^2 + 8w^2 + 9w^2 - 11w + 3w + 5w - 6 + 4 - 2$
$= 24w^2 - 3w - 4$

17. $(4a-3)(2a-1) = 8a^2 - 4a - 6a + 3$
$= 8a^2 - 10a + 3$

19. $(2+3b)(2-3b) = 2^2 - (3b)^2 = 4 - 9b^2$

21. $P = 2w + 2l$
$= 2(x-3) + 2(5x+2)$
$= 2x - 6 + 10x + 4$
$= 12x - 2$
$A = lw = (5x+2)(x-3) = 5x^2 - 13x - 6$

Cumulative Review Exercises Chapters 1–4

1. $-5 - \dfrac{1}{2}[4 - 3(-7)] = -5 - \dfrac{1}{2}[4 + 21]$
$= -5 - \dfrac{1}{2}(25)$
$= -5 - \dfrac{25}{2}$
$= -\dfrac{35}{2}$

3. $\dfrac{-3 - \sqrt{14 - (-2) + 3^2}}{-3.44 + 1.2^2} = \dfrac{-3 - \sqrt{16 + 9}}{-2}$
$= \dfrac{-3 - 5}{-2}$
$= 4$

5. $-7, \dfrac{0}{4}, 2, 0.8, \sqrt{100}$

7. $-2y - 3 = -5(y-1) + 3y$
$-2y - 3 = -5y + 5 + 3y$
$-2y - 3 = -2y + 5$
$-3 \neq 5$
No solution

9. y-axis

11. Let x = value of merchandise sold. Then,
3% of x = \$360
$0.03x = 360$
$x = \dfrac{360}{0.03} = 12000$
He sold \$12,000 worth of merchandise.

13. (a) $y = \dfrac{3}{2}(4) + 6 = 6 + 6 = 12$ in.

(b) $y = \dfrac{3}{2}(9) + 6 = \dfrac{27}{2} + 6 = 19.5$ in.

(c) $y = 14\frac{1}{4}$

$$\begin{aligned} 14\frac{1}{4} &= \frac{3}{2}x + 6 \\ 4\left(\frac{57}{4}\right) &= 4\left(\frac{3}{2}x + 6\right) \\ 57 &= 6x + 24 \\ 6x &= 33 \\ x &= \frac{33}{6} = 5.5 \text{ hours} \end{aligned}$$

(d)

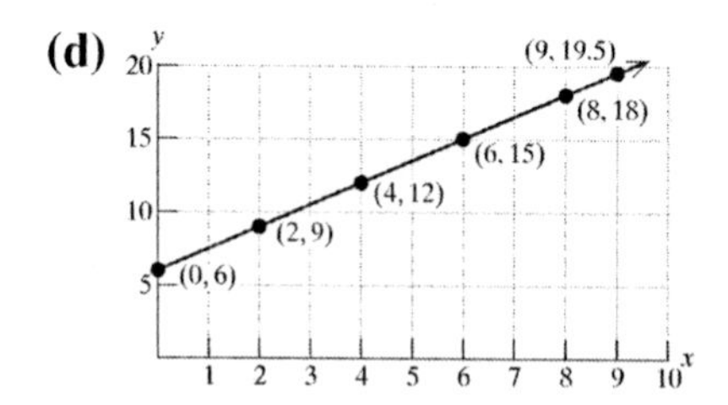

15. $(2x^2 + 3x - 7) - (-3x^2 + 12x + 8)$

$$\begin{aligned} &= 2x^2 + 3x^2 + 3x - 12x - 7 - 8 \\ &= 5x^2 - 9x - 15 \end{aligned}$$

17. $(4t - 3)^2 = (4t)^2 - 2(4t)(3) + 3^2$

$$= 16t^2 - 24t + 9$$

19. $(7x - 8) - (x + 3)^2 = 7x - 8 - (x^2 + 6x + 9)$

$$= -x^2 + x - 17$$

21. $4m^2 + 8m + 11 + \frac{24}{m - 2}$

$$\begin{array}{r} 4m^2 + 8m + 11 \\ m - 2 \overline{)\ 4m^3 + 0m^2 \quad - 5m + 2} \\ -(4m^3 - 8m^2) \qquad\qquad \\ \hline 8m^2 \quad - 5m \qquad \\ -(8m^2 - 16m) \qquad \\ \hline 11m \quad + 2 \\ -(11m - 22) \\ \hline 24 \end{array}$$

23. $\left(\frac{2c^2d^4}{8cd^6}\right)^2 = \left(\frac{c}{2d^2}\right)^2 = \frac{c^2}{4d^4}$

25. (a) $407100000 = 4.071 \times 10^8$

(b) $0.000004071 = 4.071 \times 10^{-6}$

27. $\frac{(8.2 \times 10^{-2})(6.8 \times 10^{-6})}{2.0 \times 10^{-5}} = \frac{55.76 \times 10^{-8}}{2.0 \times 10^{-5}}$

$$\begin{aligned} &= \frac{5.576 \times 10^{-7}}{2.0 \times 10^{-5}} \\ &= 2.788 \times 10^{-2} \end{aligned}$$

Chapter 5

Section 5.1 Practice Exercises

1. 7

3. 6

5. ab

7. $4w^2z$

9. $(x-y)$

11. $7(3x+1)$

13. **(a)** $3(x-2y) = 3x-3(2y) = 3x-6y$

(b) $3x-6y = 3x-3(2y) = 3(x-2y)$

15. $4p+12 = 4p+4\cdot 3 = 4(p+3)$

17. $5c^2-10c = (5c)c-(5c)(2) = 5c(c-2)$

19. $x^5+x^3 = x^3x^2+x^3 = x^3(x^2+1)$

21. $t^4-4t = tt^3-4t = t(t^3-4)$

23. $2ab+4a^3b = 2ab+2ab(2a^2)$
$= 2ab(1+2a^2)$

25. $38x^2y-19x^2y^4 = 19x^2y(2)-19x^2y(y^3)$
$= 19x^2y(2-y^3)$

27. $42p^3q^2+14pq^2-7p^4q^4$
$= 7pq^2(6p^2+2-p^3q^2)$

29. $t^5+2rt^3-3t^4+4r^2t^2$
$= t^2(t^3+2rt-3t^2+4r^2)$

31. $13(a+6)-4b(a+6) = (a+6)(13-4b)$

33. $8v(w^2-2)+(w^2-2)$
$= 8v(w^2-2)+1(w^2-2)$
$= (w^2-2)(8v+1)$

35. $21x(x+3)+7x^2(x+3) = 7x(x+3)(3+x)$
$= 7x(x+3)^2$

37. $6(z-1)^3+7z(z-1)^2-(z-1)$
$= (z-1)[6(z-1)^2+7z(z-1)-1]$
$= (z-1)[6z^2-12z+6+7z^2-7z-1]$
$= (z-1)(13z^2-19z+5)$

39. **(a)** $-2x^3-4x^2+8x = -2x(x^2+2x-4)$

(b) $-2x^3-4x^2+8x = 2x(-x^2-2x+4)$

41. $-8t^2-9t-2 = -1(8t^2+9t+2)$

43. $-4y^3+5y-7 = (-1)(4y^3-5y+7)$

45. $15p^3-30p^2 = -15p^2(p+2)$

47. $-q^4+2q^2-9q = -q(q^3-2q+9)$

49. $-7x-6y-2z = -1(7x+6y+2z)$

51. $-3(2c+5)-4c(2c+5) = -1(2c+5)(3+4c)$

53. $8a^2-4ab+6ac-3bc$
$= 4a(2a-b)+3c(2a-b)$
$= (2a-b)(4a+3c)$

55. $3q+3p+qr+pr = 3(q+p)+r(q+p)$
$= (q+p)(3+r)$

57. $6x^2+3x+4x+2 = 3x(2x+1)+2(2x+1)$
$= (2x+1)(3x+2)$

59. $2t^2+6t-5t-15 = 2t(t+3)+(-5)(t+3)$
$= (2t-5)(t+3)$

61. $6y^2-2y-9y+3 = 2y(3y-1)+(-3)(3y-1)$
$= (3y-1)(2y-3)$

63. $b^4+b^3-4b-4 = b^3(b+1)+(-4)(b+1)$
$= (b+1)(b^3-4)$

65. $3j^2k+15k+j^2+5=3k(j^2+5)+1(j^2+5)$
$=(j^2+5)(3k+1)$

67. $14w^6x^6+7w^6-2x^6-1$
$=7w^6(2x^6+1)+(-1)(2x^6+1)$
$=(2x^6+1)(7w^6-1)$

69. $15x^4+15x^2y^2+10x^3y+10xy^3$
$=5x(3x^3+3xy^2+2x^2y+2y^3)$
$=5x(3x(x^2+y^2)+2y(x^2+y^2))$
$=5x(x^2+y^2)(3x+2y)$

71. $4abx-4b^2x-4ab+4b^2$
$=4b(ax-bx-a+b)$
$=4b(x(a-b)-1(a-b))$
$=4b(a-b)(x-1)$

73. $6st^2-18st-6t^4+18t^3$
$=6t(st-3s-t^3+3t^2)$
$=6t(s(t-3)-t^2(t-3))$
$=6t(t-3)(s-t^2)$

75. $P=2w+2l$
$P=2(w+l)$

77. $S=2\pi r^2+2\pi rh$
$S=2\pi r(r+h)$

79. $\frac{1}{7}x^2+\frac{3}{7}x-\frac{5}{7}=\frac{1}{7}(x^2+3x-5)$

81. $\frac{5}{4}w^2+\frac{3}{4}w+\frac{9}{4}=\frac{1}{4}(5w^2+3w+9)$

83. $\frac{1}{12}z^2+\frac{1}{3}z+\frac{1}{2}=\frac{1}{12}z^2+\frac{4}{12}z+\frac{6}{12}$
$=\frac{1}{12}(z^2+4z+6)$

85. $\frac{5}{6}q^2+\frac{1}{3}q-2=\frac{5}{6}q^2+\frac{2}{6}q-\frac{12}{6}$
$=\frac{1}{6}(5q^2+2q-12)$

87. Answers will vary; $6x^2+9x$

89. Answers will vary; $8p^2q^2+12p^2q$

Section 5.2 Practice Exercises

1. $8p^9+24p^3=8p^3(p^6+3)$

3. $9x^2y+12xy^2-15x^2y^2$
$=3xy(3x+4y-5xy)$

5. $5x(x-2)-2(x-2)=(x-2)(5x-2)$

7. $p^2-2pq-pq+2q^2$
$=p(p-2q)-q(p-2q)$
$=(p-2q)(p-q)$

9. $6a^2+24a-12a-48$
$=6(a^2+4a-2a-8)$
$=6(a(a+4)-2(a+4))$
$=6(a+4)(a-2)$

11. A polynomial that cannot be factored is prime.

13. 12 and 1

15. -8 and -1

17. -1 and 6

19. -12 and 6

21. $3x^2+13x+4=3x^2+12x+x+4$
$=3x(x+4)+1(x+4)$
$=(x+4)(3x+1)$

23. $4w^2-9w+2=4w^2-8w-w+2$
$=4w(w-2)-1(w-2)$
$=(w-2)(4w-1)$

25. $2m^2+5m-3=2m^2+6m-m-3$
$=2m(m+3)-1(m+3)$
$=(m+3)(2m-1)$

27. $8k^2-6k-9=8k^2-12k+6k-9$
$=4k(2k-3)+3(2k-3)$
$=(2k-3)(4k+3)$

29. $4k^2-20k+25=4k^2-10k-10k+25$
$=2k(2k-5)-5(2k-5)$
$=(2k-5)(2k-5)$
$=(2k-5)^2$

31. Prime

33. $4p^2+5pq-6q^2=4p^2+8pq-3pq-6q^2$
$=4p(p+2q)-3q(p+2q)$
$=(p+2q)(4p-3q)$

35. $15m^2+mn-2n^2$
$=15m^2+6mn-5mn-2n^2$
$=3m(5m+2n)-n(5m+2n)$
$=(5m+2n)(3m-n)$

37. $3r^2-rs-14s^2=3r^2+6rs-7rs-14s^2$
$=3r(r+2s)-7s(r+2s)$
$=(r+2s)(3r-7s)$

39. Prime

41. $q^2-11q+10=q^2-10q-q+10$
$=q(q-10)-1(q-10)$
$=(q-10)(q-1)$

43. $r^2-6r-40=r^2-10r+4r-40$
$=r(r-10)+4(r-10)$
$=(r-10)(r+4)$

45. $x^2+6x-7=x^2-x+7x-7$
$=x(x-1)+7(x-1)$
$=(x-1)(x+7)$

47. $m^2-13m+42=m^2-6m-7m+42$
$=m(m-6)-7(m-6)$
$=(m-6)(m-7)$

49. $a^2+9a+20=a^2+4a+5a+20$
$=a(a+4)+5(a+4)$
$=(a+4)(a+5)$

51. Prime

53. $p^2+20pq+100q^2$
$=p^2+10pq+10pq+100q^2$
$=p(p+10q)+10q(p+10q)$
$=(p+10q)(p+10q)$
$=(p+10q)^2$

55. $x^2-xy-42y^2=x^2-7xy+6xy-42y^2$
$=x(x-7y)+6y(x-7y)$
$=(x-7y)(x+6y)$

57. $r^2+8rs+15s^2=r^2+5rs+3rs+15s^2$
$=r(r+5s)+3s(r+5s)$
$=(r+5s)(r+3s)$

59. $9z^2-21z+10=9z^2-15z-6z+10$
$=3z(3z-5)-2(3z-5)$
$=(3z-5)(3z-2)$

61. $7y^2+25y+12=7y^2+21y+4y+12$
$=7y(y+3)+4(y+3)$
$=(y+3)(7y+4)$

63. No; the factor $(2x+4)$ contains a common factor of 2.

65. $72x^2+18x-2=2(36x^2+9x-1)$
$=2(36x^2+12x-3x-1)$
$=2(12x(3x+1)-1(3x+1))$
$=2(3x+1)(12x-1)$

67. $p^3-6p^2-27p=p(p^2-6p-27)$
$=p(p^2-9p+3p-27)$
$=p(p(p-9)+3(p-9))$
$=p(p-9)(p+3)$

69. $2(3x^2+10x+7)=2(3x^2+3x+7x+7)$
$=2(3x(x+1)+7(x+1))$
$=2(x+1)(3x+7)$

71. $2p^3 - 38p^2 + 120p$
$= 2p(p^2 - 19p + 60)$
$= 2p(p^2 - 4p - 15p + 60)$
$= 2p(p(p-4) - 15(p-4))$
$= 2p(p-4)(p-15)$

73. $x^2y^2 + 14x^2y + 33x^2$
$= x^2(y^2 + 14y + 33)$
$= x^2(y^2 + 11y + 3y + 33)$
$= x^2(y(y+11) + 3(y+11))$
$= x^2(y+11)(y+3)$

75. $-k^2 - 7k - 10 = -1(k^2 + 7k + 10)$
$= -1(k^2 + 5k + 2k + 10)$
$= -1(k(k+5) + 2(k+5))$
$= -1(k+5)(k+2)$

77. $-3n^2 - 3n + 90 = -3(n^2 + n - 30)$
$= -3(n^2 + 6n - 5n - 30)$
$= -3(n(n+6) - 5(n+6))$
$= -3(n+6)(n-5)$

79. $16z^2 - 14z + 3 = 16z^2 - 8z - 6z + 3$
$= 8z(2z-1) - 3(2z-1)$
$= (2z-1)(8z-3)$

81. $b^2 - 8b + 16 = b^2 - 4b - 4b + 16$
$= b(b-4) - 4(b-4)$
$= (b-4)(b-4)$
$= (b-4)^2$

83. $-5x^2 + 25x - 30 = -5(x^2 - 5x + 6)$
$= -5(x^2 - 3x - 2x + 6)$
$= -5(x(x-3) - 2(x-3))$
$= -5(x-3)(x-2)$

85. $t^2 - t - 6 = t^2 - 3t + 2t - 6$
$= t(t-3) + 2(t-3)$
$= (t-3)(t+2)$

Section 5.3 Practice Exercises

1. $7a^9 + 28a^3 = 7a^3(a^6 + 4)$

3. $12w^2 - 4w = 4w(3w - 1)$

5. $21a^2b^2 + 12ab^2 - 15a^2b$
$= 3ab(7ab + 4b - 5a)$

7. Different (positive multiplied by a negative).

9. Both negative (their product is then positive).

11. $(x-7)(x+8)$

13. $(x+7)(x-8)$

15. Any polynomial whose only factors are 1 and itself.

17. $2y^2 - 3y - 2 = (2y+1)(y-2)$

19. $9x^2 - 12x + 4 = (3x-2)(3x-2) = (3x-2)^2$

21. $2a^2 + 7a + 6 = (2a+3)(a+2)$

23. $6t^2 + 7t - 3 = (2t+3)(3t-1)$

25. $4m^2 - 20m + 25 = (2m-5)(2m-5)$
$= (2m-5)^2$

27. $5c^2 - c + 2$ prime

29. $6x^2 - 19xy + 10y^2 = (2x-5y)(3x-2y)$

31. $12m^2 + 11mn - 5n^2 = (4m+5n)(3m-n)$

33. $6r^2 + rs - 2s^2 = (3r+2s)(2r-s)$

35. $4s^2 - 8st + t^2$ prime

37. $x^2 + 7x - 18 = (x+9)(x-2)$

39. $a^2 - 10a - 24 = (a-12)(a+2)$

41. $r^2+5r-24=(r+8)(r-3)$

43. $w^2-14w+49=(w-7)(w-7)=(w-7)^2$

45. $k^2+5k+4=(k+1)(k+4)$

47. v^2-4v+1 prime

49. $m^2-13mn+40n^2=(m-8n)(m-5n)$

51. $a^2+9ab+8b^2=(a+8b)(a+b)$

53. $x^2+9xy+20y^2=(x+5y)(x+4y)$

55. $10t^2-23t-5=(2t-5)(5t+1)$

57. $14w^2+13w-12=(7w-4)(2w+3)$

59. No; the factor $(3x+6)$ has a GCF of 3.

61. $2(m^2-6m-40)=2(m+4)(m-10)$

63. $y^3(2y^2+13y+6)=y^3(2y+1)(y+6)$

65. $d(5d^5+3d-10)$

67. $4b(b^2-b-20)=4b(b+4)(b-5)$

69. $y^2(x^2-13x+30)=y^2(x-3)(x-10)$

71. $-1(a^2+15a-34)=-1(a+17)(a-2)$

73. $-2(u^2-14u+45)=-2(u-5)(u-9)$

75. $16x^2+10x+1=(8x+1)(2x+1)$

77. $c^2-2c+1=(c-1)(c-1)=(c-1)^2$

79. $-2z^2+20z-18=-2(z^2-10z+9)$
$=-2(z-9)(z-1)$

81. $q^2-13q+42=(q-6)(q-7)$

Section 5.4 Practice Exercises

1. $3x^2-x-10=(3x-5)(x+2)$

3. $x^2yz^2+6y^2z+yz=yz(x^2z+6y+1)$

5. $12x^2-34x+10=2(6x^2-17x+5)$
$=2(3x-1)(2x-5)$

7. $ax+ab-6x-6b=a(x+b)-6(x+b)$
$=(x+b)(a-6)$

9. $x^2+6x+9=(x+3)(x+3)=(x+3)^2$

11. $(2x+3)^2=4x^2+12x+9$

13. $(6h-1)^2=36h^2-12h+1$

15. **(a)** x^2+4x+4

(b) $x^2+4x+4=(x+2)(x+2)=(x+2)^2$
$x^2+5x+4=(x+1)(x+4)$

17. **(a)** $4x^2-20x+25$

(b) $4x^2-25x+25=(x-5)(4x-25)$
$4x^2-20x+25=(2x-5)(2x-5)$
$=(2x-5)^2$

19. $y^2-10y+25=(y-5)(y-5)=(y-5)^2$

21. $m^2+6m+9=(m+3)(m+3)=(m+3)^2$

23. $r^2-2r+36$ prime

25. $49q^2-28q+4=(7q)^2-2(7q)(2)+2^2$
$=(7q-2)^2$

27. $9p^2+42p+49=(3p)^2+2(3p)(7)+7^2$
$=(3p+7)^2$

29. $25h^2+50h+16=(5h)^2+2(5h)(5)+4^2$
The middle term is not 2 times the product of $5h$ and 4. This is not a perfect square trinomial.
$(5h+2)(5h+8)$

31. $16a^2+8ab+b^2=(4a)^2+2(4a)(b)+b^2$
$=(4a+b)^2$

33. $16q^2+40qr+25r^2$
$=(4q)^2+2(4q)(5r)+(5r)^2$
$=(4q+5r)^2$

35. $a^2+2ab+b^2=(a+b)^2$

37. $k^2-k+\frac{1}{4}=k^2-2(k)\left(\frac{1}{2}\right)+\left(\frac{1}{2}\right)^2$
$=\left(k-\frac{1}{2}\right)^2$

39. $9x^2+x+\frac{1}{36}=(3x)^2+2(3x)\left(\frac{1}{6}\right)+\left(\frac{1}{6}\right)^2$
$=\left(3x+\frac{1}{6}\right)^2$

41. $x^2-5^2=x^2-25$

43. $(2w)^2-3^2=4w^2-9$

45. $x^2-36=(x+6)(x-6)$

47. $w^2-100=(w+10)(w-10)$

49. $4a^2-121b^2=(2a)^2-(11b)^2$
$=(2a+11b)(2a-11b)$

51. $49m^2-16n^2=(7m)^2-(4n)^2$
$=(7m+4n)(7m-4n)$

53. $9q^2+16$ prime

55. $c^6-25=(c^3)^2-5^2=(c^3+5)(c^3-5)$

57. $25-16t^2=5^2-(4t)^2=(5+4t)(5-4t)$

59. $p^2-\frac{1}{9}$
$p^2-\left(\frac{1}{3}\right)^2=\left(p+\frac{1}{3}\right)\left(p-\frac{1}{3}\right)$

61. $m^2+\frac{100}{81}$ prime

63. $\frac{4}{9}-w^2=\left(\frac{2}{3}\right)^2-w^2=\left(\frac{2}{3}+w\right)\left(\frac{2}{3}-w\right)$

65. **(a)** a^2 (area of outer square)
b^2 (area of inner square)
Total area $=a^2-b^2$

(b) $a^2-b^2=(a+b)(a-b)$

67. $3w^2-27=3(w^2-9)=3(w+3)(w-3)$

69. $50p^4-2=2(25p^4-1)$
$=2(5p^2+1)(5p^2-1)$

71. $2x^2+24x+72=2(x^2+12x+36)$
$=2(x+6)^2$

73. $2t^3-10t^2-2t+10=2(t^3-5t^2-t+5)$
$=2(t^2(t-5)-1(t-5))$
$=2(t-5)(t^2-1)$
$=2(t-5)(t+1)(t-1)$

75. $100y^4+25x^2=25(4y^4+x^2)$

77. $4a^2b-40ab^2+100b^3$
$=4b(a^2-10ab+25b^2)$
$=4b(a-5b)^2$

79. $2x^3+3x^2-2x-3=x^2(2x+3)-1(2x+3)$
$=(2x+3)(x^2-1)$
$=(2x+3)(x+1)(x-1)$

81. $81y^4-16=(9y^2)^2-4^2$
$=(9y^2+4)(9y^2-4)$
$=(9y^2+4)(3y+2)(3y-2)$

83. $81k^2+30k+1=(27k+1)(3k+1)$

85. $k^3+4k^2-9k-36=k^2(k+4)-9(k+4)$
$=(k+4)(k^2-9)$
$=(k+4)(k+3)(k-3)$

87. $4m^{14}-20m^7+25$
$=(2m^7)^2-2(2m^7)(5)+5^2$
$=(2m^7-5)^2$

89. $0.36x^2-0.01=(0.6x)^2-(0.1)^2$
$=(0.6x+0.1)(0.6x-0.1)$

91. $\frac{1}{4}w^2-\frac{1}{9}v^2=\left(\frac{1}{2}w\right)^2-\left(\frac{1}{3}v\right)^2$
$=\left(\frac{1}{2}w+\frac{1}{3}v\right)\left(\frac{1}{2}w-\frac{1}{3}v\right)$

93. $(y-3)^2-9=((y-3)+3)((y-3)-3)$
$=(y)(y-6)$

95. $(2p+1)^2-36=((2p+1)+6)((2p+1)-6)$
$=(2p+7)(2p-5)$

97. $16-(t+2)^2$
$=(4+(t+2))(4-(t+2))$
$=(t+6)(2-t)$ or $-1(t+6)(t-2)$

99. $100-(2b-5)^2$
$=(10+(2b-5))(10-(2b-5))$
$=(2b+5)(15-2b)$ or $-1(2b+5)(2b-15)$

Section 5.5 Practice Exercises

1. $(x-y)(x^2+xy+y^2)$
$=x^3+x^2y+xy^2-x^2y-xy^2-y^3$
$=x^3-y^3$

3. $x^3, 8, y^6, 27q^3, w^{12}, r^3s^6$

5. If the binomial is of the form a^3+b^3.

7. $a^3+b^3=(a+b)(a^2-ab+b^2)$

9. $y^3-8=y^3-2^3=(y-2)(y^2+2y+4)$

11. $1-p^3=(1-p)(1+p+p^2)$

13. $w^3+64=w^3+4^3=(w+4)(w^2-4w+16)$

15. $1000a^3+27=(10a)^3+3^3$
$=(10a+3)(100a^2-30a+9)$

17. $x^3-1000=x^3-10^3$
$=(x-10)(x^2+10x+100)$

19. $64t^3+1=(4t)^3+1^3=(4t+1)(16t^2-4t+1)$

21. $n^3-\frac{1}{8}=n^3-\left(\frac{1}{2}\right)^3=\left(n-\frac{1}{2}\right)\left(n^2+\frac{1}{2}n+\frac{1}{4}\right)$

23. $a^3+b^6=a^3+(b^2)^3$
$=(a+b^2)(a^2-ab^2+b^4)$

25. $x^9+64y^3=(x^3)^3+(4y)^3$
$=(x^3+4y)(x^6-4x^3y+16y^2)$

27. $25m^{12}+16$ prime

29. $a^2-b^2=(a+b)(a-b)$

31. $x^4-4=(x^2)^2-2^2=(x^2+2)(x^2-2)$

33. a^2+9 prime

35. $t^3+64=t^3+4^3=(t+4)(t^2-4t+16)$

37. g^3-4 prime

39. $4b^3+108=4(b^3+27)$
$=4(b+3)(b^2-3b+9)$

41. $5p^2-125=5(p^2-25)=5(p+5)(p-5)$

43. $\frac{1}{64}-8h^3=\left(\frac{1}{4}\right)^3-(2h)^3$
$=\left(\frac{1}{4}-2h\right)\left(\frac{1}{16}+\frac{1}{2}h+4h^2\right)$

45. $x^4-16=(x^2)^2-4^2$
$=(x^2+4)(x^2-4)$
$=(x^2+4)(x+2)(x-2)$

47. q^6-64
$=(q^3)^2-8^2$
$=(q^3+8)(q^3-8)$
$=(q+2)(q^2-2q+4)(q-2)(q^2+2q+4)$

49. $4b+16=4(b+4)$

51. $y^2+4y+3=(y+3)(y+1)$

53. $16z^4-81=(4z^2)^2-9^2$
$=(4z^2+9)(4z^2-9)$
$=(4z^2+9)(2z+3)(2z-3)$

55. $5r^3+5=5(r^3+1)=5(r+1)(r^2-r+1)$

57. $7p^2-29p+4=(7p-1)(p-4)$

59. $-2x^2+8x-8=-2(x^2-4x+4)$
$=-2(x-2)(x-2)$
$=-2(x-2)^2$

61. $54-2y^3=2(27-y^3)$
$=2(3-y)(9+3y+y^2)$

63. $4t^2-31t-8=(4t+1)(t-8)$

65. $2xw-10x+3yw-15y$
$=2x(w-5)+3y(w-5)$
$=(w-5)(2x+3y)$

67. $4q^2-9=(2q)^2-3^2=(2q+3)(2q-3)$

69. x^2+2x+4

71. $2x+1$

73. $\frac{64}{125}p^3-\frac{1}{8}q^3$
$=\left(\frac{4}{5}p\right)^3-\left(\frac{1}{2}q\right)^3$
$=\left(\frac{4}{5}p-\frac{1}{2}q\right)\left(\frac{16}{25}p^2+\frac{2}{5}pq+\frac{1}{4}q^2\right)$

75. $a^{12}+b^{12}=(a^4)^3+(b^4)^3$
$=(a^4+b^4)(a^8-a^4b^4+b^8)$

77. (a)

$$\begin{array}{r|l} & x^2+2x+4 \\ \hline x-2 & x^3+0x^2+0x-8 \\ & -(x^3-2x^2) \\ & \quad 2x^2+0x \\ & \quad -(2x^2-4x) \\ & \qquad 4x-8 \\ & \qquad -(4x-8) \end{array}$$

(b) $x^3-8=(x-2)(x^2+2x+4)$

79. (a)

$$\begin{array}{r|l} & m^2-m+1 \\ \hline m+1 & m^3+0m^2+0m+1 \\ & -(m^3+m^2) \\ & \quad -m^2+0m \\ & \quad -(-m^2-m) \\ & \qquad m+1 \\ & \qquad -(m+1) \end{array}$$

(b) $m^3+1=(m+1)(m^2-m+1)$

Section 5.6 Practice Exercises

1. $x^2 - 6x - 16 = (x+2)(x-8)$

3. $20b^2 - 11b - 3 = (5b+1)(4b-3)$

5. $100 - 9u^2 = 10^2 - (3u)^2 = (10+3u)(10-3u)$

7. $p^3 - 216 = p^3 - 6^3 = (p-6)(p^2+6p+36)$

9. $x^2y^3 + x^5y^2 = x^2y^2(y+x^3)$

11. $2y - 22 + 9xy - 99x = 2(y-11) + 9x(y-11)$
$= (y-11)(2+9x)$

13. $w^2 - 40w + 400 = (w-20)(w-20)$
$= (w-20)^2$

15. $x^3 - 3x^2 - 10x = x(x^2 - 3x - 10)$
$= x(x+2)(x-5)$

17. $3y^2 + 21y + 36 = 3(y^2 + 7y + 12)$
$= 3(y+3)(y+4)$

19. $p^2 + 12pq + 36q^2 = (p+6q)(p+6q)$
$= (p+6q)^2$

21. $x^2 + 3x + 8$ prime

23. $2x^2 + 13x - 24 = (2x-3)(x+8)$

25. $u^2 - 25v^2 = u^2 - (5v)^2 = (u+5v)(u-5v)$

27. $a^3b^3 - 36ab = ab(a^2b^2 - 36)$
$= ab(ab+6)(ab-6)$

29. $2x^3 - 20x^2 + 18x = 2x(x^2 - 10x + 9)$
$= 2x(x-9)(x-1)$

31. $-3a^2b^2 + 3ab^3 - 6ab^2 = -3ab^2(a-b+2)$

33. $11r(s+4) - 6(s+4) = (s+4)(11r-6)$

35. $100 + t^2$ prime

37. $x^4 - 8x = x(x^3 - 8) = x(x-2)(x^2+2x+4)$

39. $m^3 + 16$ prime

41. $5xy - 3y + 15x - 9 = y(5x-3) + 3(5x-3)$
$= (5x-3)(y+3)$

43. $2pq - 14p - 8q + 56 = 2(pq - 7p - 4q + 28)$
$= 2(p(q-7) - 4(q-7))$
$= 2(q-7)(p-4)$

45. $4x^2 - 16x + 16 = 4(x^2 - 4x + 4) = 4(x-2)^2$

47. $50 + p^2 - 15p = p^2 - 15p + 50$
$= (p-5)(p-10)$

49. $24xy + 16x^2 + 9y^2$
$= 16x^2 + 24xy + 9y^2$
$= (4x)^2 + 2(4x)(3y) + (3y)^2$
$= (4x+3y)^2$

51. $2x^5 + 6x^3 - 10x^4 - 30x^2$
$= 2x^5 - 10x^4 + 6x^3 - 30x^2$
$= 2x^2(x^3 - 5x^2 + 3x - 15)$
$= 2x^2(x^2(x-5) + 3(x-5))$
$= 2x^2(x-5)(x^2+3)$

53. $-x^2 + 16x - 63 = -1(x^2 - 16x + 63)$
$= -(x-7)(x-9)$

55. $6x^2 - 21x - 45 = 3(2x^2 - 7x - 15)$
$= 3(2x+3)(x-5)$

57. $5a^2bc^3 - 7abc^2 = abc^2(5ac - 7)$

59. $t^2 + 2t - 63 = (t+9)(t-7)$

61. $ab + ay - b^2 - by = a(b+y) - b(b+y)$
$= (b+y)(a-b)$

63. $14u^2 - 11uv + 2v^2 = (7u-2v)(2u-v)$

65. $4q^2 - 8q - 6 = 2(2q^2 - 4q - 3)$

67. $9m^2 + 16n^2$ prime

69. $6r^2 + 11r + 3 = (3r + 1)(2r + 3)$

71. $81u^2 - 90uv + 25v^2$
$= (9u)^2 - 2(9u)(5v) + (5v)^2$
$= (9u - 5v)^2$

73. $2ax - 6ay + 4bx - 12by$
$= 2(ax - 3ay + 2bx - 6by)$
$= 2(a(x - 3y) + 2b(x - 3y))$
$= 2(x - 3y)(a + 2b)$

75. $21x^4y + 41x^3y + 10x^2y$
$= x^2y(21x^2 + 41x + 10)$
$= x^2y(3x + 5)(7x + 2)$

77. $8uv - 6u + 12v - 9 = 2u(4v - 3) + 3(4v - 3)$
$= (4v - 3)(2u + 3)$

79. $12x^2 - 12x + 3 = 3(4x^2 - 4x + 1) = 3(2x - 1)^2$

81. $6n^3 + 5n^2 - 4n = n(6n^2 + 5n - 4)$
$= n(2n - 1)(3n + 4)$

83. $64 - y^2 = (8 + y)(8 - y)$

85. $x^2(x + y) - y^2(x + y) = (x + y)(x^2 - y^2)$
$= (x + y)(x + y)(x - y)$
$= (x + y)^2(x - y)$

87. $(a + 3)^4 + 6(a + 3)^5 = (a + 3)^4(1 + 6(a + 3))$
$= (a + 3)^4(1 + 6a + 18)$
$= (a + 3)^4(6a + 19)$

89. $24(3x + 5)^3 - 30(3x + 5)^2$
$= 6(3x + 5)^2[4(3x + 5) - 5]$
$= 6(3x + 5)^2[12x + 15]$
$= 6(3x + 5)^2 3(4x + 5)$
$= 18(3x + 5)^2(4x + 5)$

91. $16p^4 - q^4 = (4p^2)^2 - (q^2)^2$
$= (4p^2 + q^2)(4p^2 - q^2)$
$= (4p^2 + q^2)(2p + q)(2p - q)$

93. $y^3 + \frac{1}{64} = \left(y + \frac{1}{4}\right)\left(y^2 - \frac{1}{4}y + \frac{1}{16}\right)$

95. $6a^3 + a^2b - 6ab^2 - b^3$
$= a^2(6a + b) - b^2(6a + b)$
$= (6a + b)(a^2 - b^2)$
$= (6a + b)(a + b)(a - b)$

97. $\frac{1}{9}t^2 - \frac{1}{6}t + \frac{1}{16} = \left(\frac{1}{3}t\right)^2 + 2\left(\frac{1}{3}t\right)\left(\frac{1}{4}\right) + \left(\frac{1}{4}\right)^2$
$= \left(\frac{1}{3}t + \frac{1}{4}\right)^2$

99. $x^2 + 12x + 36 - a^2 = (x + 6)^2 - a^2$
$= (x + 6 + a)(x + 6 - a)$

101. $p^2 + 2pq + q^2 - 81 = (p + q)^2 - 9^2$
$= (p + q + 9)(p + q - 9)$

103. $b^2 - (x^2 + 4x + 4) = b^2 - (x + 2)^2$
$= (b + (x + 2))(b - (x + 2))$
$= (b + x + 2)(b - x - 2)$

105. $4 - u^2 + 2uv - v^2 = 4 - (u - v)^2$
$= (2 + (u - v))(2 - (u - v))$
$= (2 + u - v)(2 - u + v)$

107. $6ax - by + 2bx - 3ay$
$= 6ax + 2bx - by - 3ay$
$= 2x(3a + b) - y(3a + b)$
$= (3a + b)(2x - y)$

109. $u^6 - 64$
$= (u^3)^2 - (8)^2$
$= (u^3 + 8)(u^3 - 8)$
$= (u + 2)(u^2 - 2u + 4)(u - 2)(u^2 + 2u + 4)$
$= (u + 2)(u - 2)(u^2 - 2u + 4)(u^2 + 2u + 4)$

111. $x^8 - 1 = (x^4)^2 - 1^2$
$= (x^4 + 1)(x^4 - 1)$
$= (x^4 + 1)(x^2 + 1)(x^2 - 1)$
$= (x^4 + 1)(x^2 + 1)(x + 1)(x - 1)$

113. **(a)** $u^2 - 10u + 25 = (u - 5)^2$

(b) $x^4 - 10x^2 + 25 = (x^2)^2 - 10x^2 + 25$
$= (x^2 - 5)^2$

(c) $(a + 1)^2 - 10(a + 1) + 25 = ((a + 1) - 5)^2$
$= (a - 4)^2$

115. **(a)** $u^2 + 11u - 26 = (u + 13)(u - 2)$

(b) $w^6 + 11w^3 - 26 = (w^3)^2 + 11w^3 - 26$
$= (w^3 + 13)(w^3 - 2)$

(c) $(y - 4)^2 + 11(y - 4) - 26$
$= ((y - 4) + 13)((y - 4) - 2)$
$= (y + 9)(y - 6)$

117. $(5x^2 - 1)^2 - 4(5x^2 - 1) - 5$
Let $u = 5x^2 - 1$.
$u^2 - 4u - 5 = (u - 5)(u + 1)$
$(5x^2 - 1 - 5)(5x^2 - 1 + 1)$
$= (5x^2 - 6)(5x^2)$ or $(5x^2)(5x^2 - 6)$

119. $2(3w - 5)^2 - 19(3w - 5) + 35$
Let $u = 3w - 5$.
$2u^2 - 19u + 35 = (2u - 5)(u - 7)$
$(2(3w - 5) - 5)(3w - 5 - 7)$
$= (6w - 15)(3w - 12)$
$= 3(2w - 5)3(w - 4)$
$= 9(2w - 5)(w - 4)$

121. $a^2 - b^2 + a + b = (a + b)(a - b) + (a + b)$
$= (a + b)(a - b + 1)$

123. $5wx^3 + 5wy^3 - 2zx^3 - 2zy^3$
$= 5w(x^3 + y^3) - 2z(x^3 + y^3)$
$= (x^3 + y^3)(5w - 2z)$
$= (x + y)(x^2 - xy + y^2)(5w - 2z)$

Section 5.7 Practice Exercises

1. $4x - 2 + 2bx - b = 2(2x - 1) + b(2x - 1)$
$= (2x - 1)(2 + b)$

3. $4b^2 - 44b + 120 = 4(b^2 - 11b + 30)$
$= 4(b - 6)(b - 5)$

5. $16w^2 - 1 = (4w)^2 - 1^2 = (4w + 1)(4w - 1)$

7. $12k + 16 = 4(3k + 4)$

9. $2y^2 + 3y - 44 = (2y + 11)(y - 4)$

11. Linear

13. Quadratic

15. Neither

17. Quadratic

19. If $ab = 0$, then $a = 0$ or $b = 0$.

21. $(x + 3)(x - 1) = 0$
$x + 3 = 0$ or $x - 1 = 0$
$x = -3$ $\quad$ $x = 1$

23. $(2x - 7)(2x + 7) = 0$
$2x - 7 = 0$ or $2x + 7 = 0$
$2x = 7$ $\quad$ $2x = -7$
$x = \frac{7}{2}$ $\quad$ $x = -\frac{7}{2}$

25. $3(x + 5)(x + 5) = 0$
$x + 5 = 0$
$x = -5$

27. $x(3x + 1)(x + 1) = 0$
$x = 0$ or $3x + 1 = 0$ or $x + 1 = 0$
$x = 0$ $\quad$ $x = -\frac{1}{3}$ $\quad$ $x = -1$

29. $p^2 - 2p - 15 = 0$
$(p+3)(p-5) = 0$
$p + 3 = 0$ or $p - 5 = 0$
$p = -3$ $\quad p = 5$

31. $z^2 + 10z - 24 = 0$
$(z+12)(z-2) = 0$
$z + 12 = 0$ or $z - 2 = 0$
$z = -12$ $\quad z = 2$

33. $2q^2 - 7q - 4 = 0$
$(2q+1)(q-4) = 0$
$2q + 1 = 0$ or $q - 4 = 0$
$q = -\frac{1}{2}$ $\quad q = 4$

35. $0 = 9x^2 - 4$
$0 = (3x+2)(3x-2)$
$3x + 2 = 0$ or $3x - 2 = 0$
$x = -\frac{2}{3}$ $\quad x = \frac{2}{3}$

37. $2k^2 - 28k + 96 = 0$
$2(k^2 - 14k + 48) = 0$
$2(k-6)(k-8) = 0$
$k - 6 = 0$ or $k - 8 = 0$
$k = 6$ $\quad k = 8$

39. $0 = 2m^3 - 5m^2 - 12m$
$m(2m^2 - 5m - 12) = 0$
$m(2m+3)(m-4) = 0$
$m = 0$ or $2m + 3 = 0$ or $m - 4 = 0$
$m = 0$ $\quad m = -\frac{3}{2}$ $\quad m = 4$

41. To use the zero product rule to solve any equation, the equation must be factorable and equal to zero.

43. $x^2 - 10x = -16$
$x^2 - 10x + 16 = 0$
$(x-8)(x-2) = 0$
$x - 8 = 0$ or $x - 2 = 0$
$x = 8$ $\quad x = 2$

45. $4p^2 = 49$
$4p^2 - 49 = 0$
$(2p+7)(2p-7) = 0$
$2p + 7 = 0$ or $2p - 7 = 0$
$p = -\frac{7}{2}$ $\quad p = \frac{7}{2}$

47. $2(q^2 + 10q) = -50$
$2q^2 + 20q + 50 = 0$
$2(q^2 + 10q + 25) = 0$
$2(q+5)(q+5) = 0$
$q + 5 = 0$
$q = -5$

49. $-x = 3x^3 + 4x^2$
$3x^3 + 4x^2 + x = 0$
$x(3x+1)(x+1) = 0$
$x = 0$ or $3x + 1 = 0$ or $x + 1 = 0$
$x = 0$ $\quad x = -\frac{1}{3}$ $\quad x = -1$

51. $9(k-1) = -4k^2$
$9k - 9 = -4k^2$
$4k^2 + 9k - 9 = 0$
$(4k-3)(k+3) = 0$
$4k - 3 = 0$ or $k + 3 = 0$
$k = \frac{3}{4}$ $\quad k = -3$

53. $3p(p-1) = 18$
$3p^2 - 3p - 18 = 0$
$3(p^2 - p - 6) = 0$
$3(p-3)(p+2) = 0$
$p - 3 = 0$ or $p + 2 = 0$
$p = 3$ $\quad p = -2$

55. $21w^2 = 14w$

$21w^2 - 14w = 0$

$7w(3w-2) = 0$

$7w = 0$ or $3w - 2 = 0$

$w = 0$ $\quad w = \frac{2}{3}$

57. $2(4d^2 + d) = 0$

$2d(4d+1) = 0$

$2d = 0$ or $4d + 1 = 0$

$d = 0$ $\quad d = -\frac{1}{4}$

59. $t^3 + 2t^2 - 16t - 32 = 0$

$t^2(t+2) - 16(t+2) = 0$

$(t+2)(t+4)(t-4) = 0$

$t+2=0$ or $t+4=0$ or $t-4=0$

$t=-2$ $\quad t=-4$ $\quad t=4$

61. $(w+5)(w-3) = 20$

$w^2 + 2w - 15 - 20 = 0$

$w^2 + 2w - 35 = 0$

$(w+7)(w-5) = 0$

$w+7=0$ or $w-5=0$

$w=-7$ $\quad w=5$

63. $(k-6)(k-1) = -k-2$

$k^2 - 7k + 6 = -k - 2$

$k^2 - 6k + 8 = 0$

$(k-4)(k-2) = 0$

$k-4=0$ or $k-2=0$

$k=4$ $\quad k=2$

65. Let x = the number. Then,

$x + 2x^2 = 36$

$2x^2 + x - 36 = 0$

$(2x+9)(x-4) = 0$

$2x+9=0$ or $x-4=0$

$x = -\frac{9}{2}$ $\quad x = 4$

The numbers are $-\frac{9}{2}$ and 4.

67. Let x = a number. Then,

$x^2 = x + 20$

$x^2 - x - 20 = 0$

$(x+4)(x-5) = 0$

$x+4=0$ or $x-5=0$

$x=-4$ $\quad x=5$

The numbers are –4 and 5.

69. Let x = first integer. Then the next consecutive even integer is $(x + 2)$.

$x(x+2) = 48$

$x^2 + 2x - 48 = 0$

$(x+8)(x-6) = 0$

$x+8=0$ or $x-6=0$

$x=-8$ $\quad x=6$

The numbers are –8 and –6 or 6 and 8.

71. Let x = first integer. Then the next consecutive integer is $(x + 1)$.

$x^2 + (x+1)^2 = 10(x+x+1) - 9$

$x^2 + x^2 + 2x + 1 = 10(2x+1) - 9$

$2x^2 + 2x + 1 = 20x + 1$

$2x^2 - 18x = 0$

$2x(x-9) = 0$

$2x=0$ or $x-9=0$

$x=0$ $\quad x=9$

The numbers are 0 and 1 or 9 and 10.

73. Let x = length of painting and $x - 2$ be the length of the painting. Then,

$A = \text{(length)(width)}$

$120 = x(x-2)$

$120 = x^2 - 2x$

$x^2 - 2x - 120 = 0$

$(x+10)(x-12) = 0$

$x = -10$ or $x = 12$

The painting has length 12 in. and width 10 in.

75. Let x = length and $x - 7$ be the width.

(a)
$$A = lw$$
$$78 = x(x-7)$$
$$x^2 - 7x - 78 = 0$$
$$(x+6)(x-13) = 0$$
$x = -6$ or $x = 13$

The dimensions are 13 in. by 6 in.

(b) $P = 2w + 2l$
$P = 2(6) + 2(13)$
$P = 12 + 26 = 38$ in.

77. Let x = base and $3x - 5$ = the height.
$$A = \frac{1}{2}bh$$
$$125 = \frac{1}{2}x(3x-5)$$
$$250 = 3x^2 - 5x$$
$$3x^2 - 5x - 250 = 0$$
$$(3x+25)(x-10) = 0$$
$3x + 25 = 0$ or $x - 10 = 0$
$x = -\frac{25}{3}$ $\quad x = 10$

The base is 10 cm and the height is 25 cm.

79. If you let $h = 0$, then,
$$0 = -16t^2 + 64$$
$$0 = -16(t^2 - 4)$$
$$-16(t+2)(t-2) = 0$$
$t + 2 = 0$ or $t - 2 = 0$
$t = -2$ $\quad t = 2$

It will take 2 seconds to hit the ground.

81. Ground level is when $h = 0$. Then,
$$0 = -16t^2 + 64t$$
$$0 = -16t(t-4)$$
$-16t = 0$ or $t - 4 = 0$
$t = 0$ $\quad t = 4$

The times are 0 seconds and 4 seconds.

83. Given a right triangle with legs a and b and hypotenuse c, $a^2 + b^2 = c^2$.

85. Yes, $9^2 + 12^2 = 15^2$
$$81 + 144 = 225$$
$$225 = 225$$

87. No, $8^2 + 9^2 = 10^2$
$$64 + 81 = 100$$
$$145 \neq 100$$

89. $\left(\begin{matrix}\text{length}\\\text{of base}\end{matrix}\right)^2 + \left(\begin{matrix}\text{height from}\\\text{ground to top}\end{matrix}\right)^2 = 17^2$
$$x^2 + (x+7)^2 = 17^2$$
$$x^2 + x^2 + 14x + 49 = 289$$
$$2x^2 + 14x - 240 = 0$$
$$2(x+15)(x-8) = 0$$
$x = -15$ or $x = 8$

The bottom of the ladder is 8 ft from the house. The distance from the top of the ladder to the ground is 15 ft.

91. Let x = length of hypotenuse.
$$(x-4)^2 + (x-2)^2 = x^2$$
$$x^2 - 8x + 16 + x^2 - 4x + 4 = x^2$$
$$x^2 - 12x + 20 = 0$$
$$(x-10)(x-2) = 0$$
$x - 10 = 0$ or $x - 2 = 0$
$x = 10$ or $x = 2$

The length of the hypotenuse is 10 m.

93. (a) $N = \frac{x(x-3)}{2}$
$$N = \frac{4(4-3)}{2} = \frac{4}{2} = 2$$

(b) $N = \frac{x(x-3)}{2}$
$$N = \frac{5(5-3)}{2} = \frac{10}{2} = 5$$

(c) $N = \frac{x(x-3)}{2}$

$35 = \frac{x(x-3)}{2}$

$70 = x(x-3)$

$70 = x^2 - 3x$

$0 = x^2 - 3x - 70$

$0 = (x+7)(x-10)$

$x+7=0$ or $x-10=0$

$x=-7$ $\quad$ $x=10$

Ten sides.

95. $(a+2)^2 - 4(a+2) - 21 = 0$

Let $u = (a+2)$.

$u^2 - 4u - 21 = 0$

$(u+3)(u-7) = 0$

$u = -3$ or $u = 7$

$a+2=-3$ or $a+2=7$

$a=-5$ $\quad$ $a=5$

97. $2(w-1)^2 - 7(w-1) - 4 = 0$

Let $u = w - 1$.

$2u^2 - 7u - 4 = 0$

$(2u+1)(u-4) = 0$

$u = -\frac{1}{2}$ or $u = 4$

$w-1 = -\frac{1}{2}$ or $w-1=4$

$w = \frac{1}{2}$ $\quad$ $w=5$

Chapter 5 Review Exercises

1. GCF: 6

3. GCF: ab^4

5. GCF: $2c(3c-5)$

7. $6x^2 + 2x^3 - 8x = 2x(3x + x^2 - 4)$

9. $32y^2 - 48 = 16(2y^2 - 3)$

11. $-t^2 + 5t = -t(t-5)$ or $t(-t+5)$

13. $3b(b+2) - 7(b+2) = (b+2)(3b-7)$

15. $7w^2 + 14w + wb + 2b = 7w(w+2) + b(w+2)$
$= (w+2)(7w+b)$

17. $x^2 - 6x - 4x + 24 = x(x-6) - 4(x-6)$
$= (x-6)(x-4)$

19. $60y^2 - 45y - 12y + 9$
$= 3(20y^2 - 15y - 4y + 3)$
$= 3(5y(4y-3) - 1(4y-3))$
$= 3(4y-3)(5y-1)$

21. –6, 1

23. 8, 3

25. 5, –1

27. $3c^2 - 5c - 2 = 3c^2 - 6c + c - 2$
$= 3c(c-2) + 1(c-2)$
$= (c-2)(3c+1)$

29. $2t^2 + 11st + 12s^2 = 2t^2 + 8st + 3st + 12s^2$
$= 2t(t+4s) + 3s(t+4s)$
$= (t+4s)(2t+3s)$

31. $w^3 + 4w^2 - 5w = w(w^2 + 4w - 5)$
$= w(w^2 + 5w - w - 5)$
$= w(w(w+5) - 1(w+5))$
$= w(w+5)(w-1)$

33. $40v^2 + 22v - 6 = 2(20v^2 + 11v - 3)$
$= 2(20v^2 + 15v - 4v - 3)$
$= 2(5v(4v+3) - 1(4v+3))$
$= 2(4v+3)(5v-1)$

35. $x^2 + 9x - 22 = x^2 + 11x - 2x - 22$
$= x(x+11) - 2(x+11)$
$= (x+11)(x-2)$

37. $a^3b - 10a^2b^2 + 24ab^3$
$= ab(a^2 - 10ab + 24b^2)$
$= ab(a^2 - 6ab - 4ab + 24b^2)$
$= ab(a(a - 6b) - 4b(a - 6b))$
$= ab(a - 6b)(a - 4b)$

39. $3m + 9m^2 - 2 = 9m^2 + 6m - 3m - 2$
$= 3m(3m + 2) - 1(3m + 2)$
$= (3m + 2)(3m - 1)$

41. Different

43. Both positive

45. $2y^2 - 5y - 12 = (2y + 3)(y - 4)$

47. $2p^2 - 4p - 48 = 2(p^2 - 2p - 24)$
$= 2(p - 6)(p + 4)$

49. $10z^2 + 29z + 10 = (2z + 5)(5z + 2)$

51. $2p^2 - 5p + 1$ prime

53. $10w^2 - 60w - 270 = 10(w^2 - 6w - 27)$
$= 10(w + 3)(w - 9)$

55. $9c^2 - 30cd + 25d^2 = (3c - 5d)(3c - 5d)$
$= (3c - 5d)^2$

57. $v^4 - 2v^2 - 3 = (v^2)^2 - 2v^2 - 3$
$= (v^2 - 3)(v^2 + 1)$

59. $4x^2 - 20x + 25 = (2x)^2 - 2(2x)(5) + 5^2$
$= (2x - 5)^2$

61. $c^2 - 6c + 9 = c^2 - 2(c)(3) + 3^2 = (c - 3)^2$

63. $t^2 + 8t + 49$; not a perfect square trinomial. the middle term does not fit the pattern—it does not equal $2(t)(7)$.

65. $a^2 - 49 = a^2 - 7^2 = (a + 7)(a - 7)$

67. $h - 25$; not a difference of squares. Also, h is not a perfect square.

69. $100 - 81t^2 = 10^2 - (9t)^2 = (10 + 9t)(10 - 9t)$

71. $x^2 + 16$; this is a sum of squares, not a difference.

73. $2c^4 - 18 = 2(c^4 - 9) = 2(c^2 + 3)(c^2 - 3)$

75. $8x^2 + 24x + 18 = 2(4x^2 + 12x + 9)$
$= 2(2x + 3)(2x + 3)$
$= 2(2x + 3)^2$

77. $p^3 + 3p^2 - 16p - 48 = p^2(p + 3) - 16(p + 3)$
$= (p + 3)(p^2 - 16)$
$= (p + 3)(p + 4)(p - 4)$

79. $a^3 + b^3 = (a + b)(a^2 - ab + b^2)$

81. $z^3 - w^3 = (z - w)(z^2 + zw + w^2)$

83. $64 + a^3 = 4^3 + a^3 = (4 + a)(16 - 4a + a^2)$

85. $p^6 + 8 = (p^2)^3 + 2^3$
$= (p^2 + 2)(p^4 - 2p^2 + 4)$

87. $6x^3 - 48 = 6(x^3 - 8) = 6(x - 2)(x^2 + 2x + 4)$

89. v.

91. iii.

93. $216w^3 - 1 = (6w)^3 - 1^3$
$= (6w - 1)(36w^2 + 6w + 1)$

95. $128 + 2v^6 = 2(64 + v^6)$
$= 2(4^3 + (v^2)^3)$
$= 2(4 + v^2)(16 - 4v + v^2)$

97. $q^6 - 1 = (q^3)^2 - 1^2$
$= (q^3 + 1)(q^3 - 1)$
$= (q+1)(q^2 - q + 1)(q-1)(q^2 + q + 1)$
$= (q+1)(q-1)(q^2 - q + 1)(q^2 + q + 1)$

99. $3p^2 - 6p + 3 = 3(p^2 - 2p + 1) = 3(p-1)^2$

101. $k^2 - 13k + 42 = (k-6)(k-7)$

103. $q^4 - 64q = q(q^3 - 4^3)$
$= q(q-4)(q^2 + 4q + 16)$

105. $2t^2 + t + 3$ prime

107. $x^3 + 4x^2 - x - 4 = x^2(x+4) - 1(x+4)$
$= (x+4)(x^2 - 1)$
$= (x+4)(x+1)(x-1)$

109. $5p^4q - 20q^3 = 5q(p^4 - 4q^2)$
$= 5q(p^2 + 2q)(p^2 - 2q)$

111. $(y-4)^3 + 4(y-4)^2 = (y-4)^2(y-4+4)$
$= y(y-4)^2$

113. $80z + 32 + 50z^2 = 2(25z^2 + 40z + 16)$
$= 2(5z+4)^2$

115. $w^4 + w^3 - 56w^2 = w^2(w^2 + w - 56)$
$= w^2(w+8)(w-7)$

117. $14m^3 - 14 = 14(m^3 - 1)$
$= 14(m-1)(m^2 + m + 1)$

119. $a^2 - 6a + 9 - 16x^2 = (a-3)^2 - (4x)^2$
$= (a - 3 + 4x)(a - 3 - 4x)$

121. $(4x+3)^2 - 12(4x+3) + 36 = ((4x+3) - 6)^2$
$= (4x-3)^2$

123. $(4x - 1)(3x + 2) = 0$
$4x - 1 = 0$ or $3x + 2 = 0$
$x = \frac{1}{4}$ $\quad x = -\frac{2}{3}$

125. $3w(w + 3)(5w + 2) = 0$
$3w = 0$ or $w + 3 = 0$ or $5w + 2 = 0$
$w = 0$ $\quad w = -3$ $\quad w = -\frac{2}{5}$

127. $7k^2 - 9k - 10 = 0$
$(7k+5)(k-2) = 0$
$7k + 5 = 0$ or $k - 2 = 0$
$k = -\frac{5}{7}$ $\quad k = 2$

129. $q^2 - 144 = 0$
$(q+12)(q-12) = 0$
$q + 12 = 0$ or $q - 12 = 0$
$q = -12$ $\quad q = 12$

131. $5v^2 - v = 0$
$v(5v - 1) = 0$
$v = 0$ or $5v - 1 = 0$
$v = 0$ $\quad v = \frac{1}{5}$

133. $36t^2 + 60t = -25$
$36t^2 + 60t + 25 = 0$
$(6t+5)(6t+5) = 0$
$6t + 5 = 0$
$t = -\frac{5}{6}$

135. $3(y^2 + 4) = 20y$
$3y^2 + 12 = 20y$
$3y^2 - 20y + 12 = 0$
$(3y-2)(y-6) = 0$
$3y - 2 = 0$ or $y - 6 = 0$
$y = \frac{2}{3}$ $\quad y = 6$

137. Let x = height and $2x + 1$ be the base.

$$A = (\text{height})(\text{base})$$
$$78 = (x)(2x+1)$$
$$78 = 2x^2 + x$$
$$2x^2 + x - 78 = 0$$
$$(2x+13)(x-6) = 0$$
$$x = -\frac{13}{2} \text{ or } x = 6$$

The height is 6 ft and the base is 13 ft.

139. Yes; $15^2 + 20^2 = 25^2$

$$225 + 400 = 625$$
$$625 = 625$$

141. Let x = a number. Then,

$$60 - x^2 = -4$$
$$x^2 - 64 = 0$$
$$(x+8)(x-8) = 0$$

$x = -8$ or $x = 8$

The numbers are –8 and 8.

143. Let x = height, $2x + 1$ be the base.

$$A = \frac{1}{2}bh$$
$$18 = \frac{1}{2}x(2x+1)$$
$$36 = 2x^2 + x$$
$$2x^2 + x - 36 = 0$$
$$(2x+9)(x-4) = 0$$
$$x = -\frac{9}{2} \text{ or } x = 4$$

The height is 4 m and the base is 9 m.

Chapter 5 Test

1. $15x^4 - 3x + 6x^3 = 3x(5x^3 - 1 + 2x^2)$

3. $6w^2 - 43w + 7 = (6w-1)(w-7)$

5. $q^2 - 16q + 64 = (q-8)^2$

7. $3a^2 + 27ab + 54b^2 = 3(a^2 + 9ab + 18b^2)$
$= 3(a+6b)(a+3b)$

9. $xy - 7x + 3y - 21 = x(y-7) + 3(y-7)$
$= (y-7)(x+3)$

11. $-10u^2 + 30u - 20 = -10(u^2 - 3u + 2)$
$= -10(u-2)(u-1)$

13. $5y^2 - 50y + 125 = 5(y^2 - 10y + 25)$
$= 5(y-5)(y-5)$
$= 5(y-5)^2$

15. $2x^3 + x^2 - 8x - 4 = x^2(2x+1) - 4(2x+1)$
$= (2x+1)(x^2 - 4)$
$= (2x+1)(x+2)(x-2)$

17. $x^2 + 8x + 16 - y^2 = (x+4)^2 - y^2$
$= (x+4+y)(x+4-y)$

19. $12a - 6ac + 2b - bc = 6a(2-c) + b(2-c)$
$= (2-c)(6a+b)$

21. $x^2 - 7x = 0$

$x(x-7) = 0$

$x = 0$ or $x - 7 = 0$

$x = 0 \qquad x = 7$

23. $x(5x+4) = 1$

$5x^2 + 4x - 1 = 0$

$(5x-1)(x+1) = 0$

$5x - 1 = 0$ or $x + 1 = 0$

$x = \frac{1}{5} \qquad x = -1$

25. Let x = shorter leg, $3x - 3$ the longer leg, and $3x - 2$ the length of the hypotenuse.

$$(x)^2 + (3x-3)^2 = (3x-2)^2$$
$$x^2 + 9x^2 - 18x + 9 = 9x^2 - 12x + 4$$
$$x^2 - 6x + 5 = 0$$
$$(x-1)(x-5) = 0$$

$x = 1$ or $x = 5$

The shorter leg is 5 ft.

Cumulative Review Exercises Chapters 1–5

1. $\dfrac{|4-25\div(-5)\cdot 2|}{\sqrt{8^2+6^2}}=\dfrac{|4-(-5)\cdot 2|}{\sqrt{64+36}}$
$=\dfrac{|4-(-10)|}{\sqrt{100}}$
$=\dfrac{|14|}{10}$
$=\dfrac{7}{5}$

3. $-3.5-2.5x=1.5(x-3)$
$-3.5-2.5x=1.5x-4.5$
$4.0x=1.0$
$x=\dfrac{1}{4}=0.25$

5. $3x-2y=8$
$-2y=8-3x$
$\dfrac{-2y}{-2}=\dfrac{8-3x}{-2}$
$y=\dfrac{3x-8}{2}$

7. Let x = number of quarters, $(x+2)$ the number of nickels, and $(x-3)$ the number of dimes.

$$\begin{pmatrix}\text{Value}\\ \text{of}\\ \text{quarters}\end{pmatrix}+\begin{pmatrix}\text{Value}\\ \text{of}\\ \text{nickels}\end{pmatrix}+\begin{pmatrix}\text{Value}\\ \text{of}\\ \text{dimes}\end{pmatrix}=\$3.80$$

$0.25x+0.05(x+2)+0.10(x-3)=3.80$
$0.25x+0.05x+0.10+0.10x-0.30=3.80$
$0.40x=4$
$x=\dfrac{4}{0.40}$
$=10$

There are 10 quarters, 12 nickels, and 7 dimes.

9. $2\left(\dfrac{1}{3}y^3-\dfrac{3}{2}y^2-7\right)-\left(\dfrac{2}{3}y^3+\dfrac{1}{2}y^2+5y\right)$
$=\dfrac{2}{3}y^3-3y^2-14-\dfrac{2}{3}y^3-\dfrac{1}{2}y^2-5y$
$=-\dfrac{7}{2}y^2-5y-14$

11. $(2w-7)^2=(2w)^2-2(2w)(7)+7^2$
$=4w^2-28w+49$

13. $\dfrac{c^{12}c^{-5}}{c^3}=\dfrac{c^7}{c^3}=c^{7-3}=c^4$

15. $\left(\dfrac{1}{2}\right)^0-\left(\dfrac{1}{4}\right)^{-2}=1-4^2=1-16=-15$

17. $w^4-16=(w^2)^2-4^2$
$=(w^2+4)(w^2-4)$
$=(w^2+4)(w+2)(w-2)$

19. $4a^2-12a+9=(2a)^2-2(2a)(3)+3^2$
$=(2a-3)^2$

21. $y^3-27=y^3-3^3=(y-3)(y^2+3y+9)$

23. $(a-2)^2+5(a-2)+6$
Let $u=a-2$.
$u^2+5u+6=(u+2)(u+3)$
$(a-2+2)(a-2+3)=a(a+1)$

25. $4x(2x-1)(x+5)=0$
$4x=0$ or $2x-1=0$ or $x+5=0$
$x=0$ $\quad x=\dfrac{1}{2}$ $\quad x=-5$

Chapter 6

Section 6.1 Practice Exercises

1. **(a)** A number $\frac{p}{q}$ where p and q are integers and $q \neq 0$.

(b) An expression $\frac{p}{q}$ where p and q are polynomials and $q \neq 0$.

3. $x = 2;\ \frac{1}{-2-6} = \frac{1}{-8} = -\frac{1}{8}$

5. $w = 0;\ \frac{0-10}{0+6} = \frac{-10}{6} = -\frac{5}{3}$

7. $y = 8;\ \frac{8-8}{2(0)^2+0-1} = \frac{0}{-1} = 0$

9. $a = 2;$

$$\frac{(2-7)(2+1)}{(2-2)(2+5)} = \frac{(-5)(3)}{(0)(7)} = \frac{-15}{0} = \text{undefined}$$

11. **(a)** Let $x = 12$;
$$\begin{aligned} t &= \frac{24}{x} + \frac{24}{x+8} \\ &= \frac{24}{12} + \frac{24}{12+8} \\ &= 2 + \frac{24}{20} \\ &= 2 + \frac{6}{5} \\ &= 3\frac{1}{5} \text{ hr} \end{aligned}$$

(b) Let $x = 24$;
$$\begin{aligned} t &= \frac{24}{x} + \frac{24}{x+8} \\ &= \frac{24}{24} + \frac{24}{24+8} \\ &= 1 + \frac{24}{32} \\ &= 1 + \frac{3}{4} \\ &= 1\frac{3}{4} \text{ hr} \end{aligned}$$

13. $k + 2 \neq 0$

$k \neq 2;\ \{k \mid k \neq 2\}$

15. $2x - 5 \neq 0$ or $x + 8 \neq 0$

$x \neq \frac{5}{2}$ or $x \neq -8$ $\left\{x \,\middle|\, x \neq \frac{5}{2}, x \neq -8\right\}$

17. $b^2 + 5b + 6 \neq 0$

$(b+3)(b+2) \neq 0$

$b \neq -3$ or $b \neq -2$ $\{b \mid b \neq -3, b \neq -2\}$

19. Answers may vary; $\frac{1}{x-2}$

21. Answers may vary; $\frac{1}{(x+3)(x-7)}$

23. $\frac{7b^2}{21b} = \frac{\cancel{7}\cancel{b}b}{3 \cdot \cancel{7}\cancel{b}} = \frac{b}{3}$

25. $\frac{18st^5}{12st^3} = \frac{\cancel{6} \cdot 3\cancel{s}\cancel{t^3}t^2}{\cancel{6} \cdot 2\cancel{s}\cancel{t^3}} = \frac{3t^2}{2}$

27. $\frac{-24x^2y^5z}{8xy^4z^3} = \frac{-3 \cdot \cancel{8}\cancel{x} \cdot x \cdot y\cancel{y^4}\cancel{z}}{\cancel{8}\cancel{x}\cancel{y^4}\cancel{z}z^2} = -\frac{3xy}{z^2}$

29. $\frac{3(y+2)}{6(y+2)} = \frac{3}{6} = \frac{1}{2}$

31. $\frac{(p-3)(p+5)}{(p+5)(p+4)} = \frac{p-3}{p+4}$

33. $\frac{(m+11)}{4(m+11)(m-11)} = \frac{1}{4(m-11)}$

35. **(a)** $\frac{3y+6}{6y+12} = \frac{3(y+2)}{6(y+2)}$

(b) $6y + 12 = 0;\ y \neq -2\ \{y \mid y \neq -2\}$

(c) $\frac{3y+6}{6y+12} = \frac{3(y+2)}{6(y+2)} = \frac{3}{6} = \frac{1}{2}$

37. (a) $\frac{t^2-1}{t+1}=\frac{(t+1)(t-1)}{t+1}$

(b) $t+1=0$; $t\neq -1$ $\{t \mid t\neq -1\}$

(c) $\frac{t^2-1}{t+1}=\frac{(t+1)(t-1)}{(t+1)}=t+1$

39. (a) $\frac{7w}{21w^2-35w}=\frac{7w}{7w(3w-5)}$

(b) $21w^2-35w=0$

$w\neq 0$ or $w\neq \frac{5}{3}$

$\left\{w \middle| w\neq 0,\ w\neq \frac{5}{3}\right\}$

(c) $\frac{7w}{21w^2-35w}=\frac{7w}{7w(3w-5)}=\frac{1}{3w-5}$

41. (a) $\frac{9x^2-4}{6x+4}=\frac{(3x+2)(3x-2)}{2(3x+2)}$

(b) $6x+4=0$,

$x\neq -\frac{2}{3}$

$\left\{x \middle| x\neq -\frac{2}{3}\right\}$

(c) $\frac{9x^2-4}{6x+4}=\frac{(3x+2)(3x-2)}{2(3x+2)}=\frac{3x-2}{2}$

43. (a) $\frac{a^2+3a-10}{a^2+a-6}=\frac{(a+5)(a-2)}{(a+3)(a-2)}$

(b) $a^2+a-6=0$

$a\neq -3$ or $a\neq 2$

$\{a \mid a\neq -3,\ a\neq 2\}$

(c) $\frac{a^2+3a-10}{a^2+a-6}=\frac{(a+5)(a-2)}{(a+3)(a-2)}=\frac{a+5}{a+3}$

45. $\frac{5}{20a-25}=\frac{5}{5(4a-5)}=\frac{1}{4a-5}$

47. $\frac{4w-8}{w^2-4}=\frac{4(w-2)}{(w+2)(w-2)}=\frac{4}{w+2}$

49. $\frac{3x^2-6x}{9xy+18x}=\frac{3x(x-2)}{9x(y+2)}=\frac{x-2}{3(y+2)}$

51. $\frac{2x+4}{x^2-3x-10}=\frac{2(x+2)}{(x-5)(x+2)}=\frac{2}{x-5}$

53. $\frac{a^2-49}{a-7}=\frac{(a+7)(a-7)}{a-7}=a+7$

55. $\frac{q^2+25}{q+5}$ not reducible

57. $\frac{y^2+6y+9}{2y^2+y-15}=\frac{(y+3)(y+3)}{(2y-5)(y+3)}=\frac{y+3}{2y-5}$

59. $\frac{3x^2+7x-6}{x^2+7x+12}=\frac{(3x-2)(x+3)}{(x+3)(x+4)}=\frac{3x-2}{x+4}$

61. $\frac{5q^2+5}{q^4-1}=\frac{5(q^2+1)}{(q+1)(q-1)(q^2+1)}$

$=\frac{5}{(q+1)(q-1)}$

63. $\frac{ac-ad+2bc-2bd}{2ac+ad+4bc+2bd}=\frac{(c-d)(a+2b)}{(2c+d)(a+2b)}$

$=\frac{c-d}{2c+d}$

65. $\frac{49p^2-28pq+4q^2}{14p-4q}=\frac{(7p-2q)^2}{2(7p-2q)}$

$=\frac{7p-2q}{2}$

67. $\frac{2x^2-xy-3y^2}{2x^2-11xy+12y^2}=\frac{(2x-3y)(x+y)}{(2x-3y)(x-4y)}$

$=\frac{x+y}{x-4y}$

69. They are opposites.

71. $\dfrac{x-5}{5-x}=\dfrac{(-1)(5-x)}{5-x}=-1$

73. $\dfrac{-4-y}{4+y}=\dfrac{(-1)(4+y)}{4+y}=-1$

75. $\dfrac{3y-6}{12-6y}=\dfrac{3(y-2)}{-6(y-2)}=\dfrac{3}{-6}=-\dfrac{1}{2}$

77. $\dfrac{x^2-x-12}{16-x^2}=\dfrac{(x-4)(x+3)}{(-1)(x+4)(x-4)}=-\dfrac{x+3}{x+4}$

79. (a) $\dfrac{5(4)+5}{4^2-1}=\dfrac{25}{15}=\dfrac{5}{3}$

(b) $\dfrac{5}{4-1}=\dfrac{5}{3}$

81. (a) $\dfrac{3(-1)^2-2(-1)-1}{6(-1)^2-7(-1)-3}=\dfrac{3+2-1}{6+7-3}=\dfrac{4}{10}=\dfrac{2}{5}$

(b) $\dfrac{-1-1}{2(-1)-3}=\dfrac{-2}{-5}=\dfrac{2}{5}$

83. $\dfrac{w^3-8}{w^2+2w+4}=\dfrac{(w-2)(w^2+2w+4)}{w^2+2w+4}=w-2$

85. $\dfrac{(z+4)(z-4)}{(z-4)(z^2+4z+16)}=\dfrac{z+4}{z^2+4z+16}$

87. $\dfrac{(5x+4)(x^2-9)}{(x^2-9)}=5x+4$

Section 6.2 Practice Exercises

1. $\{x \mid x \neq 3, x \neq -2\}$

$\dfrac{(x+2)(x-1)}{(x-3)(x+2)}=\dfrac{x-1}{x-3}$

3. $\{a \mid a \neq 2\}$

$\dfrac{a^2-4}{a^2-4a+4}=\dfrac{(a+2)(a-2)}{(a-2)(a-2)}=\dfrac{a+2}{a-2}$

5. $\left\{t \,\middle|\, t \neq \dfrac{1}{2}\right\}$

$\dfrac{12t-6}{3-6t}=\dfrac{6(2t-1)}{-3(2t-1)}=-2$

7. $\dfrac{3}{5}\cdot\dfrac{1}{2}=\dfrac{3}{10}$

9. $\dfrac{3}{4}\div\dfrac{3}{8}=\dfrac{3}{4}\cdot\dfrac{8}{3}=\dfrac{24}{12}=2$

11. $6\cdot\dfrac{5}{12}=\dfrac{6}{1}\cdot\dfrac{5}{12}=\dfrac{30}{12}=\dfrac{5}{2}$

13. $\dfrac{\frac{21}{4}}{\frac{7}{5}}=\dfrac{21}{4}\cdot\dfrac{5}{7}=\dfrac{105}{28}=\dfrac{15}{4}$

15. $\dfrac{4x-24}{20x}\cdot\dfrac{5x}{8}=\dfrac{4(x-6)}{20x}\cdot\dfrac{5x}{8}=\dfrac{x-6}{8}$

17. $\dfrac{3y+18}{y^2}\cdot\dfrac{4y}{6y+36}=\dfrac{3(y+6)}{y^2}\cdot\dfrac{4y}{6(y+6)}=\dfrac{2}{y}$

19. $\dfrac{10}{2-a}\cdot\dfrac{a-2}{16}=\dfrac{10}{-(a-2)}\cdot\dfrac{a-2}{16}=-\dfrac{5}{8}$

21. $\dfrac{b^2-a^2}{a-b}\cdot\dfrac{a}{a^2-ab}=\dfrac{-(a+b)(a-b)}{a-b}\cdot\dfrac{a}{a(a-b)}$

$=-\dfrac{a+b}{a-b}$

23. $\dfrac{4a+12}{6a-18}\div\dfrac{3a+9}{5a-15}=\dfrac{4(a+3)}{6(a-3)}\cdot\dfrac{5(a-3)}{3(a+3)}=\dfrac{10}{9}$

25. $\dfrac{3x-21}{6x^2-42x}\div\dfrac{7}{12x}=\dfrac{3(x-7)}{6x(x-7)}\cdot\dfrac{12x}{7}=\dfrac{6}{7}$

27. $\dfrac{y^2+5y-36}{y^2-2y-8}\cdot\dfrac{y+2}{y-6}=\dfrac{(y+9)(y-4)}{(y-4)(y+2)}\cdot\dfrac{y+2}{y-6}$

$=\dfrac{y+9}{y-6}$

29. $\dfrac{t^2+4t-5}{t^2+7t+10}\cdot\dfrac{t+4}{t-1}=\dfrac{(t+5)(t-1)}{(t+5)(t+2)}\cdot\dfrac{t+4}{t-1}$
$=\dfrac{t+4}{t+2}$

31. $\dfrac{m^2-n^2}{9}\div\dfrac{3n-3m}{27m}$
$=\dfrac{(m+n)(m-n)}{9}\cdot\dfrac{27m}{-3(m-n)}$
$=-m(m+n)$

33. $\dfrac{3p+4q}{p^2+4pq+4q^2}\div\dfrac{4}{p+2q}$
$=\dfrac{3p+4q}{(p+2q)(p+3q)}\cdot\dfrac{p+2q}{4}$
$=\dfrac{3p+4q}{4(p+2q)}$

35. $(w+3)\cdot\dfrac{w}{2w^2+5w-3}$
$=\dfrac{(w+3)}{1}\cdot\dfrac{w}{(2w-1)(w+3)}$
$=\dfrac{w}{2w-1}$

37. $\dfrac{\frac{5t-10}{12}}{\frac{4t-8}{8}}=\dfrac{5(t-2)}{12}\cdot\dfrac{8}{4(t-2)}=\dfrac{5}{6}$

39. $\dfrac{q+1}{5q^2-28q-12}\cdot(5q+2)$
$=\dfrac{q+1}{(5q+2)(q-6)}\cdot\dfrac{(5q+2)}{1}$
$=\dfrac{q+1}{q-6}$

41. $\dfrac{2a^2+13a-24}{8a-12}\div(a+8)$
$=\dfrac{(2a-3)(a+8)}{4(2a-3)}\cdot\dfrac{1}{(a+8)}$
$=\dfrac{1}{4}$

43. $(5t-1)\div\dfrac{5t^2+9t-2}{3t+8}$
$=\dfrac{(5t-1)}{1}\cdot\dfrac{3t+8}{(5t-1)(t+2)}$
$=\dfrac{3t+8}{t+2}$

45. $\dfrac{x^2+2x-3}{x^2-3x+2}\cdot\dfrac{x^2+2x-8}{x^2+4x+3}$
$=\dfrac{(x+3)(x-1)}{(x-2)(x-1)}\cdot\dfrac{(x+4)(x-2)}{(x+3)(x+1)}$
$=\dfrac{x+4}{x+1}$

47. $\dfrac{\frac{w^2-6w+9}{8}}{\frac{9-w^2}{4w+12}}=\dfrac{(w-3)(w-3)}{8}\cdot\dfrac{4(w+3)}{-(w+3)(w-3)}$
$=-\dfrac{w-3}{2}$

49. $\dfrac{k^2+3k+2}{k^2+5k+4}\div\dfrac{k^2+5k+6}{k^2+10k+24}$
$=\dfrac{(k+1)(k+2)}{(k+4)(k+1)}\cdot\dfrac{(k+6)(k+4)}{(k+2)(k+3)}$
$=\dfrac{k+6}{k+3}$

51. $\frac{b^3-3b^2+4b-12}{b^4-16}\cdot\frac{3b^2+5b-2}{3b^2-10b+3}\div\frac{3}{6b-12}=\frac{(b^2+4)(b-3)}{(b+2)(b-2)(b^2+4)}\cdot\frac{(3b-1)(b+2)}{(3b-1)(b-3)}\cdot\frac{6(b-2)}{3}=\frac{6}{3}=2$

53. $\frac{a^2-5a}{a^2+7a+12}\div\frac{a^3-7a^2+10a}{a^2+9a+18}\div\frac{a+6}{a+4}=\frac{a(a-5)}{(a+3)(a+4)}\cdot\frac{(a+6)(a+3)}{a(a-5)(a-2)}\cdot\frac{a+4}{a+6}=\frac{1}{a-2}$

55. $\frac{p^3-q^3}{p-q}\cdot\frac{p+q}{2p^2+2pq+2q^2}=\frac{(p-q)(p^2+pq+q^2)}{p-q}\cdot\frac{p+q}{2(p^2pq+q^2)}=\frac{p+q}{2}$

Section 6.3 Practice Exercises

1. $\{x \mid x \neq -1, x \neq 1\}$

$\frac{3x+3}{5x^2-5}=\frac{3(x+1)}{5(x+1)(x-1)}=\frac{3}{5(x-1)}$

3. $\frac{a+3}{a+7}\cdot\frac{a^2+3a-10}{a^2+a-6}=\frac{a+3}{a+7}\cdot\frac{(a+5)(a-2)}{(a+3)(a-2)}=\frac{a+5}{a+7}$

5. $\frac{6(a+2b)}{2(a-3b)}\cdot\frac{4(a+3b)(a-3b)}{9(a+2b)(a-2b)}=\frac{4(a+3b)}{3(a-2b)}$

7. $\frac{6}{7}=\frac{36}{42}$

9. $\frac{2}{13}=\frac{6}{39}$

11. $\frac{3}{p^2q}=\frac{15p}{5p^3q}$

13. $\frac{2x}{yz}=\frac{12xyz^3}{6y^2z^4}$

15. $\frac{w+6}{w-7}=\frac{(w+6)(w+2)}{(w-7)(w+2)}=\frac{w^2+8w+12}{(w-7)(w+2)}$

17. $\frac{6}{x-3}=\frac{(-1)(6)}{(-1)(x-3)}=\frac{-6}{3-x}$

19. a, b, c, d

21. Because x^5 is the lowest power of x that has x^3, x^5, x^4 as factors.

23. The product of unique factors is $(x+3)(x-2)$.

25. Because $(b-1)$ and $(1-b)$ are opposites; they differ by a factor of -1.

27. $3^2 \cdot 5 = 45$

29. $2^4 = 16$

31. 5 or -5

33. $3^2 \cdot x^2 y^3 = 9x^2 y^3$

35. $w^2 y$

37. $(p+3)(p-1)(p+2)$

39. $9t(t+1)^2$

41. $(y-2)(y+2)(y+3)$

43. $3-x$ or $x-3$

45. LCD: $5x^2$; $\frac{6}{5x^2}, \frac{5x}{5x^2}$

47. LCD: $5 \cdot 6x^3 = 30x^3$; $\frac{24x}{30x^3}, \frac{5y}{30x^3}$

49. LCD: $12a^2b$; $\frac{10}{12a^2b}, \frac{a^3}{12a^2b}$

51. LCD: $(m+4)(m-1)$;

$$\frac{6(m-1)}{(m+4)(m-1)} = \frac{6m-6}{(m+4)(m-1)},$$
$$\frac{3}{m-1} = \frac{3(m+4)}{(m+4)(m-1)}$$

53. LCD: $(w+3)(w-8)(w-1)$;

$$\frac{6(w+1)}{(w+3)(w-8)(w-1)} = \frac{6w+6}{(w+3)(w-8)(w-1)},$$
$$\frac{w(w+3)}{(w+3)(w-8)(w-1)} = \frac{w^2+3w}{(w+3)(w-8)(w-1)}$$

55. LCD:

$$(p-2)(p+2)(p+2) = (p-2)(p+2)^2$$
$$\frac{6p(p+2)}{(p-2)(p+2)^2} = \frac{6p^2+12p}{(p-2)(p+2)^2},$$
$$\frac{3(p-2)}{(p-2)(p+2)^2} = \frac{3p-6}{(p-2)(p+2)^2}$$

57. LCD: $a-4$ or $4-a$;

$$\frac{1}{a-4}, \frac{(-1)a}{(-1)(4-a)} = -\frac{a}{a-4} \text{ or}$$
$$\frac{(-1)(1)}{(-1)(a-4)} = -\frac{1}{4-a}, \frac{a}{4-a}$$

59. LCD: $2(x-7)$ or $2(7-x)$;

$$\frac{8}{2(x-7)}, \frac{-1(y)}{(2)(x-7)} = -\frac{y}{2(x-7)} \text{ or}$$
$$\frac{-1(8)}{2(7-x)} = -\frac{8}{2(7-x)}, \frac{y}{2(7-x)}$$

61. LCD: $a+b$;

$$\frac{1}{a+b}, \frac{6}{-1(a+b)} = -\frac{6}{a+b} \text{ or}$$
$$\frac{-1}{-a-b}, \frac{6}{-a-b}$$

63. LCD: $(z+2)(z+3)(z+7)$;

$$\frac{z(z+3)}{(z+2)(z+3)(z+7)}, \frac{-3z(z+2)}{(z+2)(z+3)(z+7)},$$
$$\frac{5(z+7)}{(z+2)(z+3)(z+7)}$$

65. LCD: $(p-2)(p+2)(p^2+2p+4)$;

$$\frac{3(p+2)}{(p^2-4)(p^2+2p+4)}, \frac{p(p^2+2p+4)}{(p^2-4)(p^2+2p+4)},$$
$$\frac{5p(p^2-4)}{(p^2-4)(p^2+2p+4)}$$

Section 6.4 Practice Exercises

1. **(a)** $x = 0;\ \frac{-5}{10} = -\frac{1}{2}$

$x = 1;\ \frac{1^2 - 4(1) - 5}{1^2 - 7(1) + 10} = -2$

$x = -1;\ \frac{(-1)^2 - 4(-1) - 5}{(-1)^2 - 7(-1) + 10} = \frac{0}{18} = 0$

$x = 2;$

$\frac{(2)^2 - 4(2) - 5}{(2)^2 - 7(2) - 10} = \frac{-9}{0} = \text{undefined}$

$x = 5;\ \frac{5^2 - 4(5) - 5}{5^2 - 7(5) + 10} = \frac{0}{0} = \text{undefined}$

(b) $(x-5)(x-2);\ \{x \mid x \neq 5, x \neq 2\}$

(c) $\frac{(x-5)(x+1)}{(x-5)(x-2)} = \frac{x+1}{x-2}$

3. $\frac{2b^2 - b - 3}{2b^2 - 3b - 9} \cdot \frac{4b-12}{2b-3} \div \frac{b^2-1}{4b+6}$

$= \frac{(2b-3)(b+1)}{(2b+3)(b-3)} \cdot \frac{4(b-3)}{2b-3} \cdot \frac{2(2b+3)}{(b+1)(b-1)}$

$= \frac{8}{b-1}$

5. $\frac{7}{8} + \frac{3}{8} = \frac{10}{8} = \frac{5}{4}$

7. $\frac{9}{16} - \frac{3}{16} = \frac{6}{16} = \frac{3}{8}$

9. $\frac{5a}{a+2} - \frac{3a-4}{a+2} = \frac{5a - (3a-4)}{a+2}$

$= \frac{2a+4}{a+2}$

$= \frac{2(a+2)}{a+2}$

$= 2$

11. $\frac{5c}{c+6} + \frac{30}{c+6} = \frac{5c+30}{c+6} = \frac{5(c+6)}{c+6} = 5$

13. $\frac{5}{t-8} - \frac{2t+1}{t-8} = \frac{5-2t-1}{t-8} = \frac{4-2t}{t-8} = \frac{-2(t-2)}{t-8}$

15. $\frac{10}{3x-7} - \frac{5}{3x-7} = \frac{10-5}{3x-7} = \frac{5}{3x-7}$

17. $\frac{m^2}{m+5} + \frac{10m+25}{m+5} = \frac{m^2+10m+25}{m+5}$

$= \frac{(m+5)^2}{m+5}$

$= m+5$

19. $\frac{2a}{a+3} + \frac{6}{a+3} = \frac{2a+6}{a+3} = \frac{2(a+3)}{a+3} = 2$

21. $\frac{x^2}{x+5} - \frac{25}{x+5} = \frac{x^2-25}{x+5}$

$= \frac{(x+5)(x-5)}{x+5}$

$= x-5$

23. $P = \frac{2x}{y} + \frac{6x}{y} + \frac{7x}{y} = \frac{15x}{y}$

25. $\frac{4}{5xy^3} + \frac{2x}{15y^2} = \frac{4(3)}{5xy^3(3)} + \frac{2x(xy)}{15y^2(xy)}$

$= \frac{12 + 2x^2y}{15xy^3}$

$= \frac{2(6 + x^2y)}{15xy^3}$

27. $\frac{z}{3z-9} - \frac{z-2}{z-3} = \frac{z}{3(z-3)} - \frac{3(z-2)}{3(z-3)}$

$= \frac{z - 3z + 6}{3(z-3)}$

$= \frac{-2z+6}{3(z-3)}$

$= \frac{-2(z-3)}{3(z-3)}$

$= -\frac{2}{3}$

29. $\frac{5}{a+1}+\frac{4}{3a+3}=\frac{5(3)}{3(a+1)}+\frac{4}{3(a+1)}$
$=\frac{15+4}{3(a+1)}$
$=\frac{19}{3(a+1)}$

31. $\frac{k}{k^2-9}-\frac{4}{k-3}$
$=\frac{k}{(k+3)(k-3)}-\frac{4(k+3)}{(k+3)(k-3)}$
$=\frac{k-4k-12}{(k+3)(k-3)}$
$=\frac{-3(k+4)}{(k+3)(k-3)}$

33. $\frac{3a-7}{6a+10}-\frac{10}{3a^2+5a}$
$=\frac{a(3a-7)}{2a(3a+5)}-\frac{10(2)}{2a(3a+5)}$
$=\frac{3a^2-7a-20}{2a(3a+5)}$
$=\frac{(3a+5)(a-4)}{2a(3a+5)}$
$=\frac{a-4}{2a}$

35. $\frac{10}{3x-7}+\frac{5}{7-3x}=\frac{10}{3x-7}+\frac{5}{-1(3x-7)}$
$=\frac{10+(-5)}{3x-7}$
$=\frac{5}{3x-7}$

37. $\frac{6a}{a^2-b^2}+\frac{2a}{a^2+ab}$
$=\frac{6a}{(a+b)(a-b)}+\frac{2a}{a(a+b)}$
$=\frac{6a^2}{a(a+b)(a-b)}+\frac{2a(a-b)}{a(a+b)(a-b)}$
$=\frac{6a^2+2a^2-2ab}{a(a+b)(a-b)}$
$=\frac{8a^2-2ab}{a(a+b)(a-b)}$
$=\frac{2a(4a-b)}{a(a+b)(a-b)}$
$=\frac{2(4a-b)}{(a+b)(a-b)}$

39. $\frac{p}{3}-\frac{4p-1}{-3}=\frac{p}{3}-\frac{-(4p-1)}{3}$
$=\frac{p+4p-1}{3}$
$=\frac{5p-1}{3}$

41. $\frac{4n}{n-8}-\frac{2n-1}{8-n}=\frac{4n}{n-8}-\frac{2n-1}{-1(n-8)}$
$=\frac{4n}{n-8}+\frac{2n-1}{n-8}$
$=\frac{4n+2n-1}{n-8}$
$=\frac{6n-1}{n-8}$

43. $\frac{5}{x}+\frac{3}{x+2}=\frac{5(x+2)}{x(x+2)}+\frac{3x}{x(x+2)}$
$=\frac{5x+10+3x}{x(x+2)}$
$=\frac{8x+10}{x(x+2)}$
$=\frac{2(4x+5)}{x(x+2)}$

45. $$\frac{4w}{w^2+2w-3}+\frac{2}{1-w}=\frac{4w}{(w+3)(w-1)}+\frac{-2}{w-1}$$
$$=\frac{4w}{(w+3)(w-1)}-\frac{2(w+3)}{(w+3)(w-1)}$$
$$=\frac{4w-2w-6}{(w+3)(w-1)}$$
$$=\frac{12w-6}{(w+3)(w-1)}$$
$$=\frac{2(w-3)}{(w+3)(w-1)}$$

47. $$\frac{3a-8}{a^2-5a+6}+\frac{a+2}{a^2-6a+8}=\frac{3a-8}{(a-2)(a-3)}+\frac{a+2}{(a-2)(a-4)}$$
$$=\frac{(3a-8)(a-4)}{(a-2)(a-3)(a-4)}+\frac{(a+2)(a-3)}{(a-2)(a-3)(a-4)}$$
$$=\frac{3a^2-20a+32+a^2-a-6}{(a-2)(a-3)(a-4)}$$
$$=\frac{4a^2-21a+26}{(a-2)(a-3)(a-4)}$$
$$=\frac{(4a-13)(a-2)}{(a-2)(a-3)(a-4)}$$
$$=\frac{4a-13}{(a-3)(a-4)}$$

49. $$\frac{3x}{x^2+x-6}+\frac{x}{x^2+5x+6}=\frac{3x}{(x+3)(x-2)}+\frac{x}{(x+3)(x+2)}$$
$$=\frac{3x(x+2)}{(x+3)(x-2)(x+2)}+\frac{x(x-2)}{(x+3)(x-2)(x+2)}$$
$$=\frac{3x^2+6x+x^2-2x}{(x+3)(x-2)(x+2)}$$
$$=\frac{4x^2+4x}{(x+3)(x-2)(x+2)}$$
$$=\frac{4x(x+1)}{(x+3)(x-2)(x+2)}$$

51. $$\frac{3y}{2y^2-y-1}-\frac{4y}{2y^2-7y-4}=\frac{3y}{(2y+1)(y-1)}-\frac{4y}{(2y+1)(y-4)}$$
$$=\frac{3y(y-4)}{(2y+1)(y-1)(y-4)}-\frac{4y(y-1)}{(2y+1)(y-1)(y-4)}$$
$$=\frac{3y^2-12y-4y^2+4y}{(2y+1)(y-1)(y-4)}$$
$$=\frac{-y^2-8y}{(2y+1)(y-1)(y-4)}$$
$$=\frac{-y(y+8)}{(2y+1)(y-1)(y-4)}$$

53. $$\frac{3}{2p-1}-\frac{4p+4}{4p^2-1}=\frac{3}{2p-1}-\frac{4p+4}{(2p+1)(2p-1)}$$
$$=\frac{3(2p+1)}{(2p+1)(2p-1)}-\frac{4p+4}{(2p+1)(2p-1)}$$
$$=\frac{6p+3-4p-4}{(2p+1)(2p-1)}$$
$$=\frac{2p-1}{(2p+1)(2p-1)}$$
$$=\frac{1}{2p+1}$$

55. $$\frac{x}{x-y}-\frac{y}{y-x}=\frac{x}{x-y}-\frac{y}{-1(x-y)}=\frac{x}{x-y}+\frac{y}{x-y}=\frac{x+y}{x-y}$$

57. $$\frac{2}{a+b}+\frac{2}{a-b}-\frac{4a}{a^2-b^2}=\frac{2(a-b)}{(a+b)(a-b)}+\frac{2(a+b)}{(a+b)(a-b)}-\frac{4a}{(a+b)(a-b)}$$
$$=\frac{2a-2b+2a+2b-4a}{(a+b)(a-b)}$$
$$=0$$

59. $P = 2w + 2l$
$$= 2\left(\frac{2}{x+3}\right) + 2\left(\frac{1}{x+2}\right)$$
$$= \frac{4}{x+3} + \frac{2}{x+2}$$
$$= \frac{4(x+2)}{(x+3)(x+2)} + \frac{2(x+3)}{(x+3)(x+2)}$$
$$= \frac{4x+8+2x+6}{(x+3)(x+2)}$$
$$= \frac{6x+14}{(x+3)(x+2)}$$
$$= \frac{2(3x+7)}{(x+3)(x+2)}$$

61. $\frac{1}{n}$

63. $\frac{12}{p}$

65. Let n = the number, then $n + \left(7 \cdot \frac{1}{n}\right)$;
$$n + \frac{7}{n} = \frac{n^2}{n} + \frac{7}{n} = \frac{n^2+7}{n}$$

67. $\frac{1}{n} - \frac{2}{n}$; $\frac{1}{n} - \frac{2}{n} = -\frac{1}{n}$

69. $\left(\frac{2}{k+1} + 3\right)\left(\frac{k+1}{4k+7}\right)$
$$= \left(\frac{2}{k+1} + \frac{3(k+1)}{k+1}\right)\left(\frac{k+1}{4k+7}\right)$$
$$= \left(\frac{3k+5}{k+1}\right)\left(\frac{k+1}{4k+7}\right)$$
$$= \frac{3k+5}{4k+7}$$

71. $\left(\frac{1}{10a} - \frac{b}{10a^2}\right) \div \left(\frac{1}{10} - \frac{b}{10a}\right)$
$$= \left(\frac{1}{10a^2} - \frac{b}{10a^2}\right) \div \left(\frac{a}{10a} - \frac{b}{10a}\right)$$
$$= \left(\frac{a-b}{10a^2}\right) \cdot \left(\frac{10a}{a-b}\right)$$
$$= \frac{1}{a}$$

73. $\frac{6a^2b^3}{72ab^7c} = \frac{6aab^3}{6 \cdot 12ab^3b^4c} = \frac{a}{12b^4c}$

75. $\frac{p^2+10pq+25q^2}{p^2+6pq+5q^2} \div \frac{10p+50q}{2p^2-2q^2}$
$$= \frac{(p+5q)(p+5q)}{(p+5q)(p+q)} \cdot \frac{2(p+q)(p-q)}{10(p+5q)}$$
$$= \frac{p-q}{5}$$

77. $\frac{20x^2+10x}{4x^3+4x^2+x} = \frac{10x(2x+1)}{x(2x+1)(2x+1)} = \frac{10}{2x+1}$

79. $\frac{h^2-49}{h+1} \div \frac{h+7}{h^2-1}$
$$= \frac{(h+7)(h-7)}{h+1} \cdot \frac{(h+1)(h-1)}{h+7}$$
$$= (h-7)(h-1)$$

81. $\frac{a}{a^2-9} - \frac{3}{6a-18}$
$$= \frac{a}{(a+3)(a-3)} - \frac{3}{6(a-3)}$$
$$= \frac{2a}{2(a+3)(a-3)} - \frac{(a+3)}{2(a+3)(a-3)}$$
$$= \frac{2a-a-3}{2(a+3)(a-3)}$$
$$= \frac{a-3}{2(a+3)(a-3)}$$
$$= \frac{1}{2(a+3)}$$

83. $(t^2+5t-24)\left(\frac{t+8}{t-3}\right)=\frac{(t+8)(t-3)}{1}\left(\frac{t+8}{t-3}\right)$
$=(t+8)^2$

85. $\frac{-3}{w^3+27}-\frac{1}{w^2-9}=\frac{-3}{(w+3)(w^2-3w+9)}-\frac{1}{(w+3)(w-3)}$
$=\frac{-3(w-3)}{(w-3)(w+3)(w^2-3w+9)}-\frac{w^2-3w+9}{(w-3)(w+3)(w^2-3w+9)}$
$=\frac{-3w+9-w^2+3w-9}{(w-3)(w+3)(w^2-3+9)}$
$=\frac{-w^2}{(w-3)(w+3)(w^2-3w+9)}$

87. $\frac{2p}{p^2+5p+6}-\frac{p+1}{p^2+2p-3}+\frac{3}{p^2+p-2}$
$=\frac{2p}{(p+2)(p+3)}-\frac{p+1}{(p-1)(p+3)}+\frac{3}{(p+2)(p-1)}$
$=\frac{2p(p-1)}{(p-1)(p+2)(p+3)}-\frac{(p+1)(p+2)}{(p-1)(p+2)(p+3)}+\frac{3(p+3)}{(p-1)(p+2)(p+3)}$
$=\frac{2p^2-2p-p^2-3p-2+3p+9}{(p-1)(p+2)(p+3)}$
$=\frac{p^2-2p+7}{(p-1)(p+2)(p+3)}$

89. $\frac{3m}{m^2+3m-10}+\frac{5}{4-2m}-\frac{1}{m+5}=\frac{3m}{(m+5)(m-2)}+\frac{5}{-2(m-2)}-\frac{1}{m+5}$
$=\frac{6m}{2(m+5)(m-2)}-\frac{5(m+5)}{2(m+5)(m-2)}-\frac{2(m-2)}{2(m+5)(m-2)}$
$=\frac{6m-5m-25-2m+4}{2(m+5)(m-2)}$
$=\frac{-m-21}{2(m+5)(m-2)}$

Section 6.5 Practice Exercises

1. $\{c\mid c\neq -1, c\neq 2\}$; $\frac{(c-2)(c+3)}{(c+1)(c-2)}=\frac{c+3}{c+1}$

3. $\{x\mid x\neq 2, x\neq -2\}$; $\frac{6x+12}{3x^2-12}=\frac{6(x+2)}{3(x+2)(x-2)}=\frac{2}{x-2}$

5. $\frac{2}{w-2}+\frac{3}{w}=\frac{2w}{w(w-2)}+\frac{3(w-2)}{w(w-2)}$
$=\frac{5w-6}{w(w-2)}$

7. $\frac{p^2+2p}{2p-1}\cdot\frac{10p^2-5p}{12p^3+24p^2}$
$=\frac{p(p+2)}{2p-1}\cdot\frac{5p(2p-1)}{12p^2(p+2)}$
$=\frac{5}{12}$

9. $\left(\frac{1}{z}-\frac{1}{2z}\right)\div\left(\frac{1}{2}+\frac{1}{2z}\right)$
$=\left(\frac{2}{2z}-\frac{1}{2z}\right)\div\left(\frac{z}{2z}+\frac{1}{2z}\right)$
$=\frac{1}{2z}\cdot\frac{2z}{z+1}$
$=\frac{1}{z+1}$

11. $\frac{\frac{1}{2}+\frac{2}{3}}{5};\ \frac{\frac{1}{2}+\frac{2}{3}}{5}=\frac{\frac{3}{6}+\frac{4}{6}}{5}=\frac{\frac{7}{6}}{5}=\frac{7}{6}\cdot\frac{1}{5}=\frac{7}{30}$

13. $\frac{3}{\frac{2}{3}+\frac{3}{4}};\ \frac{3}{\frac{8}{12}+\frac{9}{12}}=\frac{3}{\frac{17}{12}}=3\cdot\frac{12}{17}=\frac{36}{17}$

15. $\frac{\frac{1}{8}+\frac{4}{3}}{\frac{1}{2}-\frac{5}{12}}=\frac{\frac{3}{24}+\frac{32}{24}}{\frac{6}{12}-\frac{5}{12}}=\frac{\frac{35}{24}}{\frac{1}{12}}=\frac{35}{24}\cdot\frac{12}{1}=\frac{35}{2}$

17. $\frac{\frac{1}{h}+\frac{1}{k}}{\frac{1}{hk}}=\frac{\frac{k}{hk}+\frac{h}{hk}}{\frac{1}{hk}}=\frac{\frac{k+h}{hk}}{\frac{1}{hk}}=\frac{k+h}{hk}\cdot\frac{hk}{1}=k+h$

19. $\frac{\frac{n+1}{n^2-9}}{\frac{2}{n+3}}=\frac{n+1}{(n+3)(n-3)}\cdot\frac{n+3}{2}=\frac{n+1}{2(n-3)}$

21. $\frac{2+\frac{1}{x}}{4+\frac{1}{x}}=\frac{\frac{2x}{x}+\frac{1}{x}}{\frac{4x}{x}+\frac{1}{x}}$
$=\frac{\frac{2x+1}{x}}{\frac{4x+1}{x}}$
$=\frac{2x+1}{x}\cdot\frac{x}{4x+1}$
$=\frac{2x+1}{4x+1}$

23. $\frac{\frac{m}{7}-\frac{7}{m}}{\frac{1}{7}+\frac{1}{m}}=\frac{(7m)\left(\frac{m}{7}-\frac{7}{m}\right)}{(7m)\left(\frac{1}{7}+\frac{1}{m}\right)}$
$=\frac{m^2-49}{m+7}$
$=\frac{(m+7)(m-7)}{m+7}$
$=m-7$

25. $\frac{\frac{1}{5}-\frac{1}{y}}{\frac{7}{10}+\frac{1}{y^2}}=\frac{10y^2\left(\frac{1}{5}-\frac{1}{y}\right)}{10y^2\left(\frac{7}{10}+\frac{1}{y^2}\right)}$
$=\frac{2y^2-10y}{7y^2+10}$
$=\frac{2y(y-5)}{7y^2+10}$

27. $\frac{\frac{8}{a+4}+2}{\frac{12}{a+4}-2}=\frac{(a+4)\left(\frac{8}{a+4}+2\right)}{(a+4)\left(\frac{12}{a+4}-2\right)}$
$=\frac{8+2a+8}{12-2a-8}$
$=\frac{2a+16}{4-2a}$
$=\frac{2(a+8)}{2(2-a)}$
$=\frac{a+8}{2-a}$

29. $\dfrac{1-\frac{4}{t^2}}{1-\frac{2}{t}-\frac{8}{t^2}}=\dfrac{t^2\left(1-\frac{4}{t^2}\right)}{t^2\left(1-\frac{2}{t}-\frac{8}{t^2}\right)}$
$=\dfrac{t^2-4}{t^2-2t-8}$
$=\dfrac{(t+2)(t-2)}{(t-4)(t+2)}$
$=\dfrac{t-2}{t-4}$

31. $\dfrac{\frac{1}{z^2-9}+\frac{2}{z+3}}{\frac{3}{z-3}}=\dfrac{(z+3)(z-3)\left(\frac{1}{(z+3)(z-3)}+\frac{2}{z+3}\right)}{(z+3)(z-3)\left(\frac{3}{z-3}\right)}$
$=\dfrac{1+2z-6}{3z+9}$
$=\dfrac{2z-5}{3(z+3)}$

33. $\dfrac{\frac{2}{x-1}+2}{\frac{2}{x+1}-2}=\dfrac{(x-1)(x+1)\left(\frac{2}{x-1}+2\right)}{(x-1)(x+1)\left(\frac{2}{x+1}-2\right)}$
$=\dfrac{2(x+1)+2(x+1)(x-1)}{2(x-1)-2(x+1)(x-1)}$
$=\dfrac{2x^2+2x}{2x-2x^2}$
$=\dfrac{2x(x+1)}{2x(1-x)}$
$=\dfrac{x+1}{1-x}$

35. (a) $R=\dfrac{1}{\frac{1}{2}+\frac{1}{3}}=\dfrac{1}{\frac{5}{6}}=\dfrac{6}{5}\ \Omega$

(b) $R=\dfrac{1}{\frac{1}{10}+\frac{1}{15}}=\dfrac{1}{\frac{5}{30}}=\dfrac{30}{5}=6\ \Omega$

37. $1+\dfrac{1}{1+1}=1+\dfrac{1}{2}=\dfrac{3}{2}$

39. $1+\dfrac{1}{1+\frac{1}{1+\frac{1}{1+1}}}=1+\dfrac{1}{\frac{5}{3}}=1+\dfrac{3}{5}=\dfrac{8}{5}$

Section 6.6 Practice Exercises

1. $\dfrac{2}{x-3}-\dfrac{3}{x^2-x-6}$
$=\dfrac{2(x+2)}{(x-3)(x+2)}-\dfrac{3}{(x-3)(x+2)}$
$=\dfrac{2x+1}{(x-3)(x+2)}$

3. $\dfrac{t^2-5t+6}{t^2-5t-6}\div\dfrac{t^2-4}{t^2+2t+1}$
$=\dfrac{(t-3)(t-2)}{(t-6)(t+1)}\cdot\dfrac{(t+1)(t+1)}{(t-2)(t+2)}$
$=\dfrac{(t-3)(t+1)}{(t-6)(t+2)}$

5. $\dfrac{h-\frac{1}{h}}{\frac{1}{5}-\frac{1}{5h}}=\dfrac{5h\left(h-\frac{1}{h}\right)}{5h\left(\frac{1}{5}-\frac{1}{5h}\right)}$
$=\dfrac{5h^2-5}{h-1}$
$=\dfrac{5(h+1)(h-1)}{(h-1)}$
$=5(h+1)$

7. $\dfrac{1}{3}z+\dfrac{2}{3}=-2z+10$
$3\left(\dfrac{1}{3}z+\dfrac{2}{3}\right)=3(-2z+10)$
$z+2=-6z+30$
$7z=28$
$z=4$

9. $\dfrac{3}{2}p+\dfrac{1}{3}=\dfrac{2p-3}{4}$
$12\left(\dfrac{3}{2}p+\dfrac{1}{3}\right)=12\left(\dfrac{2p-3}{4}\right)$
$18p+4=6p-9$
$12p=-13$
$p=-\dfrac{13}{12}$

11.
$$\frac{2x-3}{4}+\frac{9}{10}=\frac{x}{5}$$
$$20\left(\frac{2x-3}{4}+\frac{9}{10}\right)=20\left(\frac{x}{5}\right)$$
$$10x-15+18=4x$$
$$6x=-3$$
$$x=-\frac{1}{2}$$

13. (a) LCD: $4w$

(b)
$$\frac{1}{w}-\frac{1}{2}=-\frac{1}{4}$$
$$4w\left(\frac{1}{w}-\frac{1}{2}\right)=4w\left(-\frac{1}{4}\right)$$
$$4-2w=-w$$
$$w=4$$

15. (a) LCD: $(x+3)(x-1)$

(b)
$$\frac{x+1}{(x+3)(x-1)}=\frac{1}{x+3}-\frac{1}{x-1}$$
$$(x+3)(x-1)\left[\frac{x+1}{(x+3)(x-1)}\right]=(x+3)(x-1)\left(\frac{1}{x+3}-\frac{1}{x-1}\right)$$
$$x+1=x-1-x-3$$
$$x=-5$$

17.
$$\frac{1}{8}=\frac{3}{5}+\frac{5}{y}$$
$$40y\left(\frac{1}{8}\right)=40y\left(\frac{3}{5}+\frac{5}{y}\right)$$
$$5y=24y+200$$
$$-19y=200$$
$$y=-\frac{200}{19}$$

19.
$$\frac{4}{t}=\frac{3}{t}+\frac{1}{8}$$
$$8t\left(\frac{4}{t}\right)=8t\left(\frac{3}{t}+\frac{1}{8}\right)$$
$$32=24+t$$
$$t=8$$

21. $$\frac{5}{6x}+\frac{7}{x}=1$$
$$6x\left(\frac{5}{6x}+\frac{7}{x}\right)=6x(1)$$
$$5+42=6x$$
$$6x=47$$
$$x=\frac{47}{6}$$

23. $$1-\frac{2}{y}=\frac{3}{y^2}$$
$$y^2\left(1-\frac{2}{y}\right)=y^2\left(\frac{3}{y^2}\right)$$
$$y^2-2y=3$$
$$y^2-2y-3=0$$
$$(y-3)(y+1)=0$$
$$y-3=0 \quad \text{or} \quad y+1=0$$
$$y=3 \qquad y=-1$$

25. $$\frac{a+1}{a}=1+\frac{a-2}{2a}$$
$$2a\left(\frac{a+1}{a}\right)=2a\left(1+\frac{a-2}{2a}\right)$$
$$2a+2=2a+a-2$$
$$-a=-4$$
$$a=4$$

27. $$\frac{w}{5}-\frac{w+3}{w}=-\frac{3}{w}$$
$$5w\left(\frac{w}{5}-\frac{w+3}{w}\right)=5w\left(-\frac{3}{w}\right)$$
$$w^2-5w-15=-15$$
$$w^2-5w=0$$
$$w(w-5)=0$$
$w=0$ or $w=5$

$w=0$ is extraneous, $w=5$ is the solution.

29. $$\frac{2}{m+3}=\frac{5}{4m+12}-\frac{3}{8}$$
$$\frac{2}{m+3}=\frac{5}{4(m+3)}-\frac{3}{8}$$
$$8(m+3)\cdot\frac{2}{m+3}=8(m+3)\left(\frac{5}{4(m+3)}-\frac{3}{8}\right)$$
$$16=8(m+3)\left(\frac{5}{4(m+3)}\right)-8(m+3)\cdot\frac{3}{8}$$
$$16=10-3m-9$$
$$15=-3m$$
$$m=-5$$

31. $$\frac{p}{p-4}-5=\frac{4}{p-4}$$
$$(p-4)\left(\frac{p}{p-4}-5\right)=(p-4)\cdot\frac{4}{p-4}$$
$$(p-4)\cdot\frac{p}{p-4}-5(p-4)=4$$
$$p-5p+20=4$$
$$-4p=-16$$
$$p=4$$
No solution; $p=4$ is extraneous.

33. $$\frac{2t}{t+2}-2=\frac{t-8}{t+2}$$
$$(t+2\left(\frac{2t}{t+2}-2\right)=(t+2)\cdot\frac{t-8}{t+2}$$
$$(t+2)\frac{2t}{t+2}-2(t+2)=t-8$$
$$2t-2t-4=t-8$$
$$-4=t-8$$
$$t=4$$

35. $$\frac{x^2-x}{x-2}=\frac{12}{x-2}$$
$$(x-2)\cdot\frac{x^2-x}{x-2}=(x-2)\cdot\frac{12}{x-2}$$
$$x^2-x=12$$
$$x^2-x-12=0$$
$$(x-4)(x+3)=0$$
$x=4$ or $x=-3$

37. $$\frac{x^2+3x}{x-1}=\frac{4}{x-1}$$
$$(x-1)\cdot\frac{x^2+3x}{x-1}=(x-1)\cdot\frac{4}{x-1}$$
$$x^2+3x=4$$
$$x^2+3x-4=0$$
$$(x+4)(x-1)=0$$
$x+4=0$ or $x-1=0$

$x=-4$ is the solution ($x=1$ is extraneous).

39. $$\frac{2x}{x+4}-\frac{8}{x-4}=\frac{2x^2+32}{x^2-16}$$
$$\frac{2x}{x+4}-\frac{8}{x-4}=\frac{2x^2+32}{(x+4)(x-4)}$$
Multiply both sides by LCD: $(x+4)(x-4)$
$$2x(x-4)-8(x+4)=2x^2+32$$
$$2x^2-8x-8x-32=2x^2+32$$
$$-16x=64$$
$$x=-4$$
No solution ($x=-4$ is extraneous).

41. $$\frac{x}{x+6}=\frac{72}{x^2-36}+4$$
$$\frac{x}{x+6}=\frac{72}{(x+6)(x-6)}+4$$
Multiply both sides by LCD: $(x+6)(x-6)$
$$x(x-6)=72+4(x+6)(x-6)$$
$$x^2-6x=72+4x^2-144$$
$$-3x^2-6x+72=0$$
$$-3(x^2+2x-24)=0$$
$$(x+6)(x-4)=0$$
$x+6=0$ or $x-4=0$

$x=4$ is the solution ($x=-6$ is extraneous).

43. Let x = a number. Then,
$$\frac{1}{x}+3=\frac{25}{x}$$
$$x\left(\frac{1}{x}+3\right)=x\cdot\frac{25}{x}$$
$$1+3x=25$$
$$3x=24$$
$$x=8$$
The number is 8.

45. Let x = a number. Then,
$$\frac{x+5}{x-2}=\frac{3}{4}$$
$$4(x-2)\cdot\frac{x+5}{x-2}=4(x-2)\cdot\frac{3}{4}$$
$$4x+20=3x-6$$
$$x=-26$$
The number is –26.

47. $$K=\frac{ma}{F}$$
$$FK=ma$$
$$m=\frac{FK}{a}$$

49. $$K=\frac{IR}{E}$$
$$EK=IR$$
$$E=\frac{IR}{K}$$

51. $$I=\frac{E}{R+r}$$
$$I(R+r)=E$$
$$IR+Ir=E$$
$$IR=E-Ir$$
$$R=\frac{E-Ir}{I}$$

53. $$h=\frac{2A}{B+b}$$
$$h(B+b)=2A$$
$$Bh+bh=2A$$
$$Bh=2A-bh$$
$$B=\frac{2A-bh}{h}$$

55. $\frac{V}{\pi h}=r^2$

$V=r^2\pi h$

$h=\frac{V}{r^2\pi}$

57. $x=\frac{at+b}{t}$

$xt=at+b$

$xt-at=b$

$(x-a)t=b$

$t=\frac{b}{x-a}$

59. $\frac{x-y}{xy}=z$

$x-y=xyz$

$x-xyz=y$

$x(1-yz)=y$

$x=\frac{y}{1-yz}$

61. $a+b=\frac{2A}{h}$

$h(a+b)=2A$

$h=\frac{2A}{a+b}$

63. $\frac{1}{R}=\frac{1}{R_1}+\frac{1}{R_2}$

$RR_1R_2\cdot\frac{1}{R}=RR_1R_2\left(\frac{1}{R_1}+\frac{1}{R_2}\right)$

$R_1R_2=RR_2+RR_1$

$R_1R_2=R(R_2+R_1)$

$R=\frac{R_1R_2}{R_1+R_2}$

65. $v=\frac{s_2-s_1}{t_2-t_1}$

$v(t_2-t_1)=s_2-s_1$

$vt_2-vt_1=s_2-s_1$

$vt_2=s_2-s_1+vt_1$

$t_2=\frac{s_2-s_1+vt_1}{v}$

Section 6.7 Practice Exercises

1. Equation; $\frac{b}{5}+3=9$

$5\left(\frac{b}{5}+3\right)=5(9)$

$b+15=45$

$b=30$

3. Expression; $\frac{2}{a+5}+\frac{5}{a^2-25}$

$=\frac{2(a-5)}{(a+5)(a-5)}+\frac{5}{(a+5)(a-5)}$

$=\frac{2a-10+5}{(a+5)(a-5)}$

$=\frac{2a-5}{(a+5)(a-5)}$

5. Expression;

$\frac{3y+6}{20}\div\frac{4y+8}{8}=\frac{3(y+2)}{20}\cdot\frac{8}{4(y+2)}=\frac{3}{10}$

7. Equation; $\frac{3}{p+3}=\frac{12p+19}{p^2+7p+12}-\frac{5}{p+4}$

$\frac{3}{p+3}=\frac{12p+19}{(p+3)(p+4)}-\frac{5}{p+4}$

Multiply both sides by the LCD $(p+3)(p+4)$

$3(p+4)=12p+19-5(p+3)$

$3p+12=12p+19-5p-15$

$3p+12=7p+4$

$-4p=-8$

$p=2$

9. $\frac{5}{3}=\frac{a}{8}$

$24\cdot\frac{5}{3}=24\cdot\frac{a}{8}$

$40=3a$

$a=\frac{40}{3}$

11. $\frac{2}{1.9}=\frac{x}{38}$

$1.9x=76$

$x=\frac{76}{1.9}=40$

13. $\frac{y+1}{2y}=\frac{2}{3}$

$6y\left(\frac{y+1}{2y}\right)=6y\left(\frac{2}{3}\right)$

$3(y+1)=4y$

$3y+3=4y$

$3=y$

15. $\frac{9}{2z-1}=\frac{3}{z}$

$z(2z-1)\left(\frac{9}{2z-1}\right)=z(2z-1)\left(\frac{3}{z}\right)$

$9z=3(2z-1)$

$9z=6z-3$

$3z=-3$

$z=-1$

17. $\frac{8}{9a-1}=\frac{5}{3a+2}$

$8(3a+2)=5(9a-1)$

$24a+16=45a-5$

$-21a=-21$

$a=1$

19. (a) $\frac{V_i}{V_f}=\frac{T_i}{T_f}$

$V_iT_f=T_iV_f$

$V_f=\frac{V_iT_f}{T_i}$

(b) $\frac{V_i}{V_f}=\frac{T_i}{T_f}$

$V_iT_f=T_iV_f$

$T_f=\frac{T_iV_f}{V_i}$

21. Let x = her score. Then,

$\frac{22\text{ score}}{4\text{ holes}}=\frac{x\text{ score}}{18\text{ holes}}$

$\frac{22}{4}=\frac{x}{18}$

$36\cdot\frac{22}{4}=36\cdot\frac{x}{18}$

$198=2x$

$x=99$

Her score would be 99.

23. Let x = width of the garden.

$\frac{5}{3}=\frac{8}{x}$

$3x\cdot\frac{5}{3}=3x\cdot\frac{8}{x}$

$5x=24$

$x=\frac{24}{5}=4.8$ feet

25. Let x = no. of miles. Then,

$\frac{75}{1}=\frac{x}{3.5}$

$3.5\cdot75=3.5\cdot\frac{x}{3.5}$

$x=262.5$ miles

27. Let x = number of red M&Ms. Then,

$\frac{12}{80}=\frac{x}{200}$

$400\cdot\frac{12}{80}=400\cdot\frac{x}{200}$

$60=2x$

$x=30$ red M&Ms

29. Let x = the number of incorrect ballots.

$\frac{8}{5000}=\frac{x}{2600000}$

$2600000\cdot\frac{8}{5000}=2600000\cdot\frac{x}{2600000}$

$x=4160$ incorrect ballots

31. Let x = speed of current.

	Distance	Rate	Time
Downstream	66	$20 + x$	$\frac{66}{20+x}$
Upstream	54	$20 - x$	$\frac{54}{20-x}$

$$\frac{66}{20+x} = \frac{54}{20-x}$$
$$(20+x)(20-x)\frac{66}{20+x} = (20+x)(20-x)\frac{54}{20-x}$$
$$66(20-x) = 54(20+x)$$
$$1320 - 66x = 1080 + 54x$$
$$-120x = -240$$
$$x = 2$$

The speed is 2 mph.

33. Let x = speed of plane.

	Distance	Rate	Time
With wind	370	$20 + x$	$\frac{370}{20+x}$
Against wind	290	$x - 20$	$\frac{290}{x-20}$

$$\frac{370}{x+20} = \frac{290}{x-20}$$
$$(x-20)(x+20)\frac{370}{20+x} = (x-20)(x+20)\frac{290}{x-20}$$
$$370(x-20) = 290(x+20)$$
$$370x - 7400 = 290x + 5800$$
$$80x = 13200$$
$$x = 165$$

The speed of the plane is 165 mph.

35. Let x = speed of slower motorist.

	Distance	Rate	Time
Slower motorist	270	x	$\frac{270}{x}$
Faster motorist	360	$x + 15$	$\frac{360}{x+15}$

$$\frac{270}{x} = \frac{360}{x+15}$$
$$x(x+15)\frac{270}{x} = x(x+15)\frac{360}{x+15}$$
$$270(x+15) = 360x$$
$$270x + 4050 = 360x$$
$$-90x = -4050$$
$$x = 45$$

The speeds are 45 mph and 60 mph.

37. Let x = Shanelle's speed. Then, Devon's speed = $(x + 5)$.

$$\left(\frac{\text{distance}}{\text{Shanelle's speed}}\right) = \left(\frac{\text{distance}}{\text{Devon's speed}}\right)$$
$$\frac{30}{x} = \frac{45}{x+5}$$
$$45x = 30(x+5)$$
$$45x = 30x + 150$$
$$15x = 150$$
$$x = 10$$

Shanelle skis 10 km/hr and Devon skis 15 km/hr.

39. If it takes 2 hours to paint a room, then $\frac{1}{2}$ of the job is completed in 1 hour.

41. In one minute, the cold water can fill $\frac{1}{10}$ of the sink; the hot water can fill $\frac{1}{12}$ of the sink. If x = how long it would take both faucets to fill the sink together, then both faucets can fill $\frac{1}{x}$ of the sink.

$$\frac{1}{10}+\frac{1}{12}=\frac{1}{x}$$
$$60x\left(\frac{1}{10}+\frac{1}{12}\right)=60x\left(\frac{1}{x}\right)$$
$$60x\left(\frac{1}{10}\right)+60x\left(\frac{1}{12}\right)=60x\left(\frac{1}{x}\right)$$
$$6x+5x=60$$
$$11x=60$$
$$x=\frac{60}{11}=5\frac{5}{11}$$

Both faucets can fill the sink in $5\frac{5}{11}$ minutes.

43. In one minute, one printer can do $\frac{1}{50}$ of the job; the other printer can do $\frac{1}{40}$ of the job. If x = how long it takes both printers to do the job together, then $\frac{1}{x}$ of the job can be completed in 1 minute.

$$\frac{1}{50}+\frac{1}{40}=\frac{1}{x}$$
$$200x\left(\frac{1}{50}+\frac{1}{40}\right)=200x\left(\frac{1}{x}\right)$$
$$4x+5x=200$$
$$9x=200$$
$$x=\frac{200}{9}=22\frac{2}{9}$$

Together they can do the job in $22\frac{2}{9}$ minutes.

45. Let x = how long it will take Al. Then, in 1 day, Al can complete $\frac{1}{x}$, Tim can complete $\frac{1}{5}$, and together they can complete $\frac{1}{2}$ of the job.

$$\frac{1}{x}+\frac{1}{5}=\frac{1}{2}$$
$$10x\left(\frac{1}{x}+\frac{1}{5}\right)=10x\left(\frac{1}{2}\right)$$
$$10+2x=5x$$
$$3x=10$$
$$x=\frac{10}{3}=3\frac{1}{3}$$

It would take Al $3\frac{1}{3}$ days.

47. **(a)** $\frac{15}{3}=\frac{20}{x}$

$$15x=60$$
$$x=4 \text{ cm}$$

(b) $\frac{15}{3}=\frac{25}{x}$

$$15x=75$$
$$x=5 \text{ cm}$$

49. $\frac{x}{15}=\frac{3}{12}$

$$12x=45$$
$$x=3.75 \text{ cm}$$

$$\frac{y}{18}=\frac{3}{12}$$
$$12y=54$$
$$y=4.5 \text{ cm}$$

51. $\frac{x}{16.8}=\frac{1}{2.4}$

$$2.4x=16.8$$
$$x=7$$

The height of the pole is 7 m.

53. Let x = the height of the post. Then,

$$\frac{x}{54+18}=\frac{6}{18}$$
$$\frac{x}{72}=\frac{6}{18}$$
$$18x=432$$
$$x=24$$

The pole is 24 ft.

Chapter 6 Review Exercises

1. (a) $\frac{0-2}{0+9}=-\frac{2}{9};\ \frac{1-2}{1+9}=-\frac{1}{10};\ \frac{2-2}{2+9}=0;$
$\frac{-3-2}{-3+9}=-\frac{5}{6};\ \frac{-9-2}{-9+9}=\frac{-11}{0}$ undefined

(b) $t+9\neq 0,\ \{t\mid t\neq -9\}$

3. (a) $\frac{2-1}{1-2}=-1$

(b) $\frac{-1-5}{-1+5}=\frac{-6}{4}=-\frac{3}{2}$

(c) $\frac{-x-7}{x+7}=\frac{-(-1)-7}{-1+7}=-1$

(d) $\frac{(-1)^2-4}{4-(-1)^2}=-1$

a, c, d are the expressions equal to −1.

5. $\left\{h\,\middle|\,h\neq -\frac{1}{3},\ h\neq -7\right\}$
$\frac{h+7}{(3h+1)(h+7)}=\frac{1}{3h+1}$

7. $\{w\mid w\neq -4,\ w\neq 4\}$
$\frac{2w^2+11w+12}{w^2-16}=\frac{(2w+3)(w+4)}{(w+4)(w-4)}=\frac{2w+3}{w-4}$

9. $\{k\mid k\neq 0,\ k\neq 5\}$
$\frac{15-3k}{2k^2-10k}=\frac{-3(k-5)}{2k(k-5)}=-\frac{3}{2k}$

11. $\{m\mid m\neq -1\}$
$\frac{3m^2-12m-15}{9m+9}=\frac{3(m+1)(m-5)}{9(m+1)}=\frac{m-5}{3}$

13. $\{p\mid p\neq -7\}$
$\frac{p+7}{p^2+14p+49}=\frac{p+7}{(p+7)^2}=\frac{1}{p+7}$

15. $\frac{2u+10}{u}\cdot\frac{u^3}{4u+20}=\frac{2(u+5)}{u}\cdot\frac{u^3}{4(u+5)}=\frac{u^2}{2}$

17. $\frac{8}{x^2-25}\cdot\frac{3x+15}{16}=\frac{8}{(x+5)(x-5)}\cdot\frac{3(x+5)}{16}$
$=\frac{3}{2(x-5)}$

19. $\frac{q^2-5q+6}{2q+4}\div\frac{2q-6}{q+2}$
$=\frac{(q-3)(q-2)}{2(q+2)}\cdot\frac{q+2}{2(q-3)}$
$=\frac{q-2}{4}$

21. $(s^2-6s+8)\left(\frac{4s}{s-2}\right)=\frac{(s-4)(s-2)}{1}\cdot\frac{4s}{s-2}$
$=4s(s-4)$

23. $\frac{\frac{n^2+n+1}{n^2-4}}{\frac{n^2+n+1}{n+2}}=\frac{n^2+n+1}{(n+2)(n-2)}\cdot\frac{n+2}{n^2+n+1}=\frac{1}{n-2}$

25. $\frac{3m-3}{6m^2+18m+12}\cdot\frac{2m^2-8}{m^2-3m+2}\div\frac{m+3}{m+1}$
$=\frac{3(m-1)}{6(m+2)(m+1)}\cdot\frac{2(m+2)(m-2)}{(m-2)(m-1)}\cdot\frac{m+1}{m+3}$
$=\frac{1}{m+3}$

27. $\frac{4y^2-1}{1+2y}\div\frac{y^2-4y-5}{5-y}$
$=\frac{(2y+1)(2y-1)}{2y+1}\cdot\frac{-1(y-5)}{(y-5)(y+1)}$
$=-\frac{2y-1}{y+1}$

29. $\frac{y+2}{y-3}=\frac{2(y+2)}{2y-6}=\frac{2y+4}{2y-6}$

31. $\frac{2}{r}=\frac{2(r+3)}{r^2+3r}=\frac{2r+6}{r^2+3r}$

33. $\dfrac{u+1}{u+6}=\dfrac{(u+1)(u-6)}{u^2-36}=\dfrac{u^2-5u-6}{u^2-36}$

35. LCD: xy^2z^4

37. LCD: $q(q+8)$

39. LCD: $(n-3)(n+3)(n+2)$

41. LCD: $3k-1$ or $1-3k$

43. LCD: $3-x$ or $x-3$

45. $$\begin{aligned}\frac{b-6}{b-2}+\frac{b+2}{b-2}&=\frac{b-6+b+2}{b-2}\\&=\frac{2b-4}{b-2}\\&=\frac{2(b-2)}{b-2}\\&=2\end{aligned}$$

47. $$\begin{aligned}\frac{x^2}{x+7}-\frac{49}{x+7}&=\frac{x^2-49}{x+7}\\&=\frac{(x+7)(x-7)}{x+7}\\&=x-7\end{aligned}$$

49. $$\begin{aligned}\frac{3}{4-t^2}+\frac{t}{2-t}&=\frac{3}{(2+t)(2-t)}+\frac{t(2+t)}{(2+t)(2-t)}\\&=\frac{3+2t+t^2}{(2+t)(2-t)}\\&=\frac{t^2+2t+3}{(2+t)(2-t)}\end{aligned}$$

51. $$\begin{aligned}\frac{5}{2r+12}-\frac{1}{r}&=\frac{5}{2(r+6)}-\frac{1}{r}\\&=\frac{5r}{2r(r+6)}-\frac{2(r+6)}{2r(r+6)}\\&=\frac{5r-2r-12}{2r(r+6)}\\&=\frac{3r-12}{2r(r+6)}\\&=\frac{3(r-4)}{2r(r+6)}\end{aligned}$$

53. $$\begin{aligned}&\frac{3q}{q^2+7q+10}-\frac{2q}{q^2+6q+8}\\&=\frac{3q}{(q+5)(q+2)}-\frac{2q}{(q+4)(q+2)}\\&=\frac{3q(q+4)}{(q+5)(q+4)(q+2)}-\frac{2q(q+5)}{(q+5)(q+4)(q+2)}\\&=\frac{3q^2+12q-2q^2-10q}{(q+5)(q+4)(q+2)}\\&=\frac{q^2+2q}{(q+5)(q+4)(q+2)}\\&=\frac{q(q+2)}{(q+5)(q+4)(q+2)}\\&=\frac{q}{(q+5)(q+4)}\end{aligned}$$

55. $$\begin{aligned}&\frac{x}{3x+9}-\frac{3}{x^2+3x}+\frac{1}{x}\\&=\frac{x}{3(x+3)}-\frac{3}{x(x+3)}+\frac{1}{x}\\&=\frac{x^2}{3x(x+3)}-\frac{9}{3x(x+3)}+\frac{3(x+3)}{3x(x+3)}\\&=\frac{x^2-9+3x+9}{3x(x+3)}\\&=\frac{x^2+3x}{3x(x+3)}\\&=\frac{x(x+3)}{3x(x+3)}\\&=\frac{1}{3}\end{aligned}$$

57. $\dfrac{\frac{z+5}{z}}{\frac{z-5}{3}}=\dfrac{z+5}{z}\cdot\dfrac{3}{z-5}=\dfrac{3(z+5)}{z(z-5)}$

59. $\frac{\frac{2}{y}+6}{\frac{3y+1}{4}}=\frac{4y\left(\frac{2}{y}+6\right)}{4y\left(\frac{3y+1}{4}\right)}$
$=\frac{8+24y}{3y^2+y}$
$=\frac{8(1+3y)}{y(3y+1)}$
$=\frac{8}{y}$

61. $\frac{\frac{b}{a}-\frac{a}{b}}{\frac{1}{b}-\frac{1}{a}}=\frac{ab\left(\frac{b}{a}-\frac{a}{b}\right)}{ab\left(\frac{1}{b}-\frac{1}{a}\right)}$
$=\frac{b^2-a^2}{a-b}$
$=\frac{-(a+b)(a-b)}{a-b}$
$=-(a+b)$

63. $\frac{\frac{25}{k+5}+5}{\frac{5}{k+5}-5}=\frac{(k+5)\left(\frac{25}{k+5}+5\right)}{(k+5)\left(\frac{5}{k+5}-5\right)}$
$=\frac{25+5(k+5)}{5-5(k+5)}$
$=\frac{5k+50}{-5k-20}$
$=\frac{5(k+10)}{-5(k+4)}$
$=-\frac{k+10}{k+4}$

65. $\frac{1}{y}+\frac{3}{4}=\frac{1}{4}$
$4\left(\frac{1}{y}+\frac{3}{4}\right)=4\cdot\frac{1}{4}$
$\frac{4}{y}+3=1$
$\frac{4}{y}=-2$
$y=\frac{4}{-2}=-2$

67. $\frac{w}{w-1}=\frac{3}{w+1}+1$
Multiply both sides by LCD: $(w+1)(w-1)$
$w(w+1)=3(w-1)+(w+1)(w-1)$
$w^2+w=3w-3+w^2-1$
$w=3w-4$
$-2w=-4$
$w=2$

69. $\frac{4p-4}{p^2+5p-14}+\frac{2}{p+7}=\frac{1}{p-2}$
$\frac{4p-4}{(p+7)(p-2)}+\frac{2}{p+7}=\frac{1}{p-2}$
Multiply both sides by LCD: $(p+7)(p-2)$
$4p-4+2(p-2)=p+7$
$4p-4+2p-4=p+7$
$5p=15$
$p=3$

71. $\frac{y+1}{y+3}=\frac{y^2-11y}{y^2+y-6}-\frac{y-3}{y-2}$
$\frac{y+1}{y+3}=\frac{y^2-11y}{(y+3)(y-2)}-\frac{y-3}{y-2}$
Multiply both sides by LCD: $(y+3)(y-2)$
$(y-2)(y+1)=y^2-11y-(y+3)(y-3)$
$y^2-y-2=y^2-11y-y^2+9$
$y^2-y-2=-11y+9$
$y^2+10y-11=0$
$(y-1)(y+11)=0$
$y-1=0$ or $y+11=0$
$y=1$ or $y=-11$

73. $\frac{V}{h}=\frac{\pi r^2}{3}$
$3h\left(\frac{V}{h}\right)=3h\left(\frac{\pi r^2}{3}\right)$
$3V=h\pi r^2$
$h=\frac{3V}{\pi r^2}$

33. $\frac{u+1}{u+6}=\frac{(u+1)(u-6)}{u^2-36}=\frac{u^2-5u-6}{u^2-36}$

35. LCD: xy^2z^4

37. LCD: $q(q+8)$

39. LCD: $(n-3)(n+3)(n+2)$

41. LCD: $3k-1$ or $1-3k$

43. LCD: $3-x$ or $x-3$

45.
$$\begin{aligned}\frac{b-6}{b-2}+\frac{b+2}{b-2}&=\frac{b-6+b+2}{b-2}\\&=\frac{2b-4}{b-2}\\&=\frac{2(b-2)}{b-2}\\&=2\end{aligned}$$

47.
$$\begin{aligned}\frac{x^2}{x+7}-\frac{49}{x+7}&=\frac{x^2-49}{x+7}\\&=\frac{(x+7)(x-7)}{x+7}\\&=x-7\end{aligned}$$

49.
$$\begin{aligned}\frac{3}{4-t^2}+\frac{t}{2-t}&=\frac{3}{(2+t)(2-t)}+\frac{t(2+t)}{(2+t)(2-t)}\\&=\frac{3+2t+t^2}{(2+t)(2-t)}\\&=\frac{t^2+2t+3}{(2+t)(2-t)}\end{aligned}$$

51.
$$\begin{aligned}\frac{5}{2r+12}-\frac{1}{r}&=\frac{5}{2(r+6)}-\frac{1}{r}\\&=\frac{5r}{2r(r+6)}-\frac{2(r+6)}{2r(r+6)}\\&=\frac{5r-2r-12}{2r(r+6)}\\&=\frac{3r-12}{2r(r+6)}\\&=\frac{3(r-4)}{2r(r+6)}\end{aligned}$$

53.
$$\begin{aligned}&\frac{3q}{q^2+7q+10}-\frac{2q}{q^2+6q+8}\\&=\frac{3q}{(q+5)(q+2)}-\frac{2q}{(q+4)(q+2)}\\&=\frac{3q(q+4)}{(q+5)(q+4)(q+2)}-\frac{2q(q+5)}{(q+5)(q+4)(q+2)}\\&=\frac{3q^2+12q-2q^2-10q}{(q+5)(q+4)(q+2)}\\&=\frac{q^2+2q}{(q+5)(q+4)(q+2)}\\&=\frac{q(q+2)}{(q+5)(q+4)(q+2)}\\&=\frac{q}{(q+5)(q+4)}\end{aligned}$$

55.
$$\begin{aligned}&\frac{x}{3x+9}-\frac{3}{x^2+3x}+\frac{1}{x}\\&=\frac{x}{3(x+3)}-\frac{3}{x(x+3)}+\frac{1}{x}\\&=\frac{x^2}{3x(x+3)}-\frac{9}{3x(x+3)}+\frac{3(x+3)}{3x(x+3)}\\&=\frac{x^2-9+3x+9}{3x(x+3)}\\&=\frac{x^2+3x}{3x(x+3)}\\&=\frac{x(x+3)}{3x(x+3)}\\&=\frac{1}{3}\end{aligned}$$

57. $\frac{\frac{z+5}{z}}{\frac{z-5}{3}}=\frac{z+5}{z}\cdot\frac{3}{z-5}=\frac{3(z+5)}{z(z-5)}$

59. $\dfrac{\frac{2}{y}+6}{\frac{3y+1}{4}} = \dfrac{4y\left(\frac{2}{y}+6\right)}{4y\left(\frac{3y+1}{4}\right)}$
$= \dfrac{8+24y}{3y^2+y}$
$= \dfrac{8(1+3y)}{y(3y+1)}$
$= \dfrac{8}{y}$

61. $\dfrac{\frac{b}{a}-\frac{a}{b}}{\frac{1}{b}-\frac{1}{a}} = \dfrac{ab\left(\frac{b}{a}-\frac{a}{b}\right)}{ab\left(\frac{1}{b}-\frac{1}{a}\right)}$
$= \dfrac{b^2-a^2}{a-b}$
$= \dfrac{-(a+b)(a-b)}{a-b}$
$= -(a+b)$

63. $\dfrac{\frac{25}{k+5}+5}{\frac{5}{k+5}-5} = \dfrac{(k+5)\left(\frac{25}{k+5}+5\right)}{(k+5)\left(\frac{5}{k+5}-5\right)}$
$= \dfrac{25+5(k+5)}{5-5(k+5)}$
$= \dfrac{5k+50}{-5k-20}$
$= \dfrac{5(k+10)}{-5(k+4)}$
$= -\dfrac{k+10}{k+4}$

65. $\dfrac{1}{y}+\dfrac{3}{4}=\dfrac{1}{4}$
$4\left(\dfrac{1}{y}+\dfrac{3}{4}\right)=4\cdot\dfrac{1}{4}$
$\dfrac{4}{y}+3=1$
$\dfrac{4}{y}=-2$
$y=\dfrac{4}{-2}=-2$

67. $\dfrac{w}{w-1}=\dfrac{3}{w+1}+1$
Multiply both sides by LCD: $(w+1)(w-1)$
$w(w+1)=3(w-1)+(w+1)(w-1)$
$w^2+w=3w-3+w^2-1$
$w=3w-4$
$-2w=-4$
$w=2$

69. $\dfrac{4p-4}{p^2+5p-14}+\dfrac{2}{p+7}=\dfrac{1}{p-2}$
$\dfrac{4p-4}{(p+7)(p-2)}+\dfrac{2}{p+7}=\dfrac{1}{p-2}$
Multiply both sides by LCD: $(p+7)(p-2)$
$4p-4+2(p-2)=p+7$
$4p-4+2p-4=p+7$
$5p=15$
$p=3$

71. $\dfrac{y+1}{y+3}=\dfrac{y^2-11y}{y^2+y-6}-\dfrac{y-3}{y-2}$
$\dfrac{y+1}{y+3}=\dfrac{y^2-11y}{(y+3)(y-2)}-\dfrac{y-3}{y-2}$
Multiply both sides by LCD: $(y+3)(y-2)$
$(y-2)(y+1)=y^2-11y-(y+3)(y-3)$
$y^2-y-2=y^2-11y-y^2+9$
$y^2-y-2=-11y+9$
$y^2+10y-11=0$
$(y-1)(y+11)=0$
$y-1=0$ or $y+11=0$
$y=1$ or $y=-11$

73. $\dfrac{V}{h}=\dfrac{\pi r^2}{3}$
$3h\left(\dfrac{V}{h}\right)=3h\left(\dfrac{\pi r^2}{3}\right)$
$3V=h\pi r^2$
$h=\dfrac{3V}{\pi r^2}$

75. $\frac{m+2}{8}=\frac{m}{3}$

$$3(m+2)=8m$$
$$3m+6=8m$$
$$6=5m$$
$$m=\frac{6}{5}$$

77. Let x = grams of fat in 6 oz.

$$\frac{4\text{ g}}{2\text{ oz}}=\frac{x}{6\text{ oz}}$$
$$2x=24$$
$$x=12$$

12 grams in 6 oz bag.

79. Let x = time to fill pool if both pumps are working together. Then in 1 minute the first pump can fill $\frac{1}{24}$ of the pool, the second pump can fill $\frac{1}{56}$ of the pool, and together they can fill $\frac{1}{x}$ of the pool.

$$\frac{1}{24}+\frac{1}{56}=\frac{1}{x}$$
$$168x\left(\frac{1}{24}+\frac{1}{56}\right)=168x\left(\frac{1}{x}\right)$$
$$7x+3x=168$$
$$10x=168$$
$$x=16.8$$

Together both pumps can fill the pool in 16.8 hours.

Chapter 6 Test

1. **(a)** $\{x \mid x \neq 2\}$

(b) $\frac{5(x-2)(x+1)}{30(2-x)}=\frac{5(x-2)(x+1)}{-30(x-2)}=-\frac{x+1}{6}$

3. **(a)** $\frac{-1+4}{-1-4}=\frac{3}{-5}=-\frac{3}{5}$

(b) $\frac{7-2(-1)}{2(-1)-7}=\frac{9}{-9}=-1$

(c) $\frac{9(-1)^2+16}{-9(-1)^2-16}=\frac{25}{-25}=-1$

(d) $-\frac{-1+5}{-1+5}=-\frac{4}{4}=-1$

b, c and d are the expressions equal to -1.

5. $\frac{9-b^2}{5b+15}\div\frac{b-3}{b+3}=\frac{-(b+3)(b-3)}{5(b+3)}\cdot\frac{b+3}{b-3}$

$$=-\frac{b+3}{5}$$

7. $\frac{t}{t-2}-\frac{8}{t^2-4}=\frac{t}{t-2}-\frac{8}{(t+2)(t-2)}$

$$=\frac{t(t+2)}{(t+2)(t-2)}-\frac{8}{(t+2)(t-2)}$$
$$=\frac{t^2+2t-8}{(t+2)(t-2)}$$
$$=\frac{(t+4)(t-2)}{(t+2)(t-2)}$$
$$=\frac{t+4}{t+2}$$

9. $\frac{1-\frac{4}{m}}{m-\frac{16}{m}}=\frac{m\left(1-\frac{4}{m}\right)}{m\left(m-\frac{16}{m}\right)}$

$$=\frac{m-4}{m^2-16}$$
$$=\frac{m-4}{(m+4)(m-4)}$$
$$=\frac{1}{m+4}$$

11.

$$\frac{p}{p-1}+\frac{1}{p}=\frac{p^2+1}{p^2-p}$$
$$\frac{p}{p-1}+\frac{1}{p}=\frac{p^2+1}{p(p-1)}$$
$$p(p-1)\left(\frac{p}{p-1}+\frac{1}{p}\right)=p(p-1)\left(\frac{p^2+1}{p(p-1)}\right)$$
$$p^2+p-1=p^2+1$$
$$p-1=1$$
$$p=2$$

13. $$\frac{4x}{x-4}=3+\frac{16}{x-4}$$
$$(x-4)\left(\frac{4x}{x-4}\right)=(x-4)\left(3+\frac{16}{x-4}\right)$$
$$4x=3(x-4)+16$$
$$4x=3x-12+16$$
$$x=4$$
No solution ($x = 4$ does not check).

15. Let x = the number. Then,
$$\frac{3}{2}+\frac{1}{x}=\frac{2}{5}\cdot\frac{1}{x}$$
$$\frac{3}{2}+\frac{1}{x}=\frac{2}{5x}$$
$$10x\left(\frac{3}{2}+\frac{1}{x}\right)=10x\left(\frac{2}{5x}\right)$$
$$15x+10=4$$
$$15x=-6$$
$$x=-\frac{6}{15}=-\frac{2}{5}$$

17. Let x = cups of carrots. Then,
$$\frac{\frac{1}{2}\text{ cup}}{6\text{ servings}}=\frac{x}{15\text{ servings}}$$
$$6x=\frac{1}{2}\cdot 15$$
$$6x=\frac{15}{2}$$
$$x=\frac{1}{6}\cdot\frac{15}{2}=\frac{5}{4}=1\frac{1}{4}\text{ cups}$$

19. Let x = time it takes second printer to complete job. Then in one hour both printers can complete $\frac{1}{2}$ of the job, the first printer can complete $\frac{1}{6}$ of the job, and the second printer can complete $\frac{1}{x}$ of the job. Therefore,
$$\frac{1}{6}+\frac{1}{x}=\frac{1}{2}$$
$$12x\left(\frac{1}{6}+\frac{1}{x}\right)=12x\cdot\frac{1}{2}$$
$$2x+12=6x$$
$$12=4x$$
$$x=3$$
It takes the second printer 3 hours to complete the job alone.

21. (a) LCD: $15(x + 3)$

(b) LCD: $3x^2y^2$

Cumulative Review Exercises Chapters 1–6

1. $\left(\frac{1}{2}\right)^{-4}+2^4=2^4+2^4=16+16=32$

3. Rational: $\sqrt{4}=2, \sqrt{9}=3, \sqrt{16}=4, \sqrt{49}=7$
Irrational: $\sqrt{5}, \sqrt{20}$

5. $$-3(x-5)-2<-2x+5$$
$$-3x+15-2<-2x+5$$
$$-x<-8$$
$$x>8$$
$(8, \infty)$

8

7. Let x = width of the pool. Then, $(2x + 1)$ is the length.
$$P=2w+2l$$
$$104=2x+2(2x+1)$$
$$104=2x+4x+2$$
$$102=6x$$
$$x=17$$
The width is 17 m and the length is 35 m.

9. $$4x-3=3x+7$$
$$x=10$$
The angles are 37°.

11. $20^2 = 12^2 + 16^2$
$400 = 144 + 256$
$400 = 400$

13. $\left(\frac{4x^{-1}y^{-2}}{z^4}\right)^{-2}(2y^{-1}z^3)^3$
$= \left(\frac{z^4}{4x^{-1}y^{-2}}\right)^2\left(\frac{2z^3}{y}\right)^3$
$= \left(\frac{2^8}{16x^{-2}y^{-4}}\right)\left(\frac{8z^9}{y^3}\right)$
$= \left(\frac{2^8x^2y^4}{16}\right)\left(\frac{8z^9}{y^3}\right)$
$= \frac{x^2yz^{17}}{2}$

15. $(5x-3)^2 = (5x)^2 - 2(5x)(3) + 3^2$
$= 25x^2 - 30x + 9$

17. $\frac{8a^4b^2 - 2ab^3 + a^3b^2}{2ab^2}$
$= \frac{8a^2b^4}{2ab^2} - \frac{2ab^3}{2ab^2} + \frac{a^3b^2}{2ab^2}$
$= 4ab^2 - b + \frac{a^2}{2}$

19. $10cd + 5d - 6c - 3 = 5d(2c+1) - 3(2c+1)$
$= (2c+1)(5d-3)$

21. $\left\{x \middle| x \neq 5, x \neq -\frac{1}{2}\right\}$

23. $\frac{2x-6}{x^2-16} \div \frac{10x^2-90}{x^2-x-12}$
$= \frac{2(x-3)}{(x+4)(x-4)} \cdot \frac{(x+3)(x-4)}{10(x+3)(x-3)}$
$= \frac{1}{5(x+4)}$

25. $\frac{\frac{3}{4}-\frac{1}{x}}{\frac{1}{3x}-\frac{1}{4}} = \frac{12x\left(\frac{3}{4}-\frac{1}{x}\right)}{12x\left(\frac{1}{3x}-\frac{1}{4}\right)}$
$= \frac{9x-12}{4-3x}$
$= \frac{3(3x-4)}{-1(3x-4)}$
$= -3$

27. $\frac{2b-5}{6} = \frac{4b}{7}$
$7(2b-5) = 6(4b)$
$14b - 35 = 24b$
$-35 = 10b$
$b = \frac{-35}{10} = -\frac{7}{2}$

29. **(a)** linear

(b)

x	y
0	−10
2	0
1	−5

(c)

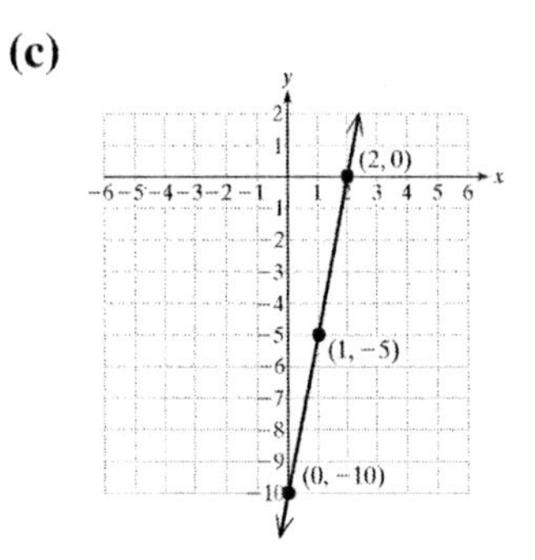

31. $2ab + a^2 + b^2 - 16$
$= a^2 + 2ab + b^2 - 16$
$= (a+b)^2 - 4^2$
$= ((a+b)+4)((a+b)-4)$
$= (a+b+4)(a+b-4)$

Chapter 7

Section 7.1 Practice Exercises

1. $4x + 2y = 8$

To find the x-intercept, substitute $y = 0$.

$$\begin{aligned} 4x + 2(0) &= 8 \\ 4x &= 8 \\ x &= 2 \end{aligned}$$

The x-intercept is (2, 0).

To find the y-intercept, substitute $x = 0$.

$$\begin{aligned} 4(0) + 2y &= 8 \\ 2y &= 8 \\ y &= 4 \end{aligned}$$

The y-intercept is (0, 4).

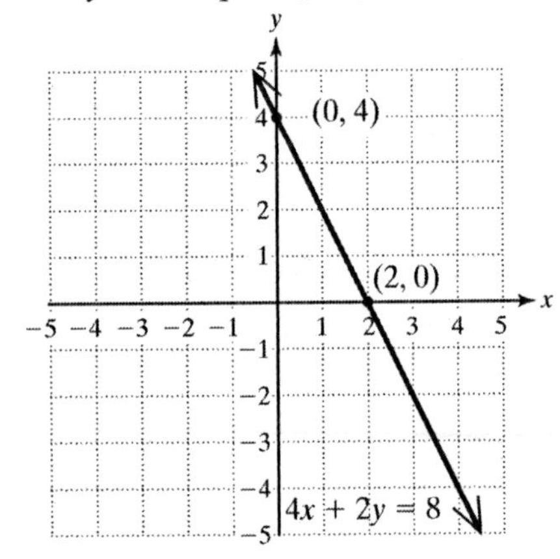

3. $3x - 4y = 6$

To find the x-intercept, substitute $y = 0$.

$$\begin{aligned} 3x - 4(0) &= 6 \\ 3x &= 6 \\ x &= 2 \end{aligned}$$

The x-intercept is (2, 0).

To find the y-intercept, substitute $x = 0$.

$$\begin{aligned} 3(0) - 4y &= 6 \\ -4y &= 6 \\ y &= \frac{6}{-4} = -\frac{3}{2} \end{aligned}$$

The y-intercept is $\left(0, -\frac{3}{2}\right)$.

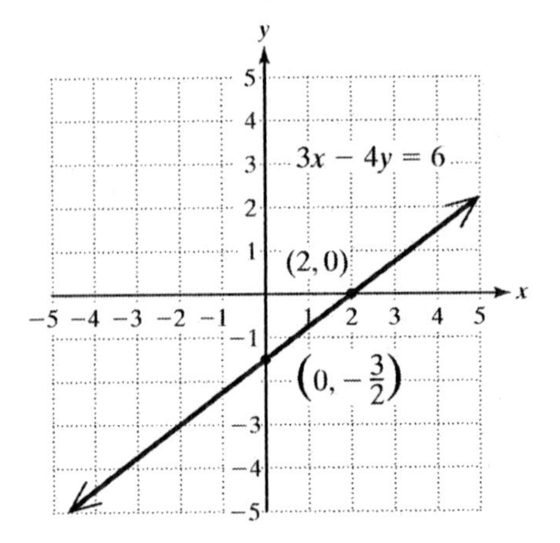

5. $6x - 2y = 0$

To find the x-intercept, substitute $y = 0$.

$$\begin{aligned} 6x - 2(0) &= 0 \\ 6x &= 0 \\ x &= 0 \end{aligned}$$

The x-intercept is (0, 0).

To find the y-intercept, substitute $x = 0$.

$$\begin{aligned} 6(0) - 2y &= 0 \\ 2y &= 0 \\ y &= \frac{0}{-2} = 0 \end{aligned}$$

The y-intercept is (0, 0).

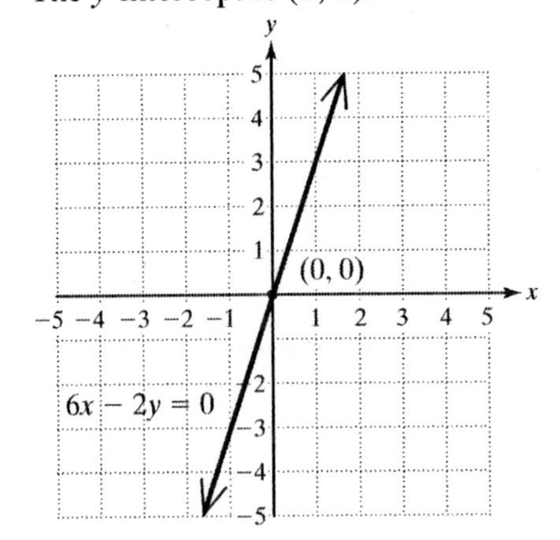

7. To write the line in slope-intercept form, solve for y.

$$4x+2y=8$$
$$2y=-4x+8$$
$$\frac{2y}{2}=\frac{-4x}{2}+\frac{8}{2}$$
$$y=-2x+4$$

slope: -2; y-intercept: $(0, 4)$

9. To write the line in slope-intercept form, solve for y.

$$9y=5x+3$$
$$\frac{9y}{9}=\frac{5x}{9}+\frac{3}{9}$$
$$y=\frac{5}{9}x+\frac{1}{3}$$

slope: $\frac{5}{9}$; y-intercept: $\left(0, \frac{1}{3}\right)$

11. To write the line in slope-intercept form, solve for y.

$$4x-3y=0$$
$$3y=4x$$
$$\frac{-3y}{-3}=\frac{-4x}{-3}$$
$$y=\frac{4}{3}x$$

slope: $\frac{4}{3}$; y-intercept: $(0, 0)$

13. To write the line in slope-intercept form, solve for y. We first multiply by the LCD, 15, to clear the equation of fractions.

$$\frac{1}{3}x-\frac{2}{5}y=1$$
$$\frac{15}{1}\left(\frac{1}{3}x-\frac{2}{5}y\right)=\frac{15}{1}(1)$$
$$\frac{15}{1}\left(\frac{1}{3}x\right)-\frac{15}{1}\left(\frac{2}{5}y\right)=15$$
$$5x-6y=15$$
$$6y=5x+15$$
$$\frac{-6y}{-6}=\frac{-5x}{-6}+\frac{15}{-6}$$
$$y=\frac{5}{6}x-\frac{5}{2}$$

slope: $\frac{5}{6}$; y-intercept: $\left(0, -\frac{5}{2}\right)$

15–19. iv, v, vi, ii, iii, i

21. Let $(x_1, y_1)=(1, -3)$ and $(x_2, y_2)=(2, 6)$.

$$m=\frac{y_2-y_1}{x_2-x_1}=\frac{6-(-3)}{2-1}=\frac{6+3}{1}=\frac{9}{1}=9$$

23. Let $(x_1, y_1)=(-3, -7)$, $(x_2, y_2)=(4, -1)$.

$$m=\frac{y_2-y_1}{x_2-x_1}=\frac{-1-(-7)}{4-(-3)}=\frac{-1+7}{4+3}=\frac{6}{7}$$

25. (a)

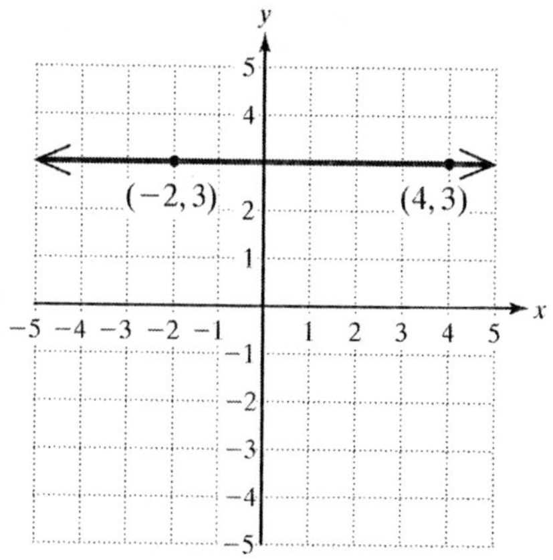

(−2, 3) (4, 3)

(b) Let $(x_1, y_1)=(-2, 3)$, $(x_2, y_2)=(4, 3)$. The slope of the line is

$$m=\frac{y_2-y_1}{x_2-x_1}=\frac{3-3}{4-(-2)}=\frac{0}{6}=0$$

(c) The slope of a horizontal line is *zero*.

27. **(a)**

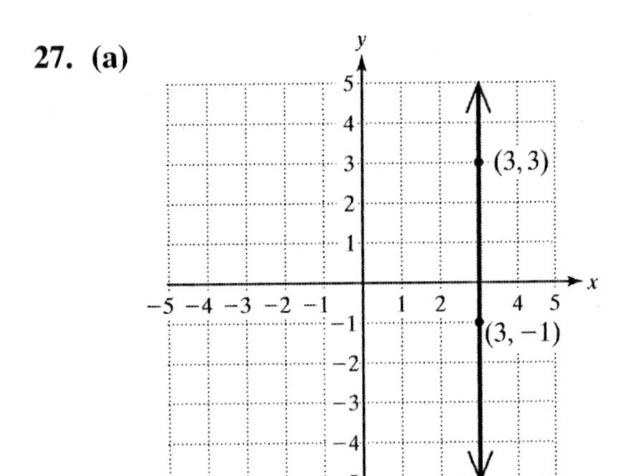

(b) Let $(x_1, y_1) = (3, -1), (x_2, y_2) = (3, 3)$.
The slope of the line is
$$m = \frac{y_2 - y_1}{x_2 - x_1} = \frac{3-(-1)}{3-3} = \frac{4}{0}$$
Since division by zero is undefined, the slope of the line is undefined.

(c) The slope of a vertical line is *undefined.*

29.
$$\begin{aligned} y - y_1 &= m(x - x_1) \\ y - 1 &= 3(x-(-2)) \\ y - 1 &= 3(x+2) \\ y - 1 &= 3x + 6 \\ y &= 3x + 7 \end{aligned}$$

31.
$$\begin{aligned} y - y_1 &= m(x - x_1) \\ y - 6 &= \frac{1}{4}(x-(-8)) \\ y - 6 &= \frac{1}{4}(x+8) \\ y - 6 &= \frac{1}{4}x + 2 \\ y &= \frac{1}{4}x + 8 \end{aligned}$$

33.
$$\begin{aligned} y - y_1 &= m(x - x_1) \\ y - (-2.2) &= 4.1(x - 5.3) \\ y + 2.2 &= 4.1x - 21.73 \\ y &= 4.1x - 23.93 \end{aligned}$$

35.
$$\begin{aligned} y - y_1 &= m(x - x_1) \\ y - (-2) &= 0(x - 3) \\ y + 2 &= 0 \\ y &= -2 \end{aligned}$$

37. $m = \frac{-6-0}{-2-1} = \frac{-6}{-3} = 2$
$$\begin{aligned} y - y_1 &= m(x - x_1) \\ y - (-6) &= 2(x-(-2)) \\ y + 6 &= 2(x+2) \\ y + 6 &= 2x + 4 \\ y &= 2x - 2 \end{aligned}$$

39. $m = \frac{-3-2}{1-(-7)} = \frac{-5}{8} = -\frac{5}{8}$
$$\begin{aligned} y - y_1 &= m(x - x_1) \\ y - (-3) &= -\frac{5}{8}(x-1) \\ y + 3 &= -\frac{5}{8}x + \frac{5}{8} \\ y &= -\frac{5}{8}x - \frac{19}{8} \end{aligned}$$

41. $m = \frac{3.1-(-5.3)}{2.2-12.2} = \frac{8.4}{-10.0} = -0.84$
$$\begin{aligned} y - y_1 &= m(x - x_1) \\ y - 3.1 &= -0.84(x - 2.2) \\ y - 3.1 &= -0.84x + 1.848 \\ y &= -0.84x + 4.948 \end{aligned}$$

43. A line parallel to a horizontal line is a horizontal line and its equation has the form y equal to a constant.
$y = 1$

45. A line perpendicular to a horizontal line is a vertical line and its equation has the form x equal to a constant.
$x = 2$

47. A line parallel to $x = 4$, a vertical line, is a vertical line and its equation has the form x equal to k, a constant.
$$x = \frac{5}{2}$$

49. A line perpendicular to $x = 0$, a vertical line, is a horizontal line and its equation has the form y equal to k, a constant.
$y = 2$

51. A line whose slope is undefined is a vertical line and its equation has the form x equal to k, a constant.
$x = -6$

53. $m = \dfrac{0-3}{-4-(-4)} = \dfrac{-3}{0}$ Undefined
A line whose slope is undefined is a vertical line and its equation has the form x equal to k, a constant.
$x = -4$

55. **(a)** $m = \dfrac{67-142}{0-15} = \dfrac{-75}{-15} = 5$

(b)
$$y - y_1 = m(x - x_1)$$
$$y - 67 = 5(x - 0)$$
$$y - 67 = 5x$$
$$y = 5x + 67$$

(c) $y = 5(25) + 67$
$y = 125 + 67$
$y = 192$
The median price of a one-family house in 2005 would be \$192,000.

57. The slope of the given line is $\frac{1}{2}$. The slope of a line perpendicular to the given line is the negative reciprocal of $\frac{1}{2}$ or -2.
$$y - y_1 = m(x - x_1)$$
$$y - 2 = -2(x - (-5))$$
$$y - 2 = -2(x + 5)$$
$$y - 2 = -2x - 10$$
$$y = -2x - 8$$

59. Write the given line in slope-intercept form to determine the slope of the line.
$$3x - y = 6$$
$$-y = -3x + 6$$
$$y = 3x - 6$$
The slope of the given line is 3. The slope of a line parallel to the given line is also 3.
$$y - y_1 = m(x - x_1)$$
$$y - 4 = 3(x - 4)$$
$$y - 4 = 3x - 12$$
$$y = 3x - 8$$

61. Write the given line in slope-intercept form to determine the slope of the line.
$$-5x + y = 4$$
$$y = 5x + 4$$
The slope of the given line is 5. The slope of a line perpendicular to the given line is the negative reciprocal of 5 or $-\frac{1}{5}$.
$$y - y_1 = m(x - x_1)$$
$$y - (-6) = -\frac{1}{5}(x - 0)$$
$$y + 6 = -\frac{1}{5}x$$
$$y = -\frac{1}{5}x - 6$$

Section 7.2 Practice Exercises

1. $y = 0.095x \qquad x \geq 0$

(a) $y = 0.095(1000)$
$y = \$95$

(b) $y = 0.095(2000)$
$y = \$190$

(c) y-intercept: $y = 0.095(0)$
$y = 0 \qquad (0, 0)$
For 0 kilowatt hours used, the cost is \$0.

(d) $m = 0.095$
The cost increases by \$0.095 for each kilowatt-hour used.

(e)

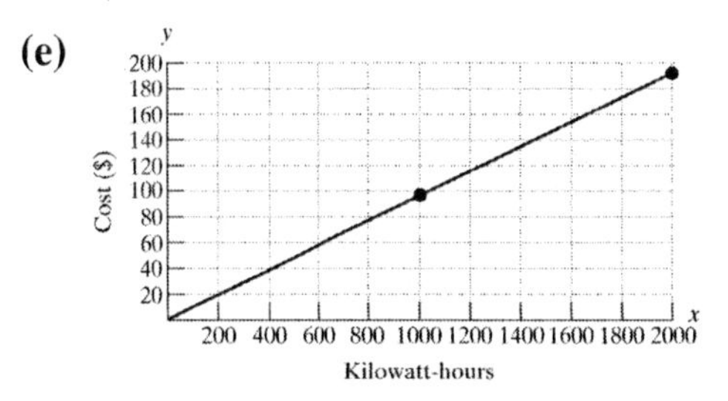

3. Male: $y = 21.5x + 28.6$
Female: $y = 3.49x + 24.5$

(a) $m = 3.49$
The number of female inmates has increased by 3.49 thousand per year between 1987 and 1997.

(b) $m = 21.5$
The number of male inmates has increased by 21.5 thousand per year between 1987 and 1997.

(c) Males; the number of male inmates is increasing at a faster rate than the number of female inmates.

5. $y = -2.333x + 124.0$

(a) y, temperature

(b) x, latitude

(c) $y = -2.333(40.0) + 124.0$
$y = -93.32 + 124.0$
$y = 30.7°$

(d) $y = -2.333(47.4) + 124.0$
$y = -110.5842 + 124.0$
$y = 13.4°$

(e) $m = -2.333$
The average temperature in January decreases 2.333° per 1° of latitude.

(f)
$$0 = -2.333x + 124.0$$
$$2.333x = 124.0$$
$$x = 53.2 \qquad (53.2, 0)$$
At 53.2° latitude, the average temperature in January is 0°.

7. (a) Using the points (0, 169) and (5, 156) the slope of the equation can be found.
$$m = \frac{169 - 156}{0 - 5} = \frac{13}{-5} = -2.6$$
Using this slope, the point (0, 169), and the point-slope formula the equation can be determined.

$$y - 169 = -2.6(x - 0)$$
$$y - 169 = -2.6x$$
$$y = -2.6x + 169$$

(b) The year 2000 corresponds to $x = 6$.
$y = -2.6(6) + 169$
$y = -15.6 + 169$
$y = 153.4$ hours

9. (a) First use the points to determine the slope.
$$m = \frac{57.75 - 82.25}{17 - 24} = \frac{-24.5}{-7} = 3.5$$
Next use the point-slope formula to determine the equation.
$$y - 57.75 = 3.5(x - 17)$$
$$y - 57.75 = 3.5x - 59.5$$
$$y = 3.5x - 1.75$$

(b) $m = 3.5$
For each additional inch in length of a person's arm, the person's height increases by 3.5 in.

(c) $y = 3.5(21.5) - 1.75$
$y = 75.25 - 1.75$
$y = 73.5$ in.
The person's height would be 73.5 in. or 6 ft $1\frac{1}{2}$ in.

11. (a) $y = 0.25x + 20$

(b) $y = 0.25(258) + 20$
$y = 64.50 + 20$
$y = 84.50$
It will cost $84.50 to rent the car and drive it 258 miles.

13. (a) $y = 25x + 20$

(b) $y = 25(20) + 20$
$y = 500 + 20$
$y = 520$
It will cost $520 for 20 tennis lessons.

15. (a) $y = 35x + 1200$

(b) $y = 35(100) + 1200$
$y = 3500 + 1200$
$y = 4700$
It will cost \$4700 to produce 100 items in one month.

17. **(a)** $y = 0.8x + 100$

(b) $y = 0.8(200) + 100$
$y = 160 + 100$
$y = 260$
It will cost \$260 to produce 200 loaves of bread in one day.

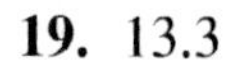

19. 13.3

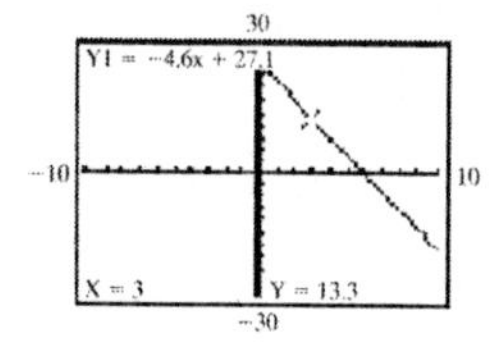

21. 345

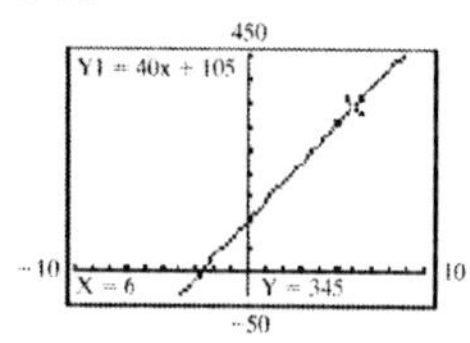

Section 7.3 Practice Exercises

1.
$$y - y_1 = m(x - x_1)$$
$$y - (-7) = -3(x - 2)$$
$$y + 7 = -3(x - 2)$$
$$y + 7 = -3x + 6$$
$$y = -3x - 1$$

3. $m = \dfrac{-3-4}{6-(-1)} = \dfrac{-7}{6+1} = \dfrac{-7}{7} = -1$
$$y - y_1 = m(x - x_1)$$
$$y - 4 = -1[x - (-1)]$$
$$y - 4 = -1(x + 1)$$
$$y - 4 = -x - 1$$
$$y = -x + 3$$

5. **(a)** The slope of the given line is $\frac{2}{3}$. The slope of a line parallel to the given line is also $\frac{2}{3}$.
$$y - y_1 = m(x - x_1)$$
$$y - 4 = \frac{2}{3}[x - (-3)]$$
$$y - 4 = \frac{2}{3}(x + 3)$$
$$y - 4 = \frac{2}{3}x + 2$$
$$y = \frac{2}{3}x + 6$$

(b) The slope of the given line is $\frac{2}{3}$. The slope of a line perpendicular to the given line is the negative reciprocal of $\frac{2}{3}$, or $-\frac{3}{2}$.
$$y - y_1 = m(x - x_1)$$
$$y - 4 = -\frac{3}{2}[x - (-3)]$$
$$y - 4 = -\frac{3}{2}(x + 3)$$
$$y - 4 = -\frac{3}{2}x - \frac{9}{2}$$
$$y = -\frac{3}{2}x - \frac{1}{2}$$

7. $\{(A, 1), (A, 2), (B, 2), (C, 3), (D, 5), (E, 4)\}$

9. {(Pregnant women, 60), (Nursing mothers, 65), (Infants under 1 year old, 14), (Children from 1 to 4 years, 16), (Adults, 50)}

11. Domain: $\{A, B, C, D, E\}$;
Range: $\{1, 2, 3, 4, 5\}$

13. Domain: {Pregnant women, Nursing mothers, Infants under 1 year old, Children from 1 to 4 years, Adults};
Range: {60, 65, 14, 16, 50}

15. **(a)** Answers will vary. For example: {(Julie, New York), (Peggy, Florida), (Stephen, Kansas), (Pat, New York)}

(b) Answers will vary. From the example in part (a), Domain: {Julie, Peggy, Stephen, Pat}; Range: {New York, Florida, Kansas}

17. **(a)** $y = 2x - 1$

(b)

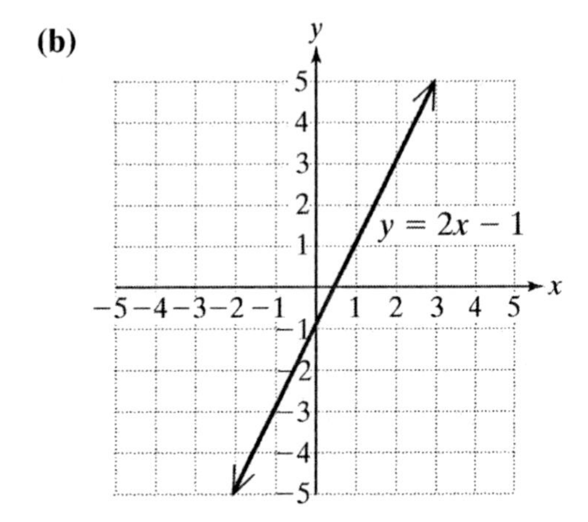

(c) Domain: $(-\infty, \infty)$; Range: $(-\infty, \infty)$

19. Domain: $[-5, 3]$; Range: $[-2.1, 2.8]$

21. Domain: $[0, 4.2]$; Range: $[-2.1, 2.1]$

23. Domain: $(-\infty, 0]$; Range: $(-\infty, \infty)$

25. Domain: $(-2, \infty)$; Range: $(2, \infty)$

27. Domain: $[-4, \infty)$; Range: $[0, \infty)$

29. Domain: $\{-3, -1, 1, 3\}$; Range: $\{0, 1, 2, 3\}$

31. Domain: $[-4, 5)$; Range: $\{-2, 1, 3\}$

33. **(a)** 2.85 inches

(b) 9.33 inches

(c) December

(d) (Nov, 2.66)

(e) (Sept, 7.63)

(f) {Jan, Feb, Mar, Apr, May, June, July, Aug, Sept, Oct, Nov, Dec}

35. **(a)** grade 9: $y = -12.64(9) + 195.22$

$y = -113.76 + 195.22$

$y = 81.46\%$

grade 10: $y = -12.64(10) + 195.22$

$y = -126.40 + 195.22$

$y = 68.82\%$

grade 11: $y = -12.64(11) + 195.22$

$y = -139.04 + 195.22$

$y = 56.18\%$

grade 12: $y = -12.64(12) + 195.22$

$y = -151.68 + 195.22$

$y = 43.54\%$

(b) No, 7 is not in the domain.

37. Domain: $(-\infty, \infty)$; Range: $[0, \infty)$

39. Domain: $(-\infty, \infty)$; Range: $[0, \infty)$

41. Domain: $(-\infty, \infty)$; Range: $[-2, \infty)$

43. The domain $(-\infty, \infty)$ and the range $[c, \infty)$ will be the same for all values of c.

45. **(a)**

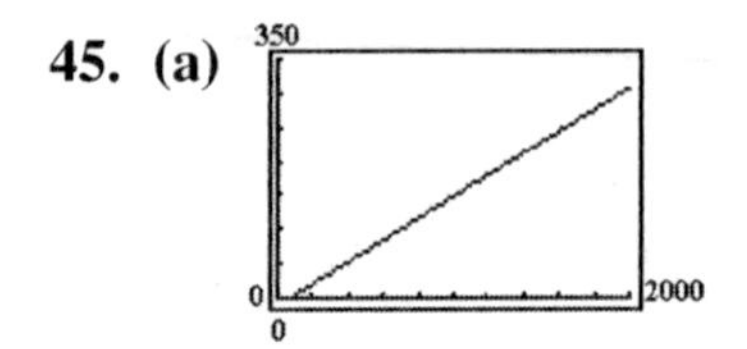

(b)

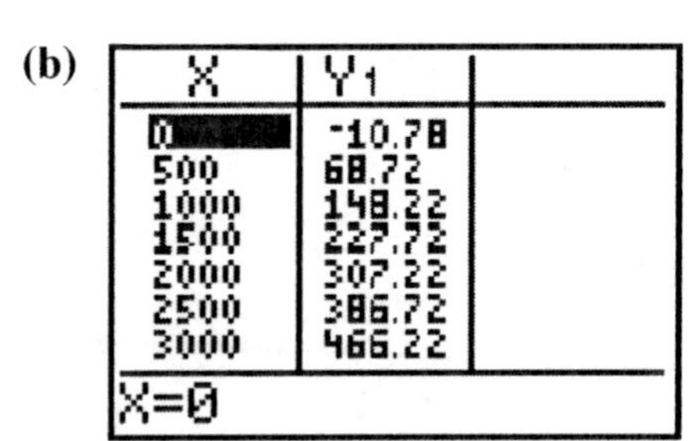

X	Y1
0	-10.78
500	68.72
1000	148.22
1500	227.72
2000	307.22
2500	386.72
3000	466.22

X=0

Section 7.4 Practice Exercises

1. **(a)** {(Doris, Mike), (Richard, Nora), (Doris, Molly), (Richard, Mike)}

(b) Domain: {Doris, Richard}

(c) Range: {Mike, Nora, Molly}

(d) The relation is not a function since Doris is paired with both Mike and Molly.

3. **(a)** $\{(3, 10), (4, 12), (5, 12), (6, 12)\}$

(b) Domain: $\{3, 4, 5, 6\}$

(c) Range: $\{10, 12\}$

(d) The relation is a function since each element in the domain is paired with exactly one element in the range.

5. Domain: $[0, 4]$; Range: $[1, 4]$

7. Domain: $\{-4\}$; Range: $(-\infty, \infty)$

9. The relation is not a function since a vertical line will cross the graph at more than one point.

11. The relation is a function since any vertical line will cross the graph at no more than one point.

13. The relation is not a function since a vertical line will cross the graph at more than one point.

15. $f(2) = 6(2) - 2 = 10$

17. $h(4) = 7$

19. $g(0) = (0)^2 - 4(0) + 1 = 1$

21. $k(0) = |0 - 2| = |-2| = 2$

23. $f(t) = 6t - 2$

25. $h(u) = 7$

27. $g(-3) = (-3)^2 - 4(-3) + 1$
$= 9 + 12 + 1$
$= 22$

29. $k(0) = |-2 - 2| = |-4| = 4$

31. $f(x+1) = 6(x+1) - 2 = 6x + 6 - 2 = 6x + 4$

33. $g(x-2) = (x-2)^2 - 4(x-2) + 1$
$= x^2 - 4x + 4 - 4x + 8 + 1$
$= x^2 - 8x + 13$

35. $g(x+h) = (x+h)^2 - 4(x+h) + 1$
$= x^2 + 2xh + h^2 - 4x - 4h + 1$

37. $h(a + b) = 7$

39. $f(-a) = 6(-a) - 2 = -6a - 2$

41. $k(-c) = |-c - 2|$

43. $f\left(\frac{1}{2}\right) = 6\left(\frac{1}{2}\right) - 2 = 3 - 2 = 1$

45. $h\left(\frac{1}{7}\right) = 7$

47. $f(-2.8) = 6(-2.8) - 2 = -18.8$

49. $p(2) = -7$

51. $p(3) = 2\pi$

53. $q(2) = -5$

55. $q(6) = 4$

57. $\left\{-3, -7, -\frac{3}{2}, 1.2\right\}$

59. $\{6,0\}$

61. $f(-3) = 5, f(1.2) = 5$
-3 and 1.2

63. $g(6) = 0, g(1) = 0$
6 and 1

65. $f(-7) = -3$

67. **(a)** $f(0) = 2$

(b) $f(3) = 1$

(c) $f(-2) = 1$

(d) $f(-3) = -3, x = -3$

(e) $f(1) = 3, x = 1$

(f) $[-3, 3]$

(g) $[-3, 3]$

69. Answers will vary. To find excluded values, set the denominator equal to zero and solve that equation. Since $x - 2 = 0$ when $x = 2$, the domain will be all real numbers except 2.

71. Set the denominator equal to zero to find any values of the variable that result in division by zero. $x - 4 = 0 \Rightarrow x \neq 4$. Thus, the domain is $(-\infty, 4) \cup (4, \infty)$.

73. Set the denominator equal to zero to find any values of the variable that result in division by zero. $t \neq 0$. Thus, the domain is $(-\infty, 0) \cup (0, \infty)$.

75. Set the denominator equal to zero to find any values of the variable that result in division by zero. $p^2 + 2 = 0$. Since $p^2 + 2$ is always a positive number, the denominator will never become 0. Thus, the domain is $(-\infty, \infty)$.

77. $t + 7 \geq 0 \Rightarrow t \geq -7$. Thus, the domain is $[-7, \infty)$.

79. $a - 3 \geq 0 \Rightarrow a \geq 3$. Thus, the domain is $[3, \infty)$.

81. $2x + 1 \geq 0 \Rightarrow x \geq -\frac{1}{2}$. Thus, the domain is $\left[-\frac{1}{2}, \infty\right)$.

83. The domain of a polynomial is all real numbers.

85. Since $q(t)$ is a polynomial function, the domain is $(-\infty, \infty)$.

87. Since $g(x)$ is a polynomial function, the domain is $(-\infty, \infty)$.

89. **(a)** $h(1) = 45.1$, $h(1.5) = 38.975$

(b) After 1 second, the height of the ball is 45.1 feet. After 1.5 seconds, the height of the ball is 38.975 feet.

91. **(a)** $d(1) = 5.9$, $d(2) = 11.8$

(b) After 1 hour, the distance is 5.9 miles. After 2 hours, the distance is 11.8 miles.

93. **(a)** $N(1) \approx 2.6$ indicates that a 1 year old has an average of 2.6 doctor's visits per year;
$N(20) \approx 1.9$ indicates that a 20 year old has an average of 1.9 doctor's visits per year;
$N(40) \approx 2.3$ indicates that a 40 year old has an average of 2.3 doctor's visits per year;
$N(75) \approx 5.6$ indicates that a 75 year old has an average of 5.6 doctor's visits per year.

(b) Label the points (1, 2.6); (20, 1.9), (40, 2.3) and (75, 5.6).

(c) About 24 years old

95. $\left(-\infty, \frac{1}{3}\right) \cup \left(\frac{1}{3}, \infty\right)$

97. $(4, \infty)$

99.

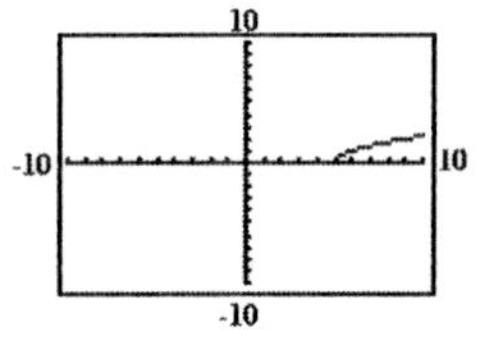

10

-10 10

-10

101.

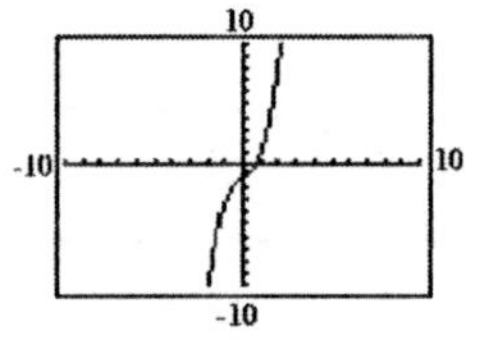

10

-10 10

-10

103. **(a)**

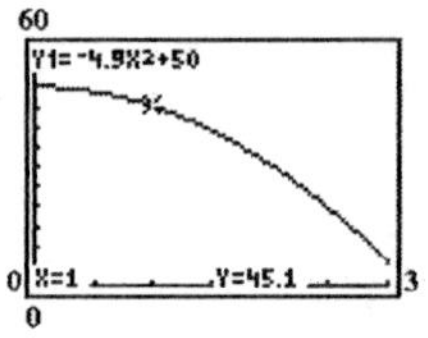

60

Y1=-4.9X2+50

X=1 Y=45.1

0 3

0

(b) $h(1) = 45.1$; $h(1.5) = 38.975$

Section 7.5 Practice Exercises

1. **(a)** It is a function since each domain value is paired with exactly one range value.

(b) {6, 5, 4, 3, 2, 1}

(c) {1, 2, 3, 4, 5, 6}

3. $k(0) = (0)^2 - 2 = 0 - 2 = -2$

5. $k(a) = (a)^2 - 2 = a^2 - 2$

7. $k(-2) = (-2)^2 - 2 = 4 - 2 = 2$

9. $k(b-1) = (b-1)^2 - 2$
$= b^2 - 2b + 1 - 2$
$= b^2 - 2b - 1$

11. Since $f(x)$ is a polynomial function, the domain is $(-\infty, \infty)$.

13. Set the radicand, $x - 6 \geq 0$. Thus, $x \geq 6$. In interval notation the domain is $[6, \infty)$.

15. **(a)** $f(3) = 3(3) = 9$
It takes 9 pounds to stretch the spring 3 inches.

(b) $f(0) = 3(0) = 0$
It takes 0 pounds to stretch the spring 0 inches.

17.

x	$g(x)$
–2	2
–1	1
0	0
1	1
2	2

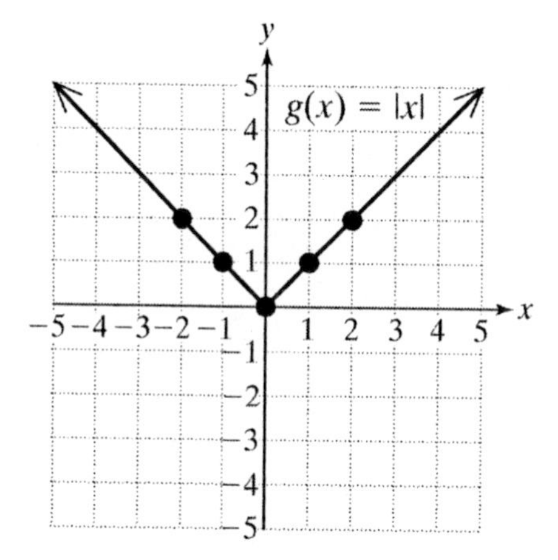

19.

x	$p(x)$
–2	–8
–1	–1
0	0
1	1
2	8

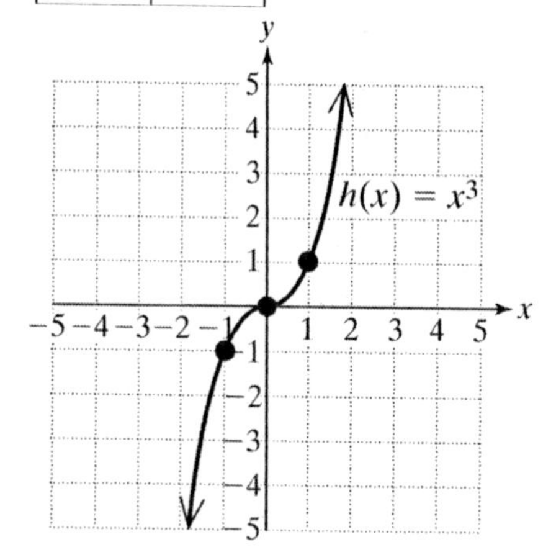

21.

x	$p(x)$
0	0
1	1
4	2
9	3
16	4

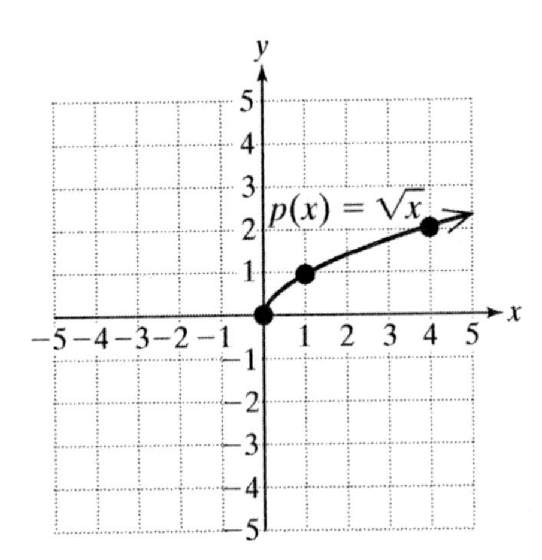

23. **(a)** Set the radicand, $x + 4 \geq 0$. Thus, $x \geq -4$. In interval notation the domain is $[-4, \infty)$.

(b) $f(0) = \sqrt{(0)+4} = \sqrt{4} = 2;$
$f(5) = \sqrt{(5)+4} = \sqrt{9} = 3;$
$f(-3) = \sqrt{(-3)+4} = \sqrt{1} = 1$

25. **(a)** Since $h(x)$ is a polynomial function, the domain is $(-\infty, \infty)$.

(b) $h(1) = -(1)^2 + 2 = -1 + 2 = 1;$
$h(-1) = -(-1)^2 + 2 = -1 + 2 = 1;$
$h(0) = -(0)^2 + 2 = 0 + 2 = 2$

27. **(a)** Set the denominator equal to zero to find any values of the variable that result in division by zero.
$x - 3 = 0 \Rightarrow x \neq 3$. Thus, the domain is $(-\infty, 3) \cup (3, \infty)$.

(b) $p(0) = \frac{2}{0-3} = \frac{2}{-3} = -\frac{2}{3};$
$p(1) = \frac{2}{1-3} = \frac{2}{-2} = -1;$
$p(2) = \frac{2}{2-3} = \frac{2}{-1} = -2;$
$p(4) = \frac{2}{4-3} = \frac{2}{1} = 2;$
$p(5) = \frac{2}{5-3} = \frac{2}{2} = 1;$
$p(6) = \frac{2}{6-3} = \frac{2}{3}$

29. $M(x) = 8x + 1$
x-intercept(s)
$0 = 8x + 1$
$-\frac{1}{8} = x$
The x-intercept is $\left(-\frac{1}{8}, 0\right)$.
y-intercept
$M(0) = 8(0) + 1 = 1$
The y-intercept is (0, 1).

31. $C(x) = -5x$
x-intercept(s)
$0 = -5x$
$0 = x$
The x-intercept is (0, 0).
y-intercept
$C(0) = -5(0) = 0$
The y-intercept is (0, 0).

33. $A(x) = (2x + 1)(x - 5)$
x-intercept(s)
$0 = (2x + 1)(x - 5)$
$2x + 1 = 0$ or $x - 5 = 0$
$x = -\frac{1}{2}$ or $x = 5$
The x-intercepts are $\left(-\frac{1}{2}, 0\right)$, (5, 0).
y-intercept
$A(0) = (2 \cdot 0 + 1)(0 - 5) = (1)(-5) = -5$
The y-intercept is (0, −5).

35. $p(x) = x^2 - 3x - 10$

<u>x-intercept(s)</u>

$0 = x^2 - 3x - 10$

$0 = (x-5)(x+2)$

$x - 5 = 0 \quad \text{or} \quad x + 2 = 0$

$x = 5 \quad \text{or} \quad x = -2$

The x-intercepts are (5, 0), (–2, 0).

<u>y-intercept</u>

$p(0) = (0)^2 - 3(0) - 10 = -10$

The y-intercept is (0, –10).

37. $g(x) = x^2 + 6x + 9$

<u>x-intercept(s)</u>

$0 = x^2 + 6x + 9$

$0 = (x+3)(x+3)$

$x + 3 = 0 \quad \text{or} \quad x + 3 = 0$

$x = -3 \quad \text{or} \quad x = -3$

The x-intercept is (–3, 0).

<u>y-intercept</u>

$g(0) = (0)^2 + 6(0) + 9 = 9$

The y-intercept is (0, 9).

39. $h(x) = 4x(x-3)(3x+2)$

<u>x-intercept(s)</u>

$0 = 4x(x-3)(3x+2)$

$0 = 4x(x-3)(3x+2)$

$4x = 0 \quad \text{or} \quad x - 3 = 0 \quad \text{or} \quad 3x + 2 = 0$

$x = 0 \quad \text{or} \quad x = 3 \quad \text{or} \quad x = -\frac{2}{3}$

The x-intercepts are (0, 0), (3, 0), $\left(-\frac{2}{3}, 0\right)$.

<u>y-intercept</u>

$h(0) = 4(0)(0-3)(3 \cdot 0 + 2) = 0$

The y-intercept is (0, 0).

41. x-intercept (–1, 0); y-intercept (0, 1)

43. x-intercepts (–2, 0), (2, 0); y-intercept (0, –2)

45. x-intercept none; y-intercept (0, 2)

47. (a)

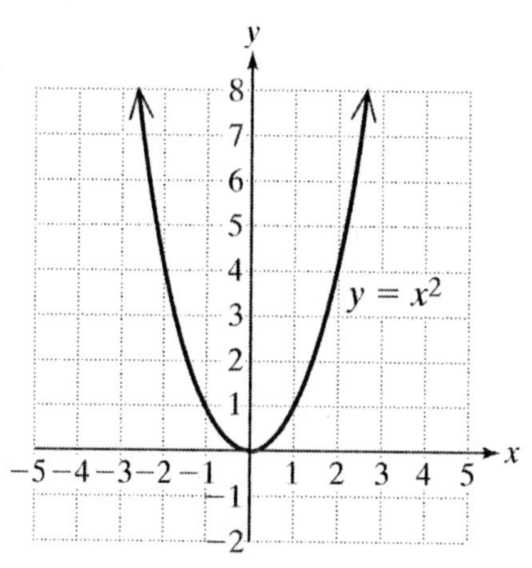

(b) $y = x^2$ is a function. The graph passes the vertical line test.

(c)

x	y
4	–2
1	–1
0	0
1	1
4	2

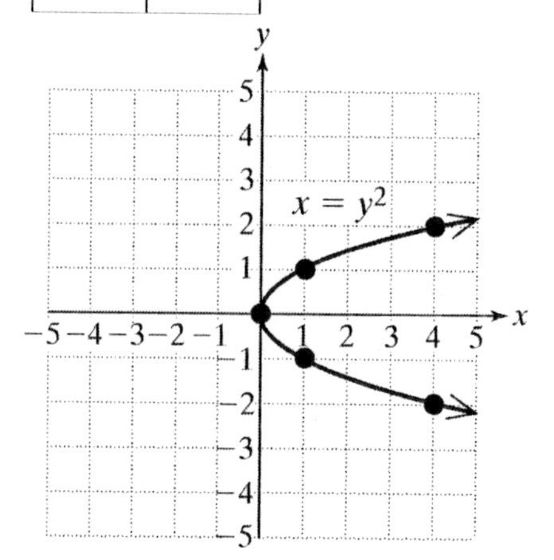

(d) $x = y^2$ is not a function. The graph does not pass the vertical line test.

49. (a) $(-\infty, \infty)$

(b) (0, 0)

(c) vi

51. **(a)** $(-\infty, \infty)$

(b) $(0, 1)$

(c) vii

53. **(a)** $[-1, \infty)$

(b) $(0, 1)$

(c) vii

55. **(a)** $(-\infty, 3) \cup (3, \infty)$

(b) $\left(0, -\frac{1}{3}\right)$

(c) ii

57. **(a)** $(-\infty, \infty)$

(b) $(0, 2)$

(c) iv

59.

X	Y1	
[illegible]	1	
0	2	
1	1	
2	-2	
3	-7	
4	-14	
5	-23	

X=-1

61.

X	Y1	
0	-.6667	
1	-1	
2	-2	
3	ERROR	
4	2	
5	1	
6	.66667	

X=0

Section 7.6 Practice Exercises

1. $f(1) = 3$

3. None

5. $[-5, \infty)$

7. y is not a function of x.

9. y is a function of x.

11. **(a)** increase

(b) decrease

13. $T = kq$

15. $W = \frac{k}{p^2}$

17. $Q = k \cdot \frac{x}{y^3}$

19. $L = kw\sqrt{v}$

21.
$$\begin{aligned} y &= kx \\ 18 &= k(4) \\ \frac{18}{4} &= k \\ \frac{9}{2} &= k \end{aligned}$$

23.
$$\begin{aligned} p &= \frac{k}{q} \\ 32 &= \frac{k}{16} \\ 512 &= k \end{aligned}$$

25.
$$\begin{aligned} y &= kwv \\ 8.75 &= k(50)(0.1) \\ 8.75 &= k(5) \\ \frac{8.75}{5} &= k \\ 1.75 &= k \end{aligned}$$

27.
$$\begin{aligned} Z &= kw^2 \\ 14 &= k(4)^2 \\ 14 &= k(16) \\ \frac{14}{16} &= k \\ \frac{7}{8} &= k \end{aligned}$$

The variation model is: $Z = \frac{7}{8}w^2$.

$$\begin{aligned} Z &= \frac{7}{8}(8)^2 \\ Z &= \frac{7}{8}(64) \\ Z &= 56 \end{aligned}$$

29. $L = ka\sqrt{b}$
$72 = k(8)\sqrt{9}$
$72 = k8(3)$
$72 = k(24)$
$3 = k$

The variation model is $L = 3a\sqrt{b}$.

$L = 3\left(\frac{1}{2}\right)\sqrt{36}$
$L = 3\left(\frac{1}{2}\right)(6)$
$L = 9$

31. $B = k \cdot \frac{m}{n}$
$20 = k \cdot \frac{10}{3}$
$\frac{3}{10} \cdot 20 = k$
$6 = k$

The variation model is $B = (6) \cdot \frac{m}{n}$.

$B = (6) \cdot \frac{15}{12}$
$B = \frac{15}{2}$

33. $A = k \cdot n$
$56800 = k \cdot (80000)$
$\frac{56800}{80000} = k$
$\frac{71}{100} = k$

The variation model is $A = \frac{71}{100} \cdot n$.

$A = \frac{71}{100} \cdot (500000)$
$A = 355{,}000$ tons

35. $d = ks^2$
$109 = k(40)^2$
$109 = k(1600)$
$\frac{109}{1600} = k$

The variation model is: $d = \frac{109}{1600}s^2$.

$d = \frac{109}{1600}(25)^2$
$d = \frac{109}{1600}(625)$
$d = 42.6$ feet

37. $C = k \cdot \frac{v}{r}$
$9 = k \cdot \frac{90}{10}$
$\frac{10}{90} \cdot 9 = k$
$1 = k$

The variation model is $C = (1) \cdot \frac{v}{r}$.

$C = (1) \cdot \frac{185}{10}$
$C = 18.5$ amps

39. $R = k \cdot \frac{1}{d^2}$
$4 = k \cdot \frac{40}{0.1^2}$
$4 = k \cdot \frac{40}{0.01}$
$4 = k \cdot (4000)$
$\frac{4}{4000} = k$
$\frac{1}{1000} = k$

The variation model is $R = \left(\frac{1}{1000}\right) \cdot \frac{1}{d^2}$.

$R = \left(\frac{1}{1000}\right) \cdot \frac{50}{0.2^2}$
$R = \left(\frac{1}{1000}\right) \cdot \frac{50}{0.04}$
$R = 1.25$ ohms

41. $W = kr^3$

$4.32 = k(3)^3$
$4.32 = k(27)$
$\frac{4.32}{27} = k$
$0.16 = k$

The variation model is $W = 0.16r^3$.

$W = 0.16(5)^3$
$W = 0.16(125)$
$W = 20$ pounds

43. $S = k \cdot \frac{wt^2}{l}$

$417 = k \cdot \frac{6(2)^2}{48}$
$417 = k \cdot \frac{6(4)}{48}$
$417 = k \cdot \frac{24}{48}$
$417 = k \cdot \frac{1}{2}$
$2 \cdot 417 = k$
$834 = k$

The variation model is $S = 834 \cdot \frac{wt^2}{l}$.

$S = 834 \cdot \frac{12(4)^2}{72}$
$S = 834 \cdot \frac{12(16)}{72}$
$S = 834 \cdot \frac{192}{72}$
$S = 834 \cdot \frac{8}{3}$
$S = 2224$ pounds

45. (a) $A = kl^2$

$A = k(2l)^2$
$A = k(4)l^2$

Area will be 4 times greater.

(b) $A = k(3l)^2$
$A = k(9)l^2$

Area will be 9 times greater.

Chapter 7 Review Exercises

1. $3x - 4y = 8$

(a) x-intercept

$3x - 4(0) = 8$
$3x = 8$
$x = \frac{8}{3}$

The x-intercept is $\left(\frac{8}{3}, 0\right)$.

y-intercept

$3(0) - 4y = 8$
$-4y = 8$
$y = -2$

The y-intercept is $(0, -2)$.

(b) To find the slope, write the line in slope-intercept form.

$3x - 4y = 8$
$-4y = -3x + 8$
$y = \frac{3}{4}x - 2$

The slope is $\frac{3}{4}$.

(c)

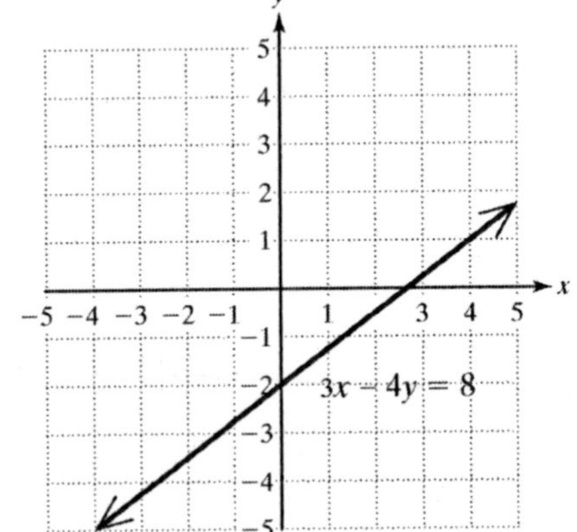

3. $x = 4$

(a) x-intercept
If $y = 0$, then $x = 4$. The x-intercept is $(4, 0)$.
y-intercept
Since x cannot be 0, there is no y-intercept.

(b) The slope is undefined.

(c)

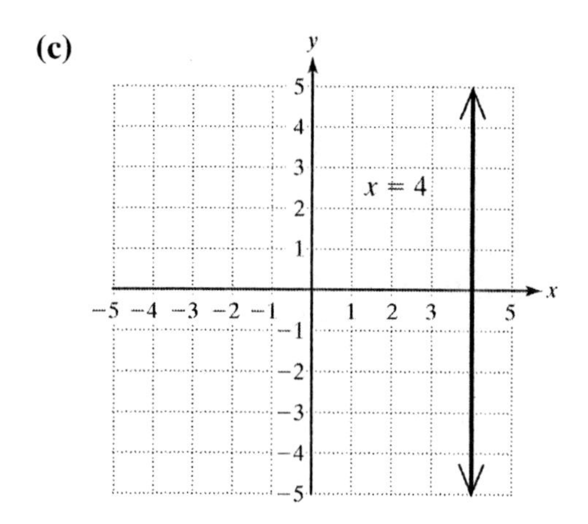

5. (a)
$$4x-2y=8$$
$$-2y=-4x+8$$
$$y=2x-4$$

(b) The slope is 2. The y-intercept is $(0, -4)$.

7. $y - y_1 = m(x - x_1)$

9.
$$y-8=-6(x-(-1))$$
$$y-8=-6(x+1)$$
$$y-8=-6-6$$
$$y=-6+2$$

11. First, use the points to determine the slope.
$$m=\frac{-4-(-2)}{0-8}=\frac{-2}{-8}=\frac{1}{4}$$
Use this value, one of the points, and the point-slope formula to determine the equation of the line.
$$y-(-4)=\frac{1}{4}(x-0)$$
$$y+4=\frac{1}{4}x$$
$$y=\frac{1}{4}x-4$$

13. The slope of the given line is 4. The slope of a line parallel to it will also be 4. Use this value, the given point, and the slope-intercept form to determine the equation of the new line.
$$y=mx+b$$
$$12=4(8)+b$$
$$12=32+b$$
$$-20=b$$
$$y=4x-20$$

15. Write the given line in slope-intercept form.
$$5x+6y=-18$$
$$6y=-5x-18$$
$$y=-\frac{5}{6}x-3$$

The slope of the given line is $-\frac{5}{6}$. The slope of a line perpendicular to the given line is the negative reciprocal, $\frac{6}{5}$. Use the point-slope formula to find the equation of the line.
$$y-y_1=m(x-x_1)$$
$$y-12=\frac{6}{5}(x-5)$$
$$y-12=\frac{6}{5}x-6$$
$$y=\frac{6}{5}x+6$$

17. Find the slope of the line.
$$m=\frac{-9-0}{0-(-1)}=\frac{-9}{1}=-9$$
Since the y-intercept is $(0, -9)$, we can use the slope-intercept form to write the equation of the line.
$$y = mx + b$$
$$y = -9x - 9$$

19. Any line perpendicular to the y-axis must be horizontal. Horizontal lines have the form $y = k$. Since the line passes through $(-2, -1)$, the equation is $y = -1$.

21. (a) $m = \frac{1.3 - 0.9}{1 - 4} = \frac{0.4}{-3} = -0.13$

(b) The number of robberies decreased by an average of 0.13 million per year between 1994 and 1999.

(c) Use the slope, one of the points, and the point-slope formula to determine the equation of the line.
$y - 1.3 = -0.13(x - 1)$
$y - 1.3 = -0.13x + 0.13$
$y = -0.13x + 1.43$

(d) $y = -0.13(5) + 1.43$
$y = -0.65 + 1.43$
$y = 0.78$
There were 780,000 robberies in 1998.

23. (a) $y = 20x + 55$

(b) $y = 20(9) + 55$
$y = 180 + 55$
$y = 235$
The total cost of renting the system for nine months is \$235.

25. For example: {(Peggy, Kent), (Charlie, Laura), (Tom, Matt), (Tom, Chris)}

27. Domain: $[-3, 9]$; Range: $[0, 60]$

29. Domain: $\{-3, -1, 0, 2, 3\}$;
Range: $\left\{-2, 1, 0, \frac{5}{2}\right\}$

31. Answers will vary. For example:

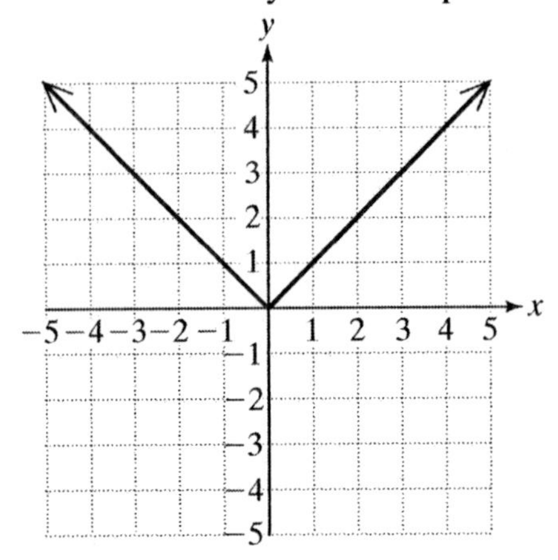

33. (a) Function

(b) Domain: $(-\infty, \infty)$

(c) Range: $(-\infty, 0.35)$

35. (a) Not a function

(b) Domain: $\{0, 4\}$

(c) Range: $\{2, 3, 4, 5\}$

37. (a) Function

(b) Domain: $\{6, 7, 8, 9\}$

(c) Range: $\{9, 10, 11, 12\}$

39. $f(1) = 6(1)^2 - 4 = 6(1) - 4 = 6 - 4 = 2$

41. $f(t) = 6(t)^2 - 4 = 6t^2 - 4$

43. $f(\pi) = 6(\pi)^2 - 4 = 6\pi^2 - 4$

45. $f(x + h) = 6(x + h)^2 - 4$
$= 6(x^2 + 2xh + h^2) - 4$
$= 6x^2 + 12xh + 6h^2 - 4$

47. Set the denominator equal to zero to find any values of the variable that result in division by zero. $x - 11 = 0 \Rightarrow x \neq 11$. Thus, the domain is $(-\infty, 11) \cup (11, \infty)$.

49. Set the radicand, $x + 2 \geq 0$. thus, $x \geq -2$. In interval notation the domain is $[-2, \infty)$.

51.

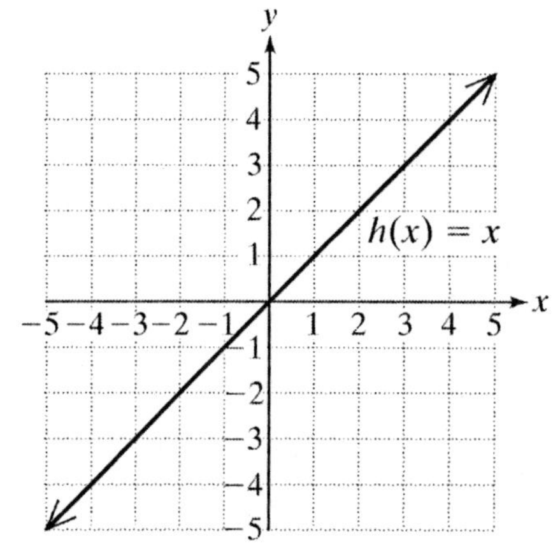

53.

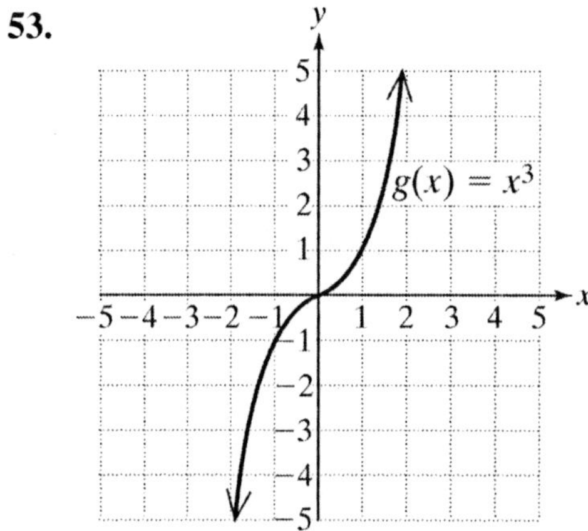

55.

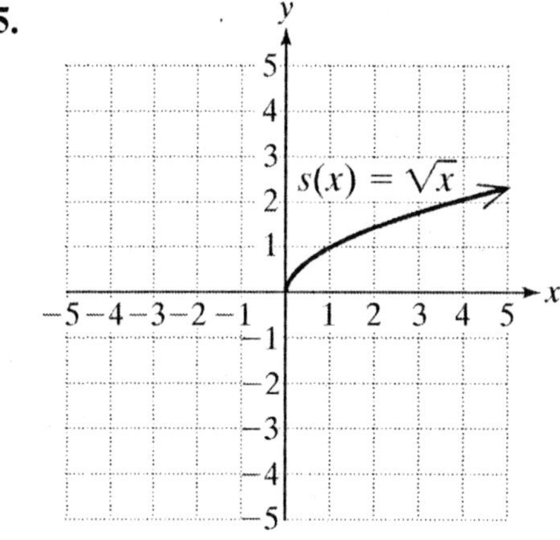

57.

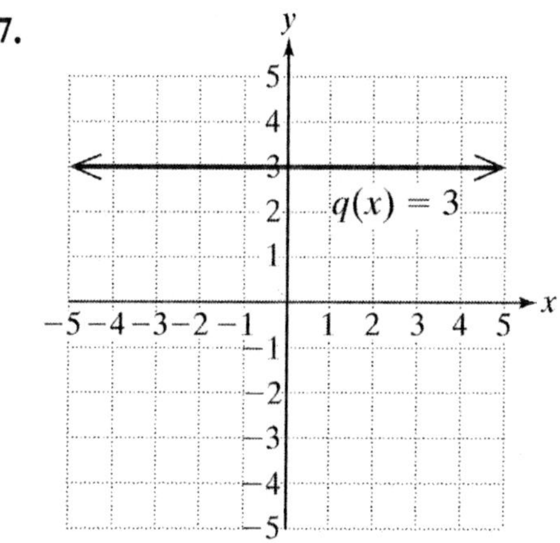

59. $p(x) = 4x - 7$

<u>x-intercept(s)</u>

$0 = 4x - 7$

$7 = 4x$

$\frac{7}{4} = x$

The x-intercept is $\left(\frac{7}{4}, 0\right)$.

<u>y-intercept</u>

$p(0) = 4(0) - 7 = -7$

The y-intercept is $(0, -7)$.

61. $F(x) = x^2 - 16$

<u>x-intercept(s)</u>

$0 = x^2 - 16$

$0 = (x-4)(x+4)$

$x - 4 = 0$ or $x + 4 = 0$

$x = 4$ or $x = -4$

The x-intercepts are $(4, 0)$, $(-4, 0)$.

<u>y-intercept</u>

$F(0) = (0)^2 - 16 = -16$

The y-intercept is $(0, -16)$.

63. $r(x) = (x - 3)(x + 2)(2x - 1)$

<u>x-intercept(s)</u>

$0 = (x - 3)(x + 2)(2x - 1)$

$x - 3 = 0$ or $x + 2 = 0$ or $2x - 1 = 0$

$x = 3$ or $x = -2$ or $x = \frac{1}{2}$

The x-intercepts are $(3, 0)$, $(-2, 0)$, $\left(\frac{1}{2}, 0\right)$.

<u>y-intercept</u>
$r(0) = (0-3)(0+2)(2\cdot 0-1)$
$= (-3)(2)(-1)$
$= 6$
The y-intercept is (0, 6).

65. (a) $s(4) = (4-2)^2 = (2)^2 = 4;$
$s(3) = (3-2)^2 = (1)^2 = 1;$
$s(2) = (2-2)^2 = (0)^2 = 0;$
$s(1) = (1-2)^2 = (1)^2 = 1;$
$s(0) = (0-2)^2 = (-2)^2 = 4$

(b) Domain: $(-\infty, \infty)$

67. (a) $h(-3) = \frac{3}{-3-3} = \frac{3}{-6} = -\frac{1}{2};$
$h(-1) = \frac{3}{-1-3} = \frac{3}{-4} = -\frac{3}{4};$
$h(0) = \frac{3}{0-3} = \frac{3}{-3} = -1;$
$h(2) = \frac{3}{2-3} = \frac{3}{-1} = -3;$
$h(4) = \frac{3}{4-3} = \frac{3}{1} = 3;$
$h(5) = \frac{3}{5-3} = \frac{3}{2};$
$h(7) = \frac{3}{7-3} = \frac{3}{4}$

(b) Domain: $(-\infty, 3) \cup (3, \infty)$

69. (a) $b(0) = 0.7(0) + 4.5 = 0 + 4.5 = 4.5$ means that in 1985 the per capita consumption of bottled water was 4.5 gallons;
$b(7) = 0.7(7) + 4.5 = 4.9 + 4.5 = 9.4$ means that in 1992 the per capita consumption of bottled water was 9.4 gallons.

(b) Slope is 0.7, the coefficient of t. The slope indicates that the consumption increased by 0.7 gallon per year.

71. $y = \frac{K}{x^3}$
$32 = \frac{K}{(2)^3}$
$32 = \frac{K}{8}$
$256 = K$
The variation model is $y = \frac{256}{x^3}$.
When $x = 4$, $y = \frac{256}{(4)^3} = \frac{256}{64} = 4.$

73. $d = k\sqrt{h}$
$26.4 = k\sqrt{16}$
$26.4 = 4k$
$6.6 = k$
The variation model is: $d = 6.6\sqrt{h}.$
$d = 6.6\sqrt{64} = 6.6(8) = 52.8$ km

Chapter 7 Test

1. $m = \frac{y_2 - y_1}{x_2 - x_1} = \frac{4-(-3)}{6-2} = \frac{7}{4}$

3. The given line is written in slope-intercept form. It has a slope of 8. The slope of a perpendicular line is the negative reciprocal, $-\frac{1}{8}$.

5. $x = 2y$
<u>x-intercept</u>
$x = 2(0)$
$x = 0$
The x-intercept is (0, 0).
<u>y-intercept</u>
$(0) = 2y$
$0 = y$
The y-intercept is (0, 0).

7. Solve for y to write the equation in slope-intercept form.

$$2x = 3y - 12$$
$$2x + 12 = 3y$$
$$\frac{2}{3}x + 4 = y$$
$$y = \frac{2}{3}x + 4$$

The slope is $\frac{2}{3}$ and the y-intercept is (0, 4).

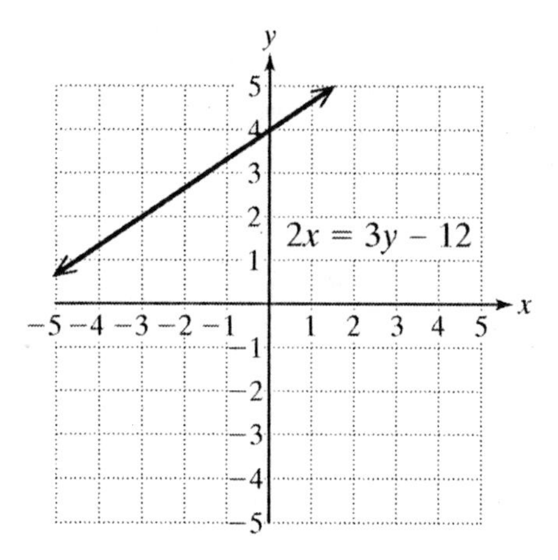

9. A line that is parallel to the x-axis is a horizontal line and has the form y equal to k, a constant. $y = -6$

11. First, write the given equation in slope-intercept form to find the slope of the given line.

$$x + 3y = 9$$
$$3y = -x + 9$$
$$y = -\frac{1}{3}x + 3$$

The slope of the given line is $-\frac{1}{3}$. The slope of a line perpendicular to the given line is the negative reciprocal of $-\frac{1}{3}$ or 3. Use this value, the point, and the point-slope formula to determine the equation of the perpendicular line.

$$y - (-1) = 3(x - (-3))$$
$$y + 1 = 3(x + 3)$$
$$y + 1 = 3x + 9$$
$$y = 3x + 8$$

13. **(a)** $y = 1.5x + 10$

(b) $y = 1.5(10) + 10$
$y = 15 + 10$
$y = 25$
The cost of attending the State Fair and going on 10 rides is $25.

15. **(a)** The relation does not define y as a function of x since 3 is paired with 1 and 3.

(b) $\{-3, -1, 1, 3\}$

(c) $\{3, -2, -1, 1\}$

17. **(a)** The relation does define y as a function of x since each x value is paired with exactly one y value.

(b) $\{1975, 1985, 1997\}$

(c) $\{47.4\%, 62.2\%, 72.1\%\}$

19. **(a)** $f(-3) = \frac{1}{3}(-3) + 6 = -1 + 6 = 5$

(b) $g(-3) = |-3| = 3$

(c) $f(0) + g(6) = \left(\frac{1}{3}(0) + 6\right) + |6|$
$= (0 + 6) + 6$
$= 6 + 6$
$= 12$

(d) $\frac{f(3)}{g(-1)} = \frac{\frac{1}{3}(3) + 6}{|-1|} = \frac{1 + 6}{1} = \frac{7}{1} = 7$

21. **(a)** $r(-2) = (-2 - 1)^3 = (-3)^3 = -27;$
$r(-1) = (-1 - 1)^3 = (-2)^3 = -8;$
$r(0) = (0 - 1)^3 = (-1)^3 = -1;$
$r(1) = (1 - 1)^3 = (0)^3 = 0;$
$r(2) = (2 - 1)^3 = (1)^3 = 1;$
$r(3) = (3 - 1)^3 = (2)^3 = 8$

(b) $(-\infty, \infty)$

23. (a) $s(0) = 1.6(0) + 36 = 0 + 36 = 36$
This means that in 1985 the per capita consumption of soft drinks in the United States was 36 gallons.
$s(7) = 1.6(7) + 36 = 11.2 + 36 = 47.2$
This means that in 1992 the per capita consumption of soft drinks in the United States was 47.2 gallons.

(b) $m = 1.6$ means that there is an increase of 1.6 gallons per year.

25. (a) ii

(b) v

(c) iv

Cumulative Review Exercises, Chapters 1–7

1.
$$\frac{1}{3}t+\frac{1}{5}=\frac{1}{10}(t-2)$$
$$\frac{1}{3}t+\frac{1}{5}=\frac{1}{10}t-\frac{1}{5}$$
$$\frac{1}{3}t-\frac{1}{10}t=-\frac{1}{5}-\frac{1}{5}$$
$$\frac{10}{30}t-\frac{3}{30}t=-\frac{2}{5}$$
$$\frac{7}{30}t=-\frac{2}{5}$$
$$\frac{30}{7}\cdot\frac{7}{30}t=\left(\frac{30}{7}\right)\left(-\frac{2}{5}\right)$$
$$y=-\frac{12}{7}$$

3. (a) $[6, \infty)$

(b) $(-\infty, 17)$

(c) $[-2, 3]$

5.
$$V=\frac{1}{3}\pi r^2 h$$
$$V=\frac{1}{3}\pi(8)^2(22)$$
$$V=\frac{1}{3}\pi(64)(22)$$
$$V=1474\text{ in}^3$$

7. The pitch of the roof is $\frac{\text{rise}}{\text{run}}$ where the rise is 13 feet and the run is $\frac{1}{2}(65)$ or 32.5.

Thus, the pitch is $\frac{13}{32.5}=\frac{2}{5}$.

9.
$$x+2x+(2x-5)=180$$
$$5x-5=180$$
$$5x=185$$
$$x=37;$$
$$2x=74;$$
$$2x-5=74-5=69$$
The measures of the angles are 37°, 74°, 69°.

11.
$$y=\frac{kx}{z}$$
$$y=\frac{k(9)}{\frac{1}{2}}$$
$$3=9k$$
$$\frac{1}{3}=k;$$
$$y=\frac{1}{3}\frac{x}{z}$$
$$y=\frac{1}{3}\cdot\frac{3}{4}$$
$$y=\frac{1}{4}$$

13.
$$(4b-3)(2b^2+1)=8b^3+4b-6b^2-3$$
$$=8b^3-6b^2+4b-3$$

15. $(6w^3-5w^2-2w)\div(2w^2)=\frac{6w^3-5w^2-2w}{2w^2}=\frac{6w^3}{2w^2}-\frac{5w^2}{2w^2}-\frac{2w}{2w^2}=3w-\frac{5}{2}-\frac{1}{w}$

17. $$\begin{aligned}\frac{2x^2+11x-21}{4x^2-10x+6}\div\frac{2x^2-98}{x^2-x+xa-a}&=\frac{2x^2+11x-21}{4x^2-10x+6}\cdot\frac{x^2-x+xa-a}{2x^2-98}\\&=\frac{(2x-3)(x+7)}{2(2x-3)(x-1)}\cdot\frac{(x-1)(x+a)}{2(x-7)(x+7)}\\&=\frac{x+a}{4(x-7)}\end{aligned}$$

19. $$\begin{aligned}\frac{x}{x^2+5x-50}-\frac{1}{x^2-7x+10}+\frac{1}{x^2+8x-20}&=\frac{x}{(x+10)(x-5)}-\frac{1}{(x-2)(x-5)}+\frac{1}{(x+10)(x-2)}\\&=\frac{x(x-2)-1(x+10)+1(x-5)}{(x+10)(x-5)(x-2)}\\&=\frac{x^2-2x-x-10+x-5}{(x+10)(x-5)(x-2)}\\&=\frac{x^2-2x-15}{(x+10)(x-5)(x-2)}\\&=\frac{(x-5)(x+3)}{(x+10)(x-5)(x-2)}\\&=\frac{x+3}{(x+10)(x-2)}\end{aligned}$$

21. $\frac{4y}{y+2}-\frac{y}{y-1}=\frac{9}{(y+2)(y-1)}$

Multiply both sides by the LCD: $(y+2)(y-1)$

$$\begin{aligned}4y(y-1)-y(y+2)&=9\\4y^2-4y-y^2-2y&=9\\3y^2-6y-9&=0\\3(y^2-2y-3)&=0\\3(y-3)(y+1)&=0\end{aligned}$$

$3=0$ or $y-3=0$ or $y+1=0$

No solution or $y=3$ or $y=-1$

The solution is $y=3$, $y=-1$.

23. $8x^2+2x-15=(2x+3)(4x-5)$

25. **(a)** The year 1996 corresponds to an x value of 1996 – 1988 = 8. Thus, $N(8)=420(8)+5260=8620$ students.

(b) $14902 = 420x + 5260$

$9642 = 420x$

$\frac{9642}{420} \approx 22.96$

$1988 + 22.96 = 2010.96$

If this linear trend continues, in the year 2011 the number of FTE will reach 14,902.

27. **(a)** $f(4) = \frac{1}{2}(4) - 1 = 2 - 1 = 1$

(b) $g(-3) = 3(-3)^2 - 2(-3)$

$= 3(9) + 6$

$= 27 + 6$

$= 33$

29. $I = krt$

$1120 = k(0.08)(2)$

$1120 = k(0.16)$

$7000 = k$

The variation model is $I = 7000rt$.

$I = 7000(0.10)(5)$

$I = \$3500$

Chapter 8

Section 8.1 Practice Exercises

1. $3x - y = 7$
$x - 2y = 4$ point: $(2, -1)$
Substitute the given point into both equations.
$3(2) - (-1) \stackrel{?}{=} 7$ ✓
$2 - 2(-1) \stackrel{?}{=} 4$ ✓

Because the given point is a solution to each equation, it is a solution to the system of equations. Yes

3. $2x - 3y = 12$
$3x + 4y = 12$ point: $(0, 4)$

Substitute the given point into both equations.
$2(0) - 3(4) \stackrel{?}{=} 12$
$3(0) + 4(4) \stackrel{?}{=} 12$

Because the given point is not a solution of either equation it is not a solution of the system of equations. No

5. $3x - 6y = 9$
$x - 2y = 3$ point: $\left(4, \frac{1}{2}\right)$
Substitute the given point into both equations.
$3(4) - 6\left(\frac{1}{2}\right) \stackrel{?}{=} 9$ ✓
$4 - 2\left(\frac{1}{2}\right) \stackrel{?}{=} 3$ ✓

Because the given point is a solution to each equation, it is a solution to the system of equations. Yes

7.(a) $y = 2x - 3$
$y = 2x + 5$

(b) $y = 2x + 1$
$y = 4x - 5$

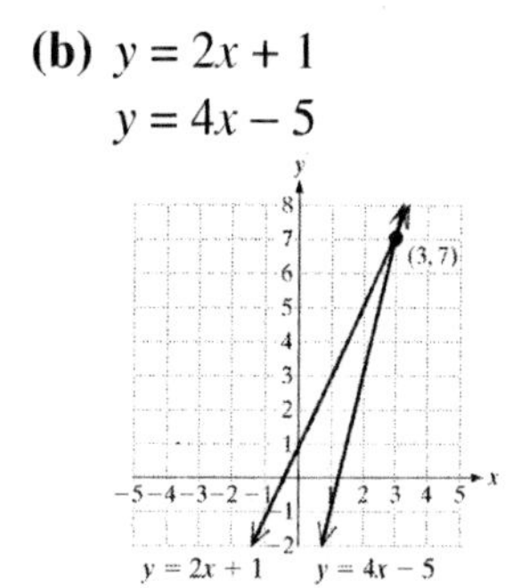

(c) $y = 3x - 5$
$y = 3x - 5$

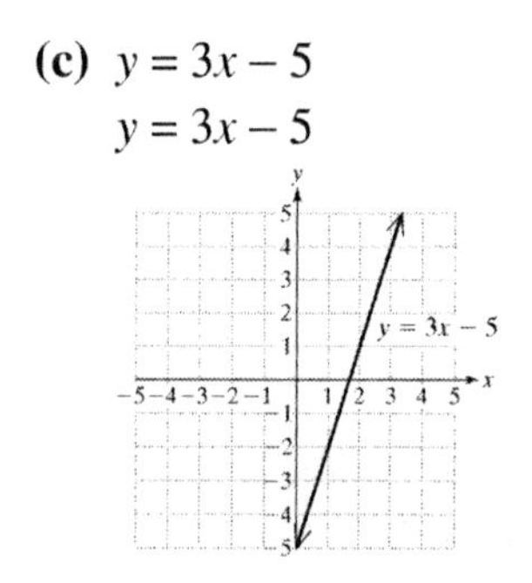

9. *c* Coinciding lines have the same slope and the same y-intercept.

11. *a* An inconsistent system graphs as parallel lines.

13. *a* The graph of the system is parallel lines.

15. *b* Intersecting lines always have different slopes.

17. *c* Coinciding lines or system with the same slopes and same y-intercept has infinitely many solution.

19. *b* Since parallel lines do not intersect, parallel lines represent a system with no solution.

21. *d* The lines intersect at the origin or $(0, 0)$.

23. $y = -x + 4$
$y = x - 2$

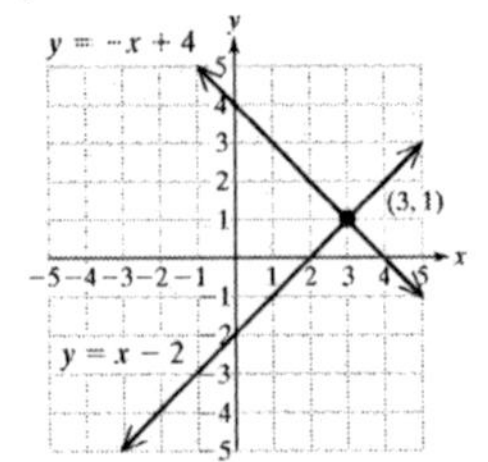

(3, 1) consistent; independent

25. $2x + y = 0$
$3x + y = 1$
Rewrite in slope-intercept form.
$y = -2x$
$y = -3x$

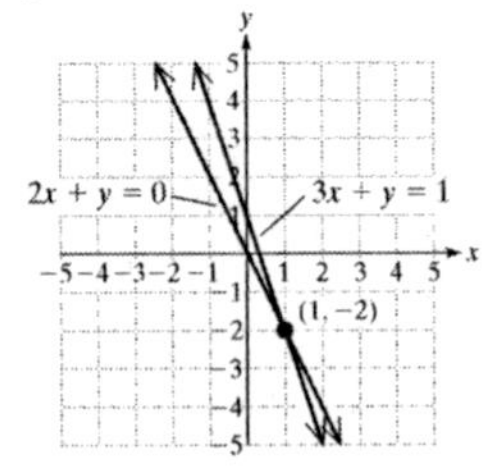

$(1, -2)$ consistent; independent

27. $2x + y = 6$
$x = 1$
Rewrite the first equation in slope-intercept form. The second equation is a vertical line through the point (1, 0).
$y = -2x + 6$

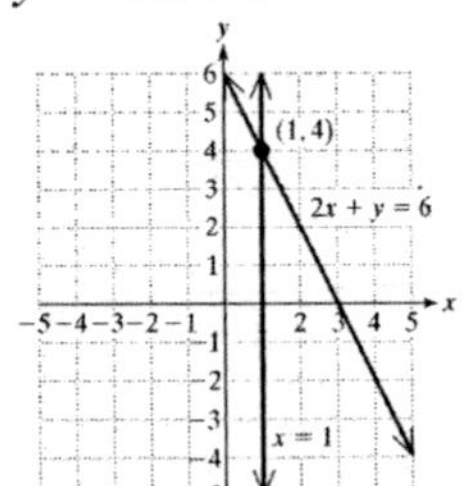

(1, 4) consistent; independent

29. $-6x - 3y = 0$
$4x + 2y = 4$
Rewrite both equations in slope-intercept form.
$y = -2x$
$y = -2x + 2$

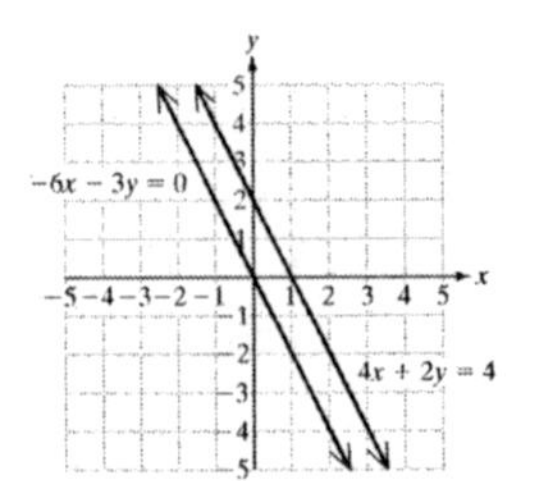

No solution; inconsistent; independent

31. $-2x + y = 3$
$6x - 3y = -9$
Rewrite both equations in slope-intercept form.
$y = 2x + 3$
$y = 2x + 3$

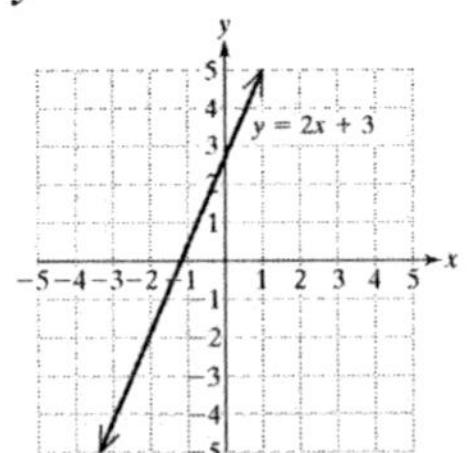

Infinitely many solutions
$\{(x, y) | y = 2x + 3\}$
consistent; dependent

33. $y = 6$
$2x + 3y = 12$
Rewrite the second equation in slope-intercept form. The first equation is a horizontal line through the point (0, 6).
$$y = -\frac{2}{3}x + 4$$

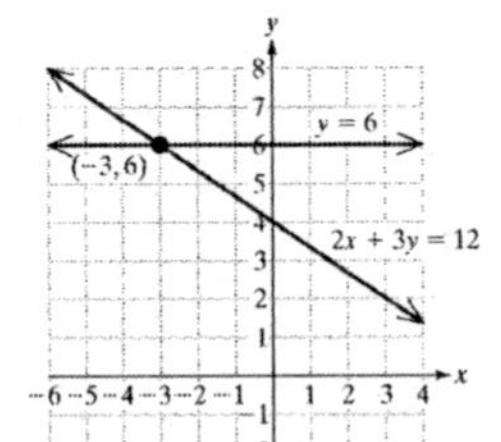

$(-3, 6)$ consistent; independent

35. $-5x + 3y = -9$
$$y = \frac{5}{3}x - 3$$
Rewrite first equation in slope-intercept form.

$y = \frac{5}{3}x - 3$

$y = \frac{5}{3}x - 3$

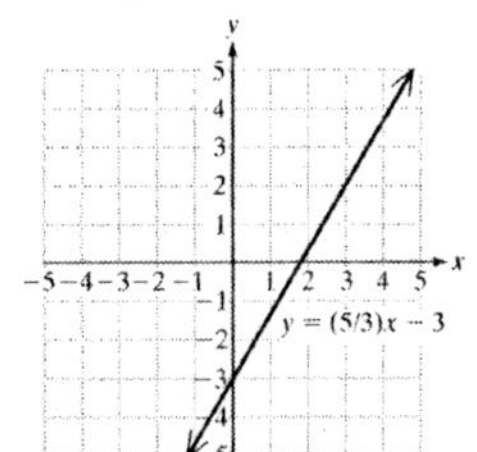

Infinitely many solutions.

$\left\{(x, y) \middle| y = \frac{5}{3}x - 3\right\}$

consistent; dependent

37. $x = 4 + y$

$3y = -3x$

Rewrite both equations in slope-intercept form.

$y = x - 4$

$y = -x$

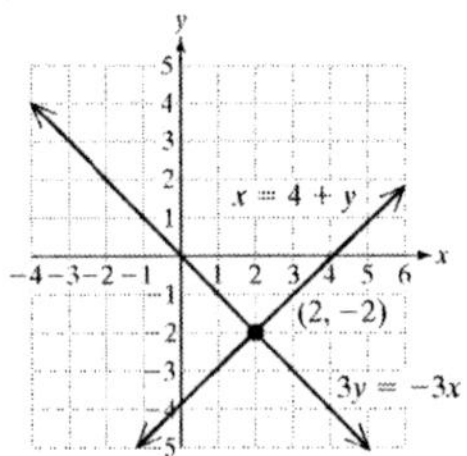

(2, –2) consistent; independent

39. $-x + y = 3$

$4y = 4x + 6$

Rewrite both equations in slope-intercept form.

$y = x + 3$

$y = x + \frac{3}{2}$

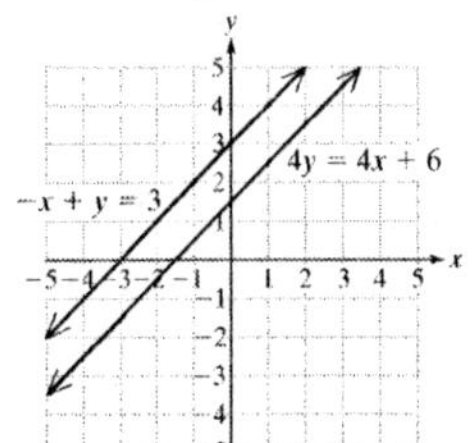

No solution; inconsistent; independent

41. $x = 4$

$2y = 4$

Solve the second equation for y.

$x = 4$

$y = 2$

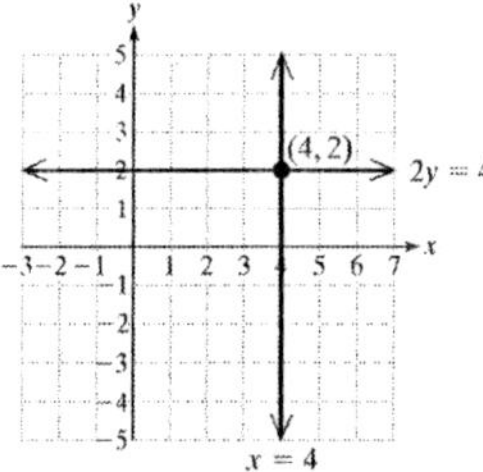

(4, 2) consistent; independent

43. $2x + 3y = 8$

$-4x - 6y = 6$

Rewrite both equations in slope-intercept form.

$y = -\frac{2}{3}x + \frac{8}{3}$

$y = -\frac{2}{3}x - 1$

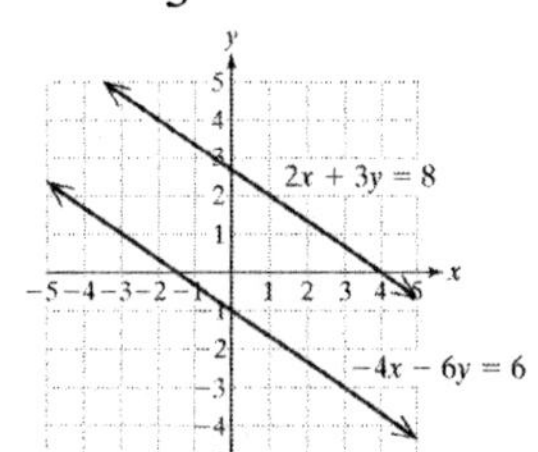

No solution; inconsistent; independent

45. $2x + y = 4$

$4x - 2y = -4$

Rewrite both equations in slope-intercept form.

$y = -2x + 4$

$y = 2x + 2$

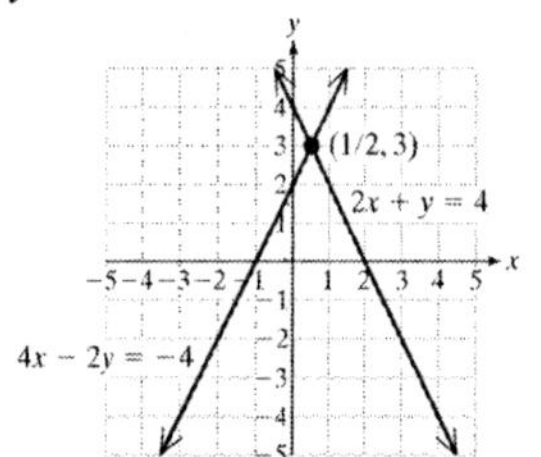

$\left(\frac{1}{2}, 3\right)$ consistent; independent

47. $y = 0.5x + 2$
$-x + 2y = 4$
Rewrite the second equation in slope-intercept form.
$y = 0.5x + 2$
$y = 0.5x + 2$

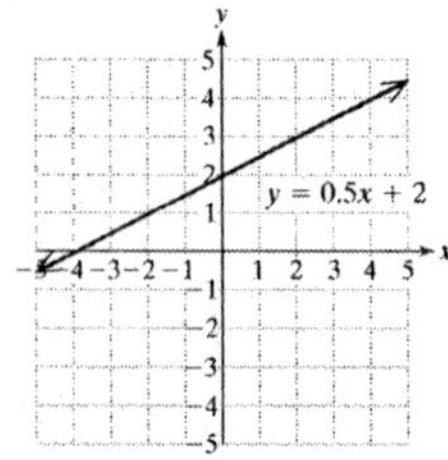

Infinitely many solutions
$\{(x, y) | y = 0.5x + 2\}$
consistent; dependent

49. The point of intersection gives the answer. Four lessons will cost \$120 for each instructor.

51. The point of intersection gives the answer. In 1997 the number of cases will be about the same.

53. For example: $4x + y = 9$
$-2x - y = -5$
These equations were found by simply putting together some combination of x and y and substituting $x = 2$ and $y = 1$ into them to determine the constant value.

55. For example: $2x + 2y = 1$
This equation was found by multiplying each side of the given equation by a different value.

57. (2, 1)

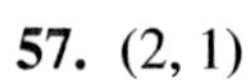

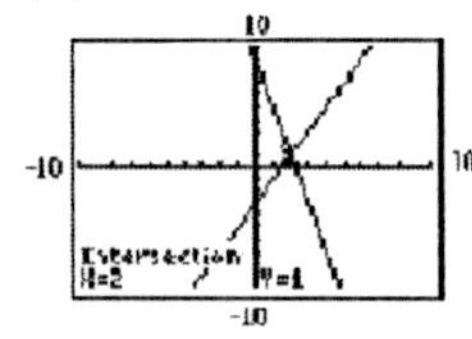

59. (3, 1)

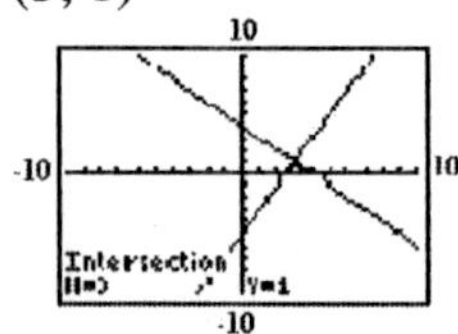

61. No solution

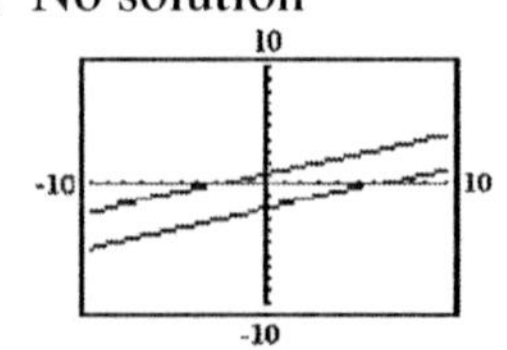

Section 8.2 Practice Exercises

1. $2x - y = 4$
$-2y = -4x + 8$
Rewrite both equations in slope-intercept form.
$y = 2x - 4$
$y = 2x - 4$
Since the slopes and the y-intercepts are equal, they are coinciding lines.

3. $2x + 3y = 6$
$x - y = 5$
Rewrite both equations in slope-intercept form.
$y = -\frac{2}{3}x + 2$
$y = x - 5$
Since the slopes are unequal, the lines intersect.

5. $2x = \frac{1}{2}y + 2$
$4x - y = 13$
Rewrite both equations in slope-intercept form.
$y = 4x - 4$
$y = 4x - 13$
Since the slopes are equal but the y-intercepts are different, the lines are parallel.

7. $3x + 2y = -3$
$y = 2x - 12$
The second equation is solved for y. Substitute this value for y into the first equation.

$$\begin{aligned} 3x+2y&=-3 \\ 3x+2(2x-12)&=-3 \\ 3x+4x-24&=-3 \\ 7x-24&=-3 \\ 7x&=21 \\ x&=3 \end{aligned}$$

$y = 2x - 12$
$y = 2(3) - 12$
$y = -6$
Solution: $(3, -6)$

9. $x=-4y+16$
$3x+5y=20$
The first equation is solved for x. Substitute this value for x into the second equation.

$$\begin{aligned} 3x+5y&=20 \\ 3(-4y+16)+5y&=20 \\ -12y+48+5y&=20 \\ -7y+48&=20 \\ -7y&=-28 \\ y&=4 \end{aligned}$$

$x = -4y + 16$
$x = -4(4) + 16$
$x = 0$
Solution: $(0, 4)$

11. $3x+5y=7$
$y=-\frac{3}{5}x+3$
The second equation is solved for y. Substitute this value for y into the first equation.

$$\begin{aligned} 3x+5y&=7 \\ 3x+\left(-\frac{3}{5}x+3\right)&=7 \\ 3x-3x+15&=7 \\ 15&=7 \quad \text{Inconsistent} \end{aligned}$$

There is no solution.

13. $x=\frac{6}{5}y+3$
$5x-6y=15$
The first equation is solved for x. Substitute this value for x into the second equation.

$$\begin{aligned} 5x-6y&=15 \\ 5\left(\frac{6}{5}y+3\right)-6y&=15 \\ 6y+15-6y&=15 \\ 15&=15 \quad \text{Identity} \end{aligned}$$

Infinitely many solutions.
Solution: $\left\{(x, y) \middle| x=\frac{6}{5}y+3\right\}$

15. **(a)** y in the second equation is easier to solve for because its coefficient is 1.

(b) $4x-2y=-6$
$3x+y=8$
Solving the second equation for y gives $y = -3x + 8$. Substitute this value for y into the first equation.

$$\begin{aligned} 4x-2y&=-6 \\ 4x-2(-3x+8)&=-6 \\ 4x+6x-16&=-6 \\ 10x-16&=-6 \\ 10x&=10 \\ x&=1 \end{aligned}$$

$$\begin{aligned} 3x+y&=8 \\ 3(1)+y&=8 \\ 3+y&=8 \\ y&=5 \end{aligned}$$

Solution: $(1, 5)$

17. $4x-y=-1$
$2x+4y=13$
Solving the first equation for y gives $y = 4x + 1$. Substitute this value for y into the second equation.

$$\begin{aligned} 2x+4y&=13 \\ 2x+4(4x+1)&=13 \\ 2x+16x+4&=13 \\ 18x+4&=13 \\ 18x&=9 \\ x&=\frac{1}{2} \end{aligned}$$

$$4x - y = -1$$
$$4\left(\frac{1}{2}\right) - y = -1$$
$$2 - y = -1$$
$$-y = -3$$
$$y = 3$$

Solution: $\left(\frac{1}{2}, 3\right)$

19. $x - 3y = -1$
$2x = 4y + 2$

Solving the first equation for x gives $x = 3y - 1$. Substitute this value for x into the second equation.

$$2x = 4y + 2$$
$$2(3y - 1) = 4y + 2$$
$$6y - 2 = 4y + 2$$
$$2y - 2 = 2$$
$$2y = 4$$
$$y = 2$$
$$x - 3y = -1$$
$$x - 3(2) = -1$$
$$x = 5$$

Solution: (5, 2)

21. $-2x + 5y = 5$
$x - 4y = -10$

Solving the second equation for x gives $x = 4y - 10$. Substitute this value for x into the first equation.

$$-2x + 5y = 5$$
$$-2(4y - 10) + 5y = 5$$
$$-8y + 20 + 5y = 5$$
$$-3y + 20 = 5$$
$$-3y = -15$$
$$y = 5$$
$$x - 4y = -1$$
$$x - 4(5) = -10$$
$$x = 10$$

Solution: (10, 5)

23. $3x + 2y = -1$
$\frac{3}{2}x + y = 4$

Solving the second equation for y gives $y = -\frac{3}{2}x + 4$. Substitute this value for y into the first equation.

$$3x + 2y = -1$$
$$3x + 2\left(-\frac{3}{2}x + 4\right) = -1$$
$$3x - 3x + 8 = -1$$
$$8 = -1$$

No solution

25. $10x - 30y = -10$
$2x - 6y = -2$

None of the variables in either equation have a coefficient of 1. You may choose either equation to solve for either variable. Solving the first equation for x gives $x = 3y - 1$. Substitute this value for x into the second equation.

$$2x - 6y = -2$$
$$2(3y - 1) - 6y = -2$$
$$6y - 2 - 6y = -2$$
$$-2 = -2 \quad \text{Identity}$$

Infinitely may solutions

Solution: $\left\{(x, y) \middle| y = \frac{1}{3}x + \frac{1}{3}\right\}$

27. $2x + y = 3$
$y = -7$

A value for been determined. To find a value for x, substitute $y = -7$ into the first equation.

$$2x + y = 3$$
$$2x - 7 = 3$$
$$2x = 10$$
$$x = 5$$

Solution: (5, –7)

29. $x + 2y = -2$
$4x = -2y - 17$

Solving the first equation for x gives $x = -2y - 2$. Substitute the value for x into the second equation.

$$4x = -2y - 17$$
$$4(-2y - 2) = -2y - 17$$
$$-8y - 8 = -2y - 17$$
$$-6y - 8 = -17$$
$$-6y = -9$$
$$y = \frac{3}{2}$$

$$x+2y=-2$$
$$x+2\left(\frac{3}{2}\right)=-2$$
$$x+3=-2$$
$$x=-5$$

Solution: $\left(-5, \frac{3}{2}\right)$

31. $y=-\frac{1}{2}x-4$

$y=4x-13$

The first equation is solved for y. Substitute this value into the second equation.

$$y=4x-13$$
$$-\frac{1}{2}x-4=4x-13$$
$$-x-8=8x-26$$
$$-9x-8=-26$$
$$-9x=-18$$
$$x=2$$
$$y=-\frac{1}{2}x-4$$
$$y=-\frac{1}{2}(2)-4$$
$$y=-5$$

Solution: $(2, -5)$

33. $y=-2x+3$

$y-4=-2(x+3)$

The first equation is solved for y. Substitute this value into the second equation.

$$y-4=-2(x+3)$$
$$-2x+1-4=-2(x+3)$$
$$-2x-3=-2x-6$$
$$-3=-6$$

No solution

35. $3x+2y=4$

$2x-3y=-6$

None of the variables in either equation have a coefficient of 1. You may choose either equation to solve for either variable. Solving the first equation for x gives $x=-\frac{2}{3}y+\frac{4}{3}$. Substitute this value for x into the second equation.

$$2x-3y=-6$$
$$2\left(-\frac{2}{3}y+\frac{4}{3}\right)-3y=-6$$
$$-\frac{4}{3}y+\frac{8}{3}-3y=-6$$
$$-4y+8-9y=-18$$
$$-13y+8=-18$$
$$-13y=-26$$
$$y=2$$
$$3x+2y=4$$
$$3x+2(2)=4$$
$$3x=0$$
$$x=0$$

Solution: $(0, 2)$

37. $y=0.25x+1$

$-x+4y=4$

The first equation is solved for y. Substitute this value into the second equation.

$$-x+4y=4$$
$$-x+4(0.25x+1)=4$$
$$-x+x+4=4$$
$$4=4 \quad \text{Identity}$$

Infinitely many solutions

Solution: $\{(x, y) | y = 0.25x + 1\}$

39. $11x+6y=17$

$5x-4y=1$

None of the variables in either equation has a coefficient of 1. You may choose either equation to solve for either variable. Solving the second equation for x gives $x=\frac{4}{5}y+\frac{1}{5}$.

Substitute this value for x into the first equation.

$$11x+6y=17$$
$$11\left(\frac{4}{5}y+\frac{1}{5}\right)+6y=17$$
$$\frac{44}{5}y+\frac{11}{5}+6y=17$$
$$44y+11+30y=85$$
$$74y+11=85$$
$$74y=74$$
$$y=1$$

$$5x-4y=1$$
$$5x-4(1)=1$$
$$5x=5$$
$$x=1$$
Solution: (1, 1)

41. $x+2y=4$
$4y=-2x-8$

Solving the first equation for x gives $x = -2y + 4$. Substitute the value for x into the second equation.

$$4y=-2x-8$$
$$4y=-2(-2y+4)-8$$
$$4y=4y-8-8$$
$$4y=4y-16$$
$$0=-16$$
Contradiction; no solution

43. $\frac{1}{3}(2x+y)=1$
$x+y=4$

Solving the second equation for y gives $y = -x + 4$. Substitute the value for y into the first equation.

$$\frac{1}{3}(2x+y)=1$$
$$\frac{1}{3}(2x+(-x+4))=1$$
$$\frac{1}{3}(2x-x+4)=1$$
$$\frac{1}{3}(x+4)=1$$
$$x+4=3$$
$$x=-1$$
$$x+y=4$$
$$-1+y=4$$
$$y=5$$
Solution: (−1, 5)

45. $\frac{a}{3}+\frac{b}{2}=-4$
$a-3b=6$

Solving the second equation for a gives $a = 3b + 6$. Substitute the value for a into the first equation.

$$\frac{a}{3}+\frac{b}{2}=-4$$
$$\frac{3b+6}{3}+\frac{b}{2}=-4$$
$$2(3b+6)+3b=-24$$
$$6b+12+3b=-24$$
$$9b+12=-24$$
$$9b=-36$$
$$b=-4$$
$$a-3b=6$$
$$a-3(-4)=6$$
$$a=-6$$
Solution: (−6, −4)

47. Let x represent one number. Let y represent the other number. The statement "two numbers have a sum of 106" translates to the equation $x + y = 106$. The statement "one number is 10 less than the other" translates to the equation $x = y - 10$.

$$x+y=106$$
$$x=y-10$$

The second equation is solved for x. Substitute this value into the first equation.

$$x+y=106$$
$$y-10+y=106$$
$$2y-10=106$$
$$2y=116$$
$$y=58$$
$$x=y-10$$
$$x=58-10$$
$$x=48$$
The numbers are 48 and 58.

49. Let x represent the measure of one angle. Let y represent the measure of the second angle. The statement "two angles are supplementary" translates to the equation $x + y = 180$. The statement "one angle is 15° more than 10 times the other angle" translates to the equation $x + 10y + 15$.

$x + y = 180$
$x = 10y + 15$

The second equation is solved for x. Substitute this value into the first equation.

$$\begin{aligned} x + y &= 180 \\ 10y + 15 + y &= 180 \\ 11y + 15 &= 180 \\ 11y &= 165 \\ y &= 15 \end{aligned}$$

$x = 10y + 15$
$x = 10(15) + 15$
$x = 165$

The measures of the angles are 165° and 15°.

51. Let x represent the measure of the first angle. Let y represent the measure of the second angle. The statement "two angles are complementary" translates to the equation $x + y = 90$. the statement "one angle is 10° more than 3 times the other angle" translates to the equation $x = 3y + 10$.

$x + y = 90$
$x = 3y + 10$

The second equation is solved for x. Substitute this value into the first equation.

$$\begin{aligned} x + y &= 90 \\ 3y + 10 + y &= 90 \\ 4y &= 80 \\ y &= 20 \end{aligned}$$

$x = 3y + 10$
$x = 3(20) + 10$
$x = 70$

The measures of the angles are 70° and 20°.

53. Let x represent the measure of one of the acute angles. Let y represent the measure of the other acute angle. The sum of the measures of the angles of a triangle is 180°. In a right triangle one of the angles is a right angle measuring 90°. Thus the sum of the measures of the other two angles is 90°. From this information comes the first equation: $x + y = 90$. The statement "one of the acute angles is 6° less than the other acute angle" translates to the equation $x = y - 6$.

$x + y = 90$
$x = y - 6$

The second equation is solved for x. Substitute this value into the first equation.

$$\begin{aligned} x + y &= 90 \\ y - 6 + y &= 90 \\ 2y - 6 &= 90 \\ 2y &= 96 \\ y &= 48 \end{aligned}$$

$x = y - 6$
$x = 48 - 6$
$x = 42$

The measures of the angles are 42° and 48°.

55. Let x represent the cost of a soft drink. Let y represent the cost of a hot dog. The statement "the total cost of a soft drink and a hot dog is \$2.50" translates to the equation $x + y = 2.50$. The statement "the price of a hot dog is \$1.00 more than the cost of the soft drink" translates to the equation $y = x + 1.00$.

$x + y = 2.50$
$y = x + 1.00$

The second equation is solved for x. Substitute this value into the first equation.

$$\begin{aligned} x + y &= 2.50 \\ x + x + 1.00 &= 2.50 \\ 2x + 1.00 &= 2.50 \\ 2x &= 1.50 \\ x &= 0.75 \end{aligned}$$

$y = x + 1.00$
$y = 0.75 + 1.00$
$y = 1.75$

A hot dog costs \$1.75 and a soft drink costs \$0.75.

57. (a) Since the point of intersection falls between grid lines, it is difficult to determine the exact number of months. It appears that the number is about 22 months.

(b) $y = 20x + 55$
$y = 22.50x$
The second equation is solved for y. Substitute this value into the first equation.

$$y = 20x + 55$$
$$22.50x = 20x + 55$$
$$2.50x = 55$$
$$x = 22$$

$y = 22.50x$
$y = 22.50(22)$
$y = 495$
(22, 495); when rented for 22 months, both companies charge $495.

(c) Company B is more expensive if the system is rented for more than 22 months. Company A is more expensive if the system is rented for less than 22 months.

59. It is stated that the system is dependent. Therefore, x can equal any real number. Pick a value for x, substitute that value into one of the equations and determine a corresponding value for y.
If $x = 0$, then $y = 3$. (0, 3)
If $x = 1$, then $y = 5$. (1, 5)
If $x = -1$, then $y = 1$. (−1, 1)

Section 8.3 Practice Exercises

1. $x + y = 8$
$y = x - 2$ Ordered pair: (5, 3)

$5 + 3 \stackrel{?}{=} 8$ ✓
$3 \stackrel{?}{=} 5 - 2$ ✓ A solution
Since the ordered pair is a solution to both equations, it is a solution of the system.

3. $x = y + 1$
$-x + 2y = 0$ Ordered pair: (3, 2)

$3 \stackrel{?}{=} 2 + 1$ ✓
$-3 + 2(2) \stackrel{?}{=} 0$ Not a solution
Since the ordered pair is not a solution to the second equation, it is not a solution of the system.

5. (a) False. Remember, the process is to have opposite coefficients. Multiply the second equation by −2.

(b) True. This will create opposite coefficients on the x-variable.

(c) True. This will create opposite coefficients on the x-variable.

7. (a) y-variable, because it has a coefficient of one in the second equation.

(b) $3x - 4y = 2$
$17x + y = 35$
Both equations are already written in standard form. There are no fractions or decimals. Neither set of coefficients are opposites. Multiplying the second equation by 4 will create opposite coefficients on y.

$$\begin{array}{r} 3x - 4y = 2 \\ \underline{68x + 4y = 140} \\ 71x = 142 \\ x = 2 \end{array}$$

$$3(2) - 4y = 2$$
$$6 - 4y = 2$$
$$-4y = -4$$
$$y = 1$$

Solution: (2, 1)

9. Since the statement is a contradiction, the system will have no solutions. The lines are parallel.

11. Since the statement is an identity, the system will have infinitely many solutions. The lines coincide.

13. $x + 2y = 8$
$5x - 2y = 4$
Both equations are already written in standard form. Coefficients on y are opposites. Add the two equations and then solve the resulting equation.

$$\begin{array}{r} x + 2y = 8 \\ \underline{5x - 2y = 4} \\ 6x = 12 \\ x = 2 \end{array}$$

$2+2y=8$
$2y=6$
$y=3$
Solution: (2, 3)

15. $a+b=3$
$3a+b=13$
Both equations are already written in standard form. Neither set of coefficients are opposites. Multiplying the first equation by −1 will result in opposite coefficients on b.

$$\begin{array}{r} -a-b=-3 \\ 3a+b=13 \\ \hline 2a=10 \\ a=5 \end{array}$$

$5+b=3$
$b=-2$
Solution: (5, −2)

17. $-3x+y=1$
$-6x-2y=-2$
Both equations are already written in standard form. Neither set of coefficients is opposites. Multiplying the first equation by 2 will result in opposite coefficients on y.

$$\begin{array}{r} -6x+2y=2 \\ -6x-2y=-2 \\ \hline -12x=0 \\ x=0 \end{array}$$

$-3(0)+y=1$
$0+y=1$
$y=1$
Solution: (0, 1)

19. $3x-5y=13$
$x-2y=5$
Both equations are already written in standard form. Neither set of coefficients is opposites. Multiplying the second equation by −3 will result in opposite coefficients on x.

$$\begin{array}{r} 3x-5y=13 \\ -3x+6y=-15 \\ \hline y=-2 \end{array}$$

$x-2(-2)=5$
$x+4=5$
$x=1$
Solution: (1, −2)

21. $-2x+y=-5$
$8x-4y=12$
Both equations are already written in standard form. Neither set of coefficients is opposites. Multiplying the first equation by 4 will result in opposite coefficients on y.

$$\begin{array}{r} -8x+4y=-20 \\ 8x-4y=12 \\ \hline 0=-8 \end{array}$$

Contradiction; no solution

23. $x+2y=2$
$-3x-6y=-6$
Both equations are already written in standard form. Neither set of coefficients is opposites. Multiplying the first equation by 3 will result in opposite coefficients on x.

$$\begin{array}{r} 3x+6y=6 \\ -3x-6y=-6 \\ \hline 0=0 \end{array} \quad \text{Identity}$$

Infinitely many solutions

Solution: $\left\{(x, y) \middle| y=-\frac{1}{2}x+1\right\}$

25. $3a+2b=11$
$7a-3b=-5$
Both equations are already written in standard form. Neither set of coefficients is opposites. However, the signs of the coefficients on b are opposites. Choose to eliminate b by multiplying the first equation by 3 and the second equation by 2 to obtain opposite coefficients on b.

$$\begin{array}{r} 9a+6b=33 \\ 14a-6b=-10 \\ \hline 23a=23 \\ a=1 \end{array}$$

$3(1)+2b=11$
$3+2b=11$
$2b=8$
$b=4$
Solution: (1, 4)

27. $3x-5y=7$
$5x-2y=-1$

Both equations are already written in standard form. Neither set of coefficients is opposites. Neither set of coefficients has opposite signs. Choose either variable to eliminate. x will be the chosen variable in this solution. To obtain opposite coefficients on x multiply the first equation by 5 and the second equation by –3.

$$\begin{aligned} 15x-25y&=35\\ -15x+6y&=3\\ \hline -19y&=38\\ y&=-2 \end{aligned}$$

$$\begin{aligned} 3x-5(-2)&=7\\ 3x+10&=7\\ 3x&=-3\\ x&=-1 \end{aligned}$$

Solution: (–1, –2)

29. $2(x+1)=-3y+9$
$3x-10=-4y$

Write both equations in standard form.
$2x+3y=7$
$3x+4y=10$

Neither set of coefficients are opposites of each other. Neither variable has coefficients with opposite signs. Choose either variable to eliminate. x will be the chosen variable in this solution. To obtain opposite coefficients on x multiply the first equation by –3 and the second equation by 2.

$$\begin{aligned} -6x-9y&=-21\\ 6x+8y&=20\\ \hline -y&=-1\\ y&=1 \end{aligned}$$

$$\begin{aligned} 3x-10&=-4(1)\\ 3x-10&=-4\\ 3x&=6\\ x&=2 \end{aligned}$$

Solution: (2, 1)

31. $4x-5y=0$
$8(x-1)=10y$

Write the second equation in standard form.
$4x-5y=0$
$8x-10y=8$

The coefficient on x in the second equation is a multiple of the coefficient on x in the first equation. Multiply the first equation by –2 to obtain opposite coefficients on x.

$$\begin{aligned} -8x+10y&=0\\ 8x-10y&=8\\ \hline 0&=8 \end{aligned}$$

Contradiction; no solution

33. $5x-2y=4$
$y=-3x+9$

Since the second equation is solved for y, use the substitution method by substituting this value for y into the first equation and solving this new equation for x.

$$\begin{aligned} 5x-2(-3x+9)&=4\\ 5x+6x-18&=4\\ 11x-18&=4\\ 11x&=22\\ x&=2 \end{aligned}$$

$y=-3(2)+9$
$y=3$
Solution: (2, 3)

35. $x+y=6$
$x-y=1$

Both equations are written in standard form. The coefficients on y are opposites. Use the addition method. Add the equations and solve the resulting equation.

$$\begin{aligned} x+y&=6\\ x-y&=1\\ \hline 2x&=7\\ x&=\frac{7}{2} \end{aligned}$$

$$\begin{aligned} \frac{7}{2}+y&=6\\ y&=\frac{5}{2} \end{aligned}$$

Solution: $\left(\frac{7}{2},\frac{5}{2}\right)$

37. $3x = 5y - 9$

$2y = 3x + 3$

Since neither equation is solved for x or y, consider using the addition method. Write both equations in standard form.

$3x - 5y = -9$

$-3x + 2y = 3$

The coefficients on x are opposites. Add the two equations and then solve the resulting equation.

$$\begin{aligned} 3x - 5y &= -9 \\ -3x + 2y &= 3 \\ \hline -3y &= -6 \\ y &= 2 \end{aligned}$$

$$\begin{aligned} 3x &= 5(2) - 9 \\ 3x &= 1 \\ x &= \frac{1}{3} \end{aligned}$$

Solution: $\left(\frac{1}{3}, 2\right)$

39. $y = -5x + 1$

$15x - 3 = -3y$

Since the first equation is solved for y, use the substitution method. Substitute the value of y into the second equation and solve the resulting equation for x.

$15x - 3 = -3(-5x + 1)$

$15x - 3 = 15x - 3$

$-3 = -3$ Identity

Infinitely many solutions

Solution: $\{(x, y) | y = -5x + 1\}$

41. $x + 2y = 4$

$x - y = -1$

Both equations are in standard form. Use the addition method. Since the signs on y are opposites, multiply the second equation by 2 to obtain opposite coefficients on y.

$$\begin{aligned} x + 2y &= 4 \\ 2x - 2y &= -2 \\ \hline 3x &= 2 \\ x &= \frac{2}{3} \end{aligned}$$

$$\begin{aligned} \frac{2}{3} - y &= -1 \\ -y &= -\frac{5}{3} \\ y &= \frac{5}{3} \end{aligned}$$

Solution: $\left(\frac{2}{3}, \frac{5}{3}\right)$

43. $8x - 16y = 24$

$2x - 4y = 0$

Since both equations are in standard form, use the addition method. The coefficient on x in the first equation is a multiple of the coefficient on x in the second equation. Multiply the second equation by –4 to obtain opposite coefficients on x.

$$\begin{aligned} 8x - 16y &= 24 \\ -8x + 16y &= 0 \\ \hline 0 &= 24 \end{aligned}$$

Contradiction; no solution

45. $\frac{m}{2} + \frac{n}{5} = \frac{13}{10}$

$3(m - n) = m - 10$

Since neither equation is solved for m and n, consider using the addition method. Write both equations in standard form.

$5m + 2n = 13$

$2m - 3n = -10$

Neither set of coefficients are opposites of each other. The n-variable has coefficients with opposite signs. To obtain opposite coefficients on n multiply the first equation by 3 and the second equation by 2.

$$\begin{aligned} 15m + 6n &= 39 \\ 4m - 6n &= -20 \\ \hline 19m &= 19 \\ m &= 1 \end{aligned}$$

$$\begin{aligned} 3(1 - n) &= 1 - 10 \\ 3 - 3n &= -9 \\ -3n &= -12 \\ n &= 4 \end{aligned}$$

Solution: $(1, 4)$

47. $2(p-3q) = p+4$
$3p+8 = 5p-q$

Since neither equation is solved for p or q, consider using the addition method. Write both equations in standard form.

$p-6q = 4$
$-2p+q = -8$

The signs of the p-variable are opposite. Multiply the first equation by 2 to obtain opposite coefficients on p.

$$\begin{array}{r} 2p-12q=8 \\ -2p+q=-8 \\ \hline -11q=0 \\ q=0 \end{array}$$

$$\begin{aligned} 3p+8&=5p-0 \\ -2p+8&=0 \\ -2p&=-8 \\ p&=4 \end{aligned}$$

Solution: (4, 0)

49. $9a-2b = 8$
$6(3a+1) = 4b+22$

Since neither equation is solved for a or b, consider using the addition method. Write both equations in standard form.

$9a-2b = 8$
$18a-4b = 16$

Neither variable has opposite signs nor opposite coefficients. The coefficient on a in the second equation is a multiple of the coefficient in the first equation. Multiply the first equation by –2 to obtain opposite coefficients on a.

$$\begin{array}{r} -18a+4b=-16 \\ 18a-4b=16 \\ \hline 0=0 \end{array} \quad \text{Identity}$$

Infinitely many solutions

Solution: $\left\{(a, b) \middle| b = \frac{9}{2}a - 4\right\}$

51. Let x represent the first positive number. Let y represent the second positive number. The statement "sum of two positive numbers is 26" translates to the equation $x + y = 26$. The statement "their difference is 14" translates to the equation $x - y = 14$.

$x+y = 26$
$x-y = 14$

The coefficients on y are opposites. Add the two equations and solve the resulting equation.

$$\begin{array}{r} x+y=26 \\ x-y=14 \\ \hline 2x=40 \\ x=20 \end{array}$$

$$\begin{aligned} 20+y&=26 \\ y&=6 \end{aligned}$$

The two positive numbers are 20 and 6.

53. Let x represent the smaller number. Let y represent the larger number. The statement "eight times the smaller of two numbers plus 2 times the larger number is 44" translates to the equation $8x + 2y = 44$. The statement "three times the smaller number minus 2 times the larger number is zero" translates to the equation $3x - 2y = 0$.

$8x+2y = 44$
$3x-2y = 0$

The coefficients on y are opposites. Add the two equations and solve the resulting equation.

$$\begin{array}{r} 8x+2y=44 \\ 3x-2y=0 \\ \hline 11x=44 \\ x=4 \end{array}$$

$$\begin{aligned} 8(4)+2y&=44 \\ 32+2y&=44 \\ 2y&=12 \\ y&=6 \end{aligned}$$

The two numbers are 4 and 6.

55. Let x represent the number of calories in a piece of cake. Let y represent the number of calories in a scoop of ice cream. The statement "number of calories in a piece of cake is 20 less than 3 times the number of calories in a scoop of ice cream" translates to the equation $x = 3y - 20$. The statement "together, the cake and ice cream have 460 calories" translates to the equation $x + y = 460$.

$$x = 3y - 20$$
$$x + y = 460$$

Since the first equation is solved for x, use the substitution method. Substitute the value for x into the second equation and solve the resulting equation for y.

$$3y - 20 + y = 460$$
$$4y - 20 = 460$$
$$4y = 480$$
$$y = 120$$

$$x = 3(120) - 20$$
$$x = 360 - 20$$
$$x = 340$$

A piece of cake has 340 calories and a scoop of ice cream has 120 calories.

57. $2x + 3y = 6$
$x - y = 5$

Multiply the second equation by -2 to obtain opposite coefficients on x.

$$\begin{array}{r} 2x + 3y = 6 \\ -2x + 2y = -10 \\ \hline 5y = -4 \end{array}$$
$$y = -\frac{4}{5}$$

To determine a value for x, perform the addition method again this time eliminating y.

Since the signs on y are opposites, multiply the second equation by 3 to obtain opposite coefficients on y.

$$\begin{array}{r} 2x + 3y = 6 \\ 3x - 3y = 15 \\ \hline 5x = 21 \end{array}$$
$$x = \frac{21}{5}$$

Solution: $\left(\frac{21}{5}, -\frac{4}{5}\right)$

59. $2x - 5y = 4$
$3x - 3y = 4$

To obtain opposite coefficients on x, multiply the first equation by -3 and the second equation by 2.

$$\begin{array}{r} -6x + 15y = -12 \\ 6x - 6y = 8 \\ \hline 9y = -4 \end{array}$$
$$y = -\frac{4}{9}$$

To determine a value for x, perform the addition method again, this time eliminating y.

$$2x - 5y = 4$$
$$3x - 3y = 4$$

To obtain opposite coefficients on y, multiply the first equation by 3 and the second equation by -5.

$$\begin{array}{r} 6x - 15y = 12 \\ -15x + 15y = -20 \\ \hline -9x = -8 \end{array}$$
$$x = \frac{8}{9}$$

Solution: $\left(\frac{8}{9}, -\frac{4}{9}\right)$

61. **(a)** Substitution method because the first equation is solved for x.

(b) $x = -2y + 5$
$2x - 4y = 10$

Substitute the value of x into the second equation and solve the resulting equation.

$$2x - 4y = 10$$
$$2(-2y + 5) - 4y = 10$$
$$-4y + 10 - 4y = 10$$
$$-8y + 10 = 10$$
$$-8y = 0$$
$$y = 0$$

$$x = -2y + 5$$
$$x = -2(0) + 5$$
$$x = 5$$

Solution: (5, 0)

63.(a) Addtion method because both equations are written in standard form and the y coefficients have opposite signs.

(b) $3x-6y=30$
$2x+3y=-22$
To obtain opposite coefficients on y, multiply the second equation by 2.
$$\begin{aligned}3x-6y&=30\\ \underline{4x+6y}&\underline{=-44}\\ 7x&=-14\\ x&=-2\end{aligned}$$

$$\begin{aligned}3(-2)-6y&=30\\ -6y&=36\\ y&=-6\end{aligned}$$
Solution: $(-2,-6)$

65. (a) Substitution method because the first equation is solved for y.

(b) $y=0.4x-0.3$
$-4x+10y=20$
Substitute the value for y into the second equation and solve the resulting equation.
$$\begin{aligned}-4x+10(0.4x-0.3)&=20\\ -4x+4x-3&=20\\ -3&=20\end{aligned}$$
No solution

67. If (1, 2) is a solution to the system, then substitute $x=1$ and $y=2$ into the equations and solve for A and B.

First equation:
$$\begin{aligned}A(1)+3(2)&=8\\ A+6&=8\\ A&=2\end{aligned}$$

Second equation:
$$\begin{aligned}1+B(2)&=-7\\ 2B&=-8\\ B&=-4\end{aligned}$$

Section 8.4 Practice Exercises

1. $-2x+y=6$
$2x+y=2$

(a) Graphing Method: The x- and y-intercepts of $-2x+y=6$ are $(-3, 0)$ and $(0, 6)$. The x- and y-intercepts of $2x+y=2$ are $(1, 0)$ and $(0, 2)$. Use these points to graph the equations.
Point of intersection: $(-1, 4)$

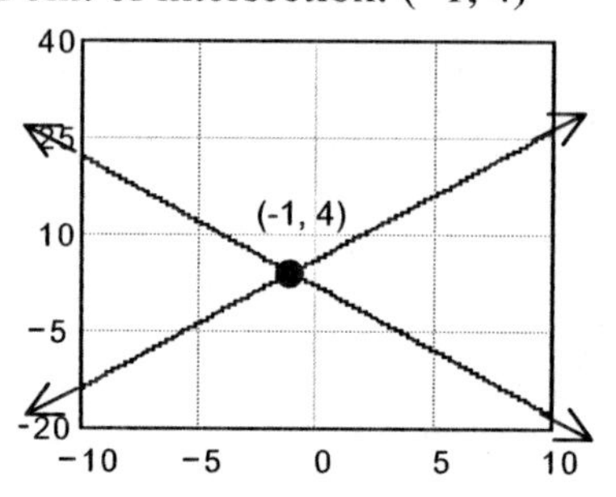

(b) Substitution method: Solve the first equation for y.
$$\begin{aligned}-2x+y&=6\\ y&=2x+6\end{aligned}$$
Substitute this value of y into the second equation and solve the resulting equation for x.
$$\begin{aligned}2x+y&=2\\ 2x+2x+6&=2\\ 4x+6&=2\\ 4x&=-4\\ x&=-1\end{aligned}$$

$$\begin{aligned}-2(-1)+y&=6\\ 2+y&=6\\ y&=4\end{aligned}$$
Solution: $(-1, 4)$

(c) Addition method: The equations of the system are written in standard form. The coefficients on x are opposites. Add the two equations together and solve the resulting equation.
$$\begin{aligned}-2x+y&=6\\ \underline{2x+y}&\underline{=2}\\ 2y&=8\\ y&=4\end{aligned}$$

$$-2x + y = 6$$
$$-2x + 4 = 6$$
$$-2x = 2$$
$$x = -1$$

Solution: $(-1, 4)$

3. $y = -2x + 6$
$4x - 2y = 8$

(a) Graphing method: The x- and y-intercepts of $y = -2x + 6$ are (3, 0) and (0, 6). the x- and y-intercepts of $4x - 2y = 8$ are (2, 0) and (0, −4). Use these points to graph the equations.

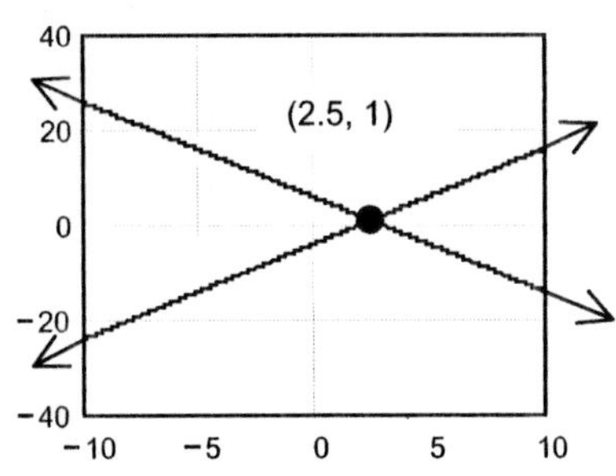

Point of intersection: $\left(\frac{5}{2}, 1\right)$

(b) Substitution method: Since the first equation is solved for y, substitute this value into the second equation and solve the resulting equation for x.

$$4x - 2y = 8$$
$$4x - 2(-2x + 6) = 8$$
$$4x + 4x - 12 = 8$$
$$8x - 12 = 8$$
$$8x = 20$$
$$x = \frac{20}{8} = \frac{5}{2}$$

$$y = -2x + 6$$
$$y = -2\left(\frac{5}{2}\right) + 6$$
$$y = 1$$

Solution: $\left(\frac{5}{2}, 1\right)$

(c) Addition method: Rewrite the equations in standard form.

$$2x + y = 6$$
$$4x - 2y = 8$$

Since the y-variables have opposite signs, multiply the first equation by 2 to obtain opposite coefficients on y.

$$4x + 2y = 12$$
$$\underline{4x - 2y = 8}$$
$$8x = 20$$
$$x = \frac{20}{8} = \frac{5}{2}$$

$$4x - 2y = 8$$
$$4\left(\frac{5}{2}\right) - 2y = 8$$
$$10 - 2y = 8$$
$$-2y = -2$$
$$y = 1$$

Solution: $\left(\frac{5}{2}, 1\right)$

5. Let x represent the measure of one of the angles. Let y represent the measure of the other angle. The statement "two angles are complementary" translates to the equation $x + y = 90$. The statement "one angle is 10° less than 9 times the other" translates to the equation $x = 9y - 10$.

$$x + y = 90$$
$$x = 9y - 10$$

Since one equation is solved for x, use the substitution method. Substitute this value for x into the first equation and solve the resulting equation for y.

$$x + y = 90$$
$$9y - 10 + y = 90$$
$$10y - 10 = 90$$
$$10y = 100$$
$$y = 10$$

$x = 9y - 10$
$x = 9(10) - 10$
$x = 80$

the measures of the angles are 80° and 10°.

7. Let d represent the number of points scored by the Dallas Cowboys. Let b represent the number of points scored by the Buffalo Bills. The statement "the Dallas Cowboys scored four more than twice the number of points scored by the Buffalo Bills" translates to the equation $d = 2b + 4$. The statement "total number of points scored by both teams was 43" translates to the equation
$d + b = 43$.

$$d = 2b + 4$$
$$d + b = 43$$

Since one equation is solved for d, use the substitution method. Substitute this value for d into the first equation and solve the resulting equation for b.

$$d + b = 43$$
$$2b + 4 + b = 43$$
$$3b + 4 = 43$$
$$3b = 39$$
$$b = 13$$

$$d = 2b + 4$$
$$d = 2(13) + 4$$
$$d = 30$$

Dallas scored 30 points, and Buffalo scored 13 points.

9. Let t represent the cost of one tape. Let c represent the cost of one CD.

$$\begin{pmatrix}\text{Cost of}\\\text{three tapes}\end{pmatrix}+\begin{pmatrix}\text{Cost of}\\\text{two CDs}\end{pmatrix}=\begin{pmatrix}\text{Total}\\\text{cost}\end{pmatrix}$$
$$\begin{pmatrix}\text{Cost of}\\\text{one tape}\end{pmatrix}+\begin{pmatrix}\text{Cost of}\\\text{four Cds}\end{pmatrix}=\begin{pmatrix}\text{Total}\\\text{cost}\end{pmatrix}$$

$$3t + 2c = 62.50$$
$$t + 4c = 72.50$$

Since both equations are in standard form, use the addition method. To obtain opposite coefficients on t, multiply the second equation by -3.

$$3t + 2c = 62.50$$
$$-3t - 12c = -217.50$$
$$-10c = -155.00$$
$$c = 15.50$$

$$t + 4c = 72.50$$
$$t + 4(15.50) = 72.50$$
$$t + 62.00 = 72.50$$
$$t = 10.50$$

Tapes cost \$10.50 each and CDs cost \$15.50 each.

11. Let t represent the cost of one share of technology stock. Let m represent the cost of one share of mutual fund.

$$\begin{pmatrix}\text{Cost of}\\\text{100 shares}\\\text{technology}\\\text{stock}\end{pmatrix}+\begin{pmatrix}\text{Cost of}\\\text{200 shares}\\\text{mutual}\\\text{fund}\end{pmatrix}=\begin{pmatrix}\text{Total}\\\text{cost}\end{pmatrix}$$
$$\begin{pmatrix}\text{Cost of}\\\text{300 shares}\\\text{technology}\\\text{stock}\end{pmatrix}+\begin{pmatrix}\text{Cost of}\\\text{50 shares}\\\text{mutual}\\\text{fund}\end{pmatrix}=\begin{pmatrix}\text{Total}\\\text{cost}\end{pmatrix}$$

$$100t + 200m = 3800$$
$$300t + 50m = 5350$$

Since the equations are written in standard form, use the addition method. To obtain opposite coefficients on t, multiply the first equation by -3.

$$-300t - 600m = -11400$$
$$300t + 50m = 5350$$
$$-550m = -6050$$
$$m = 11$$

$$100t + 200m = 3800$$
$$100t + 200(11) = 3800$$
$$100t + 2200 = 3800$$
$$100t = 1600$$
$$t = 16$$

Technology stock costs \$16 per share and the mutual fund costs \$11 per share.

13. Let x represent the amount Shanelle invested in the 10% account. Let y represent the amount Shanelle invested in the 7% account.

	10% Account	7% Account	Total
Principal invested	x	y	\$10,000
Interest earned	$0.10x$	$0.07y$	\$805

$$x + y = 10{,}000$$
$$0.1x + 0.07y = 805$$

Since the equations are in standard form and the coefficient on x in the first equation is one, use the addition method by multiplying the second equation by –0.1.

$$-0.1x - 0.1y = -1000$$
$$0.1x + 0.07y = 805$$
$$-0.03y = -195$$
$$y = 6500$$

$$x + 6500 = 10{,}000$$
$$x = 3500$$

Shanelle invested \$3500 in the 10% account and \$6500 in the 7% account.

15. Let x represent the amount invested at 7%.
Let y represent the amount invested at 4%.
The statement "twice as much money in an account earning 7% simple interest as she did in an account earning 4% simple interest" translates to the equation $x = 2y$. The statement "the total interest at the end of 1 year was \$720" translates to the equation $0.07x + 0.04y = 720$.

$$x = 2y$$
$$0.07x + 0.04y = 720$$

Since the first equation is solved for x, use the substitution method by substituting this value for x into the second equation and solving the resulting equation.

$$0.07(2y) + 0.04y = 720$$
$$0.14y + 0.04y = 720$$
$$0.18y = 720$$
$$y = 4000$$

$x = 2(4000)$
$x = 8000$
Janise invested \$8000 in the 7% account and \$4000 in the 4% account.

17.

	7.5% Account	6% Account	Total
Principal	x	y	\$12,000
Interest	$0.075x$	$0.06y$	\$840

(Amount invested at 7.5%) + (Amount invested at 6%) = (Total investment)
$\Rightarrow x + y = 12000$;
(Interest earned from 7.5% account) + (Interest earned from 6% account) = (Total Interest)
$\Rightarrow 0.075x + 0.06y = 840$

$$\begin{cases} x+y=12000 \\ 0.075x+0.06y=840 \end{cases}$$

Multiply the second equation by 1000. Solve the first equation for x and substitute this expression into the second equation.

$$\begin{aligned} x &= 12000 - y \\ 75x+60y &= 840000 \\ 75(12000-y)+60y &= 840000 \\ 900000-75y+60y &= 840000 \\ -15y &= -60000 \\ y &= 4000; \end{aligned}$$

$x = 12000 - 4000 = 8000$
\$8,000 was invested at 7.5%.
\$4,000 was invested at 6%.

19. Let x represent the amount of 50% disinfectant solution. Let y represent the amount of 40% disinfectant solution.

	50% Mixture	40% Mixture	46% Mixture
Amount of solution	x	y	25
Amount of disinfectant	$0.50x$	$0.40y$	0.46(25)

$$\begin{aligned} x+y &= 25 \\ 0.5x+0.4y &= 11.5 \end{aligned}$$

Since the equations are in standard form and the coefficient on x is one, use the addition method by multiplying the first equation by –0.5.

$$\begin{aligned} -0.5x-0.5y &= -12.5 \\ 0.5x+0.4y &= 11.5 \\ \hline -0.1y &= -1 \\ y &= 10 \end{aligned}$$

$$\begin{aligned} x+10 &= 25 \\ x &= 15 \end{aligned}$$

15 gallons of 50% mixture should be mixed with 10 gallons of 40% mixture.

21. Let x represent the amount of 45% disinfectant solution. Let y represent the amount of 30% disinfectant solution.

	45% Mixture	30% Mixture	39% Mixture
Amount of solution	x	y	20
Amount of disinfectant	$0.45(x)$	$0.30(y)$	0.39(20)

$$\begin{aligned} x+y &= 20 \\ 0.45x+0.3y &= 7.8 \end{aligned}$$

Since the equations are in standard from and the coefficient on x is one, use the addition method by multiplying the first equation by –0.45.

$$\begin{aligned} -0.45x-0.45y &= -9.0 \\ 0.45x+0.3y &= 7.8 \\ \hline -0.15y &= -1.2 \\ y &= 8 \end{aligned}$$

$$\begin{aligned} x+8 &= 20 \\ x &= 12 \end{aligned}$$

12 gallons of the 45% disinfectant solution should be mixed with 8 gallons of the 30% disinfectant mixture.

23.

	18% moisturizer	24% moisturizer	22% moisturizer
Number of ounces of face cream	x	y	12
Number of ounces of moisturizer	$0.18x$	$0.24y$	0.22(12)

(Amount of 18% moisturizer) + (Amount of 24% moisturizer) = (Total amount of moisturizer)
$\Rightarrow x + y = 12$;
(Amount of pure moisturizer in 18% cream) + (Amount of pure moisturizer in 24% cream)
= (Amount of pure moisturizer in resulting cream) $\Rightarrow 0.18x + 0.24y = 2.64$

$$\begin{cases} x + y = 12 \\ 0.18x + 0.24y = 2.64 \end{cases}$$

Multiply the second equation by 100. Solve the second equation for x and substitute this expression into the second equation.

$$\begin{aligned} x &= 12 - y \\ 18x + 24y &= 264 \\ 18(12 - y) + 24y &= 264 \\ 216 - 18y + 24y &= 264 \\ 6y &= 48 \\ y &= 8; \end{aligned}$$

$x = 12 - 8 = 4$

Combine 4 ounces of the 18% moisturizer and 8 ounces of the 24% moisturizer.

25. Let x represent the speed of the boat. Let y represent the speed of the current.

	Distance	Rate	Time
Downstream	16	$x + y$	2
Return	16	$x - y$	4

$$\begin{aligned} 2x + 2y &= 16 \\ 4x - 4y &= 16 \end{aligned}$$

Since the equations are in standard form and the coefficients on y have opposite signs, use the addition method by multiplying the first equation by 2.

$$\begin{aligned} 4x + 4y &= 32 \\ 4x - 4y &= 16 \\ \hline 8x &= 48 \\ x &= 6 \end{aligned}$$

$$2(6)+2y=16$$
$$2y=4$$
$$y=2$$

The speed of the boat in still water is 6 mph and the speed of the current is 2 mph.

27. Let x represent the speed of the plane in still air. Let y represent the speed of the wind.

	Distance	Rate	Time
Flying with the wind	800	$x+y$	2.5
Flying against the wind	700	$x-y$	2.5

$$2.5x+2.5y=800$$
$$2.5x-2.5y=700$$

Since the equations are in standard from and the coefficients on y are opposites, use the addition method.

$$2.5x+2.5y=800$$
$$2.5x-2.5y=700$$
$$5.0x=1500$$
$$x=300$$

$$2.5(300)+2.5y=800$$
$$750+2.5y=800$$
$$2.5y=50$$
$$y=20$$

The speed of the plane in still air is 300 mph and the speed of the wind is 20 mph.

29.

	Distance	Rate	Time
Jeannie	D	r	2
Juan	$20-D$	$r+2$	2

(Jeannie's Distance) = (Jeannie's rate) × (Jeannie's time) ⇒ $D = 2r$;
(Juan's Distance) = (Juan's rate) × (Juan's time) ⇒ $20 - D = 2(r + 2)$
Substitute the expression for Jeannie's time for D in Juan's equation.

$$20-2r=2(r+2)$$
$$20-2r=2r+4$$
$$-4r=-16$$
$$r=4;$$
$$r+2=6$$

Jeannie's rate is 4 mph.
Juan's rate is 6 mph.

31. Let x represent the number of dimes. Let y represent the number of nickels.

$$\begin{pmatrix}\text{number}\\ \text{of nickels}\end{pmatrix}=\begin{pmatrix}\text{number}\\ \text{of dimes}\end{pmatrix}+5$$
$$\begin{pmatrix}\text{value of}\\ \text{dimes}\end{pmatrix}+\begin{pmatrix}\text{value of}\\ \text{nickels}\end{pmatrix}=\begin{pmatrix}\text{total}\\ \text{value}\end{pmatrix}$$

$$y=x+5$$
$$0.10x+0.05y=2.80$$

Since the first equation is solved for y, use the substitution method by substituting this value into the second equation.

$$0.10x+0.05(x+5)=2.80$$
$$0.10x+0.05x+0.25=2.80$$
$$0.15x+0.25=2.80$$
$$0.15x=2.55$$
$$x=17$$

$$y=x+5$$
$$y=17+5$$
$$y=22$$

There are 17 dimes and 22 nickels.

33. (a) Let x represent the number of free throws. Let y represent the number of field goals. The statement "made 2432 baskets...some were free throws and some were field goals" translates to the equation $x + y = 2432$. The statement "the number of field goals was 762 more than the number of free-throws" translates to the equation $y = x + 762$.

$$x+y=2432$$
$$y=x+762$$

Since the second equation is solved for y, use the substitution method substituting this value into the first equation.

$$x + x + 762 = 2432$$
$$2x + 762 = 2432$$
$$2x = 1670$$
$$x = 835$$

$y = x + 762$
$y = 835 + 762$
$y = 1597$
He made 835 free-throws and 1597 field goals.

(b) $835(1) + 1597(2) = 835 + 3194 = 4029$
He scored 4029 points.

(c) $\frac{4029}{80} = 50.3625$
He averaged approximately 50 points per game.

35. Let x represent the speed of the plane in still air. Let y represent the speed of the wind.

	Distance	Rate	Time
Flying with a tailwind	350	$x + y$	$1\frac{3}{4}$
Flying with a headwind	210	$x - y$	$1\frac{3}{4}$

$$\frac{7}{4}x + \frac{7}{4}y = 350$$
$$\frac{7}{4}x - \frac{7}{4}y = 210$$

Clear fractions by multiplying both equations by 4.
$7x + 7y = 1400$
$7x - 7y = 840$

Since the equations are in standard form and the coefficients on y are opposites, use the addition method.
$$7x + 7y = 1400$$
$$7x - 7y = 840$$
$$14x = 2240$$
$$x = 160$$

$$\frac{7}{4}x + \frac{7}{4}y = 350$$
$$\frac{7}{4}(160) + y = 350$$
$$1120 + 7y = 1400$$
$$7y = 280$$
$$y = 40$$
The speed of the plane in still air is 160 mph and the wind is 40 mph.

37. Let x represent the number of pounds of candy needed. Let y represent the number of pounds of nuts needed.

	Candy	Nuts	Total
Number of	x	y	20
Value of	$1.80x$	$1.20y$	1.56(20)

$$x + y = 20$$
$$1.80x + 1.20y = 31.20$$

Since the equations are written in standard form, use the addition method by multiplying the first equation by -1.20.
$$-1.20x - 1.20y = -24.00$$
$$1.80x + 1.20y = 31.20$$
$$0.60x = 7.20$$
$$x = 12$$

$$x + y = 20$$
$$12 + y = 20$$
$$y = 8$$
12 pounds of candy should be mixed with 8 pounds of nuts.

39. Let x represent the amount invested in the first account. Let y represent the amount invested in the second account.

	First Account	Second Account	Total
Principal	x	y	60,000
Interest	$0.055x$	$0.065y$	3750

$x + y = 60{,}000$
$0.055x + 0.065y = 3750$

Since the equations are written in standard form, use the addition method by multiplying the first equation by –0.055.

$$\begin{array}{r} -0.055x - 0.055y = -3300 \\ 0.055x + 0.065y = 3750 \\ \hline 0.010y = 450 \\ y = 45{,}000 \end{array}$$

$$\begin{aligned} x + y &= 60{,}000 \\ x + 45{,}000 &= 60{,}000 \\ x &= 15{,}000 \end{aligned}$$

$15,000 is invested in the 5.5% account and $45,000 is invested in the 6.5% account.

41. Let x represent the amount of Miracle-Gro needed. Let y represent the amount of Green Light needed.

	Miracle-Gro	Green Light	Total
Amount of mixture	x	y	60
Amount of nitrogen	$0.15x$	$0.12y$	0.13(60)

$x + y = 60$
$0.15x + 0.12y = 7.8$

Since the equations are written in standard form, use the addition method by multiplying the first equation by –0.12.

$$\begin{array}{r} -0.12x - 0.12y = -7.2 \\ 0.15x + 0.12y = 7.8 \\ \hline 0.03x = 0.6 \\ x = 20 \end{array}$$

$$\begin{aligned} x + y &= 60 \\ 20 + y &= 60 \\ y &= 40 \end{aligned}$$

20 ounces of Miracle-Gro should be mixed with 40 ounces of Green Light.

43.
$$\begin{aligned} y_s &= y_d \\ x &= -10x + 500 \\ 20x &= -30x + 1500 \\ 50x &= 1500 \\ x &= 30 \end{aligned}$$

Supply equals demand when the price is $30.

45. Let x represent the number of women college students. Let y represent the number of men college students. The statement "500 college students" translates to the equation $x + y = 500$. The statement "340 said that the campus lacked adequate lighting" together with "$\frac{4}{5}$ of the women and $\frac{1}{2}$ of the men said that they thought the campus lacked adequate lighting" translates to the equation $\frac{4}{5}x + \frac{1}{2}y = 340$.

$x + y = 500$
$\frac{4}{5}x + \frac{1}{2}y = 340$

Clear fractions from the second equation by multiplying it by 10.

$x + y = 500$
$8x + 5y = 3400$

Since the equations are written in standard form, use the addition method by multiplying the first equation by –5.

$$\begin{array}{r} -5x - 5y = -2500 \\ 8x + 5y = 3400 \\ \hline 3x = 900 \\ x = 300 \end{array}$$

$$\begin{aligned} x + y &= 500 \\ 300 + y &= 500 \\ y &= 200 \end{aligned}$$

There were 300 women and 200 men in the survey.

47. Let x represent the number of 15-second commercials. Let y represent the number of 30-second commercials. The statement "there were 22 commercials" translates to the equation $x + y = 22$. the statement "some commercials were 15 seconds long and some were 30 seconds long" together with "the total playing time for commercials was 9.5 minutes" translates to the equation $15x + 30y = 570$. Remember that you must have all units the same. Therefore

$$\begin{aligned} 9.5 \text{ minutes} &= 9.5(60 \text{ seconds}) \\ &= 570 \text{ seconds} \end{aligned}$$

$$\begin{aligned} x + y &= 22 \\ 15x + 30y &= 570 \end{aligned}$$

Since the equations are written in standard form, use the addition method by multiplying the first equation by -15.

$$\begin{aligned} -15x - 15y &= -330 \\ 15x + 30y &= 570 \\ \hline 15y &= 240 \\ y &= 16 \end{aligned}$$

$$\begin{aligned} x + y &= 22 \\ x + 16 &= 22 \\ x &= 6 \end{aligned}$$

There are six 15-second commercials and sixteen 30-second commercials.

Section 8.5 Practice Exercises

1. (a) Solve the first equation for y. Substitute this expression for y into the second equation.

$$\begin{aligned} y &= 4 - 3x \\ 4x + y &= 5 \\ 4x + (4 - 3x) &= 5 \\ 4x + 4 - 3x &= 5 \\ x &= 1; \end{aligned}$$

$y = 4 - 3(1) = 1$

The solution is (1, 1).

(b) Multiply the first equation by -1.

$$\begin{aligned} -3x - y &= 4 \\ 4x + y &= 5 \\ \hline x &= 1; \end{aligned}$$

$4(1) + y = 5$

$y = 5 - 4 = 1$

The solution is (1, 1).

3. (a) Solve the second equation for x. Substitute this expression for x into the second equation.

$$x = 2 + \frac{2}{3}y$$

$$\frac{1}{2}\left(2 + \frac{2}{3}y\right) - \frac{1}{3}y = 1$$

$$1 + \frac{1}{3}y - \frac{1}{3}y = 1$$

$$1 = 1 \quad \text{Identity}$$

$$\left\{(x, y) \middle| x - \frac{2}{3}y = 2\right\}$$

(b) Multiply the first equation by 6. Multiply the second equation by 3.

$$\begin{aligned} 3x - 2y &= 6 \\ 3x - 2y &= 6 \\ \hline 0 &= 0 \quad \text{Identity} \end{aligned}$$

$$\left\{(x, y) \middle| x - \frac{2}{3}y = 2\right\}$$

5.

	Distance	Rate	Time
Car driving east	D	$r - 7$	3
Car driving west	$369 - D$	r	3

Distance of car driving east) = (Rate of car driving east) × (Time of car driving east)
$\Rightarrow D = 3(r - 7)$;
(Distance of car driving west) = (Rate of car driving west) × (Time of car driving west)
$\Rightarrow 369 - D = 3r$

$$\begin{cases} D = 3(r - 7) \\ 369 - D = 3r \end{cases}$$

Substitute the expression for D in the first equation into the second equation.

$$\begin{aligned}369-3(r-7)&=3r\\369-3r+21&=3r\\-6r&=-390\\r&=65;\end{aligned}$$

$r-7=65-7=58$

The speed of the car driving west is 65 mph.
The speed of the car driving east is 58 mph.

7. $\begin{cases}2x-y+z=10\\4x+2y-3z=10\\x-3y+2z=8\end{cases}$

(2, 1, 7) $\quad$ $\begin{aligned}2(2)-1+7&=10\\4-6&=10\\-2&\neq10\end{aligned}$

Since this point is not true in the first equation, it is not a solution to the system.

(3, –10, –6) $\quad$ $\begin{aligned}2(3)-(-10)+(-6)&=10\\6+10-6&=10\\10&=10\end{aligned}$

$$\begin{aligned}4(3)+2(-10)-3(-6)&=10\\12-20+18&=10\\10&=10\end{aligned}$$

$$\begin{aligned}3-3(-10)+2(-6)&=8\\3+30-12&=8\\21&\neq8\end{aligned}$$

Since this point is not true in the last equation, it is not a solution to the system.

(4, 0, 2) $\quad$ $\begin{aligned}2(4)-0+2&=10\\8+2&=10\\10&=10\end{aligned}$

$$\begin{aligned}4(4)+2(0)-3(2)&=10\\16+0-6&=10\\10&=10\end{aligned}$$

$$\begin{aligned}4-3(0)+2(2)&=8\\4-0+4&=8\\8&=8\end{aligned}$$

Since this point is true in all three equations, it is a solution to the system.

9. $\begin{cases}x+2y-z=5\\x-3y+z=-5\\4x+y-z=4\end{cases}$

(0, 4, 3) $\quad$ $\begin{aligned}0+2(4)-3&=5\\0+8-3&=5\\5&=5\end{aligned}$

$$\begin{aligned}0-3(4)+3&=-5\\0-12+3&=-5\\9&\neq-5\end{aligned}$$

Since this point is not true in the second equation, it is not a solution to the system.

(3, 6, 10) $\quad$ $\begin{aligned}3+2(6)-10&=5\\3+12-10&=5\\5&=5\end{aligned}$

$$\begin{aligned}3-3(6)+10&=-5\\3-18+10&=-5\\-5&=-5\end{aligned}$$

$$\begin{aligned}-2(3)+6-10&=-4\\-6+6-10&=-4\\-10&\neq-4\end{aligned}$$

Since this point is not true in the third equation, it is not a solution to the system.

(3, 3, 1) $\quad$ $\begin{aligned}3+2(3)-1&=5\\3+6-1&=5\\8&\neq5\end{aligned}$

Since this point is not true in the first equation, it is not a solution to the system.

11. A: $x+y+z=6$
B: $-x+y-z=-2$
C: $2x+3y+z=1$

Eliminate the z variable from equations A and B by adding.

$$\begin{aligned}x+y+z&=6\\\underline{-x+y-z}&\underline{=-2}\\2y&=4\\y&=2\end{aligned}$$

Eliminate the z variable from equations B and C by adding.

$$\begin{array}{r} -x+y-z=-2 \\ 2x+3y+z=11 \\ \hline x+4y=9 \end{array} \quad \text{Equation D}$$

Substitute the value for y into equation D.

$x+4(2)=9$
$x+8=9$
$x=1$

Substitute $x = 1$ and $y = 2$ into any original equation to solve for z.

$x+y+z=6$
$1+2+z=6$
$z=3$

The solution to the system is (1, 2, 3).

13. A: $-3x+y-z=8$
B: $-4x+2y+3z=-3$
C: $2x+3y-2z=-1$

Eliminate the z variable from equations A and B by multiplying equation A by 3.

$$\begin{array}{r} -9x+3y-3z=24 \\ -4x+2y+3z=-3 \\ \hline -13x+5y=21 \end{array} \quad \text{Equation D}$$

Eliminate the z variable from equations A and C by multiplying equation A by –2.

$$\begin{array}{r} 6x-2y+2z=-16 \\ 2x+3y-2z=-1 \\ \hline 8x+y=-17 \end{array} \quad \text{Equation E}$$

Use equations D and E to form a linear system in two variables. Eliminate the y variable by multiplying equation E by –5.

$$\begin{array}{r} -13x+5y=21 \\ -40x-5y=85 \\ \hline -53x=106 \\ x=-2 \end{array}$$

Substitute $x=-2$ into either equation D or E to solve for y.

$8(-2)+y=-17$
$-16+y=-17$
$y=-1$

Substitute $x=-2$ and $y=-1$ into any original equation to solve for x.

$2(-2)+3(-1)-2z=-1$
$-4-3-2z=-1$
$-2z=6$
$z=-3$

The solution to the system is (–2, –1, –3).

15. A: $2x-y+z=-1$
B: $-3x+2y-2z=1$
C: $5x+3y+3z=16$

Eliminate the y variable from equations A and B by multiplying equation A by 2.

$$\begin{array}{r} 4x-2y+2z=-2 \\ -3x+2y-2z=1 \\ \hline x=-1 \end{array}$$

Eliminate the y variable from equations A and C by multiplying equation A by 3.

$$\begin{array}{r} 6x-3y+3z=-3 \\ 5x+3y+3z=16 \\ \hline 11x+6z=13 \end{array} \quad \text{Equation D}$$

Substitute $x=-1$ into equation D to solve for z.

$11(-1)+6z=13$
$-11+6z=13$
$6z=24$
$z=4$

Substitute $x=-1$ and $z=4$ into any original equation to solve for y.

$2(-1)-y+4=-1$
$-2-y+4=-1$
$-y=-3$
$y=3$

The solution to the system is (–1, 3, 4).

17. A: $2x-3y+2z=-1$
B: $x+2y=-4$
C: $x+z=1$

Eliminate the x variable from equations A and B by multiplying equation B by –2.

$$\begin{array}{r} 2x-3y+2z=-1 \\ -2x-4y=8 \\ \hline -7y+2z=7 \end{array} \quad \text{Equation D}$$

Eliminate the x variable from equations B and C by multiplying equation B by –1.

$$\begin{array}{r} -x-2y=4 \\ x+z=1 \\ \hline -2y+z=5 \end{array} \quad \text{Equation E}$$

Use equations D and E to form a linear system in two variables. Eliminate the z variable by multiplying equation E by –2.

$$\begin{array}{r} -7y+2z=7 \\ 4y-2z=-10 \\ \hline -3y=-3 \\ y=1 \end{array}$$

Substitute $y = 1$ into either equation D or E to solve for z.

$$\begin{aligned} -7(1)+2z&=7 \\ -7+2z&=7 \\ 2z&=14 \\ z&=7 \end{aligned}$$

Substitute $z = 7$ into equation C to solve for x.

$$\begin{aligned} x+7&=1 \\ x&=-6 \end{aligned}$$

The solution to the system is (–6, 1, 7).

19. A: $4x+9y=8$
B: $8x+6z=-1$
C: $6y+6z=-1$

Eliminate the x variable from equations A and B by multiplying equation A by –2.

$$\begin{array}{r} -8x-18y=-16 \\ 8x+6z=-1 \\ \hline -18y+6z=-17 \end{array} \quad \text{Equation D}$$

Eliminate the z variable from equations C and D by multiplying equation z by –1.

$$\begin{array}{r} -6y-6z=1 \\ -18y+6z=-17 \\ \hline -24y=-16 \\ y=\frac{2}{3} \end{array}$$

Substitute $y=\frac{2}{3}$ into equation C to solve for z.

$$\begin{aligned} 6\left(\frac{2}{3}\right)+6z&=-1 \\ 4+6z&=-1 \\ 6z&=-5 \\ z&=-\frac{5}{6} \end{aligned}$$

Substitute $z=-\frac{5}{6}$ into equation B to solve for x.

$$\begin{aligned} 8x+6\left(-\frac{5}{6}\right)&=-1 \\ 8x-5&=-1 \\ 8x&=4 \\ x&=\frac{1}{2} \end{aligned}$$

The solution to the system is $\left(\frac{1}{2}, \frac{2}{3}, -\frac{5}{6}\right)$.

21. Let x represent the measure of the smallest angle.
Let y represent the measure of the middle angle.
Let z represent the measure of the largest angle.

A: $x+y+z=180$
B: $y=2x+5$
C: $z=3x-11$

A: $x+y+z=180$
B: $-2x+y=5$
C: $-3x+z=-11$

Eliminate the y variable from equations A and B by multiplying equation B by –1.

$$\begin{array}{r} x+y+z=180 \\ 2x-y=-5 \\ \hline 3x+z=175 \end{array} \quad \text{Equation D}$$

Eliminate the z variable from equations C and D by multiplying equation C by –1.

$$\begin{array}{r} 3x-z=11 \\ 3x+z=175 \\ \hline 6x=186 \\ x=31 \end{array}$$

Substitute $x = 31$ into equation B to solve for y.
$y = 2(31) + 5$
$y = 62 + 5$
$y = 67$
Substitute $x = 31$ into equation C to solve for z.
$z = 3(31) - 11$
$z = 92 - 11$
$z = 81$
The measures of the angles are 31°, 67° and 82°.

23. Let x represent the length of the shortest side.
Let y represent the length of the middle side.
Let z represent the length of the longest side.

A: $x + y + z = 54$
B: $z = x + y$
C: $x = \frac{1}{2}y$

A: $x + y + z = 54$
B: $-x - y + z = 0$
C: $x - \frac{1}{2}y = 0$

Eliminate the x variable from equations A and B by adding.

$$\begin{array}{r} x + y + z = 54 \\ \underline{-x - y + z = 0} \\ 2z = 54 \\ z = 27 \end{array}$$

Eliminate the y variable from equations A and C by multiplying equation C by 2.

$$\begin{array}{r} x + y + z = 54 \\ \underline{2x - y = 0} \\ 3x + z = 54 \end{array} \qquad \text{Equation D}$$

Substitute $z = 27$ into equation D to solve for x.
$3x + 27 = 54$
$3x = 27$
$x = 9$

Substitute $x = 9$ and $z = 27$ into equation A to solve for y.
$9 + y + 27 = 54$
$y + 36 = 54$
$y = 18$
The lengths of the sides are 9 cm, 18 cm and 27 cm.

25. Let a represent the number of adult tickets.
Let c represent the number of children's tickets.
Let s represent the number of senior tickets.

A: $a + c + s = 222$
B: $7a + 5c + 4s = 1383$
C: $a = 2(c + 2)$

A: $a + c + s = 222$
B: $7a + 5c + 4s = 1383$
C: $a - 2c - 2s = 0$

Eliminate the c variable from equations A and C by multiplying equation A by 2.

$$\begin{array}{r} 2a + 2c + 2s = 444 \\ \underline{a - 2c - 2s = 0} \\ 3a = 444 \\ a = 148 \end{array}$$

Eliminate the s variable from equations A and B by multiplying equation A by –4.

$$\begin{array}{r} -4a - 4c - 4s = -888 \\ \underline{7a + 5c + 4s = 1383} \\ 3a + c = 495 \end{array} \qquad \text{Equation D}$$

Substitute $a = 148$ into equation D to solve for c.
$3(148) + c = 495$
$444 + c = 495$
$c = 51$
Substitute $a = 148$ and $c = 51$ into equation A to solve for s.
$148 + 51 + s = 222$
$199 + s = 222$
$s = 23$
There were 148 adult tickets sold, 51 children's tickets sold and 23 senior tickets sold.

27. Let x represent the ounces of peanuts.
Let y represent the ounces of pecans.
Let z represent the ounces of cashews.

A: $x+y+z=48$
B: $x=y+z$
C: $z=2y$

A: $x+y+z=48$
B: $x-y-z=0$
C: $-2y+z=0$

Eliminate the y variable from equations A and B by adding.

$$\begin{array}{r} x+y+z=48 \\ x-y-z=0 \\ \hline 2x=48 \\ x=24 \end{array}$$

Eliminate the z variable from equation B and C by adding.

$$\begin{array}{r} x-y-z=0 \\ -2y+z=0 \\ \hline x-3y=0 \end{array} \quad \text{Equation D}$$

Substitute $x = 24$ into equation D to solve for y.

$$24-3y=0$$
$$-3y=-24$$
$$y=8$$

Substitute $y = 8$ into equation C to solve for z.

$z = 2(8)$
$z = 16$

There are 24 oz of peanuts, 8 oz of pecans and 16 oz of cashews.

29. Let x represent the enrollment at Vanderbilt.
Let y represent the enrollment at Baylor.
Let z represent the enrollment at Pace.

A: $y=2x$
B: $z=x+2800$
C: $x+y+z=27{,}200$

A: $-2x+y=0$
B: $-x+z=2800$
C: $x+y+z=27{,}200$

Eliminate the z variable in equations B and C by multiplying equation B by -1 and adding.

$$\begin{array}{r} x-z=-2800 \\ x+y+z=27{,}200 \\ \hline 2x+y=24{,}400 \end{array} \quad \text{Equation D}$$

Eliminate the x variable in equations A and D by adding.

$$\begin{array}{r} -2x+y=0 \\ 2x+y=24{,}400 \\ \hline 2y=24{,}400 \\ y=12{,}200 \end{array}$$

Substitute $y = 12{,}200$ into equation A to solve for x.

$$y=2x$$
$$12{,}200=2x$$
$$6100=x$$

Substitute $x = 6100$ into equation B to solve for z.

$z = x + 2800$
$z = 6100 + 2800 = 8900$

The enrollments for the schools are
Vanderbilt: 6100
Baylor: 12,200
Pace: 8900

31. A: $2x+y+3z=2$
B: $x-y+2z=-4$
C: $x+3y-z=1$

Eliminate the y variable from equations A and B by adding.

$$\begin{array}{r} 2x+y+3z=2 \\ x-y+2z=-4 \\ \hline 3x+5z=-2 \end{array} \quad \text{Equation D}$$

Eliminate the y variable from equations B and C by multiplying equation B by 3.

$$\begin{array}{r} 3x-3y+6z=-12 \\ x+3y-z=1 \\ \hline 4x+5z=-11 \end{array} \quad \text{Equation E}$$

Use equations D and E to form a linear system in two variables. Eliminate the z variable by multiplying equation D by -1.

$$\begin{array}{r} -3x-5z=2 \\ 4x+5z=-11 \\ \hline x=-9 \end{array}$$

Substitute $x = -9$ into either equation D or E to solve for z.

$$4(-9)+5z=-11$$
$$-36+5z=-11$$
$$5z=25$$
$$z=5$$

Substitute $x = -9$ and $z = 5$ into any original equation to solve for y.

$$2(-9)+y+3(5)=2$$
$$-18+y+15=2$$
$$y=5$$

The solution to the system is (–9, 5, 5).

33. A: $6x-2y+2z=2$
B: $4x+8y-2z=5$
C: $-2x-4y+z=-2$

Eliminate the variable z from equations A and B by adding.

$$6x-2y+2z=2$$
$$\underline{4x+8y-2z=5}$$
$$10x+6y=7 \quad \text{Equation D}$$

Eliminate the z variable from equations B and C by multiplying equation C by 2.

$$4x+8y-2z=5$$
$$\underline{-4x-8y+2z=-4}$$
$$0\neq -4$$

This is inconsistent system. There is no solution.

35. A: $\frac{1}{2}x+\frac{2}{3}y=\frac{5}{2}$
B: $\frac{1}{2}x-\frac{1}{2}z=-\frac{3}{10}$
C: $\frac{1}{3}y-\frac{1}{4}z=\frac{3}{4}$

Multiply equation A by 6, equation B by 10 and equation C by 12.

A: $3x+4y=15$
B: $2x-5z=-3$
C: $4y-3z=9$

Eliminate the x variable from equations A and B by multiplying equation A by 2 and equation B by –3.

$$6x+8y=30$$
$$\underline{-6x+15z=9}$$
$$8y+15z=39 \quad \text{Equation D}$$

Eliminate the y variable from equations C and D by multiplying equation C by –2.

$$-8y+6z=-18$$
$$\underline{8y+15z=39}$$
$$21z=21$$
$$z=1$$

Substitute $z = 1$ into equation C and solve for y.

$$4y-3(1)=9$$
$$4y-3=9$$
$$4y=12$$
$$y=3$$

Substitute $z = 1$ into equation B to solve for x.

$$2x-5(1)=-3$$
$$2x-5=-3$$
$$2x=2$$
$$x=1$$

The solution to the system is (1, 3, 1).

37. A: $2x+y-3z=-3$
B: $3x-2y+4z=1$
C: $4x+2y-6z=-6$

Eliminate the y variable from equations A and B by multiplying equation A by 2.

$$4x+2y-6z=-6$$
$$\underline{3x-2y+4z=1}$$
$$7x-2z=-5 \quad \text{Equation D}$$

Eliminate the y variable from equations B and C by adding.

$$3x-2y+4z=1$$
$$\underline{4x+2y-6z=-6}$$
$$7x-2z=-5 \quad \text{Equation E}$$

Use equations D and E to form a linear system in two variables. Eliminate the x variable by multiplying equation D by –1.

$$-7x+2z=5$$
$$\underline{7x-2z=-5}$$
$$0=0$$

The system is dependent. There are infinitely many solutions.
General solution:

Express z in terms of x in equation E.

$$-2z = -7x - 5$$
$$z = \frac{7x+5}{2}$$

Express y in terms of x by substituting $z = \frac{7x+5}{2}$ in equation C.

$$4x + 2y - 6\left(\frac{7x+5}{2}\right) = -6$$
$$4x + 2y - 3(7x+5) = -6$$
$$4x + 2y - 21x - 15 = -6$$
$$2y - 17x = 9$$
$$y = \frac{17x+9}{2}$$

When x is arbitrary, then y and z both depend on x.

$$\left\{(x,\ y,\ z) \middle| x \text{ is arbitrary, } y = \frac{17x+9}{2},\ z = \frac{7x+5}{2}\right\}$$

Express x in terms of z in equation E.

$$7x - 2z = -5$$
$$x = \frac{2z-5}{7}$$

Express y in terms of z by substituting $x = \frac{2z-5}{7}$ in equation A.

$$2\left(\frac{2z-5}{7}\right) + y - 3z = -3$$
$$2(2z-5) + 7y - 21z = -21$$
$$4z - 10 + 7y - 21z = -21$$
$$7y - 17z = -11$$
$$y = \frac{17z-11}{7}$$

When z is arbitrary, then x and y both depend on z.

$$\left\{(x,\ y,\ z) \middle| x = \frac{2z-5}{7},\ y = \frac{17z-11}{7},\ z \text{ is arbitrary}\right\}$$

Eliminate the x variable in equations A and B by multiplying equation A by 3 and equation B by -2.

$$\begin{array}{r} 6x + 3y - 9z = -9 \\ -6x + 4y - 8z = -2 \\ \hline 7y - 17z = -11 \end{array} \quad \text{Equation F}$$

Express z in terms of y in equation F.

$$-17z = -7y - 11$$
$$z = \frac{7y+11}{17}$$

Express x in terms of z by substituting $z = \frac{7y+11}{17}$ in equation A.

$$2x + y - 3\left(\frac{7y+11}{17}\right) = -3$$
$$34x + 17y - 3(7y+11) = -51$$
$$34x + 17y - 21y - 33 = -51$$
$$34x - 4y = -18$$
$$x = \frac{4y-18}{34}$$
$$x = \frac{2y-9}{17}$$

When y is arbitrary, then x and z both depend on y.

$$\left\{(x, y, z) \middle| x = \frac{2y-9}{17},\ y \text{ is arbitrary},\ z = \frac{7y+11}{17}\right\}$$

39. A: $-0.1y + 0.2z = 0.2$
B: $0.1x + 0.1y + 0.1z = 0.2$
C: $-0.1x + 0.3z = 0.2$

Multiply equations A, B and C by 10.
A: $-y + 2z = 2$
B: $x + y + z = 2$
C: $-x + 3z = 2$

Eliminate x from equations B and C by adding.

$$x + y + z = 2$$
$$-x + 3z = 2$$
$$\overline{\quad y + 4z = 4\quad}$$ Equation D

Eliminate y from equations A and D by adding.

$$-y + 2z = 2$$
$$y + 4z = 4$$
$$\overline{\quad 6z = 6\quad}$$
$$z = 1$$

Substitute $z = 1$ into equation D to solve for y.

$$y + 4(1) = 4$$
$$y = 0$$

Substitute $z = 1$ into equation C to solve for x.

$$-x + 3(1) = 2$$
$$-x = -1$$
$$x = 1$$

The solution to the system is $(1, 0, 1)$.

41. A: $2x-4y+8z=0$
B: $-x-3y+z=0$
C: $x-2y+5z=0$

Eliminate the x variable from equations A and B by multiplying equation B by 2.

$$\begin{array}{r} 2x-4y+8z=0 \\ \underline{-2x-6y+2z=0} \\ -10y+10z=0 \end{array} \quad \text{Equation D}$$

Eliminate the x variable from equations B and C by adding.

$$\begin{array}{r} -x-3y+z=0 \\ \underline{x-2y+5z=0} \\ -5y+6z=0 \end{array} \quad \text{Equation E}$$

Use equations D and E to form a linear system in two variables. Eliminate the y variable by multiplying equation E by -1.

$$\begin{array}{r} -10y+10z=0 \\ \underline{10y-6z=0} \\ 4z=0 \\ z=0 \end{array}$$

Substitute $z = 0$ into equation E to solve for y.

$$\begin{array}{r} -5y+6(0)=0 \\ -5y=0 \\ y=0 \end{array}$$

Substitute $y = 0$ and $z = 0$ into any original equation to solve for x.

$$\begin{array}{r} 2x-4(0)+8(0)=0 \\ 2x=0 \\ x=0 \end{array}$$

The solution to the system is (0, 0, 0).

43. A: $4x-2y-3z=0$
B: $-8x-y+z=0$
C: $2x-y-\frac{3}{2}z=0$

Eliminate the x variable from equations A and B by multiplying equation A by 2.

$$\begin{array}{r} 8x-4y-6z=0 \\ \underline{-8x-y+z=0} \\ -5y-5z=0 \end{array} \quad \text{Equation D}$$

Eliminate the x variable from equations A and C by multiplying equation C by -2.

$$\begin{array}{r} 4x-2y-3z=0 \\ \underline{-4x+2y+3z=0} \\ 0=0 \end{array}$$

This is a dependent system. There are infinitely many solutions.

45. Plug each of the given points (x, y) into the equation $y=ax^2+bx+c$.

$(1,-1)$: $-1=a(1)^2+b(1)+c$
$-1=a+b+c$

$(0, 3)$: $3=a(0)^2+b(0)+c$
$3=c$

$(-2, 17)$: $17=a(-2)^2+b(-2)+c$
$17=4a-2b+c$

Next, solve the resulting system of three equations in three unknowns.

A: $a+b+c=-1$
B: $c=3$
C: $4a-2b+c=17$

We can eliminate the c variable in equations A and C by substituting $c = 3$.

$a+b+3=-1$
$4a-2b+3=17$

$a+b=-4$ Equation D
$4a-2b=14$ Equation E

Eliminate the b variable from equations D and E by multiplying equation D by 2 and adding.

$$\begin{array}{r} 2a+2b=-8 \\ \underline{4a-2b=14} \\ 6a=6 \\ a=1 \end{array}$$

Substitute $a = 1$ into equation D to solve for b.

$$\begin{array}{r} a+b=-4 \\ 1+b=-4 \\ b=-5 \end{array}$$

Substitute the solution, $a = 1$, $b = -5$, $c = 3$, into the quadratic function.

$y=ax^2+bx+c$
$y=(1)x^2+(-5)x+(3)$
$y=x^2-5x+3$

47. Plug each of the given points (x, y) into the equation $y = ax^2 + bx + c$.

$(1, -5)$: $-5 = a(1)^2 + b(1) + c$
$-5 = a + b + c$

$(-1, -9)$: $-9 = a(-1)^2 + b(-1) + c$
$-9 = a - b + c$

$(3, -17)$: $-17 = a(3)^2 + b(3) + c$
$-17 = 9a + 3b + c$

Next, solve the resulting system of three equations in three unknowns.

A: $a + b + c = -5$
B: $a - b + c = -9$
C: $9a + 3b + c = -17$

Eliminate the b variable in equations A and B by adding.

$a + b + c = -5$
$a - b + c = -9$
$2a + 2c = -14$ Equation D

Eliminate the b variable from equations B and C by multiplying equation B by 3 and adding.

$3a - 3b + 3c = -27$
$9a + 3b + c = -17$
$12a + 4c = -44$ Equation E

Eliminate the c variable from equations D and E by multiplying equation D by –2 and adding.

$-4a - 4c = 28$
$12a + 4c = -44$
$8a = -16$
$a = -2$

Substitute $a = -2$ into equation D to solve for c.

$2a + 2c = -14$
$2(-2) + 2c = -14$
$-4 + 2c = -14$
$2c = -10$
$c = -5$

Substitute $a = -2$ and $c = -5$ into equation A to solve for b.

$a + b + c = -5$
$(-2) + b + (-5) = -5$
$b - 7 = -5$
$b = 2$

Substitute the solution, $a = -2$, $b = 2$, $c = -5$ into the quadratic function.

$y = ax^2 + bx + c$
$y = (-2)x^2 + (2)x + (-5)$
$y = -2x^2 + 2x - 5$

Section 8.6 Practice Exercises

1. $5x + y = 6$
$-3x + 2y = -1$

$y = -5x + 6$;
$-3x + 2(-5x + 6) = -1$
$-3x - 10x + 12 = -1$
$-13x = -13$
$x = 1$
$y = -5(1) + 6$
$y = 1$
The solution to the system is (1, 1).

3. A: $x + y - z = 8$
B: $x - 2y + z = 3$
C: $x + 3y + 2z = 7$

Eliminate the x variable from equations A and B by multiplying equation A by –1.

$-x - y + z = -8$
$x - 2y + z = 3$
$-3y + 2z = -5$ Equation D

Eliminate the x variable from equations A and C by multiplying equation A by –1.

$-x - y + z = -8$
$x + 3y + 2z = 7$
$2y + 3z = -1$ Equation E

Eliminate the y variable from equations D and E by multiplying equation D by 2 and equation E by 3.

$-6y + 4z = -10$
$6y + 9z = -3$
$13z = -13$
$z = -1$

Solve for y by substituting $z = -1$ into equation D.

$$\begin{aligned} -3y + 2(-1) &= -5 \\ -3y - 2 &= -5 \\ -3y &= -3 \\ y &= 1 \end{aligned}$$

Solve for x by substituting $y = 1$ and $z = -1$ into equation A.

$$\begin{aligned} x + 1 - (-1) &= 8 \\ x + 2 &= 8 \\ x &= 6 \end{aligned}$$

The solution to the system is (6, 1, –1).

5. An augmented matrix is one constructed from the coefficients of the variable terms and the constants.

7. The order of a matrix is the number of rows by the number of columns.

9. 4×1 order, column matrix

11. 3×3 order, square matrix

13. 1×2 order, row matrix

15. 2×4 order, none of these

17. $\left[\begin{array}{cc|c} 1 & -2 & -1 \\ 2 & 1 & -7 \end{array}\right]$

19. $\left[\begin{array}{cc|c} -9 & 13 & -5 \\ 7 & 5 & 19 \end{array}\right]$

21. $\left[\begin{array}{ccc|c} 1 & 1 & 1 & 6 \\ 1 & -1 & 1 & 2 \\ 1 & 1 & -1 & 0 \end{array}\right]$

23. $\left[\begin{array}{ccc|c} 1 & -2 & 1 & 5 \\ 2 & 6 & 3 & -2 \\ 3 & -1 & -2 & 1 \end{array}\right]$

25. **(a)** 7

(b) –2

27. $\left[\begin{array}{cc|c} 1 & \frac{1}{2} & \frac{11}{2} \\ 2 & -1 & 1 \end{array}\right]$

29. $\left[\begin{array}{cc|c} 1 & -4 & 3 \\ 5 & 2 & 1 \end{array}\right]$

31. $\left[\begin{array}{cc|c} 1 & 5 & 2 \\ 0 & 11 & 5 \end{array}\right]$

33. False; A is a 2×3 matrix

35. False

37. $x = -1$; $y = -7$

39. $x = 8$; $y = 0$; $z = -1$

41. Interchange Rows 1 and 2.

43. Multiply Row 1 by –3 and add to Row 2. Replace Row 2 with the result.

45. $\left[\begin{array}{cc|c} 1 & -2 & -1 \\ 2 & 1 & -7 \end{array}\right]$

$$-2R_1 + R_2 \Rightarrow R_2 \left[\begin{array}{cc|c} 1 & -2 & -1 \\ 0 & 5 & -5 \end{array}\right]$$

$$\frac{1}{5}R_2 \Rightarrow R_2 \left[\begin{array}{cc|c} 1 & -2 & -1 \\ 0 & 1 & -1 \end{array}\right]$$

$$2R_2 + R_1 \Rightarrow R_1 \left[\begin{array}{cc|c} 1 & 0 & -3 \\ 0 & 1 & -1 \end{array}\right]$$

The solution to the system is (–3, –1).

47. $\left[\begin{array}{cc|c} 1 & 3 & 6 \\ -4 & -9 & 3 \end{array}\right]$

$$4R_1 + R_2 \Rightarrow R_2 \left[\begin{array}{cc|c} 1 & 3 & 6 \\ 0 & 3 & 27 \end{array}\right]$$

$$\frac{1}{3}R_2 \Rightarrow R_2 \left[\begin{array}{cc|c} 1 & 3 & 6 \\ 0 & 1 & 9 \end{array}\right]$$

$$-3R_2 + R_1 \Rightarrow R_1 \left[\begin{array}{cc|c} 1 & 0 & -21 \\ 0 & 1 & 9 \end{array}\right]$$

The solution to the system is (–21, 9).

49. $\left[\begin{array}{cc|c} 1 & 3 & 3 \\ 4 & 12 & 12 \end{array}\right]$

$$-4R_1 + R_2 \Rightarrow R_2 \left[\begin{array}{cc|c} 1 & 3 & 3 \\ 0 & 0 & 0 \end{array}\right]$$

The second row of the augmented matrix represents the equation 0 = 0. Hence, the system is dependent.

$\{(x, y) | x + 3y = 3\}$

51. $\left[\begin{array}{cc|c} 1 & -1 & 4 \\ 2 & 1 & 5 \end{array}\right]$

$$-2R_1 + R_2 \Rightarrow R_2 \left[\begin{array}{cc|c} 1 & -1 & 4 \\ 0 & 3 & -3 \end{array}\right]$$

$$\frac{1}{3}R_2 \Rightarrow R_2 \left[\begin{array}{cc|c} 1 & -1 & 4 \\ 0 & 1 & -1 \end{array}\right]$$

$$R_2 + R_1 \Rightarrow R_1 \left[\begin{array}{cc|c} 1 & 0 & 3 \\ 0 & 1 & -1 \end{array}\right]$$

The solution to the system is $(3, -1)$.

53. $\left[\begin{array}{cc|c} 1 & 3 & -1 \\ -3 & -6 & 12 \end{array}\right]$

$$3R_1 + R_2 \Rightarrow R_2 \left[\begin{array}{cc|c} 1 & 3 & -1 \\ 0 & 3 & 9 \end{array}\right]$$

$$\frac{1}{3}R_2 \Rightarrow R_2 \left[\begin{array}{cc|c} 1 & 3 & -1 \\ 0 & 1 & 3 \end{array}\right]$$

$$-3R_2 + R_1 \Rightarrow R_1 \left[\begin{array}{cc|c} 1 & 0 & -10 \\ 0 & 1 & 3 \end{array}\right]$$

The solution to the system is $(-10, 3)$.

55. $\left[\begin{array}{cc|c} 3 & 1 & -4 \\ -6 & -2 & 3 \end{array}\right]$

$$\frac{1}{3}R_1 \Rightarrow R_1 \left[\begin{array}{cc|c} 1 & \frac{1}{3} & -\frac{4}{3} \\ -6 & -2 & 3 \end{array}\right]$$

$$6R_1 + R_2 \Rightarrow R_2 \left[\begin{array}{cc|c} 1 & \frac{1}{3} & -\frac{4}{3} \\ 0 & 0 & -8 \end{array}\right]$$

The second row of the augmented matrix represents the contradiction $0 = -8$. Hence, the system is inconsistent. There is not solution.

57. $\left[\begin{array}{ccc|c} 1 & 1 & 1 & 6 \\ 1 & -1 & 1 & 2 \\ 1 & 1 & -1 & 0 \end{array}\right]$

$$-1R_1 + R_2 \Rightarrow R_2 \left[\begin{array}{ccc|c} 1 & 1 & 1 & 6 \\ 0 & -2 & 0 & -4 \\ 1 & 1 & -1 & 0 \end{array}\right]$$

$$-1R_1 + R_3 \Rightarrow R_3 \left[\begin{array}{ccc|c} 1 & 1 & 1 & 6 \\ 0 & -2 & 0 & -4 \\ 0 & 0 & -2 & -6 \end{array}\right]$$

$$-\frac{1}{2}R_2 \Rightarrow R_2 \left[\begin{array}{ccc|c} 1 & 1 & 1 & 6 \\ 0 & 1 & 0 & 2 \\ 0 & 0 & -2 & -6 \end{array}\right]$$

$$-1R_2 + R_1 \Rightarrow R_1 \left[\begin{array}{ccc|c} 1 & 0 & 1 & 4 \\ 0 & 1 & 0 & 2 \\ 0 & 0 & -2 & -6 \end{array}\right]$$

$$-\frac{1}{2}R_3 \Rightarrow R_3 \left[\begin{array}{ccc|c} 1 & 0 & 1 & 4 \\ 0 & 1 & 0 & 2 \\ 0 & 0 & 1 & 3 \end{array}\right]$$

$$-1R_3 + R_1 \Rightarrow R_1 \left[\begin{array}{ccc|c} 1 & 0 & 0 & 1 \\ 0 & 1 & 0 & 2 \\ 0 & 0 & 1 & 3 \end{array}\right]$$

The solution to the system is $(1, 2, 3)$.

59. $\left[\begin{array}{ccc|c} 1 & -2 & 1 & 5 \\ 2 & 6 & 3 & -10 \\ 3 & -1 & -2 & 5 \end{array}\right]$

$$-2R_1 + R_2 \Rightarrow R_2 \left[\begin{array}{ccc|c} 1 & -2 & 1 & 5 \\ 0 & 10 & 1 & -20 \\ 3 & -1 & -2 & 5 \end{array}\right]$$

$$-3R_1 + R_3 \Rightarrow R_3 \left[\begin{array}{ccc|c} 1 & -2 & 1 & 5 \\ 0 & 10 & 1 & -20 \\ 0 & 5 & -5 & -10 \end{array}\right]$$

$$\frac{1}{10}R_2 \Rightarrow R_2 \left[\begin{array}{ccc|c} 1 & -2 & 1 & 5 \\ 0 & 1 & \frac{1}{10} & -2 \\ 0 & 5 & -5 & -10 \end{array}\right]$$

$$-2R_1 + R_2 \Rightarrow R_2 \left[\begin{array}{ccc|c} 1 & 3 & 8 & 1 \\ 1 & -9 & -18 & 9 \\ 3 & -1 & 14 & -2 \end{array}\right]$$

$$-5R_2 + R_3 \Rightarrow R_3 \left[\begin{array}{ccc|c} 1 & -2 & 1 & 5 \\ 0 & 1 & \frac{1}{3} & -2 \\ 0 & 0 & -\frac{11}{2} & 0 \end{array}\right]$$

$$2R_2 + R_1 \Rightarrow R_1 \left[\begin{array}{ccc|c} 1 & 0 & \frac{6}{5} & 1 \\ 0 & 1 & \frac{1}{10} & -2 \\ 0 & 0 & -\frac{11}{2} & 0 \end{array}\right]$$

$$-\frac{2}{11}R_3 \Rightarrow R_3 \left[\begin{array}{ccc|c} 1 & 0 & \frac{6}{5} & 1 \\ 0 & 1 & \frac{1}{10} & -2 \\ 0 & 0 & 1 & 0 \end{array}\right]$$

$$-\frac{1}{10}R_3 + R_2 \Rightarrow R_2 \left[\begin{array}{ccc|c} 1 & 0 & \frac{6}{5} & 1 \\ 0 & 1 & 0 & -2 \\ 0 & 0 & 1 & 0 \end{array}\right]$$

$$-\frac{6}{5}R_3 + R_1 \Rightarrow R_1 \left[\begin{array}{ccc|c} 1 & 0 & 0 & 1 \\ 0 & 1 & 0 & -2 \\ 0 & 0 & 1 & 0 \end{array}\right]$$

The solution to the system is (1, –2, 0).

61. $\left[\begin{array}{ccc|c} 1 & 1 & -1 & 2 \\ 2 & -1 & 1 & 1 \\ -1 & 1 & 1 & 2 \end{array}\right]$

$$-2R_1 + R_2 \Rightarrow R_2 \left[\begin{array}{ccc|c} 1 & 1 & -1 & 2 \\ 0 & -3 & 3 & -3 \\ -1 & 1 & 1 & 2 \end{array}\right]$$

$$R_1 + R_3 \Rightarrow R_3 \left[\begin{array}{ccc|c} 1 & 1 & -1 & 2 \\ 0 & -3 & 3 & -3 \\ 0 & 2 & 0 & 4 \end{array}\right]$$

$$-\frac{1}{3}R_2 \Rightarrow R_2 \left[\begin{array}{ccc|c} 1 & 1 & -1 & 2 \\ 0 & 1 & -1 & 1 \\ 0 & 2 & 0 & 4 \end{array}\right]$$

$$-1R_2 + R_1 \Rightarrow R_1 \left[\begin{array}{ccc|c} 1 & 0 & 0 & 1 \\ 0 & 1 & -1 & 1 \\ 0 & 2 & 0 & 4 \end{array}\right]$$

$$-2R_2 + R_3 \Rightarrow R_3 \left[\begin{array}{ccc|c} 1 & 0 & 0 & 1 \\ 0 & 1 & -1 & 1 \\ 0 & 0 & 2 & 2 \end{array}\right]$$

$$\frac{1}{2}R_3 \Rightarrow R_3 \left[\begin{array}{ccc|c} 1 & 0 & 0 & 1 \\ 0 & 1 & -1 & 1 \\ 0 & 0 & 1 & 1 \end{array}\right]$$

$$R_3 + R_2 \Rightarrow R_2 \left[\begin{array}{ccc|c} 1 & 0 & 0 & 1 \\ 0 & 1 & 0 & 2 \\ 0 & 0 & 1 & 1 \end{array}\right]$$

The solution to the system is (1, 2, 1).

63. $\left[\begin{array}{ccc|c} -1 & 2 & -1 & -6 \\ 1 & -2 & 1 & -5 \\ 3 & 1 & 2 & 4 \end{array}\right]$

$$R_1 \Leftrightarrow R_2 \left[\begin{array}{ccc|c} 1 & -2 & 1 & -5 \\ -1 & 2 & -1 & -6 \\ 3 & 1 & 2 & 4 \end{array}\right]$$

$$R_1 + R_2 \Rightarrow R_2 \left[\begin{array}{ccc|c} 1 & -2 & 1 & -5 \\ 0 & 0 & 0 & -11 \\ 3 & 1 & 2 & 4 \end{array}\right]$$

The second row of the augmented matrix represents the contradiction 0 = –11. Hence, the system is inconsistent. There is no solution.

65. $\left[\begin{array}{cc|c} 1 & 0 & -3 \\ 0 & 1 & -1 \end{array}\right]$

67. $\left[\begin{array}{cc|c} 1 & 3 & 3 \\ 0 & 0 & 0 \end{array}\right]$

Dependent system

69. $\left[\begin{array}{ccc|c} 1 & 0 & 0 & 1 \\ 0 & 1 & 0 & -2 \\ 0 & 0 & 1 & 0 \end{array}\right]$

Section 8.7 Practice Exercises

1. $\begin{vmatrix} -3 & 1 \\ 5 & 2 \end{vmatrix}$ $a = -3, b = 1, c = 5, d = 2$

$ad - bc = -3(2) - 1(5) = -6 - 5 = -11$

3. $\begin{vmatrix} -2 & 2 \\ -3 & -5 \end{vmatrix}$ $a = -2, b = 2, c = -3, d = -5$

$ad - bc = -2(-5) - 2(-3) = 10 + 6 = 16$

5. $\begin{vmatrix} \frac{1}{2} & 3 \\ -2 & 4 \end{vmatrix}$ $a = \frac{1}{2},\ b = 3, c = -2, d = 4$

$ad - bc = \frac{1}{2}(4) - 3(-2) = 2 + 6 = 8$

7. $\begin{vmatrix} 6 & 0 \\ 5 & 3 \end{vmatrix}$ $a = 6, b = 0, c = 5, d = 3$

$ad - bc = 6(3) - 0(5) = 18 - 0 = 18$

9. $\begin{vmatrix} -1 & 8 \\ 5 & 3 \end{vmatrix}$ $a=-1, b=8, c=5, d=3$

$ad-bc=-1(3)-8(5)=-3-40=-43$

11. $\begin{bmatrix} + & - & + \\ - & + & - \\ + & - & + \end{bmatrix}$

13. (a) $0\cdot\begin{vmatrix} -1 & 2 \\ 2 & -2 \end{vmatrix}-3\cdot\begin{vmatrix} 1 & 2 \\ 2 & -2 \end{vmatrix}+3\cdot\begin{vmatrix} 1 & 2 \\ -1 & 2 \end{vmatrix}$

$=0-3(-6)+3(4)$
$=18+12$
$=30$

(b) $-3\cdot\begin{vmatrix} 1 & 2 \\ 2 & -2 \end{vmatrix}+(-1)\cdot\begin{vmatrix} 0 & 2 \\ 3 & -2 \end{vmatrix}-2\cdot\begin{vmatrix} 0 & 1 \\ 3 & 2 \end{vmatrix}$

$=-3(-6)-1(-6)-2(-3)$
$=18+6+6$
$=30$

15. Choosing the row or column with the most zero elements simplifies the arithmetic when evaluating a determinant.

17. About the third column:

$1\cdot\begin{vmatrix} 3 & -6 \\ -2 & 8 \end{vmatrix}-0\cdot\begin{vmatrix} 5 & 2 \\ -2 & 8 \end{vmatrix}+0\cdot\begin{vmatrix} 5 & 2 \\ 3 & -6 \end{vmatrix}$

$=1(12)-0+0$
$=12$

19. About the third row:

$1\cdot\begin{vmatrix} 2 & 1 \\ -1 & 2 \end{vmatrix}-0\cdot\begin{vmatrix} 3 & 1 \\ 1 & 2 \end{vmatrix}+4\cdot\begin{vmatrix} 3 & 2 \\ 1 & -1 \end{vmatrix}$

$=1(5)-0+4(-5)$
$=5-20$
$=-15$

21. About the first column: Since all the elements are zero, the determinant will be zero.

23. $\begin{vmatrix} a & 2 \\ b & 8 \end{vmatrix}=8a-2b$

25. $x\cdot\begin{vmatrix} -2 & 6 \\ -1 & 1 \end{vmatrix}-y\begin{vmatrix} 0 & 3 \\ -1 & 1 \end{vmatrix}+z\begin{vmatrix} 0 & 3 \\ -2 & 6 \end{vmatrix}$

$=x(4)-y(3)+z(6)$
$=4x-3y+6z$

27. About the third column: Since all the elements are zero, the determinant will be zero.

29. $D=\begin{vmatrix} -3 & 8 \\ 5 & 5 \end{vmatrix}=-3(5)-8(5)=-15-40=-55$

$D_x=\begin{vmatrix} -10 & 8 \\ -13 & 5 \end{vmatrix}$
$=-10(5)-8(-13)$
$=-50+104$
$=54$

$D_y=\begin{vmatrix} -3 & -10 \\ 5 & -13 \end{vmatrix}$
$=-3(-13)-(-10)(5)$
$=39+50$
$=89$

31. $D=\begin{vmatrix} 2 & -1 \\ 3 & 1 \end{vmatrix}=2(1)-(-1)(3)=2+3=5$

$D_x=\begin{vmatrix} -1 & -1 \\ 6 & 1 \end{vmatrix}=-1(1)-(-1)(6)=-1+6=5$

$D_y=\begin{vmatrix} 2 & -1 \\ 3 & 6 \end{vmatrix}=2(6)-(-1)(3)=12+3=15$

$x=\dfrac{D_x}{D}=\dfrac{5}{5}=1$

$y=\dfrac{D_y}{D}=\dfrac{15}{5}=3$

The solution to the system is (1, 3).

33. $D=\begin{vmatrix} 7 & 3 \\ 5 & -4 \end{vmatrix}=7(-4)-3(5)=-28-15=-43$

$D_x=\begin{vmatrix} 4 & 3 \\ 9 & -4 \end{vmatrix}$
$=4(-4)-3(9)$
$=-16-27$
$=-43$

$D_y=\begin{vmatrix} 7 & 4 \\ 5 & 9 \end{vmatrix}=7(9)-4(5)=63-20=43$

$x = \frac{D_x}{D} = \frac{-43}{-43} = 1$

$y = \frac{D_y}{D} = \frac{43}{-43} = -1$

The solution to the system is $(1, -1)$.

35. $D = \begin{vmatrix} 2 & 3 \\ 6 & -12 \end{vmatrix}$
$= 2(-12) - 3(6)$
$= -24 - 18$
$= -42$

$D_x = \begin{vmatrix} 4 & 3 \\ -5 & -12 \end{vmatrix}$
$= 4(-12) - 3(-5)$
$= -48 + 15$
$= -33$

$D_y = \begin{vmatrix} 2 & 4 \\ 6 & -5 \end{vmatrix}$
$= 2(-5) - 4(6)$
$= -10 - 24$
$= -34$

$x = \frac{D_x}{D} = \frac{-33}{-42} = \frac{11}{14}$

$y = \frac{D_y}{D} = \frac{-34}{-42} = \frac{17}{21}$

The solution to the system is $\left(\frac{11}{14}, \frac{17}{21}\right)$.

37. Elimination method, substitution method or Gaussian elimination will determine if a system is inconsistent or dependent.

39. $D = \begin{vmatrix} 6 & -6 \\ 1 & -1 \end{vmatrix} = 6(-1) - (-6)(1) = -6 + 6 = 0$

Cramer's Rule not possible since $D = 0$. Use elimination method.
Eliminate the x variable by multiplying the second equation by -6.

$$\begin{array}{r} 6x - 6y = 5 \\ -6x + 6y = 48 \\ \hline 0 \neq 53 \end{array}$$

The system is inconsistent. There is no solution.

41. $D = \begin{vmatrix} -3 & -2 \\ -1 & 5 \end{vmatrix}$
$= -3(5) - (-2)(-1)$
$= -15 - 2$
$= -17$

$D_x = \begin{vmatrix} 0 & -2 \\ 0 & 5 \end{vmatrix} = 0(5) - (-2)(0) = 0$

$D_y = \begin{vmatrix} -3 & 0 \\ -1 & 0 \end{vmatrix} = -3(0) - 0(-1) = 0$

$x = \frac{D_x}{D} = \frac{0}{-17} = 0$

$y = \frac{D_y}{D} = \frac{0}{-17} = 0$

The solution to the system is $(0, 0)$.

43. $D = \begin{vmatrix} -2 & -10 \\ 1 & 5 \end{vmatrix}$
$= -2(5) - (-10)(1)$
$= -10 + 10$
$= 0$

Cramer's Rule not possible since $D = 0$. Use elimination method.
Eliminate the x variable by multiplying the second equation by 2.

$$\begin{array}{r} -2x - 10y = -4 \\ 2x + 10y = 4 \\ \hline 0 = 0 \end{array}$$

The equations are dependent. There are infinitely many solutions. The solution to the system is $\{(x, y) | x + 5y = 2\}$.

45. $D = \begin{vmatrix} 1 & 2 & 3 \\ 2 & -3 & 1 \\ 3 & -4 & 2 \end{vmatrix}$

$= -2 \cdot \begin{vmatrix} 2 & 1 \\ 3 & 2 \end{vmatrix} + (-3) \cdot \begin{vmatrix} 1 & 3 \\ 3 & 2 \end{vmatrix} - (-4) \cdot \begin{vmatrix} 1 & 3 \\ 2 & 1 \end{vmatrix}$
$= -2(4 - 3) - 3(2 - 9) + 4(1 - 6)$
$= -2(1) - 3(-7) + 4(-5)$
$= -2 + 21 - 20$
$= -1$

$$D_y = \begin{vmatrix} 1 & 8 & 3 \\ 2 & 5 & 1 \\ 3 & 9 & 2 \end{vmatrix}$$
$$= -8 \cdot \begin{vmatrix} 2 & 1 \\ 3 & 2 \end{vmatrix} + 5 \cdot \begin{vmatrix} 1 & 3 \\ 3 & 2 \end{vmatrix} - 9 \cdot \begin{vmatrix} 1 & 3 \\ 2 & 1 \end{vmatrix}$$
$$= -8(4-3) + 5(2-9) - 9(1-6)$$
$$= -8(1) + 5(-7) - 9(-5)$$
$$= -8 - 35 + 45$$
$$= 2$$
$$y = \frac{D_y}{D} = \frac{2}{-1} = -2$$

47. $$D = \begin{vmatrix} 4 & 4 & -3 \\ 8 & 2 & 3 \\ 4 & -4 & 6 \end{vmatrix}$$
$$= 4 \cdot \begin{vmatrix} 2 & 3 \\ -4 & 6 \end{vmatrix} - 8 \cdot \begin{vmatrix} 4 & -3 \\ -4 & 6 \end{vmatrix} + 4 \cdot \begin{vmatrix} 4 & -3 \\ 2 & 3 \end{vmatrix}$$
$$= 4(12+12) - 8(24-12) + 4(12+6)$$
$$= 4(24) - 8(12) + 4(18)$$
$$= 96 - 96 + 72$$
$$= 72$$
$$D_x = \begin{vmatrix} 3 & 4 & -3 \\ 0 & 2 & 3 \\ -3 & -4 & 6 \end{vmatrix}$$
$$= 3 \cdot \begin{vmatrix} 2 & 3 \\ -4 & 6 \end{vmatrix} - 0 \cdot \begin{vmatrix} 4 & -3 \\ -4 & 6 \end{vmatrix} + (-3) \cdot \begin{vmatrix} 4 & -3 \\ 2 & 3 \end{vmatrix}$$
$$= 3(12+12) - 0(24-12) - 3(12+6)$$
$$= 3(24) - 0 - 3(18)$$
$$= 72 - 54$$
$$= 18$$
$$x = \frac{D_x}{D} = \frac{18}{72} = \frac{1}{4}$$

49. $$D = \begin{vmatrix} 8 & 1 & 0 \\ 0 & 7 & 1 \\ 1 & 0 & -3 \end{vmatrix}$$
$$= -1 \cdot \begin{vmatrix} 0 & 1 \\ 1 & -3 \end{vmatrix} + 7 \cdot \begin{vmatrix} 8 & 0 \\ 1 & -3 \end{vmatrix} - 0 \cdot \begin{vmatrix} 8 & 0 \\ 0 & 1 \end{vmatrix}$$
$$= -1(0-1) + 7(-24-0) - 0(8-0)$$
$$= -1(-1) + 7(-24) - 0$$
$$= 1 - 168$$
$$= -167$$

$$D_y = \begin{vmatrix} 8 & 1 & 0 \\ 0 & 0 & 1 \\ 1 & -2 & -3 \end{vmatrix}$$
$$= -1 \cdot \begin{vmatrix} 0 & 1 \\ 1 & -3 \end{vmatrix} + 0 \cdot \begin{vmatrix} 8 & 0 \\ 1 & -3 \end{vmatrix} - (-2) \cdot \begin{vmatrix} 8 & 0 \\ 0 & 1 \end{vmatrix}$$
$$= -1(0-1) + 0(-24-0) + 2(8-0)$$
$$= -1(-1) + 0 + 2(8)$$
$$= 1 + 16$$
$$= 17$$
$$y = \frac{D_y}{D} = \frac{17}{-167} = -\frac{17}{167}$$

51. $$D = \begin{vmatrix} 4 & 0 & 1 \\ 0 & 1 & 0 \\ 1 & 0 & 1 \end{vmatrix}$$
$$= -0 \cdot \begin{vmatrix} 0 & 0 \\ 1 & 1 \end{vmatrix} + 1 \cdot \begin{vmatrix} 4 & 1 \\ 1 & 1 \end{vmatrix} - 0 \cdot \begin{vmatrix} 4 & 1 \\ 0 & 0 \end{vmatrix}$$
$$= 0 + 1(4-1) - 0$$
$$= 3$$
$$D_x = \begin{vmatrix} 7 & 0 & 1 \\ 2 & 1 & 0 \\ 4 & 0 & 1 \end{vmatrix}$$
$$= 7 \cdot \begin{vmatrix} 1 & 0 \\ 0 & 1 \end{vmatrix} - 2 \cdot \begin{vmatrix} 0 & 1 \\ 0 & 1 \end{vmatrix} + 4 \cdot \begin{vmatrix} 0 & 1 \\ 1 & 0 \end{vmatrix}$$
$$= 7(1-0) - 2(0-0) + 4(0-1)$$
$$= 7 - 0 - 4$$
$$= 3$$
$$D_y = \begin{vmatrix} 4 & 7 & 1 \\ 0 & 2 & 0 \\ 1 & 4 & 1 \end{vmatrix}$$
$$= -7 \cdot \begin{vmatrix} 0 & 0 \\ 1 & 1 \end{vmatrix} + 2 \cdot \begin{vmatrix} 4 & 1 \\ 1 & 1 \end{vmatrix} - 4 \cdot \begin{vmatrix} 4 & 1 \\ 0 & 0 \end{vmatrix}$$
$$= -7(0-0) + 2(4-1) - 4(0-0)$$
$$= 0 + 2(3) - 0$$
$$= 6$$
$$D_z = \begin{vmatrix} 4 & 0 & 7 \\ 0 & 1 & 2 \\ 1 & 0 & 4 \end{vmatrix}$$
$$= 7 \cdot \begin{vmatrix} 0 & 1 \\ 1 & 0 \end{vmatrix} - 2 \cdot \begin{vmatrix} 4 & 0 \\ 1 & 0 \end{vmatrix} + 4 \cdot \begin{vmatrix} 4 & 0 \\ 0 & 1 \end{vmatrix}$$
$$= 7(0-1) - 2(0-0) + 4(4-0)$$
$$= -7 - 0 + 4(4)$$
$$= 9$$

$$x = \frac{D_x}{D} = \frac{3}{3} = 1$$
$$y = \frac{D_y}{D} = \frac{6}{3} = 2$$
$$z = \frac{D_z}{D} = \frac{9}{3} = 3$$

The solution to the system is (1, 2, 3).

53. $D = \begin{vmatrix} -8 & 1 & 1 \\ 2 & -1 & 1 \\ 3 & 0 & -1 \end{vmatrix}$

$$\begin{aligned} &= -1 \cdot \begin{vmatrix} 2 & 1 \\ 3 & -1 \end{vmatrix} + (-1) \cdot \begin{vmatrix} -8 & 1 \\ 3 & -1 \end{vmatrix} - 0 \cdot \begin{vmatrix} -8 & 1 \\ 2 & 1 \end{vmatrix} \\ &= -1(-2-3) - 1(8-3) - 0 \\ &= -1(-5) - 1(5) \\ &= 5 - 5 \\ &= 0 \end{aligned}$$

Cramer's Rule does not apply.

55. $\begin{vmatrix} y & -2 \\ 8 & 7 \end{vmatrix} = 30$

$$\begin{aligned} 7y + 16 &= 30 \\ 7y &= 14 \\ y &= 2 \end{aligned}$$

57.

$$\begin{vmatrix} -1 & 0 & 2 \\ 4 & t & 0 \\ 0 & -5 & 3 \end{vmatrix} = -4$$

$$\begin{aligned} 0 \cdot \begin{vmatrix} 4 & 0 \\ 0 & 3 \end{vmatrix} + t \cdot \begin{vmatrix} -1 & 2 \\ 0 & 3 \end{vmatrix} - (-5) \cdot \begin{vmatrix} -1 & 2 \\ 4 & 0 \end{vmatrix} &= -4 \\ 0 + t(-3) + 5(-8) &= -4 \\ -3t - 40 &= -4 \\ -3t &= 36 \\ t &= -12 \end{aligned}$$

59.

$$\begin{aligned} \begin{vmatrix} 5 & 2 & 0 & 0 \\ 0 & 4 & -1 & 1 \\ -1 & 0 & 3 & 0 \\ 0 & -2 & 1 & 0 \end{vmatrix} &= 5 \cdot \begin{vmatrix} 4 & -1 & 1 \\ 0 & 3 & 0 \\ -2 & 1 & 0 \end{vmatrix} - 0 + (-1) \cdot \begin{vmatrix} 2 & 0 & 0 \\ 4 & -1 & 1 \\ -2 & 1 & 0 \end{vmatrix} - 0 \\ &= 5\left[0 + 3 \cdot \begin{vmatrix} 4 & 1 \\ -2 & 0 \end{vmatrix} - 0\right] - 1\left[2 \cdot \begin{vmatrix} -1 & 1 \\ 1 & 0 \end{vmatrix} - 0 + 0\right] \\ &= 5[3(0+2)] - 1[2(0-1)] \\ &= 5(6) - 1(-2) \\ &= 30 + 2 \\ &= 32 \end{aligned}$$

61. (a) $\begin{vmatrix} 1 & 0 & 1 & 1 \\ 2 & 5 & -1 & 1 \\ 2 & 0 & 0 & -1 \\ 0 & -1 & 1 & 0 \end{vmatrix} = -0 + 5\begin{vmatrix} 1 & 1 & 1 \\ 2 & 0 & -1 \\ 0 & 1 & 0 \end{vmatrix} - 0 + (-1)\begin{vmatrix} 1 & 1 & 1 \\ 2 & -1 & 1 \\ 2 & 0 & -1 \end{vmatrix}$

$$= 5\left[-1\begin{vmatrix} 2 & -1 \\ 0 & 0 \end{vmatrix} + 0 - 1\begin{vmatrix} 1 & 1 \\ 2 & -1 \end{vmatrix}\right]$$
$$= 1\left[-1\begin{vmatrix} 2 & 1 \\ 2 & -1 \end{vmatrix} + (-1)\begin{vmatrix} 1 & 1 \\ 2 & -1 \end{vmatrix} - 0\right]$$
$$= 5[-1(0-0) - 1(-1-2)] - 1[-1(-2-2) - 1(-1-2)]$$
$$= 5(0+3) - 1(4+3)$$
$$= 15 - 7$$
$$= 8$$

(b) $y = \dfrac{D_y}{D} = \dfrac{8}{2} = 4$

Chapter 8 Review Exercises

1. $x - 4y = -4$
$x + 2y = 8$ ordered pair: (4, 2)

$4 - 4(2) \stackrel{?}{=} -4$ ✓
$4 + 2(2) \stackrel{?}{=} 8$ ✓
The ordered pair (4, 2) is a solution.

3. $3x + y = 9$
$y = 3$ ordered pair: (1, 3)

$3(1) + 3 \stackrel{?}{=} 9$ No
$3 \stackrel{?}{=} 3$ ✓
The ordered pair (1, 3) is not a solution.

5. $y = -\dfrac{1}{2}x + 4$
$y = x - 1$
The slopes are unequal. The lines intersect.

7. $y = -\dfrac{4}{7}x + 3$
$y = -\dfrac{4}{7}x - 5$
The slopes are equal and the y-intercepts are unequal. The lines are parallel.

9. $y = 9x - 2$
$9x - y = 2$
Rewrite the second equation in slope-intercept form.
$y = 9x - 2$
$y = 9x - 2$
The equations are the same. The lines coincide.

11. $y=-\frac{2}{3}x-2$
$-x+3y=-6$

Rewrite the second equation in slope-intercept form.

$y=-\frac{2}{3}x-2$
$y=\frac{1}{3}x-2$

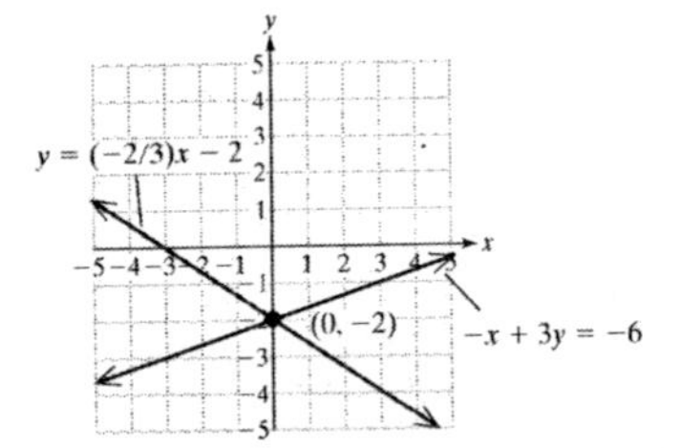

(0, –2) consistent; independent

13. $4x=-2y+10$
$2x+y=5$

Rewrite the equations in slope-intercept form.

$y=-2x+5$
$y=-2x+5$

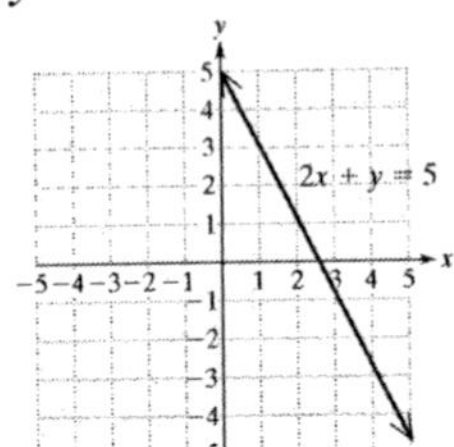

Infinitely many solutions
$\{(x, y)|y=-2x+5\}$
consistent; dependent

15. $6x-3y=9$
$y=-1$

Rewrite the first equation in slope-intercept form.

$y=2x-3$
$y=-1$

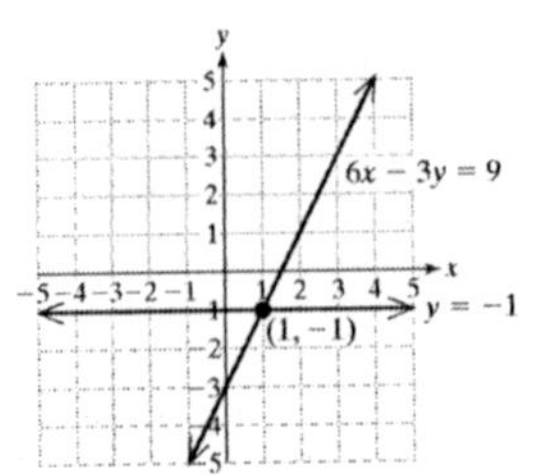

$(1,-1)$ consistent; independent

17. $x-7y=14$
$-2x+14y=14$

Rewrite the equations in slope-intercept form.

$y=\frac{1}{7}x-2$
$y=\frac{1}{7}x+1$

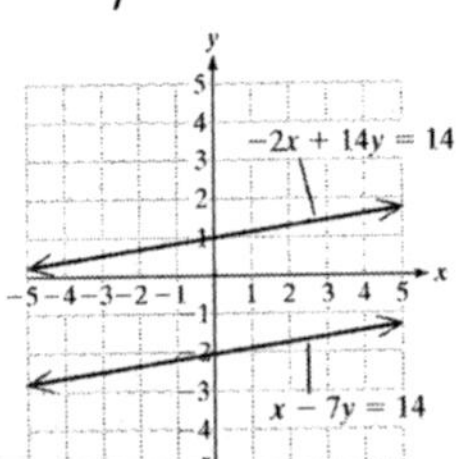

No solution; inconsistent; independent

19. The year of equality occurs at the point of intersection. 1994

21. $6x+y=2$
$y=3x-4$

Substitute the value for y into the first equation and solve for x.

$6x+3x-4=2$
$9x-4=2$
$9x=6$
$x=\frac{6}{9}=\frac{2}{3}$

$y=3x-4$
$y=3\left(\frac{2}{3}\right)-4$
$y=2-4$
$y=-2$

Solution: $\left(\frac{2}{3},-2\right)$

23. $2x+6y=10$
$x=-3y+6$

Substitute the value for x into the first equation and solve for y.
$2(-3y+6)+6y=10$
$-7y+12+6y=10$
$12=10$
Contradiction; no solution

25. (a) x in the first equation is easier to solve for because its coefficient is 1.

(b) $x+2y=11$
$5x+4y=40$

Solving the first equation for x gives $x = -2y + 11$. Substitute this value into the second equation and solve for y.
$5(-2y+11)+4y=40$
$-10y+55+4y=40$
$-6y+55=40$
$-6y=-15$
$y=\frac{-15}{-6}=\frac{5}{2}$
$x+2y=11$
$x+2\left(\frac{5}{2}\right)=11$
$x+5=11$
$x=6$
Solution: $\left(6, \frac{5}{2}\right)$

27. $3x-2y=23$
$x+5y=-15$

Solving the second equation for x gives $x = -5y - 15$. Substitute this value into the first equation and solve for y.
$3(-5y-15)-2y=23$
$-15y-45-2y=23$
$-17y-45=23$
$-17y=68$
$y=-4$

$x+5y=-15$
$x+5(-4)=-15$
$x-20=-15$
$x=5$
Solution: $(5, -4)$

29. $x-3y=9$
$5x-15y=45$

Solving the first equation for x gives $x = 3y + 9$. Substitute this value into the second equation and solve for y.
$5(3y+9)-15y=45$
$15y+45-15y=45$
$45=45$ Identity
Infinitely many solutions
$\left\{(x, y) \middle| y=-\frac{1}{3}x-3\right\}$

31. Let x represent the smaller number. Let y represent the larger number. The statement "difference of two positive number is 42" translates to the equation $y - x = 42$. The statement "the larger number is 2 more than 6 times the smaller number" translates to the equation $y = 6x + 2$.

$y-x=42$
$y=6x+2$
Since the second equation is solved for y, use the substitution method by substituting the value for y into the first equation and solving for x.
$(6x+2)-x=42$
$6x+2-x=42$
$5x+2=42$
$5x=40$
$x=8$

$y = 6x + 2$
$y = 6(8) + 2$
$y = 48 + 2$
$y = 50$
The numbers are 8 and 50.

33. Let x represent the number of rushing touchdowns. Let y represent the number of receiving touchdowns. The statement "the remaining 165 were scored rushing or receiving" translates to the equation $x + y = 165$. The statement "the number of receiving touchdowns ... was 5 more than 15 times the number of rushing touchdowns" translates to the equation $y = 15x + 5$.

$$\begin{aligned} x + y &= 165 \\ y &= 15x + 5 \end{aligned}$$

Since the second equation is solved for y, use the substitution method by substituting the value for y into the first equation and solving for x.

$$\begin{aligned} x + (15x + 5) &= 165 \\ x + 15x + 5 &= 165 \\ 16x + 5 &= 165 \\ 16x &= 160 \\ x &= 10 \end{aligned}$$

$y = 15x + 5$
$y = 15(10) + 5$
$y = 150 + 5$
$y = 155$
He scored 155 receiving touchdowns and 10 rushing touchdowns.

35. **(a)** Write both equations in standard form.

(b) Multiply one of both equations by a constant to create opposite coefficients for one of the variables.

(c) Add the equations to eliminate the variable.

(d) Solve for the remaining variable.

(e) Substitute the known variable value into an original equation to solve for the other variable.

37. **(a)** y is easier to eliminate because the signs are opposite.

(b) $9x - 2y = 14$
$4x + 3y = 14$

Multiply the first equation by 3 and the second equation by 2 to obtain opposite coefficients on y.

$$\begin{aligned} 27x - 6y &= 42 \\ 8x + 6y &= 28 \\ \hline 35x &= 70 \\ x &= 2 \end{aligned}$$

$$\begin{aligned} 9x - 2y &= 14 \\ 9(2) - 2y &= 14 \\ 18 - 2y &= 14 \\ -2y &= -4 \\ y &= 2 \end{aligned}$$

Solution: (2, 2)

39.
$$\begin{aligned} x + 3y &= 0 \\ -3x - 10y &= -2 \end{aligned}$$

Multiply the first equation by 3 to obtain opposite coefficients on x.

$$\begin{aligned} 3x + 9y &= 0 \\ -3x - 10y &= -2 \\ \hline -y &= -2 \\ y &= 2 \end{aligned}$$

$$\begin{aligned} x + 3y &= 0 \\ x + 3(2) &= 0 \\ x + 6 &= 0 \\ x &= -6 \end{aligned}$$

Solution: (−6, 2)

41. $12x = 5(y + 1)$
$5y = -1 - 4x$

Rewrite the equations in standard form.

$$\begin{aligned} 12x - 5y &= 5 \\ 4x + 5y &= -1 \\ \hline 16x &= 4 \\ x &= \frac{1}{4} \end{aligned}$$

$$5y = -1 - 4x$$
$$5y = -1 - 4\left(\frac{1}{4}\right)$$
$$5y = -1 - 1$$
$$5y = -2$$
$$y = -\frac{2}{5}$$

Solution: $\left(\frac{1}{4}, -\frac{2}{5}\right)$

43. $-8x - 4y = 16$
$10x + 5y = 5$

Multiply the first equation by 5 and the second equation by 4 to obtain opposite coefficients on y.

$$\begin{aligned} -40x - 20y &= 80 \\ 40x + 20y &= 20 \\ \hline 0 &= 100 \end{aligned} \qquad \text{Contradiction}$$

No solution

45. $0.5x - 0.2y = 0.5$
$0.4x + 0.7y = 0.4$

Multiply both equations by 10 to clear the decimals.

$5x - 2y = 5$
$4x + 7y = 4$

Multiply the first equation by 7 and the second equation by 2 to obtain opposite coefficients on y.

$$\begin{aligned} 35x - 14y &= 35 \\ 8x + 14y &= 8 \\ \hline 43x &= 43 \\ x &= 1 \end{aligned}$$

$$\begin{aligned} 0.5x - 0.2y &= 0.5 \\ 0.5(1) - 0.2y &= 0.5 \\ 0.5 - 0.2y &= 0.5 \\ -0.2y &= 0 \\ y &= 0 \end{aligned}$$

Solution: $(1, 0)$

47. (a) Addition method because the equations are written in standard form.

(b) $5x - 8y = -2$
$3x - y = -5$

Multiply the second equation by -8 to obtain opposite coefficients on y.

$$\begin{aligned} 5x - 8y &= -2 \\ -24x + 8y &= 40 \\ \hline -19x &= 38 \\ x &= -2 \end{aligned}$$

$$\begin{aligned} 3x - y &= -5 \\ 3(-2) - y &= -5 \\ -6 - y &= -5 \\ -y &= 1 \\ y &= -1 \end{aligned}$$

Solution: $(-2, -1)$

49. Let x represent the amount of money invested at 5%. Let y represent the amount of money invested at 8%.

	5% Account	8% Account	Total
Principal	x	y	20,000
Interest	$0.05x$	$0.08y$	1525

$$x + y = 20{,}000$$
$$0.05x + 0.08y = 1525$$

Multiply the second equation by 100 to eliminate the decimals.

$$x + y = 20{,}000$$
$$5x + 8y = 152{,}500$$

Multiply the first equation by -5 to obtain opposite coefficients on x.

$$\begin{aligned} 5x - 5y &= -100{,}000 \\ 5x + 8y &= 152{,}500 \\ \hline 3y &= 52{,}500 \\ y &= 17{,}500 \end{aligned}$$

$$\begin{aligned} x + y &= 20{,}000 \\ x + 17{,}500 &= 20{,}000 \\ x &= 2500 \end{aligned}$$

He invested \$2500 in the 5% account and \$17,500 in the 8% account.

51. Let b represent the speed of the boat in still water. Let c represent the speed of the current.

	Distance	Rate	Time
Downstream	80	$b+c$	4
Upstream	80	$b-c$	5

$$4b+4c=80$$
$$5b-5c=80$$

Multiply the first equation by 5 and the second equation by 4 to obtain opposite coefficients on c.

$$20b+20c=400$$
$$\underline{20b-20c=320}$$
$$40b=720$$
$$b=18$$

$$4(b+c)=80$$
$$4b+4c=80$$
$$4(18)+4c=80$$
$$72+4c=80$$
$$4c=8$$
$$c=2$$

The speed of the boat in still water is 18 mph and the speed of the current is 2 mph.

53. Let w represent the number of women voters. Let m represent the number of men voters. The statement "5700 votes were cast" translates to the equation $w + m = 5700$. The statement "3675 ... voted for the winning candidate" together with "$\frac{5}{8}$ of the women and $\frac{2}{3}$ of the men voted for the winning candidate" translates to the equation $\frac{5}{8}w+\frac{2}{3}m=3675$.

$$w+m=5700$$
$$\frac{5}{8}w+\frac{2}{3}m=3675$$

Multiply the second equation by 24 to clear the fractions.

$$w+m=5700$$
$$15w+16m=88,200$$

Multiply the first equation by -15 to obtain opposite coefficients on w.

$$-15w-15m=-85,500$$
$$\underline{15w+16m=88,200}$$
$$m=2700$$

$$w+m=5700$$
$$w+2700=5700$$
$$w=3000$$

There were 3000 women voters and 2700 men voters.

55. A: $5x+3y-z=5$
B: $x+2y+z=6$
C: $-x-2y-z=8$

Eliminate the z variable from equations A and B by adding.

$$5x+3y-z=5$$
$$\underline{x+2y+z=6}$$
$$6x+5y=11 \qquad \text{Equation D}$$

Eliminate the z variable from equations B and C by adding.

$$x+2y+z=6$$
$$\underline{-x-2y-z=8}$$
$$0=14 \qquad \text{False}$$

No solution.

57. Let x represent the length of shortest leg. Let y represent the length of middle leg. Let z represent the length of longest leg.

Shortest leg + middle leg + longest leg = perimeter $\Rightarrow x + y + z = 30$;
One leg is 2 ft more than twice the shortest $\Rightarrow y = 2x + 2$;
hypotenuse is 2 ft less than three times the shortest $\Rightarrow z = 3x - 2$

A: $x+y+z=30$
B: $-2x+y=2$
C: $-3x+z=-2$

Eliminate the z variable from equations A and C by multiplying equation C by -1.

$$\begin{array}{r} x+y+z=30 \\ 3x-z=2 \\ \hline 4x+y=32 \end{array} \quad \text{Equation D}$$

Eliminate the y variable from equations B and D by multiplying equation B by -1.

$$\begin{array}{r} 2x-y=-2 \\ 4x+y=32 \\ \hline 6x=30 \\ x=5 \end{array}$$

Substitute $x = 5$ into equation C to solve for z.

$$\begin{aligned} -3(5)+z&=-2 \\ -15+z&=-2 \\ z&=13 \end{aligned}$$

Substitute $x - 5$ into equation B to solve for y.

$$\begin{aligned} -2(5)+y&=2 \\ -10+y&=2 \\ y&=12 \end{aligned}$$

The lengths of the legs are 5 ft, 12 ft and 13 ft.

59. Order: 3×3

61. Order: 1×4

63. $\left[\begin{array}{cc|c} 1 & 1 & 3 \\ 1 & -1 & -1 \end{array}\right]$

65. $x = 9$
$y = -3$

67. **(a)** 1

(b) $\left[\begin{array}{cc|c} 1 & 3 & -11 \\ 2 & 0 & 5 \end{array}\right]$

69. $x+y=3$
$x-y=-1$

$$\left[\begin{array}{cc|c} 1 & 1 & 3 \\ 1 & -1 & -1 \end{array}\right]$$

$$-R_1+R_2 \Rightarrow R_2 \left[\begin{array}{cc|c} 1 & 1 & 3 \\ 0 & -2 & -4 \end{array}\right]$$

$$-\frac{1}{2}R_2 \Rightarrow R_2 \left[\begin{array}{cc|c} 1 & 1 & 3 \\ 0 & 1 & 2 \end{array}\right]$$

$$-R_2+R_1 \Rightarrow R_1 \left[\begin{array}{cc|c} 1 & 0 & 1 \\ 0 & 1 & 2 \end{array}\right]$$

The solution to the system is (1, 2).

71. $\begin{vmatrix} 5 & -2 \\ 2 & -3 \end{vmatrix} = -15-(-4) = -15+4 = -11$

73. $\begin{vmatrix} \frac{1}{2} & 3 \\ 1 & 8 \end{vmatrix} = \frac{1}{2}(8)-3 = 4-3 = 1$

75. $A = \begin{vmatrix} 8 & 2 & 0 \\ -1 & 4 & -2 \\ 3 & -3 & 6 \end{vmatrix}$

Element 8

$\begin{vmatrix} 4 & -2 \\ -3 & 6 \end{vmatrix} = 24-6 = 18$

77. $A = \begin{vmatrix} 8 & 2 & 0 \\ -1 & 4 & -2 \\ 3 & -3 & 6 \end{vmatrix}$

Element 2

$\begin{vmatrix} 8 & 2 \\ 3 & -3 \end{vmatrix} = -24-6 = -30$

79. $\begin{vmatrix} 2 & 1 & 0 \\ -4 & 3 & -1 \\ 3 & 0 & 1 \end{vmatrix}$ About the 3rd column

$$\begin{aligned} &= 0 \cdot \begin{vmatrix} -4 & 3 \\ 3 & 0 \end{vmatrix} - (-1) \cdot \begin{vmatrix} 2 & 1 \\ 3 & 0 \end{vmatrix} + 1 \cdot \begin{vmatrix} 2 & 1 \\ -4 & 3 \end{vmatrix} \\ &= 0 + 1(0-3) + 1(6-(-4)) \\ &= 1(-3) + 1(10) \\ &= -3+10 \\ &= 7 \end{aligned}$$

81. $\begin{vmatrix} 4 & -2 & 0 \\ 9 & 5 & 4 \\ 1 & 2 & 0 \end{vmatrix}$ About the 3rd column

$$= 0 \cdot \begin{vmatrix} 9 & 5 \\ 1 & 2 \end{vmatrix} - 4 \cdot \begin{vmatrix} 4 & -2 \\ 1 & 2 \end{vmatrix} + 0 \cdot \begin{vmatrix} 4 & -2 \\ 9 & 5 \end{vmatrix}$$
$$= 0 - 4(8 - (-2)) + 0$$
$$= -4(10)$$
$$= -40$$

83. $D = \begin{vmatrix} 3 & 2 \\ 1 & -3 \end{vmatrix} = -9 - 2 = -11$

$D_x = \begin{vmatrix} 9 & 2 \\ 1 & -3 \end{vmatrix} = -27 - 2 = -29$

$D_y = \begin{vmatrix} 3 & 9 \\ 1 & 1 \end{vmatrix} = 3 - 9 = -6$

$x = \dfrac{D_x}{D} = \dfrac{-29}{-11} = \dfrac{29}{11}$

$y = \dfrac{D_y}{D} = \dfrac{-6}{-11} = \dfrac{6}{11}$

The solution to the system is $\left(\dfrac{29}{11}, \dfrac{6}{11}\right)$.

85. $D = \begin{vmatrix} 3 & -4 \\ 2 & 3 \end{vmatrix} = 9 - (-8) = 9 + 8 = 17$

$D_x = \begin{vmatrix} 1 & -4 \\ 12 & 3 \end{vmatrix} = 3 - (-48) = 3 + 48 = 51$

$D_y = \begin{vmatrix} 3 & 1 \\ 2 & 12 \end{vmatrix} = 36 - 2 = 34$

$x = \dfrac{D_x}{D} = \dfrac{51}{17} = 3$

$y = \dfrac{D_y}{D} = \dfrac{34}{17} = 2$

The solution to the system is (3, 2).

87. $D = \begin{vmatrix} 2 & 3 & -1 \\ 1 & 0 & 3 \\ 0 & 2 & 1 \end{vmatrix}$

$$= 2 \cdot \begin{vmatrix} 0 & 3 \\ 2 & 1 \end{vmatrix} - 1 \cdot \begin{vmatrix} 3 & -1 \\ 2 & 1 \end{vmatrix} + 0$$
$$= 2(0 - 6) - 1(3 - (-2))$$
$$= 2(-6) - 1(5)$$
$$= -12 - 5$$
$$= -17$$

$D_x = \begin{vmatrix} -7 & 3 & -1 \\ 10 & 0 & 3 \\ -1 & 2 & 1 \end{vmatrix}$

$$= -7 \cdot \begin{vmatrix} 0 & 3 \\ 2 & 1 \end{vmatrix} - 10 \cdot \begin{vmatrix} 3 & -1 \\ 2 & 1 \end{vmatrix} + (-1) \cdot \begin{vmatrix} 3 & -1 \\ 0 & 3 \end{vmatrix}$$
$$= -7(0 - 6) - 10(3 - (-2)) - 1(9 - 0)$$
$$= -7(-6) - 10(5) - 1(9)$$
$$= 42 - 50 - 9$$
$$= -17$$

$D_y = \begin{vmatrix} 2 & -7 & -1 \\ 1 & 10 & 3 \\ 0 & -1 & 1 \end{vmatrix}$

$$= 2 \cdot \begin{vmatrix} 10 & 3 \\ -1 & 1 \end{vmatrix} - 1 \cdot \begin{vmatrix} -7 & -1 \\ -1 & 1 \end{vmatrix} + 0$$
$$= 2(10 - (-3)) - 1(-7 - 1)$$
$$= 2(13) - 1(-8)$$
$$= 26 + 8$$
$$= 34$$

$D_z = \begin{vmatrix} 2 & 3 & -7 \\ 1 & 0 & 10 \\ 0 & 2 & -1 \end{vmatrix}$

$$= 2 \cdot \begin{vmatrix} 0 & 10 \\ 2 & -1 \end{vmatrix} - 1 \cdot \begin{vmatrix} 3 & -7 \\ 2 & -1 \end{vmatrix} + 0$$
$$= 2(0 - 20) - 1(-3 - (-14))$$
$$= 2(-20) - 1(11)$$
$$= -40 - 11$$
$$= -51$$

$x = \dfrac{D_x}{D} = \dfrac{-17}{-17} = 1$

$y = \dfrac{D_y}{D} = \dfrac{34}{-17} = -2$

$z = \dfrac{D_z}{D} = \dfrac{-51}{-17} = 3$

The solution to the system is (1, –2, 3).

89. $D = \begin{vmatrix} 2 & -1 \\ 4 & -2 \end{vmatrix} = -4-(-4) = -4+4 = 0$

Cramer's Rule does not apply since $D = 0$. Use the elimination method.

Eliminate the x variable by multiplying the first equation by –2.

$$\begin{array}{r} -4x+2y=-2 \\ 4x-2y=2 \\ \hline 0=0 \end{array}$$

The equations are dependent. There are infinitely many solutions. The solution to the system is $\{(x, y) | y = 2x - 1\}$.

Chapter 8 Test

1. $5x+2y=-6$

$-\frac{5}{2}x - y = -3$

Rewriting each equation in slope-intercept form gives

$y = -\frac{5}{2}x - 3$

$y = -\frac{5}{2}x + 3$

Since the slopes of the two lines are equal and their y-intercepts are unequal, the lines are parallel.

3. The information needed to answer the questions occurs at the point of intersection.

(a) \$15

(b) 5,000,000 items

5. Let x represent the number of points scored by Cynthia Cooper. Let y represent the number of points scored by Sheryl Swoopes. The statement "together they scored a total 1133 points" translates to the equation $x + y = 1133$. The statement "Cooper ... scored 227 more points than ... Swoopes" translates to the equation $x = y + 227$.

$$x+y=1133$$
$$x=y+227$$

Since the second equation is solved for x, substitute this value for x into the first equation and solve for y.

$$\begin{aligned} x+y &= 1133 \\ x+227+y &= 1133 \\ 2y+227 &= 1133 \\ 2y &= 906 \\ y &= 453 \end{aligned}$$

$x = y + 227$

$x = 453 + 227$

$x = 680$

Cooper scored 680 points and Swoopes scored 453 points.

7. Let x represent the number of milliliters of 50% acid solution. Let y represent the number of milliliters of 20% acid solution.

	50% Solution	20% Solution	30% Solution
Amount of solution	x	y	36
Amount of acid	$0.5x$	$0.2y$	0.3(36)

$$x+y=36$$
$$0.5x+0.2y=10.8$$

Since the equations are written in standard form, use the addition method. Multiply the first equation by –0.5 to obtain opposite coefficients on x.

$$\begin{array}{r} -0.5x-0.5y=-18 \\ 0.5x+0.2y=10.8 \\ \hline -0.3y=-7.2 \\ y=24 \end{array}$$

$$\begin{aligned} x+y &= 36 \\ x+24 &= 36 \\ x &= 12 \end{aligned}$$

12 milliliters of the 50% acid solution should be mixed with 24 milliliters of the 20% acid solutions.

9. $\frac{1}{3}x + y = \frac{7}{3}$

$x = \frac{3}{2}y - 11$

Since the second equation is solved for x, use the substitution method by substituting this value for x into the first equation and solving for y.

$$\frac{1}{3}x + y = \frac{7}{3}$$
$$\frac{1}{3}\left(\frac{3}{2}y - 11\right) + y = \frac{7}{3}$$
$$\frac{1}{2}y - \frac{11}{3} + y = \frac{7}{3}$$

Clear the fractions by multiplying both sides of the equation by 6.

$$3y - 22 + 6y = 14$$
$$9y - 22 = 14$$
$$9y = 36$$
$$y = 4$$

$$x = \frac{3}{2}y - 11$$
$$x = \frac{3}{2}(4) - 11$$
$$x = 6 - 11$$
$$x = -5$$

Solution: $(-5, 4)$

11. $-0.25 - 0.05y = 0.2$

$10x + 2y = -8$

Clear the decimals from the first equation by multiplying it by 100.

$$-25x - 5y = 20$$
$$10x + 2y = -8$$

The equations are now in standard form. Use the addition method by multiplying the first equation by 2 and the second equation by 5 to obtain opposite coefficients on x. Add the equations and solve for y.

$$-50x - 10y = 40$$
$$50x + 10y = -40$$
$$0 = 0 \quad \text{Identity}$$

Infinitely many solutions

$\{(x, y) | y = -5x - 4\}$

13. Let x represent the measure of one of the acute angles. Let y represent the measure of the other acute angle. In a right triangle the sum of the measures of the acute angles is 90. This translates to the equation $x + y = 90$. The statement "one of the angles is 9° less than twice the other acute angle" translates to the equation $y = 2x - 9$.

$$x + y = 90$$
$$y = 2x - 9$$

Since the second equation is solved for y, use the substitution method by substituting this value into the first equation and solving for x.

$$x + y = 90$$
$$x + 2x - 9 = 90$$
$$3x - 9 = 90$$
$$3x = 99$$
$$x = 33$$

$$x + y = 90$$
$$33 + y = 90$$
$$y = 57$$

The measures of the acute angles are 33° and 57°.

15. Let x represent the amount invested in the 9% account. Let y represent the amount invested in the 11% account. The statement "five thousand dollars less was invested in an account earning 9% simple interest than in an account earning 11% simple interest" translates to the equation $x = y - 5000$. The statement "the total interest ... is \$1950" translates to the equation $0.09x + 0.11y = 1950$.

$$x = y - 5000$$
$$0.09x + 0.11y = 1950$$

Since the first equation is solved for x, use the substitution method by substituting this value for x in the second equation and solving for y.

$$0.09x + 0.11y = 1950$$
$$0.09(y - 5000) + 0.11y = 1950$$
$$0.09y - 450 + 0.11y = 1950$$
$$0.20y - 450 = 1950$$
$$0.20y = 2400$$
$$y = 12{,}000$$

$x = y - 5000$
$x = 12{,}000 - 5000$
$x = 7000$
\$7000 was invested in the 9% account and \$12,000 was invested in the 11% account. \$19,000 was the total amount invested.

17. Let p represent the speed of the plane in still air. Let w represent the speed of the wind.

	Rate	Time	Distance
With the wind	$p + w$	2	1000
Against the wind	$p - w$	2	880

$2(p + w) = 1000$
$2(p - w) = 880$

Rewrite the equations in standard form.

$2p + 2w = 1000$
$2p - 2w = 880$

Since the coefficients on w are opposites, add the equations and solve for p.

$$\begin{aligned} 2p + 2w &= 1000 \\ 2p - 2w &= 880 \\ \hline 4p &= 1880 \\ p &= 470 \end{aligned}$$

$$\begin{aligned} 2(p + w) &= 1000 \\ 2p + 2w &= 1000 \\ 2(470) + 2w &= 1000 \\ 940 + 2w &= 1000 \\ 2w &= 60 \\ w &= 30 \end{aligned}$$

The speed of the plane is 470 mph and the speed of the wind is 30 mph.

19. Let j represent the number of orders that Joanne can process.
Let k represent the number of orders that Kent can process.
Let g represent the number of orders that Geoff can process.

A: $g + j + k = 504$
B: $k = j + 20$
C: $g = j + k - 104$

or

A: $g + j + k = 504$
B: $-j + k = 20$
C: $g - j - k = -104$

Eliminate the variable g from equations A and C by multiplying equation C by -1.

$$\begin{aligned} g + j + k &= 504 \\ -g + j + k &= 104 \\ \hline 2j + 2k &= 608 \end{aligned} \qquad \text{Equation D}$$

Eliminate the j variable from equations B and D by multiplying equation B by 2.

$$\begin{aligned} -2j + 2k &= 40 \\ 2j + 2k &= 608 \\ \hline 4k &= 648 \\ k &= 162 \end{aligned}$$

Substitute $k = 162$ into equation D to find j.

$$\begin{aligned} 2j + 2(162) &= 608 \\ 2j + 324 &= 608 \\ 2j &= 284 \\ j &= 142 \end{aligned}$$

Substitute $k = 162$ and $j = 142$ into equation A to solve for G.

$$\begin{aligned} g + 142 + 162 &= 504 \\ g &= 504 - 304 \\ g &= 200 \end{aligned}$$

Joanne can process 142 orders per day.
Kent can process 162 orders per day.
Geoff can process 200 orders per day.

21. (a) $\left[\begin{array}{ccc|c} 1 & 2 & 1 & -3 \\ 0 & -8 & -3 & 10 \\ -5 & -6 & 3 & 0 \end{array}\right]$

(b) $\left[\begin{array}{ccc|c} 1 & 2 & 1 & -3 \\ 0 & -8 & -3 & 10 \\ 0 & 4 & 8 & -15 \end{array}\right]$

23. $\begin{vmatrix} 2 & -3 \\ 1 & 2 \end{vmatrix} = 4-(-3) = 4+3 = 7$

25. $\begin{vmatrix} 0 & 5 & -2 \\ 0 & 0 & 2 \\ 2 & 3 & 1 \end{vmatrix}$ About the 1st column

$= 0 - 0 + 2 \cdot \begin{vmatrix} 5 & -2 \\ 0 & 2 \end{vmatrix}$

$= 2(10-0)$

$= 20$

27. $D = \begin{vmatrix} 6 & -5 \\ -2 & -2 \end{vmatrix} = -12-10 = -22$

$D_y = \begin{vmatrix} 6 & 13 \\ -2 & 9 \end{vmatrix} = 54+26 = 80$

$y = \frac{D_y}{D} = \frac{80}{-22} = -\frac{40}{11}$

29. $D = \begin{vmatrix} 6 & -2 \\ 3 & -1 \end{vmatrix} = -6+6 = 0$

$D_x = \begin{vmatrix} 0 & -2 \\ 0 & -1 \end{vmatrix} = 0-0 = 0$

$D_y = \begin{vmatrix} 6 & 0 \\ 3 & 0 \end{vmatrix} = 0-0 = 0$

Since $D = 0$, $D_x = 0$, $D_y = 0$, the system is dependent.
The solution to the system is $\{(x, y) | y = 3x\}$.

Cumulative Review Exercises, Chapter 1–8

1. $\frac{|2-5|+10\div 2+3}{\sqrt{10^2-8^2}} = \frac{|2-5|+10\div 2+3}{\sqrt{100-64}}$

$= \frac{|-3|+5+3}{\sqrt{36}}$

$= \frac{3+5+3}{6}$

$= \frac{11}{6}$

3. $\frac{1}{3}x - \frac{3}{4} = \frac{1}{2}(x+2)$

Clear fractions by multiplying both sides by 12.

$4x - 9 = 6(x+2)$

$4x - 9 = 6x + 12$

$4x = 6x + 21$

$-2x = 21$

$x = -\frac{21}{2}$

5. $3x - 2y = 6$

$-2y = -3x + 6$

$y = \frac{3}{2}x - 3$

7. 180°

9. Let x represent the rate of the first hiker. Let y represent the rate of the second hiker. The statement "one hiker ... averages 2 mph faster than the other hiker" translates to the equation $y = x + 2$. Remember that $d = rt$. So the distance the first hiker walks is $3x$. The distance the second hiker walks is $3y$. Together they walked 18 miles which translates to the equation $3x + 3y = 18$.

$3x + 3y = 18$

$y = x + 2$

Since the second equation is solved for y, use the substitution method by substituting this value into the first equation and solving for x.

$3x + 3y = 18$

$3x + 3(x+2) = 18$

$3x + 3x + 6 = 18$

$6x + 6 = 18$

$6x = 12$

$x = 2$

$y = x + 2$

$y = 2 + 2$

$y = 4$

The rates of the hikers are 2 mph and 4 mph.

11. $(3.0\times 10^4)(6.0\times 10^8) = 18\times 10^{12} = 1.8\times 10^{13}$

13. $\left(\frac{1}{2}w^2-\frac{3}{5}w+\frac{1}{2}\right)-\left(\frac{3}{2}w^2+\frac{1}{10}w-2\right)=\frac{1}{2}w^2-\frac{3}{5}w+\frac{1}{2}-\frac{3}{2}w^2-\frac{1}{10}w+2=-w^2-\frac{7}{10}w+\frac{5}{2}$

15. $5x^2-125=5(x^2-25)=5(x-5)(x+5)$

17. $5y(y-3)(2y+1)=0$

$$\begin{aligned} 5y&=0 \quad \text{or} \quad & y-3&=0 \\ y&=0 & y&=3 \end{aligned}$$

$$\begin{aligned} \text{or } 2y+1&=0 \\ 2y&=-1 \\ y&=-\frac{1}{2} \end{aligned}$$

Solution: $y=0$ or $y=3$ or $y=-\frac{1}{2}$

19. **(a)** Nonlinear

(b) The x-intercepts occur when $y=0$.

$$\begin{aligned} 0&=x^2-x-12 \\ 0&=(x-4)(x+3) \end{aligned}$$

$$\begin{aligned} x-4&=0 \quad \text{or} \quad & x+3&=0 \\ x&=4 & x&=-3 \end{aligned}$$

The x-intercepts are (4, 0) and (–3, 0).

The y-intercepts occur when $x=0$.

$$\begin{aligned} y&=0^2-0-12 \\ y&=-12 \end{aligned}$$

The y-intercept is (0, –12).

21.
$$\begin{aligned} \frac{2a}{a-3}+\frac{28}{a^2-9}+\frac{6}{a+3}&=\frac{2a(a+3)}{(a+3)(a-3)}+\frac{28}{(a+3)(a-3)}+\frac{6(a-3)}{(a+3)(a-3)} \\ &=\frac{2a^2+6a+28+6a-18}{(a+3)(a-3)} \\ &=\frac{2a^2+12a+10}{(a+3)(a-3)} \\ &=\frac{2(a^2+6a+5)}{(a+3)(a-3)} \\ &=\frac{2(a+5)(a+1)}{(a+3)(a-3)} \end{aligned}$$

23. In Problem 21 you must change the fractions to equivalent fractions with a common denominator. In Problem 22 you must clear the denominators.

25. **(a)** The slope of a line parallel to the given line is the same as the given line: $-\frac{2}{3}$.

(b) The slope of a line perpendicular to the given line is the negative reciprocal of the given line: $\frac{3}{2}$

27. **(a) & (b)**

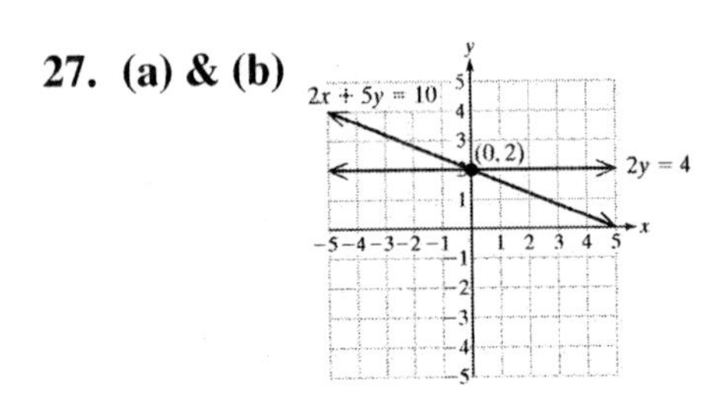

(c) Point of intersection: (0, 2)

$$2(0)+5(2) \stackrel{?}{=} 10 \ \checkmark$$
$$2(2) \stackrel{?}{=} 4 \ \checkmark$$

29. $2x + y = 3$

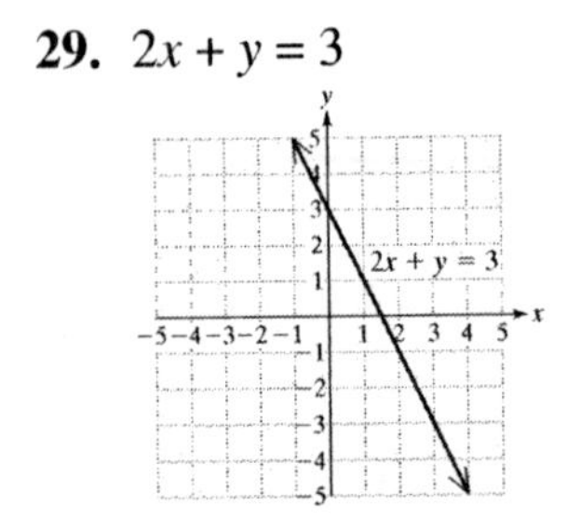

31.
$$x+y=90$$
$$x+(3y-36)=180$$

Solving the first equation for x gives $x = 90 - y$. Use the substitution method by substituting this value into the second equation to solve for y.

$$90-y+3y-36=180$$
$$2y+54=180$$
$$2y=126$$
$$y=63$$

$$x+y=90$$
$$x+63=90$$
$$x=27$$

x is 27° and y is 63°.

33. A: $3x+2y+3z=3$
B: $4x-5y+7z=1$
C: $2x+3y-2z=6$

Eliminate the x variable from equations A and C by multiplying equation A by 2 and equation C by –3.

$$6x+4y+6z=6$$
$$-6x-9y+6z=-18$$
$$-5y+12z=-12 \quad \text{Equation D}$$

Eliminate the x variable from equations B and C by multiplying equation C by –2.

$$4x-5y+7z=1$$
$$4x-6y+4z=-12$$
$$-11y+11z=-11 \quad \text{Equation E}$$

Eliminate the y variable from equations D and E by multiplying equation D by 11 and equation E by –5.

$$-55y+132z=-132$$
$$55y-55z=55$$
$$77z=-77$$
$$z=-1$$

Find y by substituting $z=-1$ into equation D.

$$-5y+12(-1)=-12$$
$$-5y=0$$
$$y=0$$

Find x by substituting $y = 0$ and $z = -1$ into any original equation.

$$3x+2(0)+3(-1)=3$$
$$3x=6$$
$$x=2$$

The solution to the system is (2, 0, –1).

Chapter 9

Section 9.1 Practice Exercises

1. $6u+5-5>2-5$
$$6u>-3$$
$$u>-\frac{1}{2}$$
Interval notation: $\left(-\frac{1}{2},\infty\right)$

3. $\left(-\frac{4}{3}\right)\left(-\frac{3}{4}p\right)\le 12\cdot\left(-\frac{4}{3}\right)$
$$p\ge -16$$
Interval notation: $[-16,\infty)$

5. $-1.5+8.1<0.1x-8.1+8.1$
$$6.6<0.1x$$
$$66<x$$
Interval notation: $(66,\infty)$

7. **(a)** $\{10, 20, 30\}$

(b) $\{5, 10, 15, 20, 25, 30, 35, 40, 50\}$

9. **(a)** $\{a, e, j, k, l, m, n, o, p, q, t, y\}$

(b) $\{j, o\}$

11. **(a)** $\{1, 2, 3, 4, 5, 6, 7, 8, 9, 10\}$

(b) $\{\ \}$, the empty set

13. **(a)** $x>2$

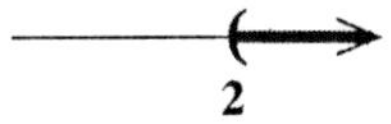

2

(b) $2+x-2<6-2$
$$x<4$$

4

(c) $4x>8$ and $2+x<6$
$$x>2 \text{ and } x<4$$
$$2<x<4$$

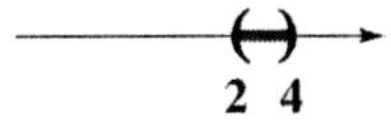

2 4

15. **(a)** $-5\le 3x-4$
$$-1\le 3x$$
$$-\frac{1}{3}\le x$$
$$x\ge -\frac{1}{3}$$

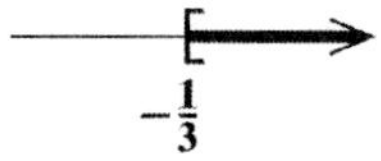

$-\frac{1}{3}$

(b) $3x-4\le 8$
$$3x\le 12$$
$$x\le 4$$

4

(c) $-5\le 3x-4\le 8$
$$-1\le 3x\le 12$$
$$-\frac{1}{3}\le x\le 4$$

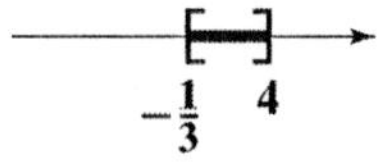

$-\frac{1}{3}$ 4

17. **(a)** $4x-7<1$
$$4x<8$$
$$x<2$$

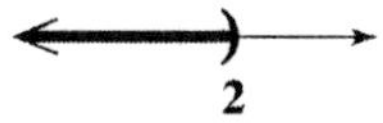

2

(b) $-3x+7>-8$
$$-3x>-15$$
$$x<5$$

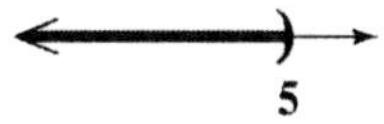

5

(c) $4x-7<1$ and $-3x+7>-8$
$$x<2 \text{ and } x<5$$
$$x<2$$

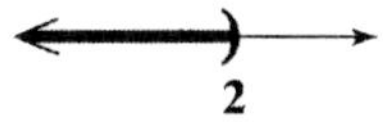

2

19. $y-7\ge -9$ and $y+2\le 5$
$$y\ge -2 \text{ and } y\le 3$$
$$-2\le y\le 3$$
Interval notation: $[-2, 3]$

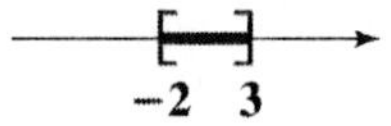

−2 3

21. $2t+7<19$ and $5t+13>28$

$2t<12$ and $5t>15$

$t<6$ and $t>3$

$3<t<6$

Interval notation: (3, 6)

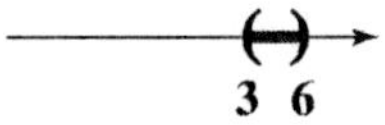

3 6

23. $0\le 2b-5<9$

$5\le 2b<14$

$\frac{5}{2}\le b<7$

Interval notation: $\left[\frac{5}{2}, 7\right)$

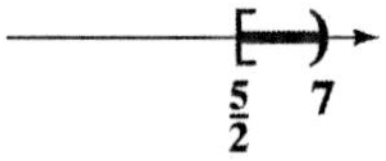

$\frac{5}{2}$ 7

25. $21k-11\le 6k+19$ and $3k-11<-k+7$

$15k-11\le 19$ and $4k-11<7$

$15k\le 30$ and $4k<18$

$k\le 2$ and $k<\frac{9}{2}$

Interval notation: $(-\infty, 2]$

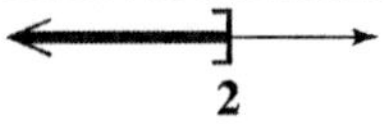

2

27. $-1<\frac{a}{6}\le 1$

$-6<a\le 6$

Interval notation: $(-6, 6]$

−6 6

29. $-\frac{2}{3}<\frac{y-4}{-6}<\frac{1}{3}$

$4>y-4>-2$

$8>y>2$

Interval notation: (2, 8)

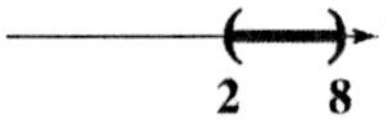

2 8

31. $\frac{2}{3}(2p-1)\ge 10$ and $\frac{4}{5}(3p+4)\ge 20$

$2p-1\ge 15$ and $3p+1\ge 25$

$2p\ge 16$ and $3p\ge 21$

$p\ge 8$ and $p\ge 7$

$p\ge 8$

Interval notation: $[8, \infty)$

8

33. **(a)** $-2x+7>9$

$-2x>2$

$x<-1$

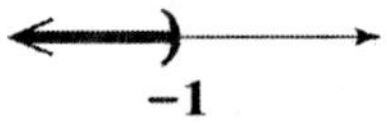

−1

(b) $3x+1<-14$

$3x<-15$

$x<-5$

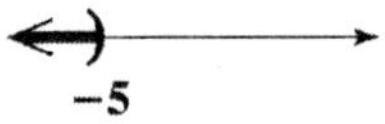

−5

(c) $x<-1$

−1

35. **(a)** $5x+8\le 23$

$5x\le 15$

$x\le 3$

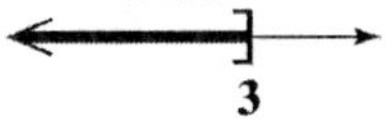

3

(b) $2x-15\ge 1$

$2x\ge 16$

$x\ge 8$

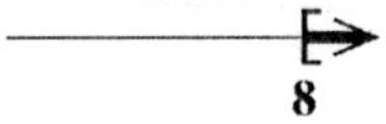

8

(c) $x\le 3$ or $x\ge 8$

3 8

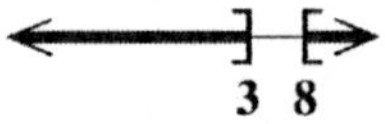

37. **(a)** $-3x-5\le -11$

$-3x\le -6$

$x\ge 2$

2

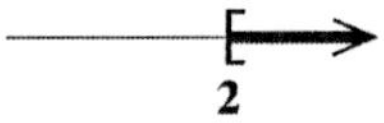

(b) $-7 \geq 5x - 2$
$-5 \geq 5x$
$-1 \geq x$
$x \leq -1$

−1

(c) $x \geq 2$ or $x \leq -1$

−1 2

39. $h + 4 < 0$ or $6h > -12$
$h < -4$ or $h > -2$

Interval notation: $(-\infty, -4) \cup (-2, \infty)$

−4 −2

41. $2y - 1 \geq 3$ or $y < -2$
$2y \geq 4$ or $y < -2$
$y \geq 2$ or $y < -2$

Interval notation: $(-\infty, -2) \cup [2, \infty)$

−2 2

43. $\frac{5}{3}v \geq 5$ or $-v - 6 > 1$
$v \geq 3$ or $-v > 7$
$v \geq 3$ or $v < -7$

Interval notation: $(-\infty, -7) \cup [3, \infty)$

−7 3

45. $5(x-1) \geq -5$ or $5 - x \leq 11$
$5x - 5 \geq -5$ or $-x \leq 6$
$5x \geq 0$ or $x \geq -6$
$x \geq 0$ or $x \geq -6$
$x \geq -6$

Interval notation: $[-6, \infty)$

−6

47. $\frac{3t-1}{10} > \frac{1}{2}$ or $\frac{3t-1}{1} < -\frac{1}{2}$
$3t - 1 > 5$ or $3t - 1 < -5$
$3t > 6$ or $3t < -4$
$t > 2$ or $t < -\frac{4}{3}$

Interval notation: $\left(-\infty, -\frac{4}{3}\right) \cup (2, \infty)$

$-\frac{4}{3}$ 2

49. $0.5w + 5 < 2.5w - 4$ or $0.3w \leq -0.1w - 1.6$
$-2w + 5 < -4$ or $0.4w \leq -1.6$
$-2 < -9$ or $w \leq -4$
$w > \frac{9}{2}$ or $w \leq -4$

Interval notation: $(-\infty, -4] \cup \left(\frac{9}{2}, \infty\right)$

−4 $\frac{9}{2}$

51. (a) $3x - 5 < 19$ and $-2x + 3 < 23$
$3x < 24$ and $-2x < 20$
$x < 8$ and $x > -10$
$-10 < x < 8$

−10 8

(b) $3x - 5 < 19$ or $-2x + 3 < 23$
$3x < 24$ or $-2x < 20$
$x < 8$ or $x > -10$
All real numbers

53. (a) $8x - 4 \geq 6.4$ or $0.3(x+6) \leq -0.6$
$8x \geq 10.4$ or $0.3x + 1.8 \leq -0.6$
$x \geq 1.3$ or $0.3x \leq -2.4$
$x \geq 1.3$ or $x \leq -8$

−8 1.3

(b) $8x - 4 \geq 6.4$ and $0.3(x+6) \leq -0.6$
$8x \geq 10.4$ and $0.3x + 1.8 \leq -0.6$
$x \geq 1.3$ and $0.3x \leq -2.4$
$x \geq 1.3$ and $x \leq -8$
No solution

55. **(a)** $4800 \le x \le 10{,}800$

(b) $x < 4800$ or $x > 10{,}800$

57. **(a)** $13 \le x \le 16$

(b) $x < 13$ or $x > 16$

59. $-3 < 2x < 12$

$-\frac{3}{2} < x < 6$

All real numbers between $-\frac{3}{2}$ and 6

61. $1+2x > 5$ or $1+2x < -1$

$2x > 4$ or $2x < -2$

$x > 2$ or $x < -1$

All real numbers greater than 2 or less than -1

Section 9.2 Practice Exercises

1. $6x-10 > 8$ or $8x+2 < 5$

$6x > 18$ or $8x < 3$

$x > 3$ or $x < \frac{3}{8}$

Interval notation: $\left(-\infty, \frac{3}{8}\right) \cup (3, \infty)$

3. $5(k-2) > -25$ and $7(1-k) > 7$

$5k-10 > -25$ and $7-7k > 7$

$5k > -15$ and $-7k > 0$

$k > -3$ and $k < 0$

$-3 < k < 0$

Interval notation: $(-3, 0)$

5. $2t-7 > 3$ and $-3t < 4$

$2t > 10$ and $t > -\frac{4}{3}$

$t > 5$ and $t > -\frac{4}{3}$

$t > 5$

Interval notation: $(5, \infty)$

7. $\frac{3}{2}h-1 < 0$ or $h+2 > \frac{7}{4}$

$\frac{3}{2}h < 1$ or $h > -\frac{1}{4}$

$h < \frac{2}{3}$ or $h > -\frac{1}{4}$

Interval notation: $(-\infty, \infty)$

9. **(a)** $(-2, 0) \cup (3, \infty)$

(b) $(-\infty, -2) \cup (0, 3)$

(c) $(-\infty, -2] \cup [0, 3]$

(d) $[-2, 0] \cup [3, \infty)$

11. **(a)** $(-1, 1]$

(b) $(-\infty, -1) \cup [1, \infty)$

(c) $(-\infty, -1) \cup (1, \infty)$

(d) $(-1, 1)$

13. **(a)** $3(2b-4)-b = 5-b$

$6b-12-b = 5-b$

$5b-12 = 5-b$

$6b-12 = 5$

$6b = 17$

$b = \frac{17}{6}$

(b) $3(2b-4)-b < 5-b$

$6b-12-b < 5-b$

$5b-12 < 5-b$

$6b-12 < 5$

$6b < 17$

$b < \frac{17}{6}$

Interval notation: $\left(-\infty, \frac{17}{6}\right)$

(c) $3(2b-4)-b>5-b$
$6b-12-b>5-b$
$5b-12>5-b$
$6b-12>5$
$6b>17$
$b>\frac{17}{6}$

Interval notation: $\left(\frac{17}{6}, \infty\right)$

15. (a) $\frac{1}{2}y+3=\frac{2}{3}y$ (Multiply by 6)
$3y+18=4y$
$-y+18=0$
$-y=-18$
$y=18$

(b) $\frac{1}{2}y+3\le\frac{2}{3}y$ (Multiply by 6)
$3y+18\le 4y$
$-y+18\le 0$
$-y\le -18$
$y\ge 18$

Interval notation: $[18, \infty)$

(c) $\frac{1}{2}y+3\ge\frac{2}{3}y$ (Multiply by 6)
$3y+18\ge 4y$
$-y+18\ge 0$
$-y\ge -18$
$y\le 18$

Interval notation: $(-\infty, 18]$

17. (a) $3w(w+4)=10-w$
$3w^2+12w=10-w$
$3w^2+13w-10=0$
$(3w-2)(w+5)=0$
$3w-2=0$ or $w+5=0$
$3w=2$ or $w=-5$
$w=\frac{2}{3}$ or $w=-5$

(b) The boundary points are $\frac{2}{3}$ and -5.

Plot the boundary points on the number line and test a point from each region.

<u>Test $w=-6$</u>: $3(-6)(-6+4)<10-(-6)$
$-18(-2)<16$
$36<16$ False

<u>Test $w=0$</u>: $3(0)(0+4)<10-0$
$0(4)<10$
$0<10$ True

<u>Test $w=1$</u>: $3(1)(1+4)<10-1$
$3(5)<9$
$15<10$ False

Since the test point $w=0$ is true, the solution is $\left(-5, \frac{2}{3}\right)$. The endpoints are not included in the solution because the symbol "<" does not include equality.

(c) The boundary points are $\frac{2}{3}$ and -5.

Plot the boundary points on the number line and test a point from each region.

<u>Test $w=-6$</u>: $3(-6)(-6+4)>10--6$
$-18(-2)>16$
$36>16$ True

<u>Test $w=0$</u>: $3(0)(0+4)>10-0$
$0(4)>10$
$0>10$ False

<u>Test $w=1$</u>: $3(1)(1+4)>10-1$
$3(5)>9$
$15>10$ True

Since the test points $w=-6$ and $w=1$ are true, the solution is $(-\infty, -5)\cup\left(\frac{2}{3}, \infty\right)$. The endpoints are not included in the solution because the symbol ">" does not include equality.

19. (a) $q^2-4q=5$
$q^2-4q-5=0$
$(q-5)(q+1)=0$
$1-5=0$ or $q+1=0$
$q=5$ or $q=-1$

(b) The boundary points are 5 and −1. Plot the boundary points on the number line and test a point from each region.

Test $q=-2$: $(-2)^2-4(-2)<5$
$4+8<5$
$12<5$ False

Test $q=0$: $(0)^2-4(0)<5$
$0-0<5$
$0<5$ True

Test $q=6$: $(6)^2-4(6)<5$
$36-24<5$
$12<5$ False

Since the test point $q=0$ is true, the solution is $[-1, 5]$. The endpoints are included in the solution because the symbol "≤" does include equality.

(c) The boundary points are 5 and −1. Plot the boundary points on the number line and test a point from each region.

Test $q=-2$: $(-2)^2-4(-2)>5$
$4+8>5$
$12>5$ True

Test $q=0$: $(0)^2-4(0)>5$
$0-0>5$
$0>5$ False

Test $q=6$: $(6)^2-4(6)>5$
$36-24>5$
$12>5$ True

Since the test points $q=-2$ and $q=6$ are true, the solution is $(-\infty, -1] \cup [5, \infty)$. The endpoints are included in the solution because the symbol "≥" does include equality.

21. $(t-7)(t+1)<0$
Find the boundary points:
$t-7=0$ or $t+1=0$
$t=7$ or $t=-1$
Plot the boundary points on the number line and test a point from each region.

Test $t=-2$: $(-2-7)(-2+1)<0$
$-9(-1)<0$
$9<0$ False

Test $t=0$: $(0-7)(0+1)<0$
$-7(1)<0$
$-7<0$ True

Test $t=8$: $(8-7)(8+1)<0$
$1(9)<0$
$9<0$ False

Since the test point $t=0$ is true, the solution is $(-1, 7)$. The endpoints are not included in the solution because the symbol "<" does not include equality.

23. $(5y-3)(y-8)>0$
Find the boundary points:
$5y-3=0$ or $y-8=0$
$y=\frac{3}{5}$ or $y=8$
Plot the boundary points on the number line and test a point from each region.

Test $y=0$: $(5(0)-3)(0-8)>0$
$-3(-8)>0$
$24>0$ True

Test $y=1$: $(5(1)-3)(1-8)>0$
$2(-7)>0$
$-14>0$ False

Test $y=9$: $(5(9)-3)(9-8)>0$
$42(1)>0$
$42>0$ True

Since the test points $y=0$ and $y=9$ are true, the solution is $\left(-\infty, \frac{3}{5}\right) \cup (8, \infty)$. The endpoints are not included in the solution because the symbol ">" does not include equality.

25. $a^2-12a\le -32$
Find the boundary points:
$a^2-12a+32=0$
$(a-8)(a-4)=0$
$a-8=0$ or $a-4=0$
$a=8$ or $a=4$
Plot the boundary points on the number line and test a point from each region.

Test $a=0$: $(0)^2-12(0)<-32$
$0<-32$ False

$\underline{\text{Test } a = 5}$: $(5)^2 - 12(5) < -32$
$$25 - 60 < -32$$
$$-35 < -32 \quad \text{True}$$

$\underline{\text{Test } a = 9}$: $(9)^2 - 12(9) < -32$
$$81 - 108 < -32$$
$$-27 < -32 \quad \text{False}$$

Since the test point $a = 0$ is true, the solution is [4, 8]. The endpoints are included in the solution because the symbol "≤" does include equality.

27. $b^2 - 121 < 0$
Find the boundary points:
$$b^2 - 121 = 0$$
$$(b + 11)(b - 11) = 0$$
$$b + 11 = 0 \quad \text{or} \quad b - 11 = 0$$
$$b = -11 \quad \text{or} \quad b = 11$$
Plot the boundary points on the number line and test a point from each region.

$\underline{\text{Test } b = -12}$: $(-12)^2 - 121 < 0$
$$144 - 121 < 0$$
$$23 < 0 \quad \text{False}$$

$\underline{\text{Test } b = 0}$: $(0)^2 - 121 < 0$
$$0 - 121 < 0$$
$$-121 < 0 \quad \text{True}$$

$\underline{\text{Test } b = 12}$: $(12)^2 - 121 < 0$
$$144 - 121 < 0$$
$$23 < 0 \quad \text{False}$$

Since the test point $b = 0$ is true, the solution is (−11, 11). The endpoints are not included in the solution because the symbol "<" does not include equality.

29. $3p^2 - 8p - 3 \geq 0$
Find the boundary points:
$$3p^2 - 8p - 3 = 0$$
$$(3p + 1)(p - 3) = 0$$
$$3p + 1 = 0 \quad \text{or} \quad p - 3 = 0$$
$$3p = -1 \quad \text{or} \quad p = 3$$
$$p = -\frac{1}{3} \quad \text{or} \quad p = 3$$
Plot the boundary points on the number line and test a point from each region.

$\underline{\text{Test } p = -1}$: $3(-1)^2 - 8(-1) - 3 > 0$
$$3 + 8 - 3 > 0$$
$$8 > 0 \quad \text{True}$$

$\underline{\text{Test } p = 0}$: $3(0)^2 - 8(0) - 3 > 0$
$$0 - 0 - 3 > 0$$
$$-3 > 0 \quad \text{False}$$

$\underline{\text{Test } p = 4}$: $3(4)^2 - 8(4) - 3 > 0$
$$48 - 32 - 3 > 0$$
$$13 > 0 \quad \text{True}$$

Since the test points $p = -1$ and $p = 4$ are true, the solution is $\left(-\infty, -\frac{1}{3}\right] \cup [3, \infty)$. The endpoints are included in the solution because the symbol "≥" does include equality.

31. $2x(x - 4)(3x + 1) > 0$
Find the boundary points:
$$2x(x - 4)(3x + 1) = 0$$
$$2x = 0 \quad \text{or} \quad x - 4 = 0 \quad \text{or} \quad 3x + 1 = 0$$
$$x = 0 \quad \text{or} \quad x = 4 \quad \text{or} \quad 3x = -1$$
$$x = 0 \quad \text{or} \quad x = 4 \quad \text{or} \quad x = -\frac{1}{3}$$
Plot the boundary points on the number line and test a point from each region.

$\underline{\text{Test } x = -1}$:
$$2(-1)(-1 - 4)(3(-1) + 1) > 0$$
$$-2(-5)(-2) > 0$$
$$-20 > 0 \quad \text{False}$$

$\underline{\text{Test } x = -0.25}$:
$$2(-0.25)(-0.25 - 4)(3(-0.25) + 1) > 0$$
$$-0.5(-4.25)(0.25) > 0$$
$$0.53125 > 0 \quad \text{True}$$

$\underline{\text{Test } x = 1}$: $2(1)(1 - 4)(3(1) + 1) > 0$
$$2(-3)(4) > 0$$
$$-24 > 0 \quad \text{False}$$

$\underline{\text{Test } x = 5}$: $2(5)(5 - 4)(3(5) + 1) > 0$
$$10(1)(16) > 0$$
$$160 > 0 \quad \text{True}$$

Since the test points $x = -0.25$ and $x = 5$ are true, the solution is $\left(-\frac{1}{3}, 0\right) \cup (4, \infty)$. The endpoints are not included in the solution because the symbol ">" does not include equality.

33. $x^3 - x^2 \le 12x$

Find the boundary points:

$$x^3 - x^2 - 12x = 0$$
$$x(x^2 - x - 12) = 0$$
$$x(x-4)(x+3) = 0$$

$x = 0$ or $x - 4 = 0$ or $x + 3 = 0$

$x = 0$ or $x = 4$ or $x = -3$

Plot the boundary points on the number line and test a point from each region.

Test $x = -4$: $(-4)^3 - (-4)^2 < 12(-4)$
$-64 - 16 < -48$
$-80 < -48$ True

Test $x = -1$: $(-1)^3 - (-1)^2 < 12(-1)$
$-1 - 1 < -12$
$-2 < -12$ False

Test $x = 1$: $(1)^3 - (1)^2 < 12(1)$
$1 - 1 < 12$
$0 < 12$ True

Test $x = 5$: $(5)^3 - (5)^2 < 12(5)$
$125 - 25 < 60$
$100 < 60$ False

Since the test points $x = -4$ and $x = 1$ are true, the solution is $(-\infty, -3] \cup [0, 4]$. The endpoints are included in the solution because the symbol "≤" does include equality.

35. $w^3 + w^2 > 4w + 4$

Find the boundary points:

$$w^3 + w^2 = 4w + 4$$
$$w^3 + w^2 - 4w - 4 = 0$$
$$w^2(w+1) - 4(w+1) = 0$$
$$(w+1)(w^2 - 4) = 0$$
$$(w+1)(w-2)(w+2) = 0$$

$w + 1 = 0$ or $w - 2 = 0$ or $w = -2$

$w = -1$ or $w = 2$ or $w = -2$

Plot the boundary points on the number line and test a point from each region.

Test $w = -3$: $(-3)^3 + (-3)^2 > 4(-3) + 4$
$-27 + 9 > -12 + 4$
$-18 > -8$ False

Test $w = -1.5$:
$(-1.5)^3 + (-1.5)^2 > 4(-1.5) + 4$
$-3.375 + 2.25 > -6.5 + 4$
$-1.125 > -2.5$ True

Test $w = 0$: $(0)^3 + (0)^2 > 4(0) + 4$
$0 + 0 > 0 + 4$
$0 > 4$ False

Test $w = 3$: $(3)^3 + (3)^2 > 4(3) + 4$
$27 + 9 > 12 + 4$
$36 > 16$ True

Since the test points $w = -1.5$ and $w = 3$ are true, the solution is $(-2, -1) \cup (2, \infty)$. The endpoints are not included in the solution because the symbol ">" does not include equality.

37. (a) $\dfrac{10}{x-5} = 5$

$$10 = 5(x-5)$$
$$10 = 5x - 25$$
$$35 = 5x$$
$$7 = x$$

(b) $\dfrac{1}{x-5} < 5$

The inequality is undefined for $x = 5$ and the solution to the related equation is 7. Therefore, the boundary points are 7 and 5.

Plot the boundary points on the number line and test a point from each region.

Test $x = 0$: $\dfrac{10}{0-5} < 5$
$-2 < 5$ True

Test $x = 6$: $\dfrac{10}{6-5} < 5$
$10 < 5$ False

Test $x = 10$: $\dfrac{10}{10-5} < 5$
$2 < 5$ True

Since the test points $x = 0$ and $x = 10$ are true, the solution is $(-\infty, 5) \cup (7, \infty)$. The endpoints are not included in the solution because the symbol "<" does not include equality.

(c) $\frac{1}{x-5} > 5$

The inequality is undefined for $x = 5$ and the solution to the related equation is 7. Therefore, the boundary points are 7 and 5.

Plot the boundary points on the number line and test a point from each region.

Test $x = 0$: $\frac{10}{0-5} > 5$

$-2 > 5$ False

Test $x = 6$: $\frac{10}{6-5} > 5$

$10 > 5$ True

Test $x = 10$: $\frac{10}{10-5} > 5$

$2 > 5$ False

Since the test point $x = 6$ is true, the solution is (5, 7). The endpoints are not included in the solution because the symbol ">" does not include equality.

39. (a) $\frac{z+2}{z-6} = -3$

$$z+2 = -3(z-6)$$
$$z+2 = -3z+18$$
$$4z = 16$$
$$z = 4$$

(b) $\frac{z+2}{z-6} \le -3$

The inequality is undefined for $z = 6$ and the solution to the related equation is 4. Therefore, the boundary points are 6 and 4.

Plot the boundary points on the number line and test a point from each region.

Test $z = 0$: $\frac{0+2}{0-6} < -3$

$-\frac{1}{3} < -3$ False

Test $z = 5$: $\frac{5+2}{5-6} < -3$

$-7 < -3$ True

Test $z = 7$: $\frac{7+2}{7-6} < -3$

$9 < -3$ False

Since the test point $z = 5$ is true, the solution is [4, 6). The endpoint of 4 is included in the solution because the symbol "≤" does include equality. Since the inequality is not defined for $z = 6$, it is not included in the solution.

(c) $\frac{z+2}{z-6} \ge -3$

The inequality is undefined for $z = 6$ and the solution to the related equation is 4. Therefore, the boundary points are 6 and 4.

Plot the boundary points on the number line and test a point from each region.

Test $z = 0$: $\frac{0+2}{0-6} > -3$

$-\frac{1}{3} > -3$ True

Test $z = 5$: $\frac{5+2}{5-6} > -3$

$-7 > -3$ False

Test $z = 7$: $\frac{7+2}{7-6} > -3$

$9 > -3$ True

Since the test points $z = 0$ and $z = 7$ are true, the solution is $(-\infty, 4] \cup (6, \infty)$. The endpoint of 4 is included in the solution because the symbol "≥" does include equality. Since the inequality is not defined for $z = 6$, it is not included in the solution.

41. $\frac{2}{x-1} \ge 0$

The inequality is undefined for $x = 1$ and there is no solution to the related equation. Therefore, the boundary point is 1.

Plot the boundary point on the number line and test a point from each region.

Test $x = 0$: $\frac{2}{0-1} > 0$

$-2 > 0$ False

Test $x = 2$: $\frac{2}{2-1} > 0$

$2 > 0$ True

Since the test point $x = 2$ is true, the solution is $(1, \infty)$. Since the inequality is not defined for $x = 1$, it is not included in the solution.

43. $\frac{a+1}{a-3}<0$

Find the boundary points:

$$\frac{a+1}{a-3}=0$$
$$a+1=0(a-3)$$
$$a+1=0$$
$$a=-1$$

The inequality is undefined for $a = 3$ and the solution to the related equation is -1. Therefore, the boundary points are 3 and -1. Plot the boundary points on the number line and test a point from each region.

Test $a=-2$: $\frac{-2+1}{-2-3}<0$

$\frac{1}{5}<0$ False

Test $a=0$: $\frac{0+1}{0-3}<0$

$-\frac{1}{3}<0$ True

Test $a=4$: $\frac{4+1}{4-3}<0$

$5<0$ False

Since the test point $a = 0$ is true, the solution is $(-1, 3)$. The endpoint -1 is not included in the solution because the symbol "$<$" does not include equality. Since the inequality is not defined for $a = 3$, it is not included in the solution.

45. $\frac{3}{2x-7}<-1$

Find the boundary points.

$$\frac{3}{2x-7}=-1$$
$$3=-1(2x-7)$$
$$3=-2x+7$$
$$-4=-2x$$
$$2=x$$

The inequality is undefined for $x=\frac{7}{2}$ and the solution to the related equation is 2. Therefore, the boundary points are $\frac{7}{2}$ and 2. Plot the boundary points on the number line and test a point from each region.

Test $x=0$: $\frac{3}{2(0)-7}<-1$

$-\frac{3}{7}<-1$ False

Test $x=3$: $\frac{3}{2(3)-7}<-1$

$-3<-1$ True

Test $x=4$: $\frac{3}{2(4)-7}<-1$

$3<-1$ False

Since the test point $x = 3$ is true, the solution is $\left(2, \frac{7}{2}\right)$. The endpoint 2 is not included in the solution because the symbol "$<$" does not include equality. Since the inequality is not defined for $x=\frac{7}{2}$, it is not included in the solution.

47. $\frac{x+1}{x-5}\geq 4$

Find the boundary points:

$$\frac{x+1}{x-5}=4$$
$$x+1=4(x-5)$$
$$x+1=4x-20$$
$$-3x=-21$$
$$x=7$$

The inequality is undefined for $x = 5$ and the solution to the related equation is 7. Therefore, the boundary points are 5 and 7. Plot the boundary points on the number line and test a point from each region.

Test $x=0$: $\frac{0+1}{0-5}>4$

$-\frac{1}{5}>0$ False

Test $x=6$: $\frac{6+1}{6-5}>4$

$7>4$ True

Test $x=8$: $\frac{8+1}{8-5}>4$

$3>4$ False

Since the test point $x = 6$ is true, the solution is $(5, 7]$. The endpoint 7 is included in the solution because the symbol "≥" does include equality. Since the inequality is not defined for $x = 5$, it is not included in the solution.

49. $\frac{1}{x} \le 2$

Find the boundary points:

$\frac{1}{x} = 2$

$1 = 2(x)$

$\frac{1}{2} = x$

The inequality is undefined for $x = 0$ and the solution to the related equation is $\frac{1}{2}$.

Therefore, the boundary points are 0 and $\frac{1}{2}$.

Plot the boundary point on the number line and test a point from each region.

Test $x = -1$: $\frac{1}{-1} < 2$

$-1 < 2$ True

Test $x = 0.25$: $\frac{1}{0.25} < 2$

$4 < 2$ False

Test $x = 1$: $\frac{1}{1} < 2$

$1 < 2$ True

Since the test points $x = -1$ and $x = 1$ are true, the solution is $(-\infty, 0) \cup \left[\frac{1}{2}, \infty\right)$. The endpoint $\frac{1}{2}$ is included in the solution because the symbol "≤" does include equality. Since the inequality is not defined for $x = 0$, it is not included in the solution.

51. $\frac{(x+2)^2}{x} > 0$

Find the boundary points:

$\frac{(x+2)^2}{x} = 0$

$(x+2)^2 = 0(x)$

$(x+2)^2 = 0$

$x + 2 = 0$

$x = -2$

The inequality is undefined for $x = 0$ and the solution to the related equation is -2. Therefore, the boundary points are 0 and -2. Plot the boundary point on the number line and test a point from each region.

Test $x = -3$: $\frac{(-3+2)^2}{-3} > 0$

$-\frac{1}{3} > 0$ False

Test $x = -1$: $\frac{(-1+2)^2}{-1} > 0$

$-1 > 0$ False

Test $x = 1$: $\frac{(1+2)^2}{1} > 0$

$9 > 0$ True

Since the test point $x = 1$ is true, the solution is $(0, \infty)$. Since the inequality is not defined for $x = 0$, it is not included in the solution.

53. $x^2 + 10x + 25 \ge 0$

$(x+5)^2 \ge 0$

The quantity $(x+5)^2$ is greater than or equal to zero for all real numbers. The solution is all real numbers, $(-\infty, \infty)$.

55. $x^2 + 2x + 1 < 0$

$(x+1)^2 < 0$

The quantity $(x+1)^2$ is greater than or equal to zero for all real numbers. There is no solution.

57. The expression x^2+4 is positive for all real numbers, x. The expression x^2 is positive or 0 for all real numbers, x. Thus, the ratio $\dfrac{x^2}{x^2+4}$ can never be negative. There is no solution.

59. The expression x^4+3x^2 is greater than zero for all real numbers, x, except 0 which makes the expression equal to 0. The solution is $\{0\}$ since the inequality is "≤."

61. $x^2+4x+4>0$

$(x+2)^2>0$

The quantity $(x+2)^2$ is greater than zero for all real numbers except $x=-2$, for which it is zero. The solution is all real numbers except -2, $(-\infty, -2) \cup (-2, \infty)$.

63. $x^2+4x+4\le 0$

$(x+2)^2\le 0$

The quantity $(x+2)^2$ is greater than zero for all real numbers except $x=-2$, for which it is zero. The solution is $\{-2\}$.

65. Interval notation: $(-\infty, 0) \cup (2, \infty)$

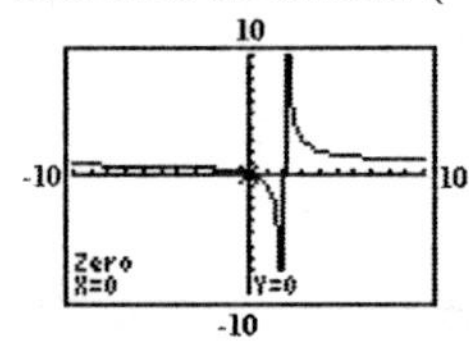

67. Interval notation: $(-1, 1)$

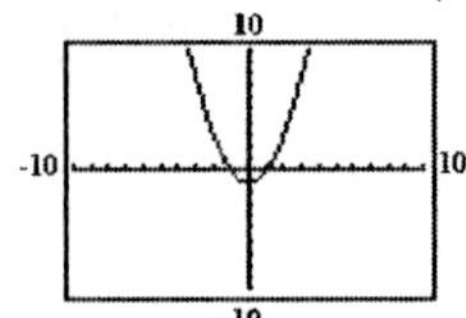

69. Enter y_1 as x^2 + 10x + 25 and determine where the graph is below or on the x-axis. Solution: $\{-5\}$.

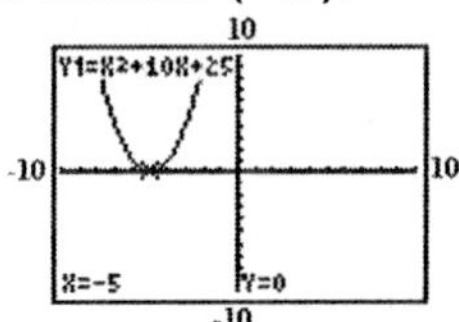

71. Enter y_1 as 8/(x^2 + 2) and determine where the graph is below the x-axis. No solution.

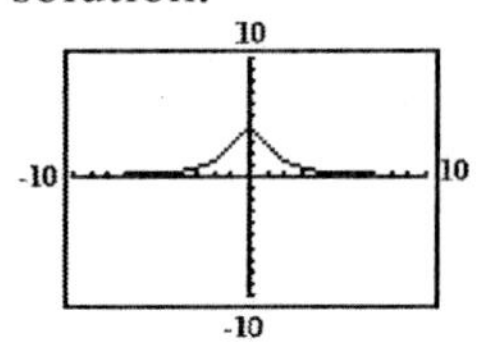

Section 9.3 Practice Exercises

1.
$$\begin{aligned}
3(a+2)-6>2 &\quad\text{and}\quad -2(a+3)+14>-3\\
3a+6-6>2 &\quad\text{and}\quad -2a+6+14>-3\\
3a>2 &\quad\text{and}\quad -2a+20>-3\\
a>\frac{2}{3} &\quad\text{and}\quad -2a>-23\\
a>\frac{2}{3} &\quad\text{and}\quad a<\frac{23}{2}
\end{aligned}$$

$$\frac{2}{3}<a<\frac{2}{3}$$

Interval notation: $\left(\dfrac{2}{3}, \dfrac{23}{2}\right)$

3. $\dfrac{4}{y-4}\ge 3$

Find the boundary points:

$$\begin{aligned}
\frac{4}{y-4}&=3\\
4&=3(y-4)\\
4&=3y-12\\
16&=3y\\
\frac{16}{3}&=y
\end{aligned}$$

The inequality is undefined for $y = 4$ and the solution to the related equation is $\dfrac{16}{3}$.

Therefore, the boundary points are 4 and $\frac{16}{3}$.

Plot the boundary points on the number line and test a point from each region.

Test $y = 0$: $\frac{4}{0-4} > 3$

$-1 > 3$ False

Test $y = 5$: $\frac{4}{5-4} > 3$

$4 > 3$ True

Test $y = 6$: $\frac{4}{6-4} > 3$

$2 > 3$ False

Since the test point $y = 5$ is true, the solution is $\left(4, \frac{16}{3}\right]$. The endpoint $\frac{16}{3}$ is included in the solution because the symbol "≥" does include equality. Since the inequality is not defined for $y = 4$, it is not included in the solution.

5. Let n represent the number.

$-1 < n + 6 < 13$

$-7 < n < 7$

All real numbers between –7 and 7.

7. $3(x - 2)(x + 4)(2x - 1) < 0$

Find the boundary points:

$3(x - 2)(x + 4)(2x - 1) = 0$

$3 \neq 0$ or $x - 2 = 0$ or $x + 4 = 0$ or $2x - 1 = 0$

$3 \neq 0$ or $x = 2$ or $x = -4$ or $2x = 1$

$x = 2$ or $x = -4$ or $x = \frac{1}{2}$

The boundary points are 2, –4, and $\frac{1}{2}$.

Plot the boundary points on the number line and test a point from each region.

Test $x = -5$: $3(-5-2)(-5+4))(2(-5)-1) < 0$

$-231 < 0$ True

Test $x = 0$: $3(0-2)(0+4)(2(0)-1) < 0$

$24 < 0$ False

Test $x = 1$: $3(1-2)(1+4)(2(1)-1) < 0$

$-15 < 0$ True

Test $x = 3$: $3(3-2)(3+4)(2(3)-1) < 0$

$105 < 0$ False

Since the test points $x = -5$ and $x = 1$ are true, the solution is $(-\infty, -4) \cup \left(\frac{1}{2}, 2\right)$.

The endpoints are not included in the solution because the symbol "<" does not include equality.

9. $|p| = 7$

$p = -7$ or $p = 7$

11. $|x|+5=11$
$|x|=6$
$x=-6 \text{ or } x=6$

13. $|y|=\sqrt{2}$
$y=-\sqrt{2} \text{ or } y=\sqrt{2}$

15. $|w|-3=-5$
$|w|=-2$
No solution

17. $|3q|=0$
$3q=0$
$q=0$

19. $\left|3x-\frac{1}{2}\right|=\frac{1}{2}$
$3x-\frac{1}{2}=-\frac{1}{2}$ or $3x-\frac{1}{2}=\frac{1}{2}$
$3x=0$ or $3x=1$
$x=0$ or $x=\frac{1}{3}$

21. $|4x-2|=|-8|$
$4x-2=-8$ or $4x-2=8$
$4x=-6$ or $4x=10$
$x=-\frac{3}{2}$ or $x=\frac{5}{2}$

23. $\left|\frac{7z}{3}-\frac{1}{3}\right|+3=6$
$\left|\frac{7z}{3}-\frac{1}{3}\right|=3$
$\frac{7z}{3}-\frac{1}{3}=-3$ or $\frac{7z}{3}-\frac{1}{3}=3$
$7z-1=-9$ or $7z-1=9$
$7z=-8$ or $7z=10$
$z=-\frac{8}{7}$ or $z=\frac{10}{7}$

25. $\left|\frac{5y+2}{2}\right|=6$
$\frac{5y+2}{2}=-6$ or $\frac{5y+2}{2}=6$
$5y+2=-12$ or $5y+2=12$
$5y=-14$ or $5y=10$
$y=-\frac{14}{5}$ or $y=2$

27. $|0.2x-3.5|=-5.6$
No solution

29. $|4w+3|=|2w-5|$
$4w+3=-(2w-5)$ or $4w+3=2w-5$
$4w+3=-2w+5$ or $2w+3=-5$
$6w+3=5$ or $2w=-8$
$6w=2$ or $w=-4$
$w=\frac{1}{3}$ or $w=-4$

31. $|2y+5|=|7-2y|$
$2y+5=-(7-2y)$ or $2y+5=7-2y$
$2y+5=-7+2y$ or $4y+5=7$
$5\neq-7$ or $4y=2$
$y=\frac{1}{2}$

33. $1=-4+\left|2-\frac{1}{4}w\right|$
$5=\left|2-\frac{1}{4}w\right|$
$2-\frac{1}{4}w=-5$ or $2-\frac{1}{4}w=5$
$-\frac{1}{4}w=-7$ or $-\frac{1}{4}w=3$
$w=28$ or $w=-12$

35. $10=4+|2y+1|$
$6=|2y+1|$
$2y+1=-6$ or $2y+1=6$
$2y=-7$ or $2y=5$
$y=-\frac{7}{2}$ or $y=\frac{5}{2}$

37. $|3b-7|-9=-9$

$$3b-7=0$$
$$3b=7$$
$$b=\frac{7}{3}$$

39. $|4w-1|=|2w+3|$

$$4w-1=-(2w+3) \quad \text{or} \quad 4w-1=2w+3$$
$$4w-1=-2w-3 \quad \text{or} \quad 2w-1=3$$
$$6w-1=-3 \quad \text{or} \quad 2w=4$$
$$6w=-2 \quad \text{or} \quad w=2$$
$$w=-\frac{1}{3} \quad \text{or} \quad w=2$$

41. $-2|x+3|=5$

$$|x+3|=-\frac{5}{2}$$

No solution

43. $|6x-9|=0$

$$6x-9=0$$
$$6x=9$$
$$x=\frac{3}{2}$$

45. $|2h-6|=|2h+5|$

$$2h-6=-(2h+5) \quad \text{or} \quad 2h-6=2h+5$$
$$2h-6=-2h-5 \quad \text{or} \quad -6\neq 5$$
$$4h-6=-5$$
$$4h=1$$
$$h=\frac{1}{4}$$

47. $\left|-\frac{1}{5}-\frac{1}{2}k\right|=\frac{9}{5}$

$$-\frac{1}{5}-\frac{1}{2}k=-\frac{9}{5} \quad \text{or} \quad -\frac{1}{5}-\frac{1}{2}k=\frac{9}{5}$$
$$-\frac{1}{2}k=-\frac{8}{5} \quad \text{or} \quad -\frac{1}{2}k=2$$
$$k=\frac{16}{5} \quad \text{or} \quad k=-4$$

49. $|3.5m-1.2|=|8.5m+6|$

$$\begin{aligned} 3.5m-1.2&=-(8.5m+6) &\text{ or }\quad 3.5m-1.2&=8.5m+6\\ 3.5m-1.2&=-8.5m-6 &\text{ or }\quad 3.5m-1.2&=8.5m+6\\ 12m-1.2&=-6 &\text{ or }\quad -5m-1.2&=6\\ 12m&=-4.8 &\text{ or }\quad -5m&=7.2\\ m&=-0.4 &\text{ or }\quad m&=-1.44 \end{aligned}$$

51. $|x|=6$

53. $|x|=\frac{4}{3}$

55. $x=2$ or $x=-\frac{1}{2}$

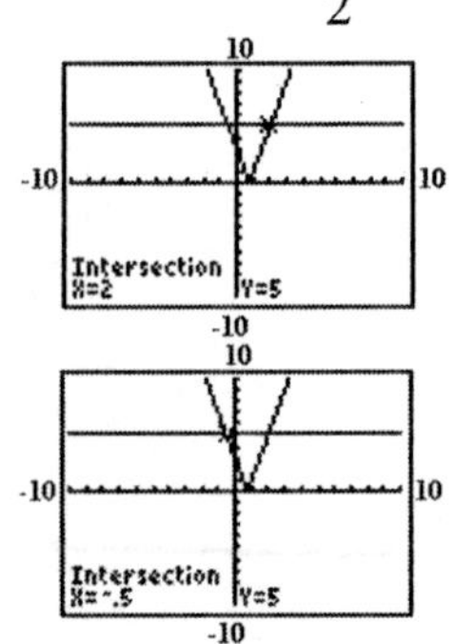

57. No solution

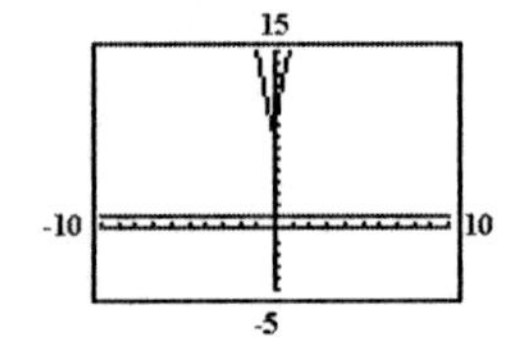

59. $x=\frac{1}{2}$

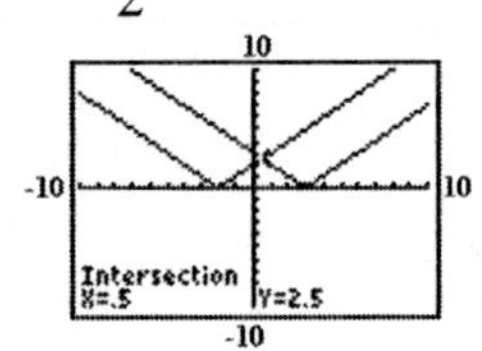

61. $x=\frac{4}{3}$ or $x=-2$

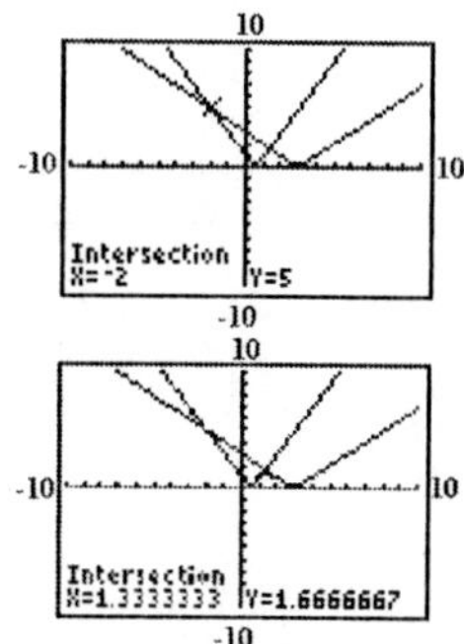

Section 9.4 Practice Exercises

1. $|10x-6|=-5$

No solution

3. $|6x|=|9x+5|$

$6x=-(9x+5)$ or $6x=9x+5$

$6x=-9x-5$ or $-3x=5$

$15x=-5$ or $x=-\frac{5}{3}$

$x=-\frac{1}{3}$ or $x=-\frac{5}{3}$

5. $-15<3w-6\le -9$

$-9<3w\le -3$

$-3<w\le -1$

Interval notation: $(-3, -1]$

−3 −1

7. $m-7\le -5$ or $m-7\ge -10$

$m\le 2$ or $m\ge -3$

All real numbers

Interval notation: $(-\infty, \infty)$

9. **(a)** $|x|=5$

$x=-5$ or $x=5$

(b) $|x|>5$

$x<-5$ or $x>5$

Interval notation: $(-\infty, -5)\cup(5, \infty)$

−5 5

(c) $|x|<5$

$-5<x<5$

Interval notation: $(-5, 5)$

−5 5

11. **(a)** $|p|=-2$

No solution

(b) $|p|>-2$

All real numbers

Interval notation $(-\infty, \infty)$

(c) $|p|<-2$

No solution

13. **(a)** $|x-3|=7$

$x-3=-7$ or $x-3=7$

$x=-4$ or $x=10$

(b) $|x-3|>7$

$x-3<-7$ or $x-3>7$

$x<-4$ or $x>10$

Interval notation: $(-\infty, -4)\cup(10, \infty)$

−4 10

(c) $|x-3|<7$

$-7<x-3<7$

$-4<x<10$

Interval notation: $(-4, 10)$

−4 10

15. **(a)** $|y+1|=-6$

No solution

(b) $|y+1|>-6$

All real numbers

Interval notation: $(-\infty, \infty)$

(c) $|y+1|<-6$

No solution

17. $|x| > 6$

$x < -6$ or $x > 6$

Interval notation: $(-\infty, -6) \cup (6, \infty)$

−6 6

19. $|t| \le 3$

$-3 \le t \le 3$

Interval notation: $[-3, 3]$

−3 3

21. $|y+2| \ge 0$

$y+2 \le 0$ or $y+2 \ge 0$

$y \le -2$ or $y \ge -2$

All real numbers

Interval notation: $(-\infty, \infty)$

23. $5 \le |2x-1|$

$|2x-1| \ge 5$

$2x-1 \le -5$ or $2x-1 \ge 5$

$2x \le -4$ or $2x \ge 6$

$x \le -2$ or $x \ge 3$

−2 3

Interval notation: $(-\infty, -2] \cup [3, \infty)$

25. $|k-7| < -3$

No solution

27. $\left|\frac{w-2}{3}\right| - 3 \le 1$

$\left|\frac{w-2}{3}\right| \le 4$

$-4 \le \frac{w-2}{3} \le 4$

$-12 \le w-2 \le 12$

$-10 \le w \le 14$

Interval notation: $[-10, 14]$

−10 14

29. $|9-4y| \ge 14$

$9-4y \le -14$ or $9-4y \ge 14$

$-4y \le -23$ or $-4y \ge 5$

$y \ge \frac{23}{4}$ or $y \le -\frac{5}{4}$

Interval notation: $\left(-\infty, -\frac{5}{4}\right] \cup \left[\frac{23}{4}, \infty\right)$

$-\frac{5}{4}$ $\frac{23}{4}$

31. $\left|\frac{2x+1}{4}\right| < 5$

$-5 < \frac{2x+1}{4} < 5$

$-20 < 2x+1 < 20$

$-21 < 2x < 19$

$-\frac{21}{2} < x < \frac{19}{2}$

Interval notation: $\left(-\frac{21}{2}, \frac{19}{2}\right)$

$-\frac{21}{2}$ $\frac{19}{2}$

33. $8 < |4-3x| + 12$

$|4-3x| + 12 > 8$

$|4-3x| > -4$

All real numbers

Interval notation: $(-\infty, \infty)$

35. $5 - |2m+1| > 5$

$-|2m+1| > 0$

$|2m+1| < 0$

No solution

37. $|p+5| \le 0$

$p+5 = 0$

$p = -5$

$\{-5\}$

−5

39. $|z-6|+5 \geq 5$
$|z-6| \geq 0$
All real numbers
Interval notation: $(-\infty, \infty)$

41. $|x|>7$

43. $|x-2| \leq 13$

45. $|x-32| \leq 0.05$

47. $\left|x-6\frac{3}{4}\right| \leq \frac{1}{8}$

49. The midpoint of the two boundary points is $\frac{-3+5}{2}=1$. The distance from the midpoint to either boundary point is $|1-5|=4$, or $|1-(-3)|=4$. Thus, the graph consists of all points whose distance from 1 is greater than 4. The inequality is $|x-1|>4$, b.

51. The midpoint of the two boundary points is $\frac{-2+6}{2}=2$. The distance from the midpoint to either boundary point is $|2-6|=4$, or $|2-(-2)|=4$. Thus, the graph consists of all points whose distance from 2 is less than 4. The inequality is $|x-2|<4$, a.

53. $(-\infty, -6) \cup (2, \infty)$

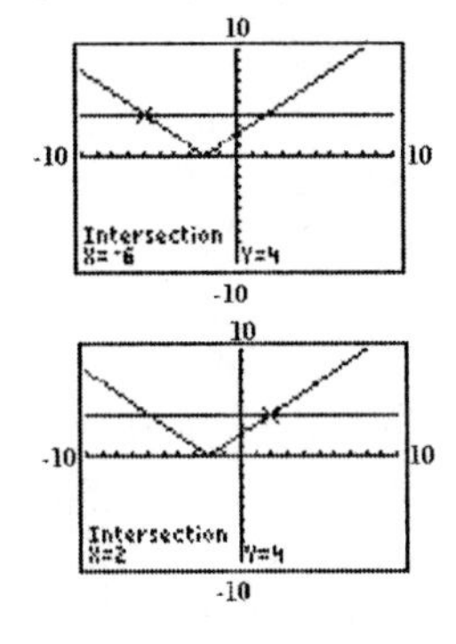

55. $(-7, 5)$

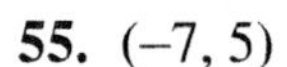

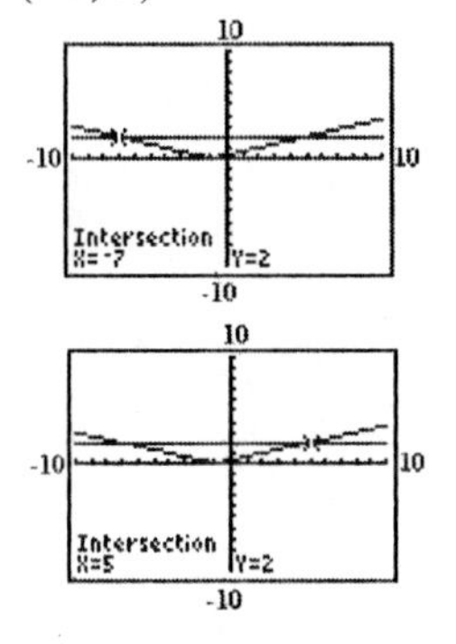

57. No solution

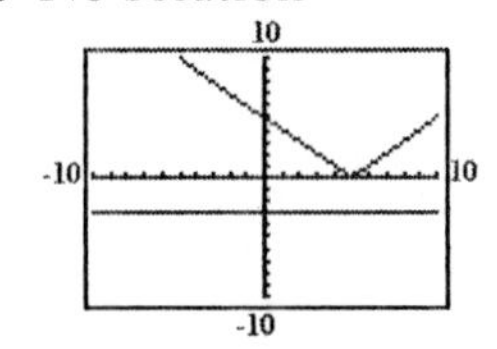

59. $(-\infty, \infty)$

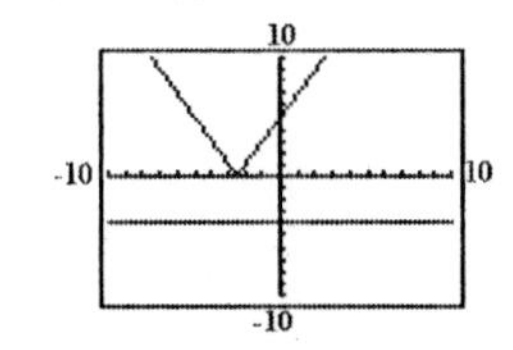

61. $x=-\frac{1}{6}$

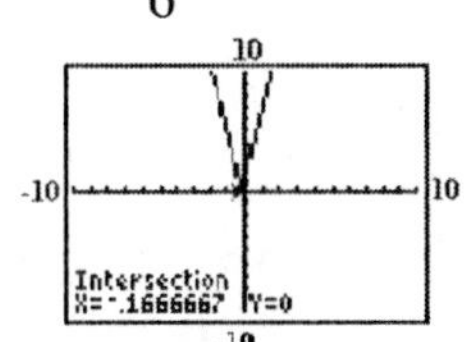

Section 9.5 Practice Exercises

1. Absolute value represents the distance from 0 on the number line. When solving $|x+3|>4$, the goal is to find the values of $x+3$ that are more than 4 units from 0 on the number line in either direction. Solve the inequality: $x+3<-4$ or $x+3>4$ or use the test point method.

3. $-3<2k-5<3$
$2<2k<8$
$1<k<4$
Interval notation: $(1, 4)$

5. $|6a-1|-4 \le 2$

$|6a-1| \le 6$

$-6 \le 6a-1 \le 6$

$-5 \le 6a \le 7$

$\frac{-5}{6} \le a \le \frac{7}{6}$

Interval notation: $\left[-\frac{5}{6}, \frac{7}{6}\right]$

7. $|2t+1|+4 \ge 7$

$|2t+1| \ge 3$

$2t+1 \le -3$ or $2t+1 \ge 3$

$2t \le -4$ or $2t \ge 2$

$t \le -2$ or $t \ge 1$

$(-\infty, -2] \cup [1, \infty)$

9. **(a)** $(3, -5) \Rightarrow x = 3, y = -5$.
Substitute these values in $2x - y > 8$.

$2(3)-(-5) > 8$

$6+5 > 8$

$11 > 8$ True

Thus, $(3, -5)$ is a solution to $2x - y > 8$.

(b) $(-1, -10) \Rightarrow x = -1, y = -10$.
Substitute these values in $2x - y > 8$.

$2(-1)-(-10) > 8$

$-2+10 > 8$

$8 > 8$ False

Thus, $(-1, -10)$ is not a solution to $2x - y > 8$. Note: $(-1, -10)$ lies on the line $2x - y = 8$.

(c) $(4, -2) \Rightarrow x = 4, y = -2$.
Substitute these values in $2x - y > 8$.

$2(4)-(-2) > 8$

$8+2 > 8$

$10 > 8$ True

Thus, $(4, -2)$ is a solution to $2x - y > 8$.

(d) $(0, 0) \Rightarrow x = 0, y = 0$.
Substitute these values in $2x - y > 8$.

$2(0)-(0) > 8$

$0+0 > 8$

$0 > 8$ False

Thus, $(4, -2)$ is not a solution to $2x - y > 8$.

11. **(a)** $(5, -3) \Rightarrow x = 5, y = -3$.
Substitute these values in $y \le -2$.

$-3 \le -2$ True

Thus, $(5, -3)$ is a solution to $y \le -2$.

(b) $(-4, -2) \Rightarrow x = -4, y = -2$.
Substitute these values in $y \le -2$.

$-2 \le -2$ True

Thus, $(-4, -2)$ is a solution to $y \le -2$.
Note: $(-4, -2)$ actually lies on the line $y = -2$.

(c) $(0, 0) \Rightarrow x = 0, y = 0$.
Substitute these values in $y \le -2$.

$0 \le -2$ False

Thus, $(0, 0)$ is not a solution to $y \le -2$.

(d) $(3, 2) \Rightarrow x = 3, y = 2$.
Substitute these values in $y \le -2$.

$2 \le -2$ False

Thus, $(3, 2)$ is not a solution to $y \le -2$.

13. To choose the correct inequality symbol, three observations must be made. First, notice the shading occurs below the line. Second, since the coefficient of y is negative in the given statement, the direction of the inequality will change. third, the boundary line is dashed indicating no equality. Thus, use the symbol $>$ for the inequality $x - y > 2$.

15. To choose the correct inequality symbol, three observations must be made. First, notice the shading occurs above the line. Second, since the coefficient of y is positive in the given statement, the direction of the inequality will not change. Third, the boundary line is solid indicating equality. Thus, use the symbol $\ge$ for the inequality $y \ge -4$.

17. The graph of $x \le 0$ includes Quadrant II and Quadrant III. The graph of $y \ge 0$ includes Quadrant I and Quadrant II. The intersection of the graphs occurs in Quadrant II. Thus, the statements are $x \le 0$ and $y \ge 0$.

19. Graph the related equation $x - 2y = 4$.

$$-2y = -x + 4$$
$$y = \frac{1}{2}x - 2$$

with y-intercept $(0, -2)$ and slope $\frac{1}{2}$.

Since the given symbol ">" contains no equality symbol, use a dashed boundary line. Test points on each side of the line to decide whether to shade above or below the line.

Test Point Above: (0, 0)

$$x - 2y > 4$$
$$0 - 2(0) > 4$$
$$0 > 4 \quad \text{False}$$

The test point (0, 0) is not a solution to the original inequality.

Test Point Below: (0, −3)

$$x - 2y > 4$$
$$0 - 2(-3) > 4$$
$$6 > 4 \quad \text{True}$$

The test point (0, −3) is a solution to the original inequality. Shade the region below the boundary.

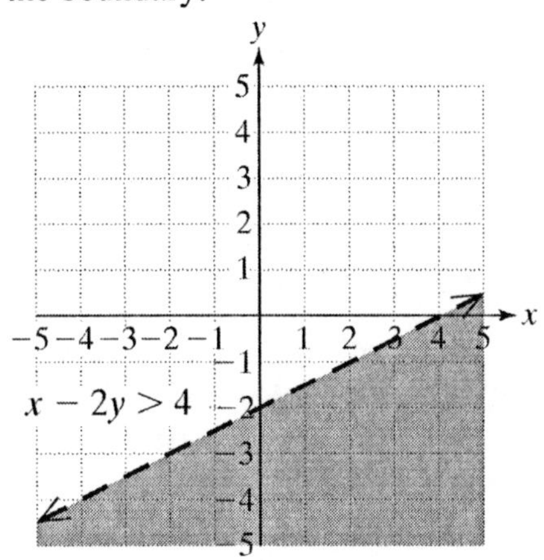

21. Graph the related equation, $5x - 2y = 10$ with y-intercept $(0, -5)$ and slope $\frac{5}{2}$.

Since the given symbol "<" contains no equality symbol, use a dashed boundary line. Test points on each side of the line to decide whether to shade above or below the line.

Test Point Above: (0, 0)

$$5x - 2y < 10$$
$$5(0) - 2(0) < 10$$
$$0 < 10 \quad \text{True}$$

The test point (0, 0) is a solution to the original inequality. Shade the region above the boundary. Since the test point above is true, any test point below would result in a false statement.

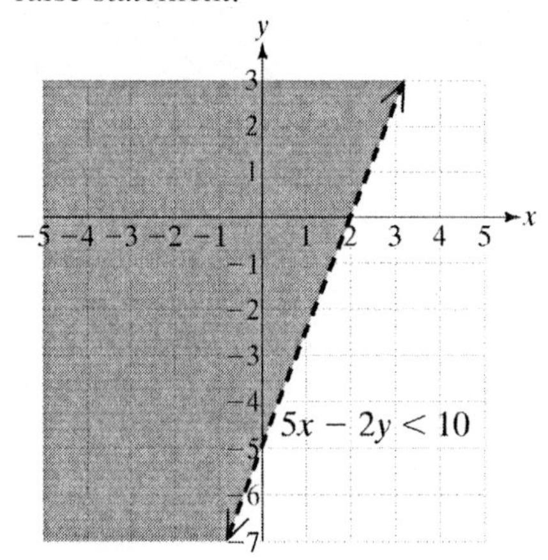

23. Graph the related equation $2x + 6y = 12$.

$$6y = -2x + 12$$
$$y = -\frac{1}{3}x + 2$$

with y-intercept $(0, 2)$ and slope $\frac{-1}{3}$.

Since the given symbol "≤" contains the equality symbol, use a solid boundary line. Test points on each side of the line to decide whether to shade above or below the line.

Test Point Above: (0, 3)

$$2x + 6y \le 12$$
$$0(0) + 6(3) \le 12$$
$$18 \le 12 \quad \text{False}$$

The test point (0, 3) is not a solution to the original inequality.

Test Point Below: (0, 0)

$$2x + 6y \le 12$$
$$0(0) + 6(0) \le 12$$
$$0 \le 12 \quad \text{True}$$

The test point (0, 0) is a solution to the original inequality. Shade the region below the boundary.

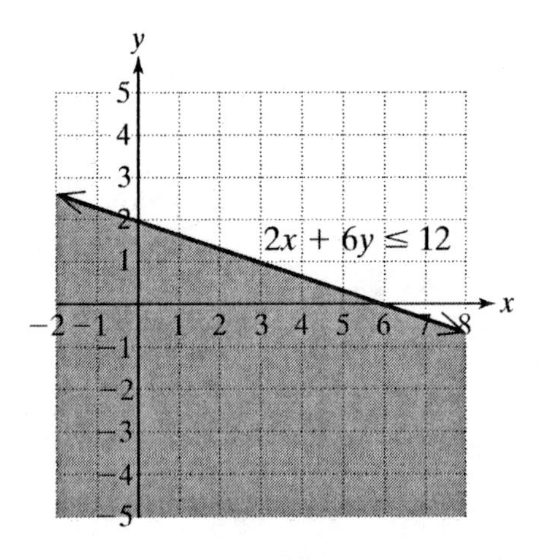

25. Graph the related equation $y = -2$ with y-intercept $(0, -2)$ and slope 0.
Since the given symbol "≥" contains the equality symbol, use a solid boundary line. Test points on each side of the line to decide whether to shade above or below the line.
Test Point Above: (0, 0)

$y \geq -2$

$0 \geq -2$ True

The test point (0, 0) is a solution to the original inequality. Shade the region above the boundary. Since the test point above is true, any test point below would result in a false statement.

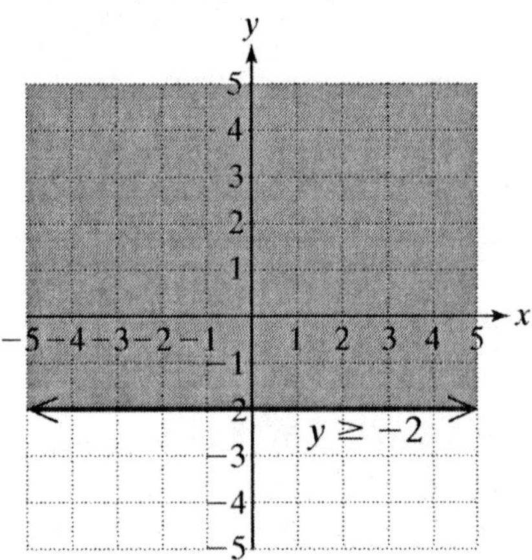

27. Graph the related equation $4x = 5$.

$x = \frac{4}{5}$

with no y-intercept and undefined slope. Since the given symbol "<" contains no equality symbol, use a dashed boundary line. Test points on each side of the line to decide whether to shade to the right or to the left of the line.
Test Point to Right (2, 0)

$4x < 5$

$4(2) < 5$

$8 < 5$ False

The test point (2, 0) is not a solution to the original inequality.
Test Point to Left: (0, 0)

$4x < 5$

$4(0) < 5$

$0 < 5$ True

The test point (0, 0) is a solution to the original inequality. Shade the region to the left of the boundary.

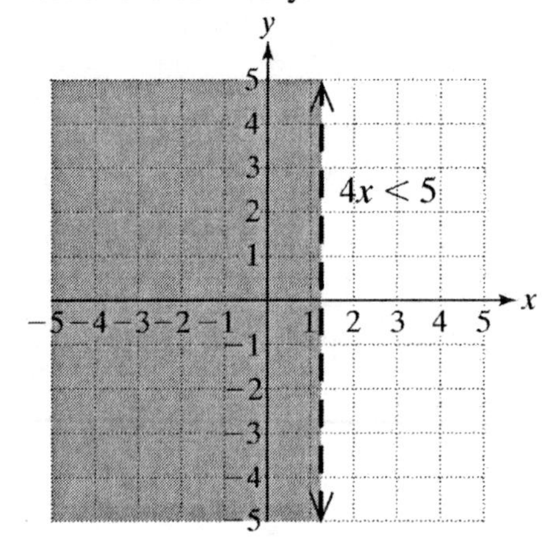

29. Graph the related equation $y = \frac{2}{5}x - 4$ with y-intercept $(0, -4)$ and slope $\frac{2}{5}$.

Since the given symbol "≥" contains the equality symbol, use a solid boundary line. Test points on each side of the line to decide whether to shade above or below the line.
Test Point Above: (0, 0)

$y \geq \frac{2}{5}x - 4$

$0 \geq \frac{2}{5}(0) - 4$

$0 \geq -4$ True

The test point (0, 0) is a solution to the original inequality. Shade the region above the boundary. Since the test point above is true, any test point below would result in a false statement.

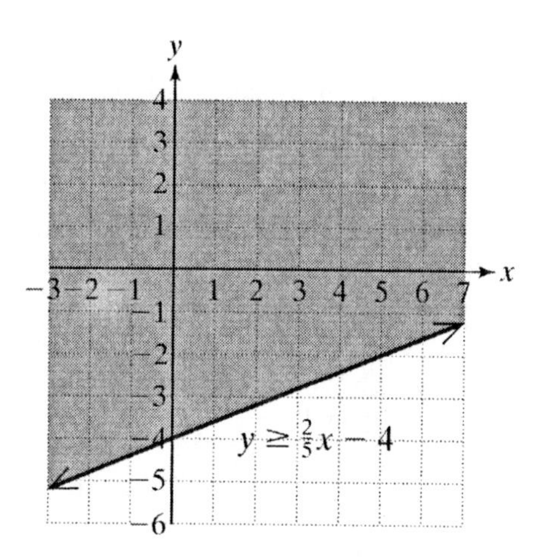

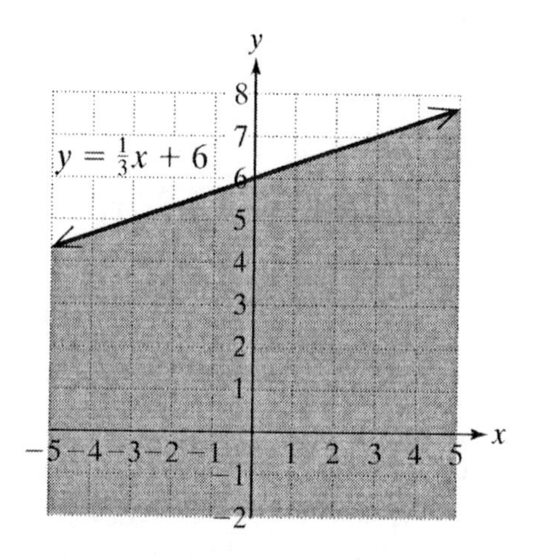

31. Graph the related equation $y = \frac{1}{3} + 6$ with y-intercept (0, 6) and slope $\frac{1}{3}$.

Since the given symbol "≤" contains the equality symbol, use a solid boundary line. Test points on each side of the line to decide whether to shade above or below the line.

Test Point Above: (0, 7)

$y < \frac{1}{3}x + 6$

$7 < \frac{1}{3}(0) + 6$

$7 < 6$ False

The test point (0, 7) is not a solution to the original inequality.

Test Point Below: (0, 0)

$y < \frac{1}{3}x + 6$

$0 < \frac{1}{3}(0) + 6$

$0 < 6$ True

The test point (0, 0) is a solution to the original inequality. Shade the region below the boundary.

33. Graph the related equation $y = 5x$ with y-intercept (0, 0) and slope 5.

Since the given symbol ">" contains no equality symbol, use a dashed boundary line. Test points on each side of the line to decide whether to shade above or below the line.

Test Point Above: (0, 2)

$y > 5x$

$2 > 5(0)$

$2 > 0$ True

The test point (0, 2) is a solution to the original inequality. Shade the region above the boundary. Since the test point above is true, any test point below would result in a false statement.

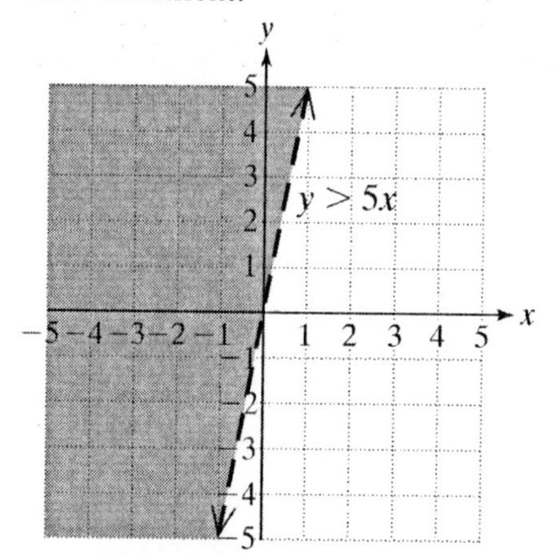

35. First, graph the related equation $\frac{x}{5}+\frac{y}{4}=1$

$$\frac{y}{4}=\frac{-x}{5}+1$$
$$5y=-4x+20$$
$$y=\frac{-4}{5}x+4$$

with y-intercept (0, 4) and slope $\frac{-4}{5}$.

Since the inequality symbol "<" contains no equality symbol, use a dashed boundary line. Test points on each side of the line to decide whether to shade above or below the line.

Test Point Above: (0, 8)

$$\frac{x}{5}+\frac{y}{4}<1$$
$$\frac{0}{5}+\frac{8}{4}<1$$
$$2<1 \quad \text{False}$$

The test point (0, 8) is not a solution to the original inequality.

Test Point Below: (0, 0)

$$\frac{x}{5}+\frac{y}{4}<1$$
$$\frac{0}{5}+\frac{0}{4}<1$$
$$0<1 \quad \text{True}$$

The test point (0, 0) is a solution to the original inequality. Shade the region below the boundary.

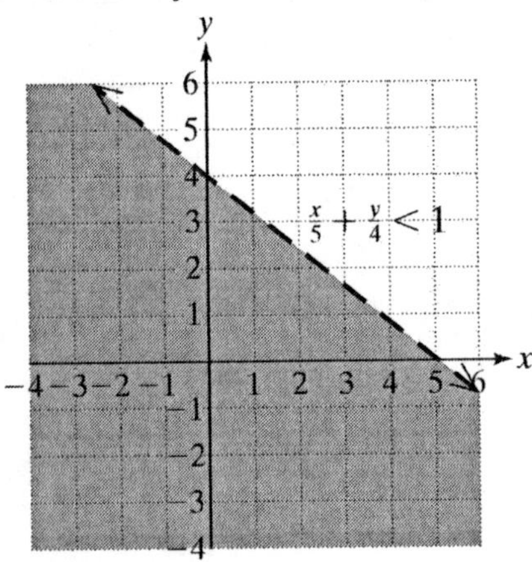

37. Graph the related equation $0.1x + 0.2y = 0.6$.

$$0.2y=-0.1x+0.6$$
$$y=-0.5x+3$$

with y-intercept (0, 3) and slope −0.5.

Since the inequality symbol "≤" contains the equality symbol, use a solid boundary line. Test points on each side of the line to decide whether to shade above or below the line.

Test Point Above: (0, 4)

$$0.1x+0.2y\le 0.6$$
$$0.1(0)+0.2(4)\le 0.6$$
$$0.8\le 0.6 \quad \text{False}$$

The test point (0, 4) is not a solution to the original inequality.

Test Point Below: (0, 0)

$$0.1x+0.2y\le 0.6$$
$$0.1(0)+0.2(0)\le 0.6$$
$$0\le 0.6 \quad \text{True}$$

The test point (0, 0) is a solution to the original inequality. Shade the region below the boundary.

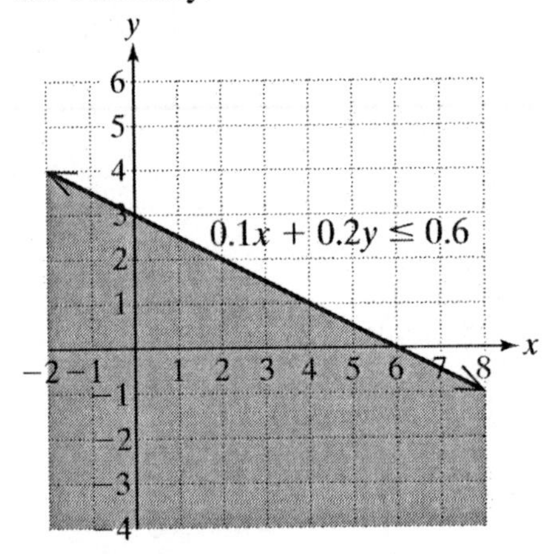

39. Solve $y \le 4$ and $y \ge -x + 2$

First Inequality

$y \le 4$

$y = 4$ (related equation)

Second Inequality

$y \ge -x + 2$

$y = -x + 2$ (related equation)

For each inequality, draw the boundary line. Then pick test points above and below the line to determine the appropriate region to shade.

Test Point Above the line $y = 4$: (0, 5)
$y \le 4$
$5 \le 4$ False
The test point (0, 5) is not a solution to the inequality.
Test Point Below the line $y = 4$: (0, 0)
$y \le 4$
$0 \le 4$ True
The test point (0, 0) is a solution to the inequality. Shade the region below the boundary.

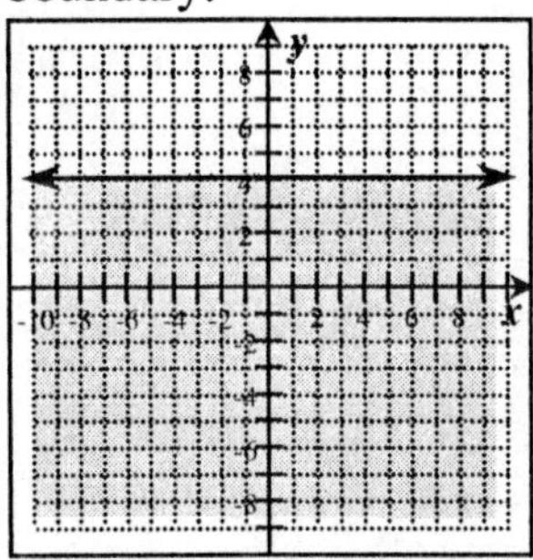

Test Point Above the line $y = -x + 2$: (3, 0)
$y \ge -x + 2$
$0 \ge -3 + 2$
$0 \ge -1$ True
The test point (3, 0) is a solution to the inequality. Shade the region above the boundary. Since the test point above is true, any test point below would result in a false statement.

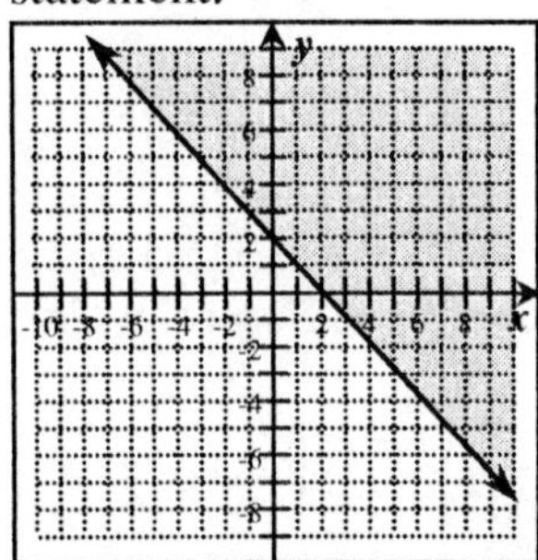

The solution is the intersection of the two individual solution sets. Therefore, the solution is the region of the plane below the line $y = 4$ and above the line $y = -x + 2$ as shaded. Include the portions of the lines $y = 4$ and $y = -x + 2$ that are in the intersection.

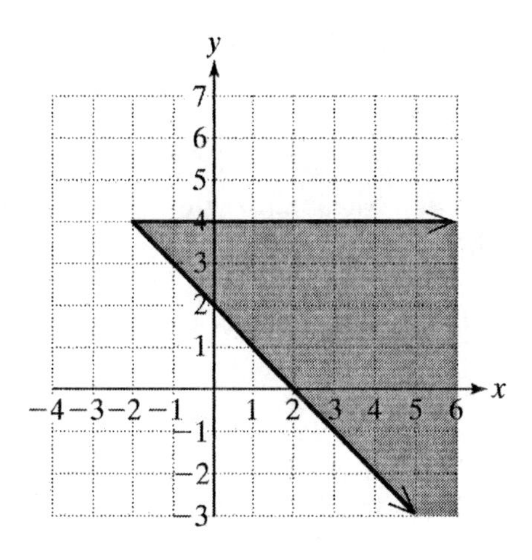

41. Solve $2x + y < 5$ or $x > 3$
First Inequality
$2x + y < 5$
$2x + y = 5$ (related equation)
$y = -2x + 5$
Second Inequality
$x > 3$
$x = 3$ (related equation)

For each inequality, draw the boundary line. Then test points on each side of the line to determine the appropriate region to shade.

Test Point Above the line $2x + y = 5$: (0, 6)
$2x + y < 5$
$2(0) + 6 < 5$
$6 < 5$ False
The test point (0, 6) is not a solution to the inequality.
Test Point Below the line $2x + y = 5$: (0, 0)
$2x + y < 5$
$2(0) + 0 < 5$
$0 < 5$ True
The test point (0, 0) is a solution to the inequality. Shade the region below the boundary.

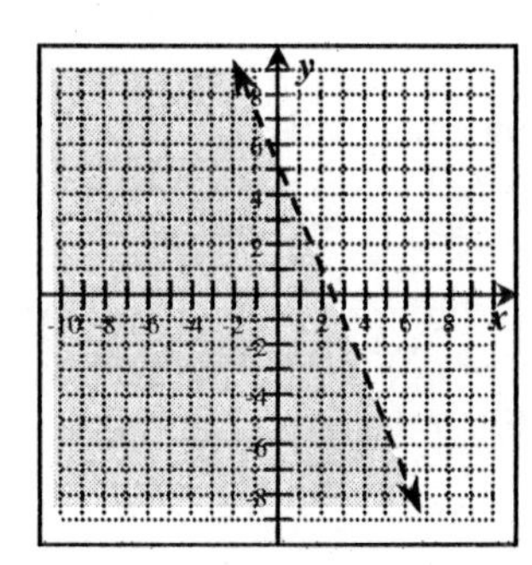

Test Point to Right of the line $x = 3$: (4, 0)

$x > 3$

$4 \geq -1$ True

The test point (4, 0) is a solution to the inequality. Shade the region to the right of the boundary. Since the test point to the right is true, any test point to the left would result in a false statement.

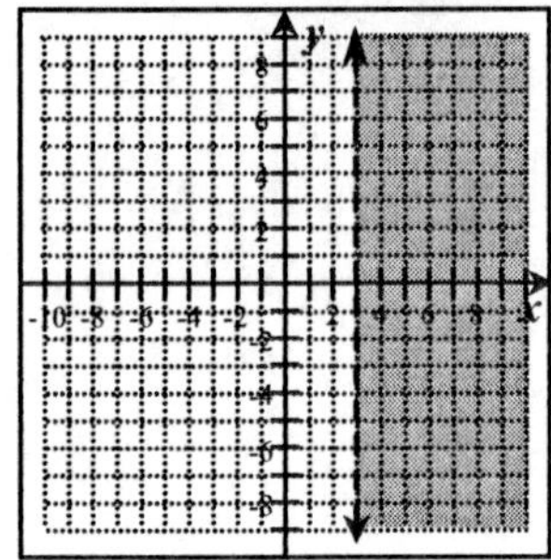

The solution is the union of the two individual solution sets. Therefore, the solution is the region of the plane below $2x + y = 5$ or to the right of $x = 3$ as shaded.

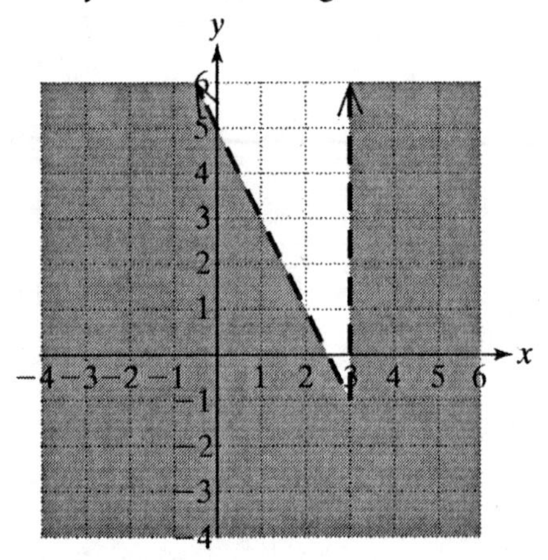

43. Solve $x + y \leq 3$ and $4x + y < 6$

First Inequality

$x + y \leq 3$

$x + y = 3$ (related equation)

$y = -x + 3$

Second Inequality

$4x + y < 6$

$4x + y = 6$ (related equation)

$y = -4x + 6$

For each inequality, draw the boundary line. Then pick test points above and below the line to determine the appropriate region to shade.

Test Point Above the line $x + y = 3$: (0, 4)

$x + y < 3$

$0 + 4 < 3$

$4 < 3$ False

The test point (0, 4) is not a solution to the inequality.

Test Point Below the line $x + y = 3$: (0, 0)

$x + y < 3$

$0 + 0 < 3$

$0 < 3$ True

The test point (0, 0) is a solution to the inequality. Shade the region below the boundary.

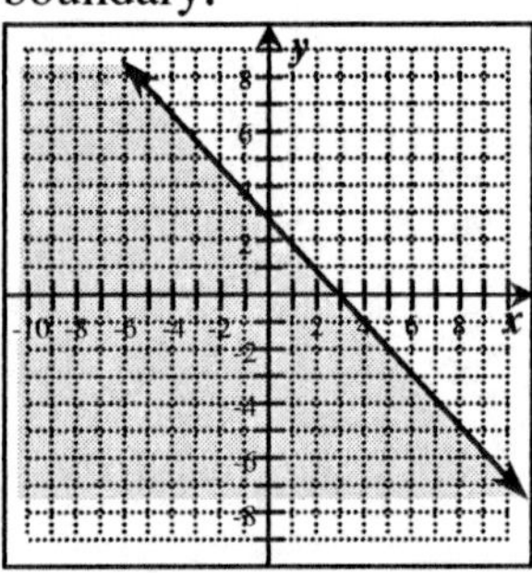

Test Point Above the line $4x + y = 6$: (0, 7)

$4x + y < 6$

$4(0) + 7 < 6$

$7 < 6$ False

The test point (0, 7) is not a solution to the inequality.

Test Point Below the line $4x + y = 6$: (0, 0)

$4x + y < 6$

$4(0) + 0 < 6$

$0 < 6$ True

The test point (0, 0) is a solution to the inequality. Shade the region below the boundary.

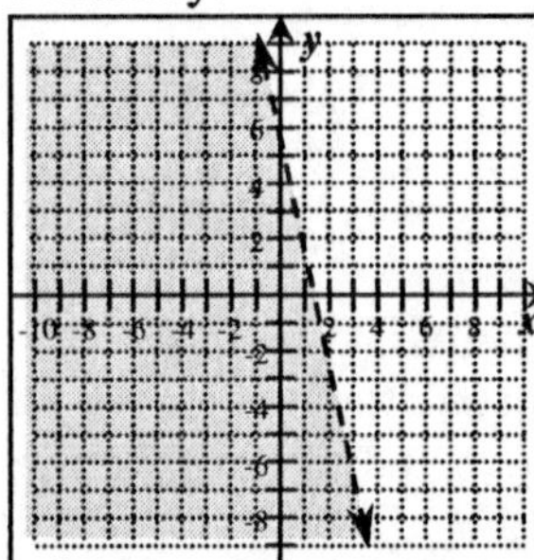

The solution is the intersection of the two individual solution sets. Therefore, the solution is the region of the plane below the line $x + y = 3$ and below the line $4x + y = 6$

as shaded. Include the portion of the line $x + y = 3$ that is in the intersection.

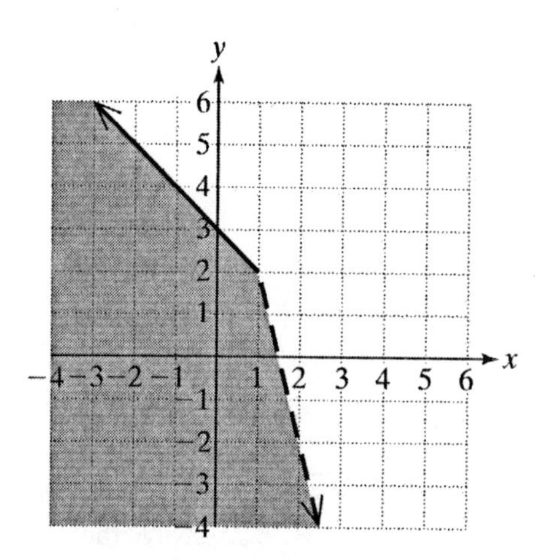

45. Solve $2x - y \le 2$ or $2x + 3y > 6$
First Inequality
$2x - y \le 2$
$2x - y = 2$ (related equation)
$-y = -2x + 2$
Second Inequality
$2x + 3y > 6$
$2x + 3y = 6$ (related equation)
$3y = -2x + 6$
$y = -\frac{2}{3}x + 2$

For each inequality, draw the boundary line. Then pick test points above and below the line to determine the appropriate region to shade.
Test Point Above the line $2x - y = 2$: (0, 0)
$2x - y < 2$
$2(0) - (0) < 2$
$0 < 2$ True
The test point (0, 0) is a solution to the inequality. Shade the region above the boundary. Since the test point above is true, any test point below would result in a false statement.

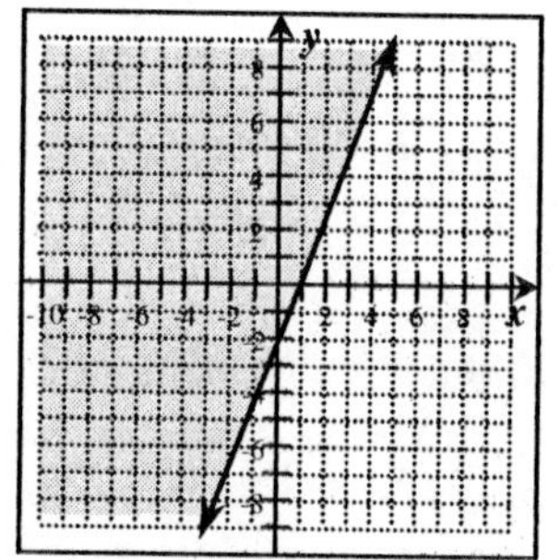

Test Point Above the line $2x + 3y = 6$: (0, 3)
$2x + 3y > 6$
$2(0) + 3(3) > 6$
$9 > 6$ True
The test point (0, 3) is a solution to the inequality. Shade the region above the boundary. Since the test point above is true, any test point below would result in a false statement.

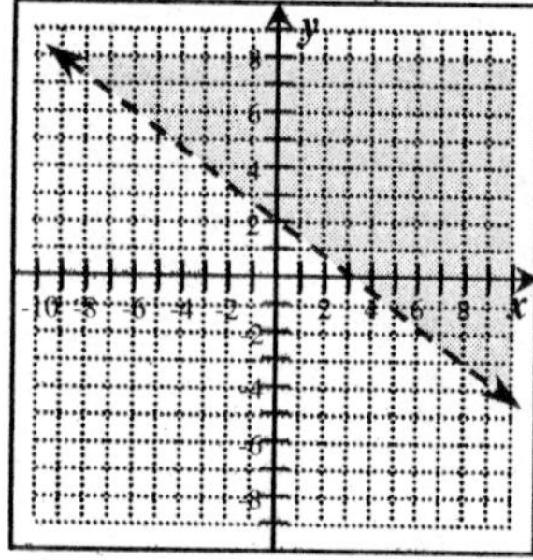

The solution is the union of the two individual solution sets. Therefore, the solution is the region of the plane above the line $2x - y = 2$ or above the line $2x + 3y = 6$ as shaded. The dashed portion of the line $2x + 3y = 6$ is not included in the solution.

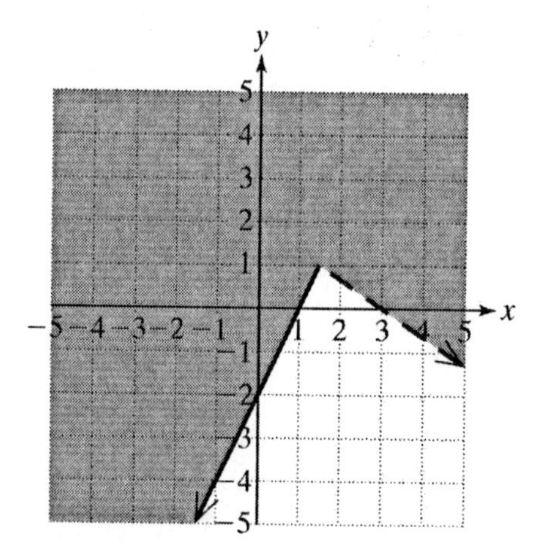

47. Solve $x \ge 4$ and $y \le 2$
First Inequality

$x \geq 4$
$x = 4$ (related equation)
Second Inequality
$y \leq 2$
$y = 2$ (related equation
For each inequality, draw the boundary line. Then test points on each side of the line to determine the appropriate region to shade.
Test Point to the Right of the line $x = 4$: (5, 0)
$x \geq 4$
$5 \geq 4$ True
The test point (5, 0) is a solution to the inequality. Shade the region to the right of the boundary. Since the test point to the right is true, any test point to the left would result in a false statement.

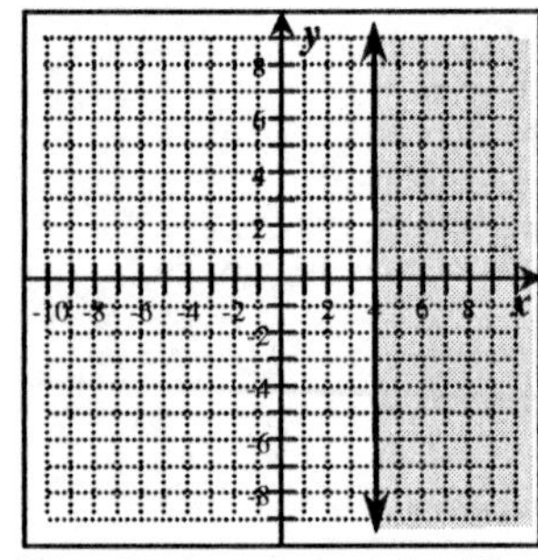

Test Point above the line $y = 2$: (0, 3)
$y \leq 2$
$3 \leq 2$ False
The test point (0, 3) is not a solution to the inequality.

Test Point Below the line $y = 2$: (0, 0)
$y \leq 2$
$0 \leq 2$ True
The test point (0, 0) is a solution to the inequality. Shade the region below the boundary.

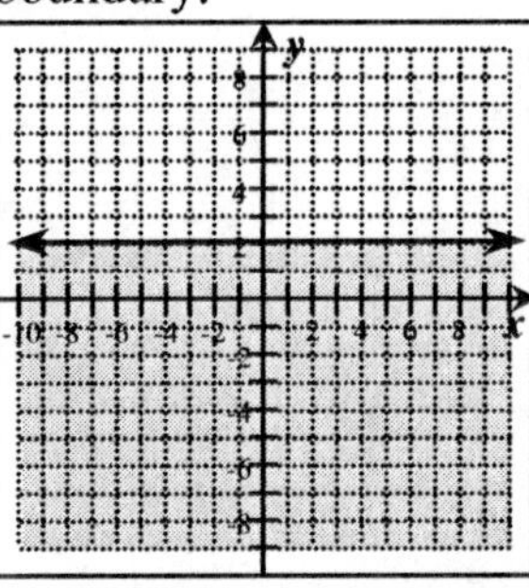

The solution is the intersection of the two individual solution sets. Therefore, the solution is the region of the plane below the line $y = 2$ and to the right of the line $x = 4$ as shaded. Include the portion of the lines $y = 2$ and $x = 4$ that are in the intersection.

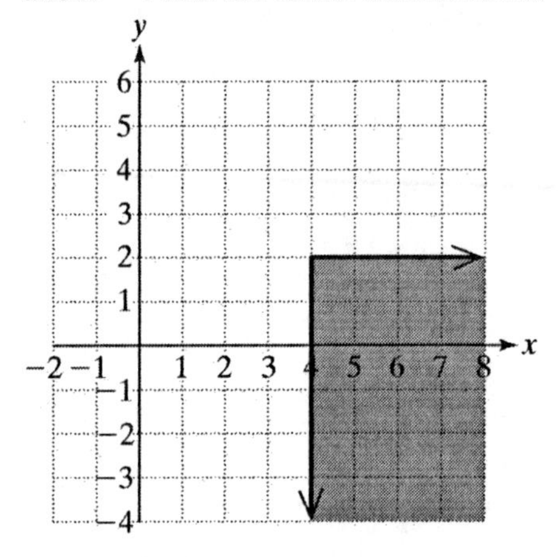

49. Solve $x \leq -2$ or $y \leq 0$
First Inequality
$x \leq -2$
$x = -2$ (related equation)
Second Inequality
$y \leq 0$
$y = 0$ (related equation)

For each inequality, draw the boundary line. Then test points on each side of the line to determine the appropriate region to shade.

Test Point to the Right of the line $x = -2$: (0, 0)

$x \le -2$

$0 \le -2$ False

The test point (0, 0) is not a solution to the inequality.

Test Point to the Left of the line $x = -2$: (–3, 0)

$x \le -2$

$-3 \le -2$ True

The test point (–3, 0) is a solution to the inequality. Shade the region below the boundary.

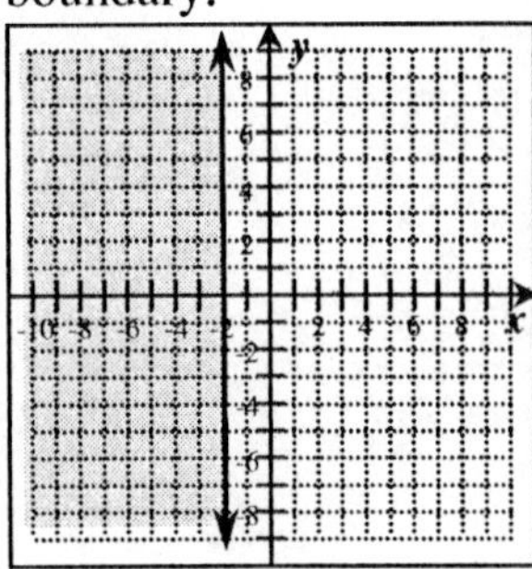

Test Point Above the line $y = 0$: (0, 1)

$y \le 0$

$1 \le 0$ False

The test point (0, 1) is not a solution to the inequality.

Test Point Below the line $y = 0$: (0, –1)

$y \le 0$

$-1 \le 0$ True

The test point (0, –1) is a solution to the inequality. Shade the region below the boundary.

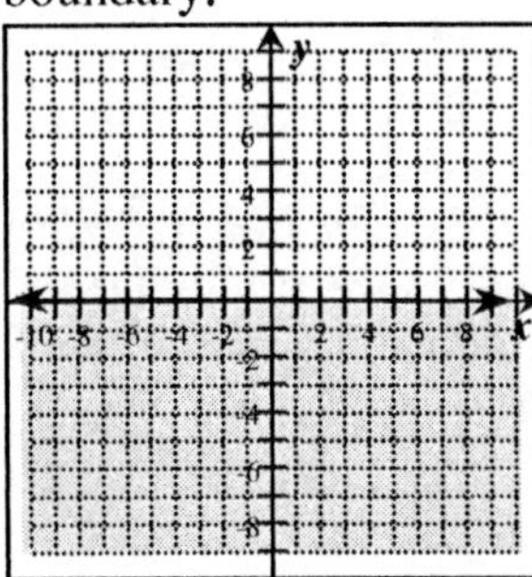

The solution is the union of the two individual solution sets. Therefore, the solution is the region of the plane to the left of the line $x = -2$ or below the line $y = 0$ as shaded. Include the lines $x = -2$ and $y = 0$.

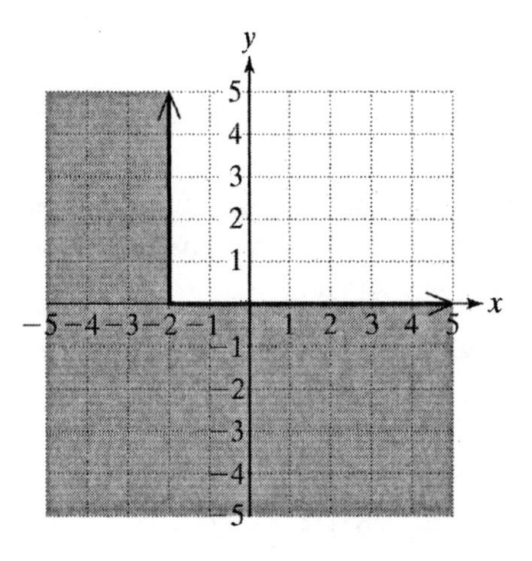

51. Solve $x \ge 0$ and $x + y < 6$

First Inequality

$x \ge 0$

$x = 0$ (related equation)

Second Inequality

$x + y < 6$

$y = -x + 6$ (related equation)

For each inequality, draw the boundary line. Then test points on each side of the line to determine the appropriate region to shade.

Test Point to the Right of the line $x = 0$: (2, 0)

$x \ge 0$

$2 \ge 0$ True

The test point (0, 0) is a solution to the inequality. Shade the region to the right of the boundary.

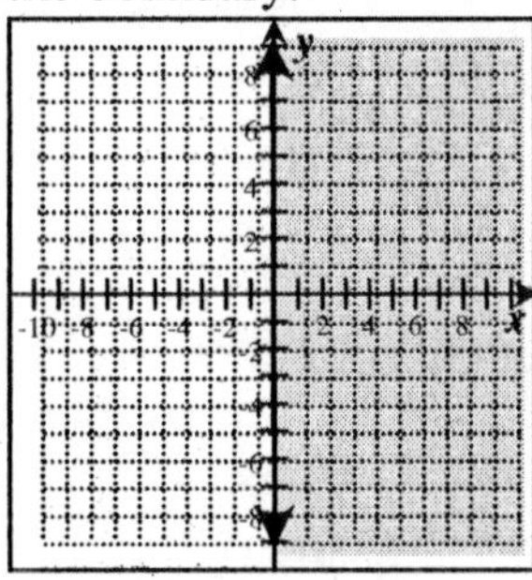

Test Point Above the line $x + y = 6$: (0, 7)

$y < -x + 6$

$7 < 0 + 6$

$7 < 6$ False

The test point (0, 7) is not a solution to the inequality.

Test Point Below the line $x + y = 6$: (0, 0)

$y < -x + 6$

$0 < 0 + 6$

$0 < 6$ True

The test point (0, 0) is a solution to the inequality. Shade the region below the boundary.

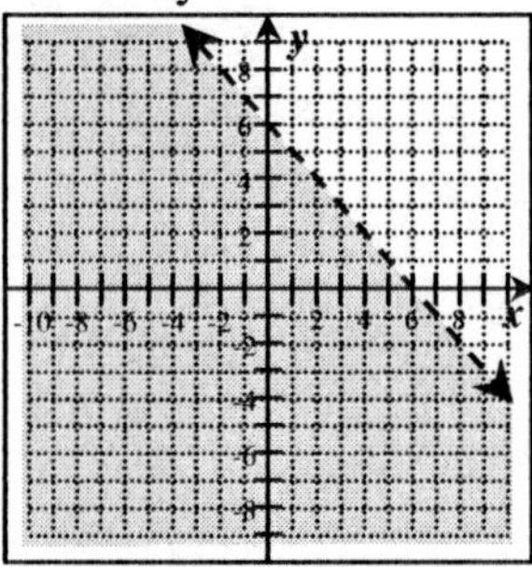

The solution is the intersection of the two individual solution sets. Therefore, the solution is the region of the plane to the right of the line $x = 0$ and below the line $y = -x + 6$ as shaded.

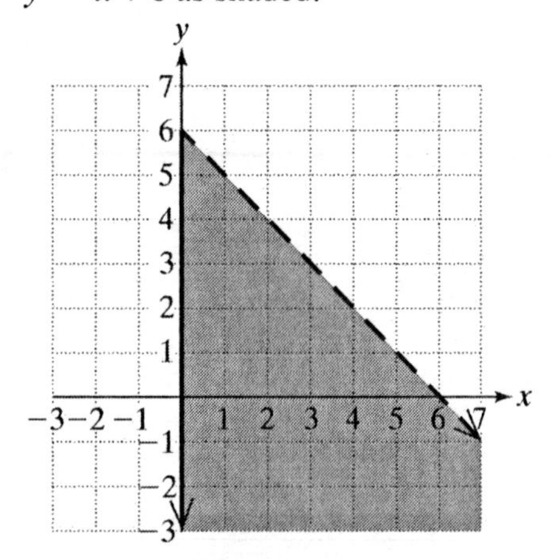

53. Solve $y \le 0$ or $x - y < -4$

First Inequality

$y \ge 0$

$y = 0$ (related equation)

Second Inequality

$x - y < -4$

$x - y = -4$ (related equation)

$-y = -x - 4$

$y = x + 4$

For each inequality, draw the boundary line. Then pick test points above and below the line to determine the appropriate region to shade.

Test Point Above the line $y = 0$: (0, 3)

$y \le 0$

$3 \le 0$ False

The test point (0, 3) is not a solution to the inequality.

Test Point Below the line $y = 0$: (0, –1)

$y \le 0$

$-1 < 0$ True

The test point (0, –1) is a solution to the inequality. Shade the region below the boundary.

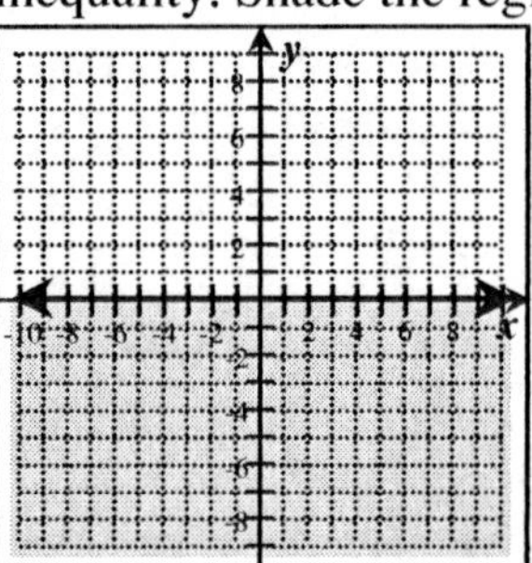

Test Point Above the line $x - y = -4$: (0, 5)

$x - y < -4$

$0 - (5) < -4$

$-5 < -4$ True

The test point (0, 5) is a solution to the inequality. Shade the region above the boundary. Since the test point above is true, any test point below would result in a false statement.

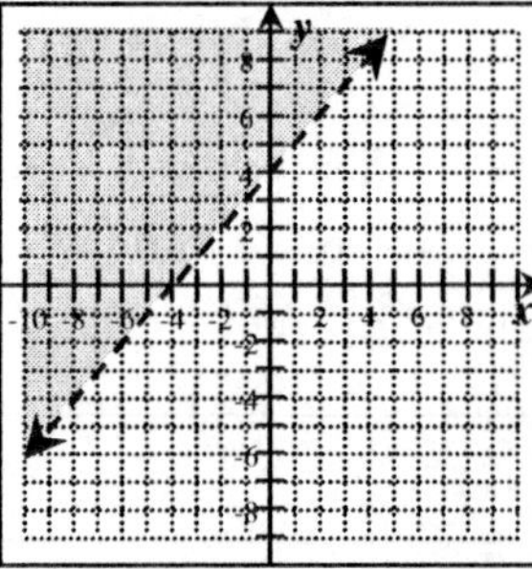

The solution is the union of the two individual solution sets. Therefore, the solution is the region of the plane above the line $x - y = -4$ or below the line $y = 0$ as shaded. The dashed portion of the line $x - y = -4$ is not included in the solution.

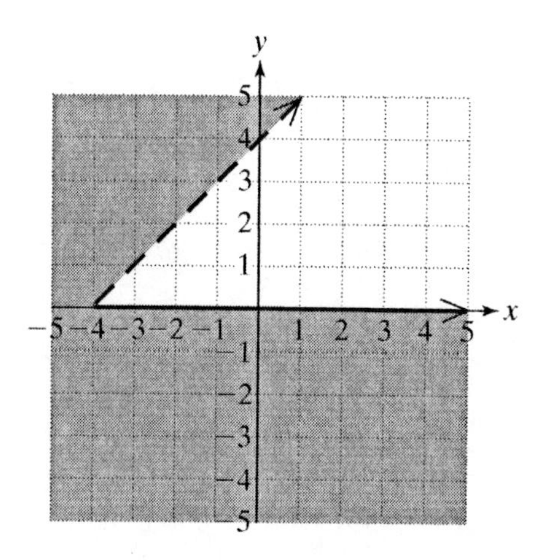

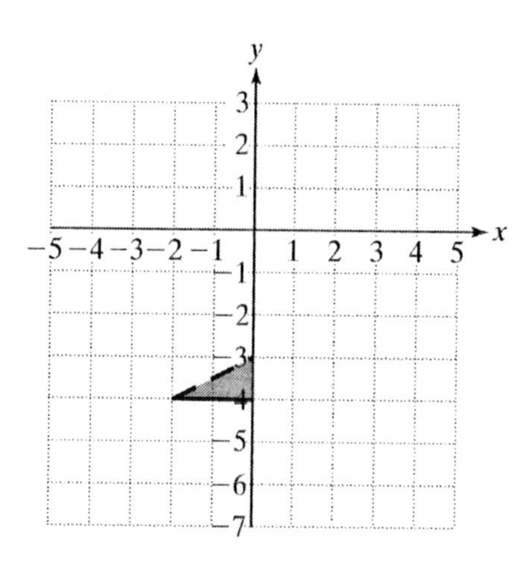

55. The two conditions, $x \geq 0$ and $y \geq 0$, represent the set of points in the first quadrant. The third condition, $x + y \leq 3$, represents the set of points below and including the line $x + y = 3$.

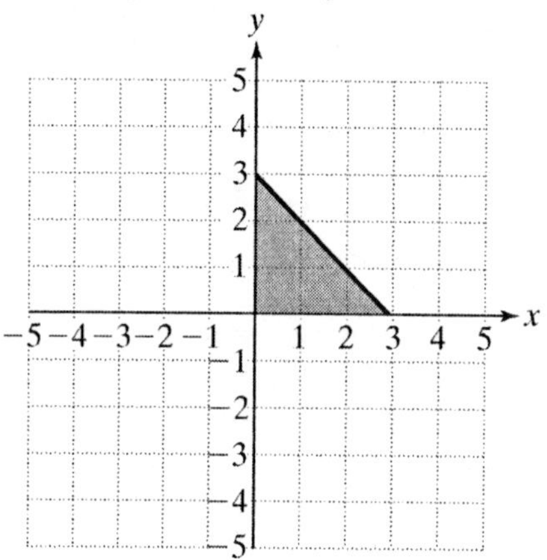

57. The condition $x \leq 0$ represents the set of points in the Quadrants II and III. The condition $y \geq -4$ represents the set of points above or on the line $y = -4$. The third condition, $y < \frac{1}{2}x - 3$, represents the set of points below the line $y = \frac{1}{2}x - 3$.

59. The first two conditions, $x \geq 0$ and $y \geq 0$, represent the set of points in the first quadrant. The third condition, $x + y \leq 8$, represents the set of points below and including the line $x + y = 8$ ($y = -x + 8$). The fourth condition, $3x + 5y \leq 30$, represents the set of points below and including the line $3x + 5y = 30$, $\left(y = -\frac{3}{5}x + 6 \right)$.

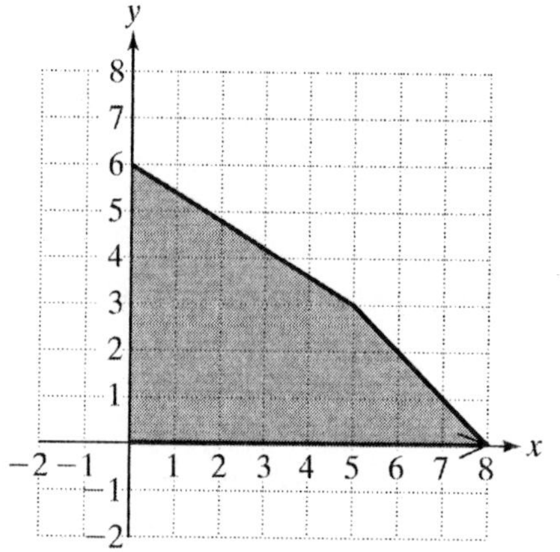

61. **(a)** $x \geq 0, y \geq 0$

(b) $x \leq 40, y \leq 40$

(c) $x + y \geq 65$

(d)

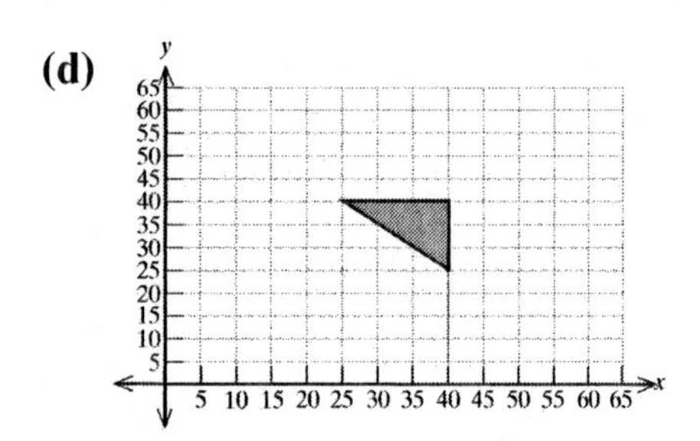

Chapter 9 Review Exercises

1. $4m > -11$ and $4m - 3 \le 13$

$m > -\frac{11}{4}$ and $4m \le 16$

$m > -\frac{11}{4}$ and $m \le 4$

Interval notation: $\left(-\frac{11}{4}, 4\right]$

3. $-3y + 1 \ge 10$ and $-2y - 5 \le -15$

$-3y \ge 9$ and $-2y \le -10$

$y \le -3$ and $y \ge 5$

No solution

5. $\frac{2}{3}t - 3 \le 1$ or $\frac{3}{4}t - 2 > 7$

$2t - 9 \le 3$ or $3t - 8 > 28$

$2t \le 12$ or $3t > 36$

$t \le 6$ or $t > 12$

Interval notation: $(-\infty, 6] \cup (12, \infty)$

7. $-7 < -7(2w + 3)$ or $-2 < -4(3w - 1)$

$-7 < -14w - 21$ or $-2 < -12w + 4$

$14 < -14w$ or $-6 < -12w$

$-1 > w$ or $\frac{1}{2} > w$

$w < -1$ or $w < \frac{1}{2}$

$w < \frac{1}{2}$

Interval notation: $\left(-\infty, \frac{1}{2}\right)$

9. $2 \ge -(b - 2) - 5b \ge -6$

$2 \ge -b + 2 - 5b \ge -6$

$2 \ge -6b + 2 \ge -6$

$0 \ge -6b \ge -8$

$0 \le b \le \frac{4}{3}$

Interval notation: $\left[0, \frac{4}{3}\right]$

11. Let n represent the number.

$-1 < \frac{1}{3}(n + 3) < 5$

$-1 < \frac{1}{3}n + 1 < 5$

$-3 < n + 3 < 15$

$-6 < n < 12$

All real numbers between -6 and 12.

13. **(a)** $125 \le x \le 200$

(b) $x < 125$ or $x > 200$

15. **(a)** The solution is the intersection of the two inequalities. Answer: $-2 \le x \le 5$

(b) The solution is the union of the two inequalities. Answer: All real numbers.

17. **(a)** $x^2 - 4 = 0$

$(x + 2)(x - 2) = 0$

$x + 2 = 0$ or $x - 2 = 0$

$x = -2$ or $x = 2$

(b) From (a) the boundary points are -2 or 2. Place the boundary points on a number line and test points within each interval.

Test point $x = -3$: $(-3)^2 - 4 < 0$

$9 - 4 < 0$

$5 < 0$ False

Test point $x = 0$: $(0)^2 - 4 < 0$

$0 - 4 < 0$

$-4 < 0$ True

Test point $x = 3$: $(3)^2 - 4 < 0$

$9 - 4 < 0$

$5 < 0$ False

Since the test point $x = 0$ is true the solution is $-2 < x < 2$. Interval notation: $(-2, 2)$. The boundary numbers are not included since the inequality "<" does not contain equality.

(c) From (a) the boundary points are -2 or 2. Place the boundary points on a number line and test points within each interval.

Test point $x = -3$: $(-3)^2 - 4 > 0$
$9 - 4 > 0$
$5 > 0$ True

Test point $x = 0$: $(0)^2 - 4 > 0$
$0 - 4 > 0$
$-4 > 0$ False

Test point $x = 3$: $(3)^2 - 4 > 0$
$9 - 4 > 0$
$5 > 0$ True

Since the test points $x = -3$ and $x = 3$ are true the solution is $x < -2$ or $x > 2$. Interval notation $(-\infty, -2) \cup (2, \infty)$. The boundary numbers are not included since the inequality ">" does not contain equality.

Part (a) represents the x-intercepts.
Part (b) represents the part of the graph below the x-axis.
Part (c) represents the part of the graph above the x-axis.

19. $2^2 - 4w - 12 < 0$
Find the boundary points.

$w^2 - 4w - 12 = 0$
$(w-6)(w+2) = 0$
$w - 6 = 0$ or $w + 2 = 0$
$w = 6$ or $w = -2$

The boundary points are 6 and -2.
Plot the boundary points on the number line and test a point from each region.

Test $w = -3$: $(-3)^2 - 4(-3) - 12 < 0$
$9 + 12 - 12 < 0$
$9 < 0$ False

Test $w = 0$: $(0)^2 - 4(0) - 12 < 0$
$0 - 0 - 12 < 0$
$-12 < 0$ True

Test $w = 7$: $(7)^2 - 4(7) - 12 < 0$
$49 - 28 - 12 < 0$
$9 < 0$ False

Since the test point $w = 0$ is true, the solution is $(-2, 6)$. The endpoints are not included in the solution because the symbol "<" does not include equality.

21. Find the boundary points.

$\frac{12}{x+2} = 6$
$12 = 6(x+2)$
$12 = 6x + 12$
$0 = 6x$
$0 = x$

The inequality is undefined for $x = -2$ and the solution to the related equation is 0. Therefore, the boundary points are -2 and 0. Plot the boundary points on the number line and test a point from each region.

Test $x = -3$: $\frac{12}{-3+2} > 6$
$-12 > 6$ False

Test $x = -1$: $\frac{12}{-1+2} > 6$
$12 > 6$ True

Test $x = 1$: $\frac{12}{1+2} > 6$
$4 > 6$ False

Since the test point $x = -1$ is true, the solution is $(-2, 0]$. The endpoint $x = 0$ is included in the solution because the symbol "≥" does include equality. The endpoint $x = -2$ is not included since the function is undefined for that value.

23. $3y(y-5)(y+2) = 0$

$3y = 0$ or $y - 5 = 0$ or $y + 2 = 0$
$y = 0$ or $y = 5$ or $y = -2$

Plot the boundary points on the number line and test a point from each region.

Test $y=-3$: $3(-3)(-3-5)(-3+2)>0$
$-9(-8)(-1)>0$
$-72>0$ False

Test $y=-1$: $3(-1)(-1-5)(-1+2)>0$
$-3(-6)(1)>0$
$18>0$ True

Test $y=1$: $3(1)(1-5)(1+2)>0$
$3(-4)(3)>0$
$-36>0$ False

Test $y=6$: $3(6)(6-5)(6+2)>0$
$18(1)(8)>0$
$144>0$ True

Since the test points $y=-1$ and $y=6$ are true, the solution is $(-2, 0) \cup (5, \infty)$. The endpoints are not included in the solution because the symbol ">" does not include equality.

25. $-x^2-4x \geq 4$
Find the boundary points.
$$-x^2-4x=4$$
$$-x^2-4x-4=0$$
$$-1(x^2+4x+4)=0$$
$$-1(x+2)(x+2)=0$$
$-1 \neq 0$ or $x+2=0$ or $x+2=0$
$x=-2$
The boundary point is -2.
Plot the boundary point on the number line and test a point from each region.

Test $x=-3$: $-(-3)^2-4(-3)>4$
$-9+12>4$
$3>4$ False

Test $x=0$: $-(0)^2-4(0)>4$
$0-0>4$
$0>4$ False

Since both intervals are false but the symbol "≥" does include equality, the solution is $\{-2\}$.

27. Find the boundary points.
$$\frac{w+1}{w-3}=1$$
$$w+1=1(w-3)$$
$$w+1=w-3$$
$$1 \neq -3$$
The inequality is undefined for $w=3$ but there is no solution to the related equation. Therefore, the boundary point is 3.
Plot the boundary points on the number line and test a point from each region.

Test $w=0$: $\frac{0+1}{0-3}>1$
$-\frac{1}{3}>1$ False

Test $w=4$: $\frac{4+1}{4-3}>1$
$5>1$ True

Since the test point $w=4$ is true, the solution is $(3, \infty)$. The endpoint $w=3$ is not included since the function is undefined for that value.

29. $t^2+10t+25 \leq 0$
Find the boundary points.
$$t^2+10t+25=0$$
$$(t+5)(t+5)=0$$
$t+5=0$ or $t+5=0$
$t=-5$
The boundary point is -5.
Plot the boundary point on the number line and test a point from each region.

Test $t=-6$: $(-6)^2+10(-6)+25<0$
$36-60+25<0$
$1<0$ False

Test $t=0$: $(0)^2+10(0)+25<0$
$0+0+25<0$
$25<0$ False

Since both intervals are false but the symbol "≤" does include equality, the solution is $\{-5\}$.

31. $|x|=10$
$x=-10$ or $x=10$

33. $|y+6|=\frac{1}{2}$
$y+6=-\frac{1}{2}$ or $y+6=\frac{1}{2}$
$y=-\frac{13}{2}$ or $y=-\frac{11}{2}$

35. $|8.7-2x|=6.1$

$8.7-2x=-6.1$ or $8.7-2x=6.1$

$-2x=-14.8$ or $-2x=-2.6$

$x=7.4$ or $x=1.3$

37. $16=|x+2|+9$

$7=|x+2|$

$x+2=-7$ or $x+2=7$

$x=-9$ or $x=5$

39. $|4x-1|+6=4$

$|4x-1|=-2$

No solution

41. $|7x-3|=0$

$7x-3=0$

$7x=3$

$x=\frac{3}{7}$

43. $|3x-5|=|2x+1|$

$3x-5=-(2x+1)$ or $3x-5=2x+1$

$3x-5=-2x-1$ or $x-5=1$

$5x-5=-1$ or $x=6$

$5x=4$ or $x=6$

$x=\frac{4}{5}$ or $x=6$

45. Both expressions give the distance between 3 and −2 on the number line.

$|3-(-2)|=|3+2|=|5|=5$; $|-2-3|=|-5|=5$

47. $|x|<4$

49. $|x+6|\geq 8$

$x+6\leq -8$ or $x+6\geq 8$

$x\leq -14$ or $x\geq 2$

Interval notation: $(-\infty, -14] \cup [2, \infty)$

−14 2

51. $|7x-1|>0$

$7x-1<0$ or $7x-1>0$

$7x<1$ or $7x>1$

$x<\frac{1}{7}$ or $x>\frac{1}{7}$

Interval notation: $\left(-\infty, \frac{1}{7}\right)\cup\left(\frac{1}{7}, \infty\right)$

$\frac{1}{7}$

53. $|3x+4|-6\leq -4$

$|3x+4|\leq 2$

$-2\leq 3x+4\leq 2$

$-6\leq 3x\leq -2$

$-2\leq x\leq -\frac{2}{3}$

Interval notation: $\left[-2, -\frac{2}{3}\right]$

-2 $-\frac{2}{3}$

55. $\left|\frac{x}{2}-6\right|<5$

$-5<\frac{x}{2}-6<5$

$-10<x-12<10$

$2<x<22$

Interval notation: (2, 22)

2 22

57. $|2x-4|+2>8$

$|2x-4|>6$

$2x-4<-6$ or $2x-4>6$

$2x<-2$ or $2x>10$

$x<-1$ or $x>5$

Interval notation: $(-\infty, -1) \cup (5, \infty)$

−1 5

59. $|5.2x-7.8|>-13$

The absolute value of any real number is nonnegative. The solution is all real numbers.

Interval notation: $(-\infty, \infty)$

61. $|3x-8|<-1$

No solution

63. If an absolute value is less than a negative number there will be no solution.

65. Graph the related equation $2x + y = 5$.

$y = -2x + 5$

with y-intercept (0, 5) and slope -2.

Since the given symbol "<" contains no equality symbol, use a dashed boundary line. Test points on each side of the line to decide whether to shade above or below the line.

Test Point Above: (0, 7)

$2x+y<5$

$2(0)+7<5$

$7<5$ False

The test point (0, 7) is not a solution to the original inequality.

Test Point Below: (0, 0)

$2x+y<5$

$2(0)+0<5$

$0<5$ True

The test point (0, 0) is a solution to the original inequality. Shade the region below the boundary.

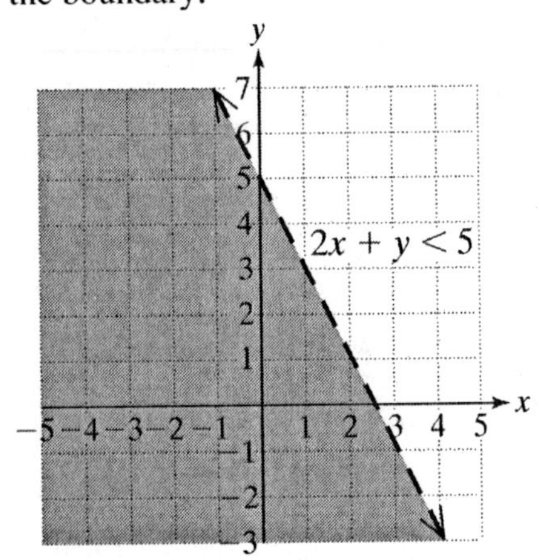

67. Graph the related equation $y=-\frac{2}{3}x+3$

with y-intercept (0, 3) and slope $-\frac{2}{3}$.

Since the given symbol "≥" contains an equality symbol, use a solid boundary line. Test points on each side of the line to decide whether to shade above or below the line.

Test Point Above: (0, 4)

$y>-\frac{2}{3}x+3$

$4>-\frac{2}{3}(0)+3$

$4>3$ True

The test point (0, 4) is a solution to the original inequality. Since the test point above is true, any test point below would result in a false statement. Shade above the line.

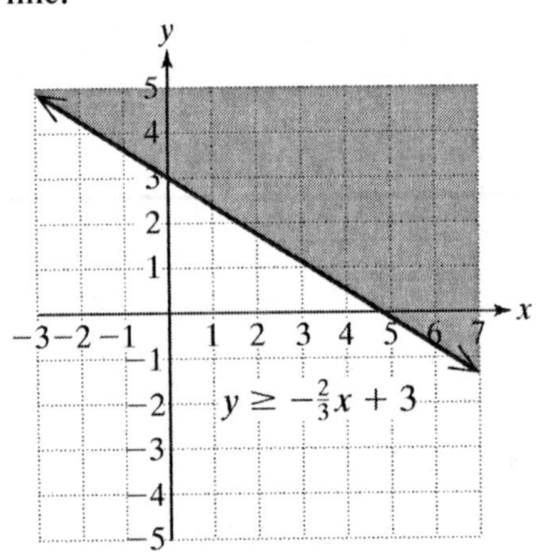

69. Graph the related equation $x=-3$ with no y-intercept and undefined slope. Since the given symbol ">" contains no equality symbol, use a dashed boundary line. Test points on each side of the line to decide whether to shade to the right or to the left of the line.

Test Point to Right: (0, 0)

$x>-3$

$0>-3$ True

The test point (0, 0) is a solution to the original inequality. Since the test point to the right is true, any test point to the left would result in a false statement. Shade to the right of the line.

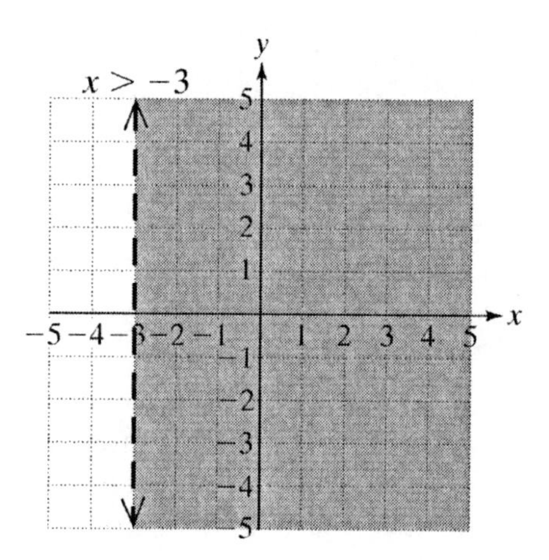

71. Graph the related equation $y = 4\frac{1}{3}$ with y-intercept $\left(0,\ 4\frac{1}{3}\right)$ and slope 0.
Since the given symbol "<" contains no equality symbol, use a dashed boundary line. Test points on each side of the line to decide whether to shade above or below the line.
Test Point Above: (0, 5)

$y < 4\frac{1}{3}$

$5 < 4\frac{1}{3}$ False

The test point (0, 5) is not a solution to the original inequality.
Test Point to Below: (0, 0)

$y < 4\frac{1}{3}$

$0 < 4\frac{1}{3}$ True

The test point (0, 0) is a solution to the original inequality. Shade the region below the boundary.

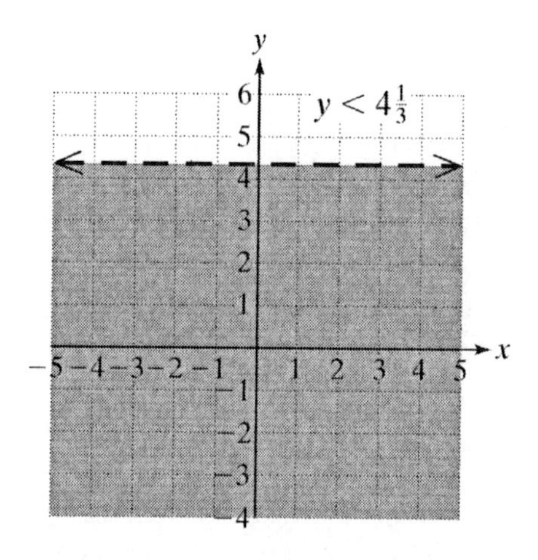

73. Graph the related equation $y = 2x$ with y-intercept (0, 0) and slope 2.
Since the given symbol "≤" contains the equality symbol, use a solid boundary line. Test points on each side of the line to decide whether to shade above or below the line.
Test Point Above: (0, 2)

$y < 2x$

$2 < 2(0)$

$2 < 0$ False

The test point (0, 2) is not a solution to the original inequality.
Test Point Below: (0, −1)

$y < 2x$

$-1 < 2(0)$

$-1 < 0$ True

The test point (0, −1) is a solution to the original inequality. Shade the region below the boundary.

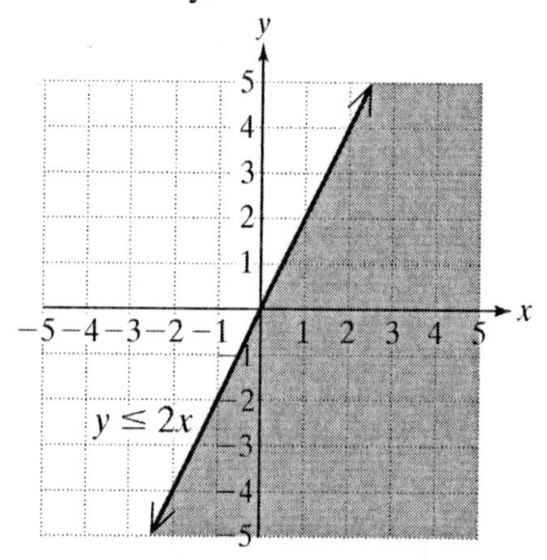

75. Solve $2x - y > -2$ and $2x - y \le 2$.

First Inequality

$2x - y > -2$

$2x - y = -2$ (related equation)

$-y = -2x - 2$

$y = 2x + 2$

Second Inequality

$2x - y \le 2$

$2x - y = 2$ (related equation)

$-y = -2x + 2$

$y = 2x - 2$

For each inequality, draw the boundary line. Then pick test points above and below the line to determine the appropriate region to shade.

Test Point Above the line $2x - y = -2$: (0, 3)

$2x - y > -2$

$2(0) - 3 > -2$

$-3 > -2$ False

The test point (0, 3) is not a solution to the inequality.

Test Point Below the line $2x - y = -2$: (0, 0)

$2x - y > -2$

$2(0) - 0 > -2$

$0 > -2$ True

The test point (0, 0) is a solution to the inequality. Shade the region below the boundary.

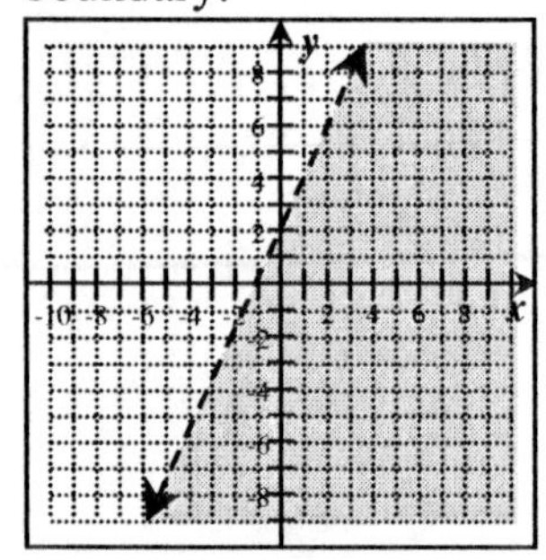

Test Point Above the line $2x - y = 2$: (0, 0)

$2x - y < 2$

$2(0) - 0 < 2$

$0 < 2$ True

The test point (0, 0) is a solution to the inequality. Shade the region above the boundary. Since the test point above is true, any test point below would result in a false statement.

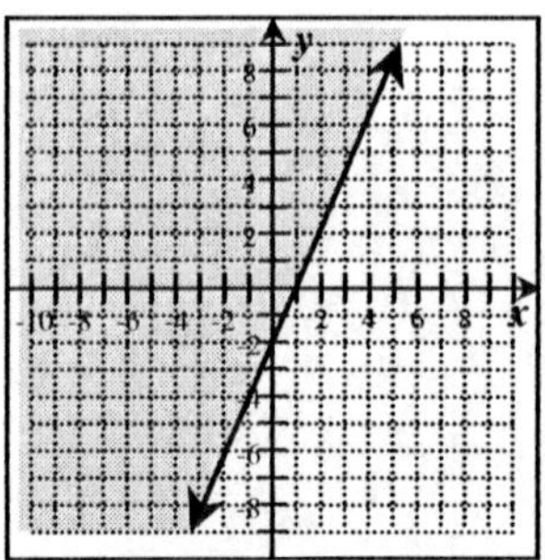

The solution is the intersection of the two individual solution sets. Therefore, the solution is the region of the plane below the line $2x - y = -2$ and above the line $2x - y = 2$ as shaded.

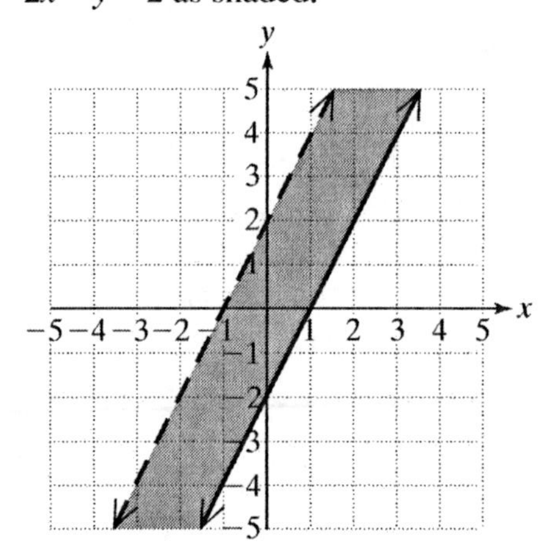

77. The two conditions, $x \ge 0$ and $y \ge 0$, represent the set of points in the first quadrant. The third condition, $y \ge -\frac{3}{2}x + 4$, represents the set of points above and including the line $y = -\frac{3}{2}x + 4$.

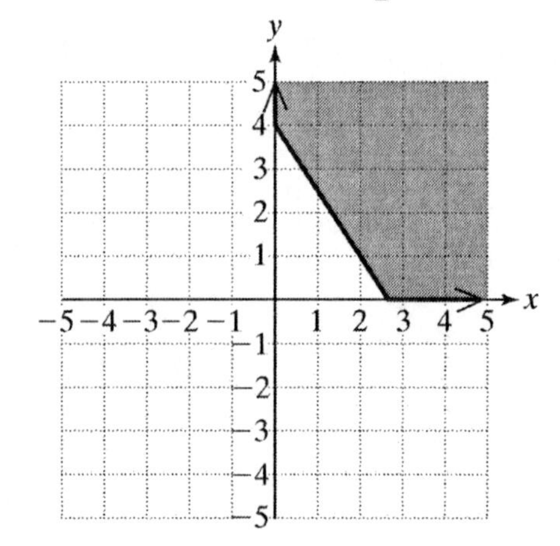

79. The condition $y \geq 0$ represents the set of points in the first and fourth quadrants. The second condition, $-2x + y \leq 4$, represents the set of points below and including the line $y = 2x + 4$. The third condition, $y \leq -x + 6$, represents the set of points below and including the line $y = -x + 6$.

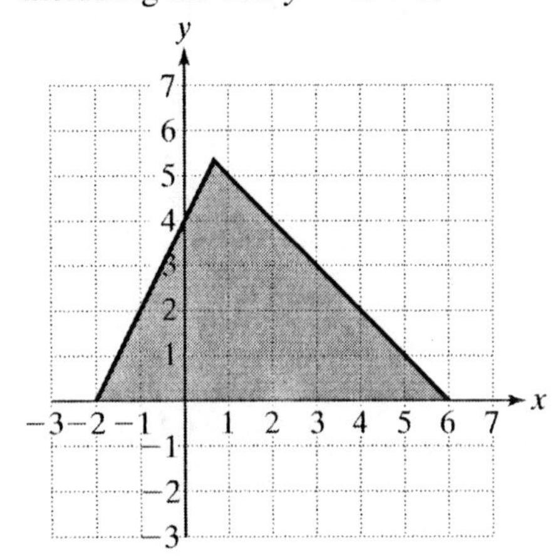

Chapter 9 Test

1. (a) $-2 \leq 3x - 1 \leq 5$

$$-1 \leq 3x \leq 6$$

$$-\frac{1}{3} \leq x \leq 2$$

Interval notation: $\left[-\frac{1}{3}, 2\right]$

(b) $-\frac{3}{5}x - 1 \leq 8$ or $-\frac{2}{3}x \geq 16$

$-\frac{3}{5}x \leq 9$ or $x \leq -24$

$x \geq -15$ or $x \leq -24$

Interval notation: $(-\infty, -24] \cup [-15, \infty)$

3. (a) $9 \leq x \leq 33$

(b) $x < 9$ or $x > 33$

(c) $x \geq 40$

5. $50 - 2a^2 > 0$

Find the boundary points:

$$50 - 2a^2 = 0$$

$$2(25 - a^2) = 0$$

$$2(5 - a)(5 + a) = 0$$

$2 \neq 0$ or $5 - a = 0$ or $5 + a = 0$

$5 = a$ or $a = -5$

The boundary points are –5 and 5.

Plot the boundary points on the number line and test a point from each region.

Test $a = -6$: $50 - 2(-6)^2 > 0$

$$50 - 72 > 0$$

$$-22 > 0 \quad \text{False}$$

Test $a = 0$: $50 - 2(0)^2 > 0$

$$50 - 0 > 0$$

$$50 > 0 \quad \text{True}$$

Test $a = 6$: $50 - 2(6)^2 > 0$

$$50 - 72 > 0$$

$$-22 > 0 \quad \text{False}$$

Since the test point $a = 0$ is true, the solution is (–5, 5). The endpoints are not included in the solution because the symbol “>” does not include equality.

7. $\frac{3}{w+3} > 2$

Find the boundary points:

$$\frac{3}{w+3} = 2$$

$$3 = 2(w + 3)$$

$$3 = 2w + 6$$

$$-3 = 2w$$

$$-\frac{3}{2} = w$$

The inequality is undefined for $w = -3$ and the solution to the related equation is $-\frac{3}{2}$.

Therefore, the boundary points are –3 and $-\frac{3}{2}$.

Plot the boundary points on the number line and test a point from each region.

Test $w = -4$: $\frac{3}{-4+3} > 2$

$$-3 > 2 \quad \text{False}$$

Test $w = -2$: $\frac{3}{-2+3} > 2$

$3 > 2$ True

Test $w = 0$: $\frac{3}{0+3} > 2$

$1 > 2$ False

Since the test point $w = -2$ is true, the solution is $\left(-3, -\frac{3}{2}\right)$. The endpoint $-\frac{3}{2}$ is not included in the solution because the symbol ">" does not include equality. Since the inequality is not defined for $w = -3$, it is not included in the solution.

9. $t^2 + 22t + 121 \leq 0$

Find the boundary points:

$t^2 + 22t + 121 = 0$

$(t+11)(t+11) = 0$

$t + 11 = 0$

$t = -11$

The boundary point is -11.

Plot the boundary point on the number line and test a point from each region.

Test $t = -12$:

$(-12)^2 + 22(-12) + 121 < 0$

$144 - 264 + 121 < 0$

$1 < 0$ False

Test $t = 0$: $(0)^2 + 22(0) + 121 < 0$

$0 + 0 + 121 < 0$

$121 < 0$ False

Since no test point is true, the solution is $\{-11\}$ since the symbol "≤" does include equality.

11. (a) $|x-3| - 4 = 0$

$|x-3| = 4$

$x - 3 = -4$ or $x - 3 = 4$

$x = -1$ or $x = 7$

(b) $|x-3| - 4 < 0$

$|x-3| < 4$

$-4 < x - 3 < 4$

$-1 < x < 7$

(c) $|x-3| - 4 > 0$

$|x-3| > 4$

$x - 3 < -4$ or $x - 3 > 4$

$x < -1$ or $x > 7$

Part (a) represents the boundary, Part (b) represents the portion of the graph below the x-axis, and Part (c) represents the portion of the graph above the x-axis.

13. $|3x - 8| > 9$

$3x - 8 < -9$ or $3x - 8 > 9$

$3x < -1$ or $3x > 17$

$x < -\frac{1}{3}$ or $x > \frac{17}{3}$

Interval notation: $\left(-\infty, -\frac{1}{3}\right) \cup \left(\frac{17}{3}, \infty\right)$

15. $|7 - 3x| + 1 > -3$

$|7 - 3x| > -4$

Interval notation: $(-\infty, \infty)$

17. $2x - 5y \leq 10$

Graph the related equation $2x - 5y = 10$ using the intercepts (5, 0) and (0, –2). Since the inequality symbol "≤" contains the equality symbol, use a solid boundary line. Test points on each side of the boundary line to decide whether to shade above or below the line.

Test point above: (0, 0)

$2x - 5y \leq 10$

$2(0) - 5(0) \leq 10$

$0 \leq 10$ True

Test point below: (6, 0)

$2x - 5y \leq 10$

$2(6) - 5(0) \leq 10$

$12 \leq 10$ False

Since the test point (0, 0) above the boundary is true in the original inequality, shade the region above the line.

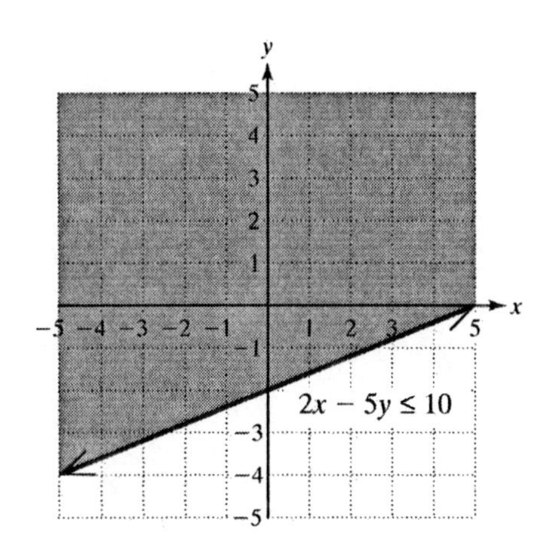

19. **(a)** $x \ge 0, y \ge 0$

(b) $300x + 400y \ge 1000$

(c)

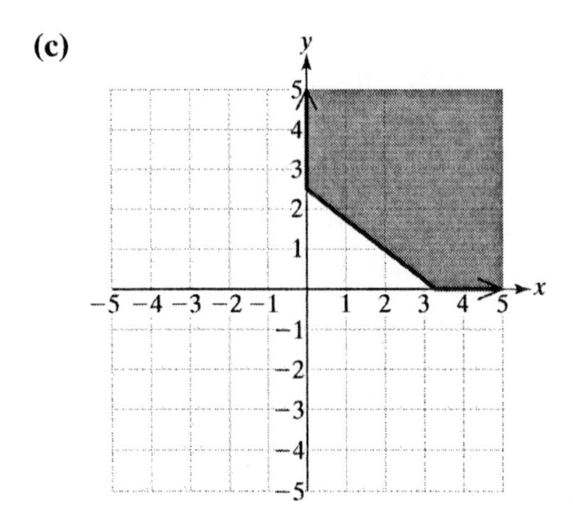

Cumulative Review Exercises, Chapters 1–9

1. $(2x-3)(x-4)-(x-5)^2$
$= 2x^2 - 11x + 12 - (x^2 - 10x + 25)$
$= 2x^2 - 11x + 12 - x^2 + 10x - 25$
$= x^2 - x - 13$

3. **(a)** $2|3-p|-4=2$
$2|3-p|=6$
$|3-p|=3$
$3-p=-3$ or $3-p=3$
$-p=-6$ or $-p=0$
$p=6$ or $p=0$

(b) $2|3-p|-4<2$
$2|3-p|<6$
$|3-p|<3$
$-3<3-p<3$
$-6<-p<0$
$6>p>0$
Interval notation: $(0, 6)$

(c) $2|3-p|-4>2$
$2|3-p|>6$
$|3-p|>3$
$3-p<-3$ or $3-p>3$
$-p<-6$ or $-p>0$
$p>6$ or $p<0$
$0<p<6$
Interval notation: $(-\infty, 0) \cup (6, \infty)$

5. Graph the related equation $4x - y = 12$.
$-y = -4x + 12$
$y = 4x - 12$
with y-intercept $(0, -12)$ and slope 4.
Since the given symbol ">" contains no equality symbol, use a dashed boundary line. Test points on each side of the line to decide whether to shade above or below the line.
Test Point Above: (0, 0)
$4x - y > 12$
$4(0) - 0 > 12$
$0 > 12$ False
The test point (0, 0) is not a solution to the original inequality.
Test Point Below: (4, 0)
$4x - y > 12$
$4(4) - 0 > 12$
$16 > 12$ True
The test point (4, 0) is a solution to the original inequality. Shade the region below the boundary.

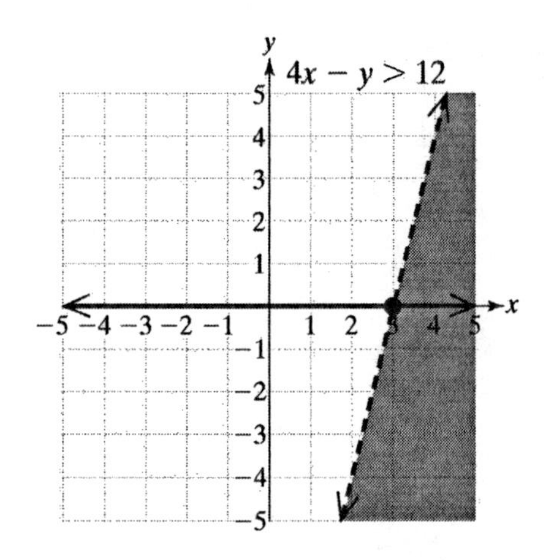

7. (a) $2x^2 + x - 10 \geq 0$

Find the boundary points:

$$2x^2 + x - 10 = 0$$
$$(2x+5)(x-2) = 0$$
$$2x+5=0 \quad \text{or} \quad x-2=0$$
$$2x=-5 \quad \text{or} \quad x=2$$
$$x=-\frac{5}{2} \quad \text{or} \quad x=2$$

The boundary points are $-\frac{5}{2}$ and 2.

Plot the boundary points on the number line and test a point from each region.

<u>Test $x=-3$</u>: $2(-3)^2 + (-3) - 10 > 0$
$$18 - 3 - 10 > 0$$
$$5 > 0 \quad \text{True}$$

<u>Test $x=0$</u>: $2(0)^2 + (0) - 10 > 0$
$$0 + 0 - 10 > 0$$
$$-10 > 0 \quad \text{False}$$

<u>Test $x=3$</u>: $2(3)^2 + (3) - 10 > 0$
$$18 + 3 - 10 > 0$$
$$11 > 0 \quad \text{True}$$

Since the test points $x=-3$ and $x=3$ are true, the solution is $\left(-\infty, -\frac{5}{2}\right] \cup [2, \infty)$. The endpoints are included in the solution because the symbol "≥" does include equality.

(b) The solution represents the portions of the graph that are above the x-axis.

9.
$$\begin{aligned} 2-3(x-5)+2[4-(2x+6)] &= 2-3(x-5)+2[4-2x-6] \\ &= 2-3(x-5)+2(-2x-2) \\ &= 2-3x+15-4x-4 \\ &= (-3x-4x)+(2+15-4) \\ &= -7x+13 \end{aligned}$$

11. (a)

$$\begin{array}{r} 2x^2 - 5x + 12 \\ x^2 + 2x - 1 \overline{\smash{)}\, 2x^4 - x^3 + 0x^2 + 5x - 7} \\ \underline{-(2x^4 + 4x^3 - 2x^2)} \\ -5x^3 + 2x^2 + 5x \\ \underline{-(-5x^3 - 10x^2 + 5x)} \\ 12x^2 + 0x - 7 \\ \underline{-(12x^2 + 24x - 12)} \\ -24x + 5 \end{array}$$

(b) $(x^2 + 2x - 1)(2x^2 - 5x + 12) + (-24x + 5) = 2x^4 - x^3 + 5x - 7$

(c) No, the remainder is not 0.

13.

$$r = \frac{k}{t}$$
$$60 = \frac{k}{10}$$
$$600 = k$$
$$r = \frac{600}{t}$$
$$r = \frac{600}{8}$$
$$r = 75 \text{ mph}$$

15. Let x represent the amount invested at 5% interest.
Let y represent the amount invested at 6.5% interest.

$$\begin{cases} x = y - 3000 \\ 0.05x + 0.065y = 770 \end{cases}$$

Substitute $y - 3000$ in the second equation for x.

$$0.05(y - 3000) + 0.065y = 770$$
$$0.05y - 150 + 0.065y = 770$$
$$0.115y - 150 = 770$$
$$0.115y = 20$$

$y = 8000 \Rightarrow x = 8000 - 3000 = 5000$

\$8000 is invested at 6.5%, \$5000 is invested at 5%.

17. (a)

$$3x + 5 = 8$$
$$3x = 3$$
$$x = 1$$

This is a vertical line with x-intercept (1, 0). There is no y-intercept. The slope is undefined.

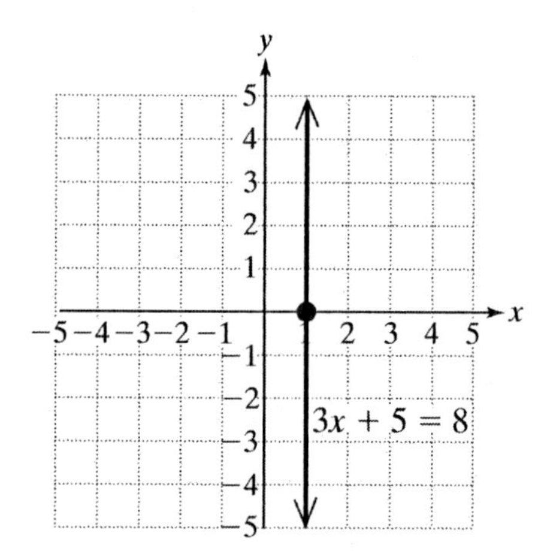

(b) $\frac{1}{2}x + y = 4$

$$y = -\frac{1}{2}x + 4$$

To find the x-intercept let $y = 0$ in the original equation, $\frac{1}{2}x + 0 = 4$, $\frac{1}{2}x = 4$, $x = 8$.

The x-intercept is (8, 0).

Compare the equation to $y = mx + b$.

The y-intercept is (0, 4).

The slope is $-\frac{1}{2}$.

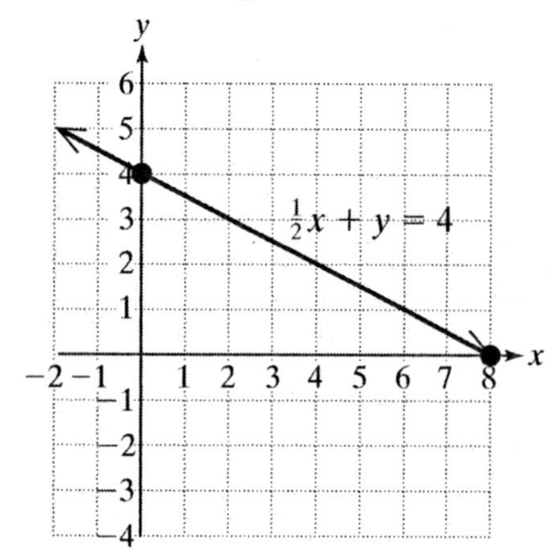

19. A: $3x + y = z + 2$
B: $y = 1 - 2x$
C: $3z = -2y$

A: $3x + y - z = 3$
B: $2x + y = 1$
C: $2y + 3z = 0$

Eliminate the y variable from equations A and B by multiplying equation A by −1 and then adding.

$$\begin{array}{r} -3x - y + z = -2 \\ 2x + y = 1 \\ \hline -x + z = -1 \end{array} \quad \text{Equation D}$$

Eliminate the y variable from equations A and C by multiplying equation A by −2.

$$\begin{array}{r} -6x - 2y + 2z = -4 \\ 2y + 3z = 0 \\ \hline -6x + 5z = -4 \end{array} \quad \text{Equation E}$$

Eliminate the x variable from equations D and E by multiplying equation D by −6.

$$\begin{array}{r} 6x - 6z = 6 \\ -6x + 5z = -4 \\ \hline -z = 2 \\ z = -2 \end{array}$$

Substitute $z = -2$ into equation C and solve for y.

$$3(-2) = -2y$$
$$-6 = -2y$$
$$3 = y$$

Substitute $y = 3$ into equation B to solve for x.

$$3 = 1 - 2x$$
$$2 = -2x$$
$$-1 = x$$

The solution to the system is (−1, 3, −2).

21. Let x represent the speed of the plane in still air.

Let y represent the wind speed.

$$\begin{cases} 6240 = 13(x - 6) \\ 6240 = 12(x + y) \end{cases}$$

$$\begin{cases} 6240 = 13x - 13y \\ 6240 = 12x + 12y \end{cases}$$

Multiply the first equation by 12 and the second equation by 13 to eliminate y.

$$\begin{cases} 74880 = 156x - 156y \\ 81120 = 156x + 156y \end{cases}$$

Adding the two equations gives
$156000 = 312x$.
$x = 500$ mph.
Thus, $6240 = 13(500 - y)$.

$$6240 = 6500 - 13y$$
$$-260 = -13y$$
$$20 = y$$

The wind speed is 20 mph. The speed of the plane in still air is 500 mph.

23. $50 - x \geq 0$

$$-x \geq -50$$
$$x \leq 50$$

Set notation: $\{x | x \leq 50\}$
Interval notation: $(-\infty, 50]$

25. $\dfrac{a^3 + 64}{16 - a^2} \div \dfrac{a^3 - 4a^2 + 16a}{a^2 - 3a - 4}$

$$= \frac{a^3 + 64}{16 - a^2} \cdot \frac{a^2 - 3a - 4}{a^3 - 4a^2 + 16a}$$
$$= \frac{(a+4)(a^2 - 4a + 16)}{(4+a)(4-a)} \cdot \frac{(a-4)(a+1)}{a(a^2 - 4a + 16)}$$
$$= -\frac{a+1}{a}$$

Chapter 10

Section 10.1 Practice Exercises

1. **(a)** 8, –8

(b) 8

(c) There are two square roots for every positive number. $\sqrt{64}$ identifies the positive square root.

3. **(a)** 9

(b) –9

5. There is no real number b such that $b^2 = -36$.

7. $\sqrt{25} = 5$

9. $\sqrt[3]{-27} = -3$

11. $\sqrt[4]{16} = 2$

13. $\sqrt[3]{\frac{1}{8}} = \frac{1}{2}$

15. $\sqrt[6]{64} = 2$

17. $\sqrt[3]{64} = 4$

19. $\sqrt[4]{-81}$; not a real number.

21. $\sqrt[6]{1{,}000{,}000} = 10$

23. $-\sqrt[3]{0.008} = -0.2$

25. $-\sqrt{0.0144} = -0.12$

27. $\sqrt{69} \approx 8.3066$

29. $7\sqrt[4]{25} \approx 15.6525$

31. $2 + \sqrt[3]{5} \approx 3.7100$

33. $\dfrac{3-\sqrt{19}}{11} \approx -0.1235$

35. **(a)** $h(-1) = \sqrt{-1-2} = \sqrt{-3}$; not a real number.

(b) $h(0) = \sqrt{0-2} = \sqrt{-2}$; not a real number.

(c) $h(1) = \sqrt{1-2} = \sqrt{-1}$; not a real number.

(d) $h(2) = \sqrt{2-2} = 0$

(e) $h(3) = \sqrt{3-2} = 1$

(f) $h(4) = \sqrt{4-2} \approx 1.41$

(g) $h(5) = \sqrt{5-2} \approx 1.73$

(h) $h(6) = \sqrt{6-2} = 2$

Domain: $[2, \infty)$

37. **(a)** $g(-1) = \sqrt[3]{-1-2} \approx -1.44$

(b) $g(0) = \sqrt[3]{0-2} \approx -1.26$

(c) $g(1) = \sqrt[3]{1-2} = -1$

(d) $g(2) = \sqrt[3]{2-2} = 0$

(e) $g(3) = \sqrt[3]{3-2} = 1$

(f) $g(4) = \sqrt[3]{4-2} \approx 1.26$

(g) $g(5) = \sqrt[3]{5-2} \approx 1.44$

(h) $g(6) = \sqrt[3]{6-2} \approx 1.59$

Domain: $(-\infty, \infty)$

39. $x - 1 \geq 1$

$x \geq 1$

$[1, \infty)$

41. Since the index is odd, the domain is all real numbers; $(-\infty, \infty)$.

43. b

45. d

47. $|a|$

49. a

51. $|a|$

53. x^2

55. $|xz^5|y^2$

57. $-\frac{x}{y}$

59. $\frac{2}{|x|}$

61. -9

63. -2

65. xy^2

67. $\frac{a^3}{b}$

69. $-\frac{5}{q}$

71. $3xy^2z$

73. $\frac{hk^2}{4}$

75. $-\frac{t}{3}$

77. $2y^2$

79. $2p^2q^3$

81. $q+p^2$

83. $\frac{6}{\sqrt[4]{x}}$

85. The sum of the square of a and the square root of b.

87. The quotient of 1 and the square of the quantity c plus d.

89.
$$A = s^2$$
$$64 = s^2$$
$$\sqrt{64} = \sqrt{s^2}$$
$$8 = s$$
The length of the sides is 8 in.

91.
$$A = s^2$$
$$97 = s^2$$
$$\sqrt{97} = \sqrt{s^2}$$
$$9.8 \approx s$$
The length of the sides is 9.8 in.

93.
$$a^2 + b^2 = c^2$$
$$12^2 + b^2 = 15^2$$
$$144 + b^2 = 225$$
$$b^2 = 81$$
$$b = 9$$
The length of the third side is 9 cm.

95.
$$a^2 + b^2 = c^2$$
$$5^2 + 12^2 = c^2$$
$$25 + 144 = c^2$$
$$169 = c^2$$
$$13 = c$$
The length of the third side is 13 ft.

97.
$$(a+b)^2 = c^2 + 4\left(\frac{1}{2}ab\right)$$
$$a^2 + 2ab + b^2 = c^2 + 2ab$$
$$a^2 + b^2 = c^2$$

99.

101. 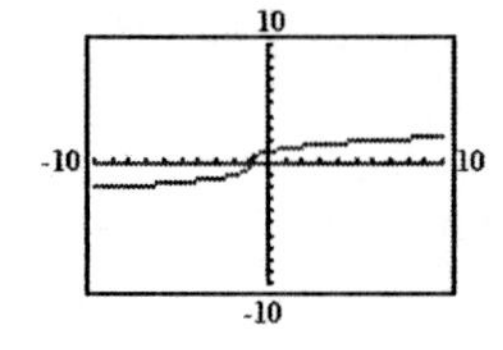

Section 10.2 Practice Exercises

1. **(a)** 3

(b) 27

3. 5

5. 3

7. Not a real number.

9. $a+1$

11. The numerator of the exponent represents the power of the base. The denominator of the exponent represents the index of the root. $a^{m/n}=\sqrt[n]{a^m}$

13. $\sqrt{25}=5$

15. $\sqrt[3]{8}=2$

17. $\sqrt[4]{81}=3$

19. $\sqrt[3]{-8}=-2$

21. $-\sqrt[3]{8}=-2$

23. $\left(\frac{1}{4}\right)^{1/2}=\sqrt{\frac{1}{4}}=\frac{1}{2}$

25. $\left(\frac{1}{27}\right)^{2/3}=\left(\sqrt[3]{\frac{1}{27}}\right)^2=\left(\frac{1}{3}\right)^2=\frac{1}{9}$

27. $(36)^{1/2}=\sqrt{36}=6$

29. $(1000)^{1/3}=\sqrt[3]{1000}=10$

31. $\sqrt[3]{\frac{1}{8}}^{\,2}+\sqrt{\frac{1}{4}}=\frac{1}{2}^2+\frac{1}{2}=\frac{1}{4}+\frac{2}{4}=\frac{3}{4}$

33. $16^{1/4}-49^{1/2}=\sqrt[4]{16}-\sqrt{49}=2-7=-5$

35. $\sqrt{\frac{1}{4}}+64^{1/3}=\sqrt{\frac{1}{4}}+\sqrt[3]{64}=\frac{1}{2}+4=\frac{9}{2}$

37. $x^{\frac{1}{4}+\frac{3}{4}}=x^{4/4}=x$

39. $p^{\frac{5}{3}-\frac{2}{3}}=p^{3/3}=p$

41. $y^{10/5}=y^2$

43. $6^{-\frac{1}{5}+\frac{6}{5}}=6^{5/5}=6$

45. $4\left(t^{\frac{1}{2}-\left(-\frac{1}{2}\right)}\right)=4\left(t^{\frac{1}{2}+\frac{1}{2}}\right)=4t$

47. $a^4\cdot a^3=a^{4+3}=a^7$

49. $5^2a^4c^{-1}d=\frac{25a^4d}{c}$

51. $\frac{x^{-8}}{y^{-9}}=\frac{y^9}{x^8}$

53. $(8w^{-3}z^9)^{1/3}=8^{1/3}w^{-1}z^3=\frac{2z^3}{w}$

55. $25^{1/2}xy^2z^3=5xy^2z^3$

57. $\frac{x^{24}y^{-4}z^8}{x^8y^3z^{12}}=x^{16}y^{-7}z^{-4}=\frac{x^{16}}{y^7z^4}$

59. $\frac{x^3y^2}{z^5}$

61. $\sqrt[3]{2x^2y}$

63. $\sqrt{\dfrac{2}{y}}$

65. $x^{1/3}$

67. $5x^{1/2}$

69. 3

71. 0.3761

73. 2.9240

75. 31.6228

77. (a) $s = A^{1/2}$
$s = 100^{1/2}$
$s = 10$
The length of the sides is 10 inches.

(b) $s = A^{1/2}$
$s = 72^{1/2}$
$s \approx 8.5$
The length of the sides is approximately 8.5 inches.

79. $r = \left(\dfrac{A}{P}\right)^{1/t} - 1$

(a) $r = \left(\dfrac{16802}{10000}\right)^{1/5} - 1$
$r = (1.6802)^{1/5} - 1$
$r \approx 1.109 - 1$
$r \approx 0.109$
The interest rate is approximately 10.9%.

(b) $r = \left(\dfrac{18000}{10000}\right)^{1/7} - 1$
$r = (1.8)^{1/7} - 1$
$r \approx 1.088 - 1$
$r \approx 0.088$
The interest rate is approximately 8.8%.

(c) The account in part (a) produced a higher average yearly return.

81. $\sqrt[6]{x}$

83. $\sqrt[8]{y}$

85. $\sqrt[15]{w}$

Section 10.3 Practice Exercises

1. $(ab^{-2}) \cdot \left(\dfrac{a}{b^{-3}}\right) = a^2 b$

3. $r^{-2}s^6 = \dfrac{s^6}{r^2}$

5. $\sqrt[7]{x^4}$

7. $y^{9/2}$

9. (a) $\dfrac{2-\sqrt{y}}{4} = \dfrac{2-\sqrt{3}}{4} \approx 0.07$

(b) $\dfrac{2-\sqrt{y}}{4} = \dfrac{2-\sqrt{5}}{4} \approx -0.06$

11. $\sqrt{4x} \cdot \sqrt{25x} = \sqrt{100x^2} = 10x$

13. $\sqrt[3]{2x} \cdot \sqrt[3]{4x^2} = \sqrt[3]{8x^3} = 2x$

15. $\sqrt[4]{(x+2)^3} \cdot \sqrt[4]{(x+2)} = \sqrt[4]{(x+2)^4} = x+2$

17. $\sqrt[5]{2y^2} \cdot \sqrt[5]{16y^3} = \sqrt[5]{32y^5} = 2y$

19. $\dfrac{\sqrt{27y}}{\sqrt{3y}} = \sqrt{\dfrac{27y}{3y}} = \sqrt{9} = 3$

21. $\dfrac{\sqrt[4]{2^5 a^3 b^7}}{\sqrt[4]{2a^3b^3}} = \sqrt[4]{\dfrac{2^5 a^3 b^7}{2a^3b^3}} = \sqrt[4]{2^4 b^4} = 2b$

23. $\dfrac{\sqrt[3]{(3x+1)^5}}{\sqrt[3]{(3x+1)^2}} = \sqrt[3]{\dfrac{(3x+1)^5}{(3x+1)^2}}$
$= \sqrt[3]{(3x+1)^3}$
$= 3x+1$

25. $\dfrac{\sqrt{(a+b)}}{\sqrt{(a+b)^3}} = \sqrt{\dfrac{(a+b)}{(a+b)^3}} = \sqrt{\dfrac{1}{(a+b)^2}} = \dfrac{1}{a+b}$

27. $\sqrt{28} = \sqrt{4 \cdot 7} = \sqrt{4} \cdot \sqrt{7} = 2\sqrt{7}$

29. $\sqrt{80} = \sqrt{16 \cdot 5} = \sqrt{16} \cdot \sqrt{5} = 4\sqrt{5}$

31. $5\sqrt{18} = 5\sqrt{9 \cdot 2}$
$= 5\sqrt{9} \cdot \sqrt{2}$
$= 5(3) \cdot \sqrt{2}$
$= 15\sqrt{2}$

33. $\sqrt[3]{54} = \sqrt[3]{27 \cdot 2} = \sqrt[3]{27} \cdot \sqrt[3]{2} = 3\sqrt[3]{2}$

35. $\sqrt{25x^4y^3} = \sqrt{25x^4y^2 \cdot y}$
$= \sqrt{25x^4y^2} \cdot \sqrt{y}$
$= 5x^2y\sqrt{y}$

37. $\sqrt[3]{27x^2y^3z^4} = \sqrt[3]{27y^3z^3 \cdot x^2z}$
$= \sqrt[3]{27y^3z^3} \cdot \sqrt[3]{x^2z}$
$= 3yz\sqrt[3]{x^2z}$

39. $\sqrt[3]{\dfrac{16a^2b}{2a^2b^4}} = \sqrt[3]{\dfrac{8}{b^3}} = \dfrac{2}{b}$

41. $\sqrt[5]{\dfrac{32x}{y^{10}}} = \sqrt[5]{\dfrac{32}{y^{10}} \cdot x} = \dfrac{2\sqrt[5]{x}}{y^2}$

43. $\dfrac{\sqrt{50x^3y}}{\sqrt{9y^4}} = \sqrt{\dfrac{50x^3y}{9y^4}}$
$= \sqrt{\dfrac{25x^2}{9y^4} \cdot 2xy}$
$= \dfrac{5x\sqrt{2xy}}{3y^2}$

45. $\sqrt{2^3a^{14}b^8c^{31}d^{22}} = \sqrt{2^2a^{14}b^8c^{30}d^{22} \cdot 2c}$
$= 2a^7b^4c^{15}d^{11}\sqrt{2c}$

47. $\dfrac{1}{\sqrt[3]{w^6}} = \dfrac{1}{w^2}$

49. $\sqrt{k^3} = k\sqrt{k}$

51. $a^2 + b^2 = c^2$
$10^2 + 8^2 = c^2$
$100 + 64 = c^2$
$164 = c^2$
$\sqrt{164} = c$
$\sqrt{4 \cdot 41} = c$
$2\sqrt{41} = c$
The third side is $2\sqrt{41}$ ft.

53. $a^2 + b^2 = c^2$
$12^2 + b^2 = 18^2$
$144 + b^2 = 324$
$b^2 = 324 - 144$
$b^2 = 180$
$b = \sqrt{180}$
$b = \sqrt{36 \cdot 5}$
$b = 6\sqrt{5}$
The third side is $6\sqrt{5}$ meters.

55.
$$a^2+b^2=c^2$$
$$90^2+90^2=c^2$$
$$8100+8100=c^2$$
$$16200=c^2$$
$$\sqrt{16200}=c$$
$$127.3\approx c$$
The distance from home plate to second base is approximately 127.3 feet.

57.
$$a^2+b^2=c^2$$
$$40^2+b^2=50^2$$
$$1600+b^2=2500$$
$$b^2=2500-1600$$
$$b^2=900$$
$$b=\sqrt{900}$$
$$b=30$$
The path from A to B and B to C would take
$$\frac{70 \text{ miles}}{55 \text{ mph}}=1.3 \text{ hours.}$$
The path from A to C would take
$$\frac{50 \text{ miles}}{35 \text{ mph}}=1.4 \text{ hours.}$$
The path from A to B and B to C is faster.

Section 10.4 Practice Exercises

1. $\sqrt[3]{-16s^4t^9}=\sqrt[3]{-8s^3t^9\cdot 2s}=-2st^3\sqrt[3]{2s}$

3. $\sqrt{3p^2}\cdot\sqrt{12p^4}=\sqrt{36p^6}=6p^3$

5. $(4x^2)^{3/2}=\left(\sqrt{4x^2}\right)^3=(2x)^3=8x^3$

7. $y^{2/3}\cdot y^{1/4}=y^{\frac{2}{3}+\frac{1}{4}}=y^{11/12}$

9. $(2.718)^{2/3}\approx 1.95$

11. **(a)** Both expressions can be simplified using the distributive property.
$7\sqrt{5}+4\sqrt{5}=(7+4)\sqrt{5}=11\sqrt{5}$
$7x+4x=(7+4)x=11x$

(b) Neither expression can be simplified because they do not contain like terms or like radicals.

13. $3\sqrt{5}+6\sqrt{5}=(3+6)\sqrt{5}=9\sqrt{5}$

15. $3\sqrt[3]{t}-2\sqrt[3]{t}=(3-2)\sqrt[3]{t}=\sqrt[3]{t}$

17. $6\sqrt{10}-\sqrt{10}=(6-1)\sqrt{10}=5\sqrt{10}$

19.
$$\sqrt[4]{3}+7\sqrt[4]{3}-\sqrt[4]{14}=(1+7)\sqrt[4]{3}-\sqrt[4]{14}$$
$$=8\sqrt[4]{3}-\sqrt[4]{14}$$

21.
$$8\sqrt{x}+2\sqrt{y}-6\sqrt{x}=(8-6)\sqrt{x}+2\sqrt{y}$$
$$=2\sqrt{x}+2\sqrt{y}$$

23. $\sqrt[3]{ab}+a\sqrt[3]{b}$ cannot be simplified further.

25. $\sqrt{2t}+\sqrt[3]{2t}$ cannot be simplified further.

27. $\frac{5}{6}z\sqrt[3]{6}+\frac{7}{9}z\sqrt[3]{6}=\left(\frac{5}{6}z+\frac{7}{9}z\right)\sqrt[3]{6}=\frac{29}{18}z\sqrt[3]{6}$

29.
$$0.81x\sqrt{y}-0.11x\sqrt{y}=(0.81x-0.11x)\sqrt{y}$$
$$=0.70x\sqrt{y}$$

31. Simplify each radical: $3\sqrt{2}+35\sqrt{2}$. Then add like radicals: $38\sqrt{2}$.

33. $\sqrt{36}+\sqrt{81}=6+9=15$

35.
$$2\sqrt{12}+\sqrt{48}=2\sqrt{4\cdot 3}+\sqrt{16\cdot 3}$$
$$=2\cdot 2\sqrt{3}+4\sqrt{3}$$
$$=(4+4)\sqrt{3}$$
$$=8\sqrt{3}$$

37.
$$4\sqrt{7}+\sqrt{63}-2\sqrt{28}=4\sqrt{7}+\sqrt{9\cdot 7}-2\sqrt{4\cdot 7}$$
$$=4\sqrt{7}+3\sqrt{7}-2\cdot 2\sqrt{7}$$
$$=(4+3-4)\sqrt{7}$$
$$=3\sqrt{7}$$

39. $3\sqrt{2a}-\sqrt{8a}-\sqrt{72a}=3\sqrt{2a}-\sqrt{4\cdot 2a}-\sqrt{36\cdot 2a}=3\sqrt{2a}-2\sqrt{2a}-6\sqrt{2a}=(3-2-6)\sqrt{2a}=-5\sqrt{2a}$

41. $2s^2\sqrt[3]{s^2t^6}+3t^2\sqrt[3]{8s^8}=2s^2\cdot t^2\sqrt[3]{s^2}+3t^2\cdot 2s^2\sqrt[3]{s^2}=(2s^2t^2+6s^2t^2)\sqrt[3]{s^2}=8s^2t^2\sqrt[3]{s^2}$

43. $7\sqrt[3]{x^4}-x\sqrt[3]{x}=7\sqrt[3]{x^3\cdot x}-x\sqrt[3]{x}=7x\sqrt[3]{x}-x\sqrt[3]{x}=(7x-x)\sqrt[3]{x}=6x\sqrt[3]{x}$

45. $5p\sqrt{20p^2}+p^2\sqrt{80}=5p\sqrt{4p^2\cdot 5}+p^2\sqrt{16\cdot 5}=5p\cdot 2p\sqrt{5}+4p^2\sqrt{5}=(10p^2+4p^2)\sqrt{5}=14p^2\sqrt{5}$

47.
$$\begin{aligned}\frac{3}{2}ab\sqrt{24a^3}+\frac{4}{3}\sqrt{54a^5b^2}-a^2b\sqrt{150a}&=\frac{3}{2}ab\sqrt{4a^2\cdot 6a}+\frac{4}{3}\sqrt{9a^4b^2\cdot 6a}-a^2b\sqrt{25\cdot 6a}\\&=\frac{3}{2}ab\cdot 2a\sqrt{6a}+\frac{4}{3}\cdot 3a^2b\sqrt{6a}-5a^2b\sqrt{6a}\\&=3a^2b\sqrt{6a}+4a^2b\sqrt{6a}-5a^2b\sqrt{6a}\\&=(3a^2b+4a^2b-5a^2b)\sqrt{6a}\\&=2a^2b\sqrt{6a}\end{aligned}$$

49.
$$\begin{aligned}x\sqrt[3]{16}-2\sqrt[3]{27x}+\sqrt[3]{54x^3}&=x\sqrt[3]{8\cdot 2}-2\sqrt[3]{27x}+\sqrt[3]{27x^3\cdot 2}\\&=2\cdot x\sqrt[3]{2}-2\cdot 3\sqrt[3]{x}+3x\sqrt[3]{2}\\&=(2x+3x)\sqrt[3]{2}-6\sqrt[3]{x}\\&=5x\sqrt[3]{2}-6\sqrt[3]{x}\end{aligned}$$

51. False; for example:
$$\begin{aligned}\sqrt{9}+\sqrt{16}&\neq\sqrt{9+16}\\7&\neq 5\end{aligned}$$

53. True

55. False; for example:
$$\begin{aligned}\sqrt{y}+\sqrt{y}&=\sqrt{2y}\\2\sqrt{y}&\neq\sqrt{2y}\end{aligned}$$

57.
$$\begin{aligned}\sqrt{48}+\sqrt{12}&=\sqrt{16\cdot 3}+\sqrt{4\cdot 3}\\&=4\sqrt{3}+2\sqrt{3}\\&=6\sqrt{3}\end{aligned}$$

59. $5\sqrt[3]{x^6}-x^2=5x^2-x^2=4x^2$

61. The difference of the square root of 18 and the square of 5.

63. The sum of the 4th root of x and the cube of y.

65. (a) (0, 6) and (6, 9)

$$6^2 + 3^2 = c^2$$
$$45 = c^2$$
$$3\sqrt{5} = c$$

(0, 6) and (2, 2)

$$4^2 + 2^2 = c^2$$
$$20 = c^2$$
$$2\sqrt{5} = c$$

(2, 2) and (4, 1)

$$1^2 + 2^2 = c^2$$
$$5 = c^2$$
$$\sqrt{5} = c$$

(4, 1) and (7, 7)

$$6^2 + 3^2 = c^2$$
$$45 = c^2$$
$$3\sqrt{5} = c$$

(7, 7) and (6, 9)

$$1^2 + 2^2 = c^2$$
$$5 = c^2$$
$$\sqrt{5} = c$$

The perimeter is
$3\sqrt{5} + 2\sqrt{5} + \sqrt{5} + 3\sqrt{5} + \sqrt{5} = 10\sqrt{5}$ yd.

(b) 22.36 yards

(c) 22.36(3)(1.49) = 99.95(1.06) = \$105.95

Section 10.5 Practice Exercises

1. $f(x) = \sqrt{-3x+1}$

(a) $f(-1) = \sqrt{-3(-1)+1} = \sqrt{4} = 2$

(b) $f(-5) = \sqrt{-3(-5)+1} = \sqrt{16} = 4$

3. $\sqrt[3]{(x-y)^3} = x - y$

5. $\sqrt[3]{-16x^5y^6z^7} = \sqrt[3]{-8x^3y^6z^6 \cdot 2x^2z}$
$= -2xy^2z^2\sqrt[3]{2x^2z}$

7. $9^{1/2} = \sqrt{9} = 3$

9. $x^{1/3}y^{1/4}x^{-1/6}y^{1/3} = x^{\frac{1}{3}-\frac{1}{6}}y^{\frac{1}{4}+\frac{1}{3}} = x^{1/6}y^{7/12}$

11. $\dfrac{a^{2/3}}{a^{1/2}} = a^{\frac{2}{3}-\frac{1}{2}} = a^{1/6}$

13. $-2\sqrt[3]{7} + 4\sqrt[3]{7} = (-2+4)\sqrt[3]{7} = 2\sqrt[3]{7}$

15. $\sqrt{2} \cdot \sqrt{10} = \sqrt{20} = \sqrt{4 \cdot 5} = 2\sqrt{5}$

17. $\sqrt[4]{16} \cdot \sqrt[4]{64} = \sqrt[4]{16 \cdot 64}$
$= \sqrt[4]{16 \cdot 16 \cdot 4}$
$= 2 \cdot 2\sqrt[4]{4}$
$= 4\sqrt[4]{4}$

19. $\left(2\sqrt{5}\right)\left(\sqrt{7}\right) = 6\sqrt{35}$

21. $\left(8a\sqrt{b}\right)\left(-3\sqrt{ab}\right) = -24a\sqrt{ab^2} = -24ab\sqrt{a}$

23. $\sqrt{3}\left(4\sqrt{3} - 6\right) = 4\sqrt{9} - 6\sqrt{3}$
$= 4 \cdot 3 - 6\sqrt{3}$
$= 12 - 6\sqrt{3}$

25. $\sqrt{2}\left(\sqrt{6} - \sqrt{3}\right) = \sqrt{12} - \sqrt{6}$
$= \sqrt{4 \cdot 3} - \sqrt{6}$
$= 2\sqrt{3} - \sqrt{6}$

27. $-3\sqrt{x}\left(\sqrt{x} + 7\right) = -3\sqrt{x^2} - 21\sqrt{x}$
$= -3x - 21\sqrt{x}$

29. $\left(\sqrt{3} + 2\sqrt{10}\right)\left(4\sqrt{3} - \sqrt{10}\right)$
$= 4\sqrt{9} - \sqrt{30} + 8\sqrt{30} - 2\sqrt{100}$
$= 12 + 7\sqrt{30} - 20$
$= -8 + 7\sqrt{30}$

31. $(\sqrt{x}+4)(\sqrt{x}-9)=\sqrt{x^2}-9\sqrt{x}+4\sqrt{x}-36$
$=x-5\sqrt{x}-36$

33. $(\sqrt[3]{y}+2)(\sqrt[3]{y}-3)=\sqrt[3]{y^2}-3\sqrt[3]{y}+2\sqrt[3]{y}-6$
$=\sqrt[3]{y^2}-\sqrt[3]{y}-6$

35. $(\sqrt{a}-3\sqrt{b})(9\sqrt{a}-\sqrt{b})$
$=9\sqrt{a^2}-\sqrt{ab}-27\sqrt{ab}+3\sqrt{b^2}$
$=9a-28\sqrt{ab}+3b$

37. $(\sqrt{7}+3)(\sqrt{7}+\sqrt{2}-5)$
$=\sqrt{49}+\sqrt{14}-5\sqrt{7}+3\sqrt{7}+3\sqrt{2}-15$
$=7+\sqrt{14}-2\sqrt{7}+3\sqrt{2}-15$
$=-8+\sqrt{14}-2\sqrt{7}+3\sqrt{2}$

39. $(\sqrt{p}+2\sqrt{q})(8+3\sqrt{p}-\sqrt{q})$
$=8\sqrt{p}+3\sqrt{p^2}-\sqrt{pq}+16\sqrt{q}+6\sqrt{pq}-2\sqrt{q^2}$
$=8\sqrt{p}+3p+5\sqrt{pq}+16\sqrt{q}-2q$

41. $\sqrt{x}\cdot\sqrt[4]{x}=x^{1/2}\cdot x^{1/4}$
$=x^{\frac{1}{2}+\frac{1}{4}}$
$=x^{\frac{2}{4}+\frac{1}{4}}$
$=x^{3/4}$
$=\sqrt[4]{x^3}$

43. $\sqrt[5]{2z}\cdot\sqrt[3]{2z}=(2z)^{1/5}(2z)^{1/3}$
$=(2z)^{\frac{1}{5}+\frac{1}{3}}$
$=(2z)^{\frac{3}{15}+\frac{5}{15}}$
$=(2z)^{8/15}$
$=\sqrt[15]{(2z)^8}$

45. $\sqrt[3]{p^2}\cdot\sqrt{p^3}=p^{2/3}\cdot p^{3/2}$
$=p^{\frac{2}{3}+\frac{3}{2}}$
$=p^{\frac{4}{6}+\frac{9}{6}}$
$=p^{13/6}$
$=\sqrt[6]{p^{13}}$
$=\sqrt[6]{p^{12}\cdot p}$
$=p^2\sqrt[6]{p}$

47. $\dfrac{\sqrt{u^3}}{\sqrt[3]{u}}=\dfrac{u^{3/2}}{u^{1/3}}$
$=u^{\frac{3}{2}-\frac{1}{3}}$
$=u^{\frac{9}{6}-\frac{2}{6}}$
$=u^{7/6}$
$=\sqrt[6]{u^7}$
$=\sqrt[6]{u^6\cdot u}$
$=u\sqrt[6]{u}$

49. $\dfrac{\sqrt{(a+b)}}{\sqrt[3]{(a+b)}}=\dfrac{(a+b)^{1/2}}{(a+b)^{1/3}}$
$=(a+b)^{\frac{1}{2}-\frac{1}{3}}$
$=(a+b)^{\frac{3}{6}-\frac{2}{6}}$
$=(a+b)^{1/6}$
$=\sqrt[6]{(a+b)}$

51. **(a)** $(x+y)(x-y)=x^2-y^2$

(b) $(x+5)(x-5)=x^2-25$

53. $(\sqrt{3}+x)(\sqrt{3}-x)=\sqrt{3}^2-x^2=3-x^2$

55. $(\sqrt{6}+\sqrt{2})(\sqrt{6}-\sqrt{2})=\sqrt{6}^2-\sqrt{2}^2$
$=6-2$
$=4$

57. $\left(8\sqrt{x}+2\sqrt{y}\right)\left(8\sqrt{x}-2\sqrt{y}\right)$
$=\left(8\sqrt{x}\right)^2-\left(2\sqrt{y}\right)^2$
$=64\sqrt{x}^2-4\sqrt{y}^2$
$=64x-4y$

59. $\left(\sqrt{13}+4\right)^2=\sqrt{13}^2+2\cdot 4\sqrt{13}+4^2$
$=13+8\sqrt{13}+16$
$=29+8\sqrt{13}$

61. $\left(\sqrt{p}-\sqrt{7}\right)^2=\sqrt{p}^2-2\cdot\sqrt{p}\sqrt{7}+\sqrt{7}^2$
$=p-2\sqrt{7p}+7$

63. $\left(\sqrt{2a}-3\sqrt{b}\right)^2$
$=\sqrt{2a}^2-2\cdot 3\sqrt{2a}\sqrt{b}+\left(3\sqrt{b}\right)^2$
$=2a-6\sqrt{2ab}+9b$

65. True

67. False; $\left(x-\sqrt{5}\right)^2=x^2-2x\sqrt{5}+5$

69. False; 5 is multiplied only with the 3.

71. True

73. $A=lw$
$A=\left(\sqrt{40}\right)\left(3\sqrt{2}\right)$
$A=3\sqrt{80}$
$A=3\sqrt{16\cdot 5}$
$A=3\cdot 4\sqrt{5}$
$A=12\sqrt{5}$
The area is $12\sqrt{5}$ ft^2.

75. $A=\frac{1}{2}bh$
$A=\frac{1}{2}\left(6\sqrt{12}\right)\left(3\sqrt{5}\right)$
$A=\frac{1}{2}\left(18\sqrt{60}\right)$
$A=9\sqrt{4\cdot 15}$
$A=9\cdot 2\sqrt{15}$
$A=18\sqrt{15}$
The area is $18\sqrt{15}$ in^2.

77. $A=bh$
$A=\left(5\sqrt{2}\right)\left(\sqrt{32}\right)$
$A=5\sqrt{64}$
$A=5\cdot 8$
$A=40$
The area is 40 m^2.

79. $\left(\sqrt[3]{a}+\sqrt[3]{b}\right)\left(\sqrt[3]{a^2}-\sqrt[3]{ab}+\sqrt[3]{b^2}\right)$
$=\sqrt[3]{a^3}-\sqrt[3]{a^2b}+\sqrt[3]{ab^2}+\sqrt[3]{a^2b}-\sqrt[3]{ab^2}+\sqrt[3]{b^3}$
$=a+b$

81. $\sqrt[3]{x}\cdot\sqrt[6]{y}=x^{1/3}\cdot y^{1/6}$
$=x^{2/6}\cdot y^{1/6}$
$=(x^2y)^{1/6}$
$=\sqrt[6]{x^2y}$

83. $\sqrt[4]{8}\cdot\sqrt{3}=8^{1/4}\cdot 3^{1/2}$
$=8^{1/4}\cdot 3^{2/4}$
$=(8\cdot 3^2)^{1/4}$
$=\sqrt[4]{8\cdot 3^2}$ or $\sqrt[4]{72}$

85. $\sqrt[4]{6}\cdot\sqrt{2}=6^{1/4}\cdot 2^{1/2}$
$=6^{1/4}\cdot 2^{2/4}$
$=2^{1/4}\cdot 3^{1/4}\cdot 2^{2/4}$
$=2^{3/4}\cdot 3^{1/4}$
$=\sqrt[4]{2^3\cdot 3}$ or $\sqrt[4]{24}$

87. $\sqrt[5]{p}\cdot\sqrt[3]{q} = p^{1/5}\cdot q^{1/3}$
$= p^{3/15}\cdot q^{5/15}$
$= (p^3q^5)^{1/15}$
$= \sqrt[15]{p^3q^5}$

Section 10.6 Practice Exercises

1. $2y\sqrt{45}+3\sqrt{20y^2} = 2y\sqrt{9\cdot 5}+3\sqrt{4y^2\cdot 5}$
$= 2y\cdot 3\sqrt{5}+3\cdot 2y\sqrt{5}$
$= 6y\sqrt{5}+6y\sqrt{5}$
$= 12y\sqrt{5}$

3. $\left(-6\sqrt{y}+3\right)\left(3\sqrt{y}+1\right)$
$= -18\sqrt{y^2}-6\sqrt{y}+9\sqrt{y}+3$
$= -18y+3\sqrt{y}+3$

5. $4\sqrt{3}+\sqrt{5}\cdot\sqrt{15} = 4\sqrt{3}+\sqrt{75}$
$= 4\sqrt{3}+\sqrt{25\cdot 3}$
$= 4\sqrt{3}+5\sqrt{3}$
$= 9\sqrt{3}$

7. $\left(8-\sqrt{t}\right)^2 = 8^2-2\cdot 8\sqrt{t}+\sqrt{t}^2$
$= 64-16\sqrt{t}+t$

9. $\left(\sqrt{2}+\sqrt{7}\right)\left(\sqrt{2}-\sqrt{7}\right) = \sqrt{2}^2-\sqrt{7}^2$
$= 2-7$
$= -5$

11. $\dfrac{x}{\sqrt{5}} = \dfrac{x}{\sqrt{5}}\cdot\dfrac{\sqrt{5}}{\sqrt{5}} = \dfrac{x\sqrt{5}}{\sqrt{5^2}} = \dfrac{x\sqrt{5}}{5}$

13. $\dfrac{7}{\sqrt[3]{x}} = \dfrac{7}{\sqrt[3]{x}}\cdot\dfrac{7\sqrt[3]{x^2}}{\sqrt[3]{x^2}} = \dfrac{7\sqrt[3]{x^2}}{\sqrt[3]{x^3}} = \dfrac{7\sqrt[3]{x^2}}{x}$

15. $\dfrac{8}{\sqrt{3z}} = \dfrac{8}{\sqrt{3z}}\cdot\dfrac{\sqrt{3z}}{\sqrt{3z}} = \dfrac{8\sqrt{3z}}{\sqrt{(3z)^2}} = \dfrac{8\sqrt{3z}}{3z}$

17. $\dfrac{1}{\sqrt[4]{2a^2}} = \dfrac{1}{\sqrt[4]{2a^2}}\cdot\dfrac{\sqrt[4]{2^3a^2}}{\sqrt[4]{2^3a^2}} = \dfrac{\sqrt[4]{8a^2}}{\sqrt[4]{(2a)^4}} = \dfrac{\sqrt[4]{8a^2}}{2a}$

19. $\dfrac{1}{\sqrt{3}} = \dfrac{1}{\sqrt{3}}\cdot\dfrac{\sqrt{3}}{\sqrt{3}} = \dfrac{\sqrt{3}}{\sqrt{(3)^2}} = \dfrac{\sqrt{3}}{3}$

21. $\dfrac{10}{\sqrt{5}} = \dfrac{10}{\sqrt{5}}\cdot\dfrac{\sqrt{5}}{\sqrt{5}} = \dfrac{10\sqrt{5}}{\sqrt{(5)^2}} = \dfrac{10\sqrt{5}}{5} = 2\sqrt{5}$

23. $\dfrac{1}{\sqrt{x}} = \dfrac{1}{\sqrt{x}}\cdot\dfrac{\sqrt{x}}{\sqrt{x}} = \dfrac{\sqrt{x}}{\sqrt{(x)^2}} = \dfrac{\sqrt{x}}{x}$

25. $\dfrac{6}{\sqrt{2y}} = \dfrac{6}{\sqrt{2y}}\cdot\dfrac{\sqrt{2y}}{\sqrt{2y}}$
$= \dfrac{6\sqrt{2y}}{\sqrt{(2y)^2}}$
$= \dfrac{6\sqrt{2y}}{2y}$
$= \dfrac{3\sqrt{2y}}{y}$

27. $\dfrac{-2a}{\sqrt{a}} = \dfrac{-2a}{\sqrt{a}}\cdot\dfrac{\sqrt{a}}{\sqrt{a}}$
$= \dfrac{-2a\sqrt{a}}{\sqrt{a^2}}$
$= \dfrac{-2a\sqrt{a}}{a}$
$= -2\sqrt{a}$

29. $\dfrac{7}{\sqrt[3]{4}} = \dfrac{7}{\sqrt[3]{4}}\cdot\dfrac{\sqrt[3]{2}}{\sqrt[3]{2}} = \dfrac{7\sqrt[3]{2}}{\sqrt[3]{2^3}} = \dfrac{7\sqrt[3]{2}}{2}$

31. $\dfrac{4}{\sqrt{w^3}} = \dfrac{4}{w\sqrt{w}} = \dfrac{4}{w\sqrt{w}}\cdot\dfrac{\sqrt{w}}{\sqrt{w}} = \dfrac{4\sqrt{w}}{w\cdot w} = \dfrac{4\sqrt{w}}{w^2}$

33. $\sqrt[4]{\frac{16}{3}} = \frac{\sqrt[4]{16}}{\sqrt[4]{3}}$
$= \frac{\sqrt[4]{2^4}}{\sqrt[4]{3}}$
$= \frac{2}{\sqrt[4]{3}} \cdot \frac{\sqrt[4]{3^3}}{\sqrt[4]{3^3}}$
$= \frac{2\sqrt[4]{3^3}}{\sqrt[4]{3^4}}$
$= \frac{2\sqrt[4]{3^3}}{3}$
$= \frac{2\sqrt[4]{27}}{3}$

35. $\frac{1}{\sqrt{x^7}} = \frac{1}{x^3\sqrt{x}} = \frac{1}{x^3\sqrt{x}} \cdot \frac{\sqrt{x}}{\sqrt{x}} = \frac{\sqrt{x}}{x^3 x} = \frac{\sqrt{x}}{x^4}$

37. $\frac{2}{\sqrt{8x^5}} = \frac{2}{2x^2\sqrt{2x}}$
$= \frac{1}{x^2\sqrt{2x}}$
$= \frac{1}{x^2\sqrt{2x}} \cdot \frac{\sqrt{2x}}{\sqrt{2x}}$
$= \frac{\sqrt{2x}}{x^2\sqrt{(2x)^2}}$
$= \frac{\sqrt{2x}}{x^2(2x)}$
$= \frac{\sqrt{2x}}{2x^3}$

39. $\sqrt[3]{\frac{16x^3}{y}} = \frac{\sqrt[3]{16x^3}}{\sqrt[3]{y}}$
$= \frac{\sqrt[3]{8x^3 \cdot 2}}{\sqrt[3]{y}}$
$= \frac{2x\sqrt[3]{2}}{\sqrt[3]{y}} \cdot \frac{\sqrt[3]{y^2}}{\sqrt[3]{y^2}}$
$= \frac{2x\sqrt[3]{2y^2}}{\sqrt[3]{y^3}}$
$= \frac{2x\sqrt[3]{2y^2}}{y}$

41. $\frac{\sqrt{x^4y^5}}{\sqrt{10x}} = \frac{x^2y^2\sqrt{y}}{\sqrt{10x}}$
$= \frac{x^2y^2\sqrt{y}}{\sqrt{10x}} \cdot \frac{\sqrt{10x}}{\sqrt{10x}}$
$= \frac{x^2y^2\sqrt{10xy}}{\sqrt{(10x)^2}}$
$= \frac{x^2y^2\sqrt{10xy}}{10x}$
$= \frac{xy^2\sqrt{10xy}}{10}$

43. $\sqrt{2} + \sqrt{6}$

45. $\sqrt{x} - 23$

47. $(\sqrt{2}+3)(\sqrt{2}+3) = (\sqrt{2})^2 - 3^2 = 2 - 9 = -7$

49. $(\sqrt{5}-\sqrt{2})(\sqrt{5}+\sqrt{2}) = (\sqrt{5})^2 - (\sqrt{2})^2$
$= 5 - 2$
$= 3$

51. $\frac{4}{\sqrt{2}+3}=\frac{4}{\sqrt{2}+3}\cdot\frac{\sqrt{2}-3}{\sqrt{2}-3}$
$=\frac{4(\sqrt{2}-3)}{(\sqrt{2})^2-3^2}$
$=\frac{4\sqrt{2}-12}{2-9}$
$=\frac{4\sqrt{2}-12}{-7}$ or $\frac{-4\sqrt{2}+12}{7}$

53. $\frac{1}{\sqrt{5}-\sqrt{2}}=\frac{1}{\sqrt{5}-\sqrt{2}}\cdot\frac{\sqrt{5}+\sqrt{2}}{\sqrt{5}+\sqrt{2}}$
$=\frac{\sqrt{5}+\sqrt{2}}{(\sqrt{5})^2-(\sqrt{2})^2}$
$=\frac{\sqrt{5}+\sqrt{2}}{5-2}$
$=\frac{\sqrt{5}+\sqrt{2}}{3}$

55. $\frac{\sqrt{7}}{\sqrt{3}+2}=\frac{\sqrt{7}}{\sqrt{3}+2}\cdot\frac{\sqrt{3}-2}{\sqrt{3}-2}$
$=\frac{\sqrt{7}(\sqrt{3}-2)}{(\sqrt{3})^2-2^2}$
$=\frac{\sqrt{21}-2\sqrt{7}}{3-4}$
$=\frac{\sqrt{21}-2\sqrt{7}}{-1}$
$=-\sqrt{21}+2\sqrt{7}$

57. $\frac{-1}{\sqrt{p}+\sqrt{q}}=\frac{-1}{\sqrt{p}+\sqrt{q}}\cdot\frac{\sqrt{p}-\sqrt{q}}{\sqrt{p}-\sqrt{q}}$
$=\frac{-\sqrt{p}+\sqrt{q}}{(\sqrt{p})^2-(\sqrt{q})^2}$
$=\frac{-\sqrt{p}+\sqrt{q}}{p-q}$

59. $\frac{2\sqrt{3}+\sqrt{7}}{3\sqrt{3}-\sqrt{7}}=\frac{2\sqrt{3}+\sqrt{7}}{2\sqrt{3}-\sqrt{7}}\cdot\frac{3\sqrt{3}+\sqrt{7}}{3\sqrt{3}+\sqrt{7}}$
$=\frac{6\cdot 3+2\sqrt{21}+3\sqrt{21}+7}{(3\sqrt{3})^2-(\sqrt{7})^2}$
$=\frac{18+5\sqrt{21}+7}{9\cdot 3-7}$
$=\frac{25+5\sqrt{21}}{20}$
$=\frac{5(5+\sqrt{21})}{20}$
$=\frac{5+\sqrt{21}}{4}$

61. $\frac{\sqrt{5}+4}{2-\sqrt{5}}=\frac{\sqrt{5}+4}{2-\sqrt{5}}\cdot\frac{2+\sqrt{5}}{2+\sqrt{5}}$
$=\frac{2\sqrt{5}+\sqrt{25}+8+4\sqrt{5}}{2^2-(\sqrt{5})^2}$
$=\frac{6\sqrt{5}+5+8}{4-5}$
$=\frac{6\sqrt{5}+13}{-1}$
$=-13-6\sqrt{5}$

63. $\frac{16}{\sqrt[3]{4}}=\frac{16}{\sqrt[3]{2^2}}\cdot\frac{\sqrt[3]{2}}{\sqrt[3]{2}}=\frac{16\sqrt[3]{2}}{\sqrt[3]{2^3}}=\frac{16\sqrt[3]{2}}{2}=8\sqrt[3]{2}$

65. $\frac{4}{x-\sqrt{2}}=\frac{4}{x-\sqrt{2}}\cdot\frac{x+\sqrt{2}}{x+\sqrt{2}}$
$=\frac{4x+4\sqrt{2}}{x^2-(\sqrt{2})^2}$
$=\frac{4x+4\sqrt{2}}{x^2-2}$

67. $T(x)=2\pi\sqrt{\frac{x}{32}}$

(a) $T(2)=2\pi\sqrt{\frac{2}{32}}$
$\approx 2\pi(0.25)$
≈ 1.57 seconds

(b) $T(1) = 2\pi\sqrt{\frac{1}{32}}$
$\approx 2\pi(0.176)$
≈ 1.11 seconds

(c) $T(0.5) = 2\pi\sqrt{\frac{0.5}{32}}$
$\approx 2\pi(0.125)$
≈ 0.79 seconds

69. $$\frac{\sqrt{6}}{2}+\frac{1}{\sqrt{6}}=\frac{\sqrt{6}}{2}\cdot\frac{\sqrt{6}}{\sqrt{6}}+\frac{1}{\sqrt{6}}\cdot\frac{\sqrt{6}}{\sqrt{6}}$$
$$=\frac{\left(\sqrt{6}\right)^2+2}{2\sqrt{6}}$$
$$=\frac{6+2}{2\sqrt{6}}$$
$$=\frac{8}{2\sqrt{6}}$$
$$=\frac{4}{\sqrt{6}}$$
$$=\frac{4}{\sqrt{6}}\cdot\frac{\sqrt{6}}{\sqrt{6}}$$
$$=\frac{4\sqrt{6}}{6}$$
$$=\frac{2\sqrt{6}}{3}$$

71. $$\sqrt{15}-\sqrt{\frac{3}{5}}+\sqrt{\frac{5}{3}}$$
$$=\sqrt{15}-\frac{\sqrt{3}}{\sqrt{5}}+\frac{\sqrt{5}}{\sqrt{3}}$$
$$=\sqrt{15}\cdot\frac{\sqrt{15}}{\sqrt{15}}-\frac{\sqrt{3}}{\sqrt{5}}\cdot\frac{\sqrt{3}}{\sqrt{3}}+\frac{\sqrt{5}}{\sqrt{3}}\cdot\frac{\sqrt{5}}{\sqrt{5}}$$
$$=\frac{\left(\sqrt{15}\right)^2-\left(\sqrt{3}\right)^2+\left(\sqrt{5}\right)^2}{\sqrt{15}}$$
$$=\frac{15-3+5}{\sqrt{15}}$$
$$=\frac{17}{\sqrt{15}}$$
$$=\frac{17}{\sqrt{15}}\cdot\frac{\sqrt{15}}{\sqrt{15}}$$
$$=\frac{17\sqrt{15}}{15}$$

73. $$\sqrt[3]{25}+\frac{3}{\sqrt[3]{5}}=\sqrt[3]{25}\cdot\frac{\sqrt[3]{5}}{\sqrt[3]{5}}+\frac{3}{\sqrt[3]{5}}$$
$$=\frac{\sqrt[3]{5^2}\sqrt[3]{5}+3}{\sqrt[3]{5}}$$
$$=\frac{\sqrt[3]{5^3}+3}{\sqrt[3]{5}}$$
$$=\frac{5+3}{\sqrt[3]{5}}$$
$$=\frac{8}{\sqrt[3]{5}}\cdot\frac{\sqrt[3]{5^2}}{\sqrt[3]{5^2}}$$
$$=\frac{8\sqrt[3]{25}}{\sqrt[3]{5^3}}$$
$$=\frac{8\sqrt[3]{25}}{5}$$

75. $$\frac{\sqrt{3}+6}{2}=\frac{\sqrt{3}+6}{2}\cdot\frac{\sqrt{3}-6}{\sqrt{3}-6}$$
$$=\frac{3-36}{2\sqrt{3}-12}$$
$$=\frac{-33}{2\sqrt{3}-12}$$

77. $\dfrac{\sqrt{a}-\sqrt{b}}{\sqrt{a}+\sqrt{b}}=\dfrac{\sqrt{a}-\sqrt{b}}{\sqrt{a}+\sqrt{b}}\cdot\dfrac{\sqrt{a}+\sqrt{b}}{\sqrt{a}+\sqrt{b}}$
$=\dfrac{a-b}{a+2\sqrt{ab}+b}$

Section 10.7 Practice Exercises

1. $\sqrt{48}=\sqrt{16\cdot 3}=4\sqrt{3}$

3. $\sqrt{\dfrac{9w^3}{16}}=\dfrac{\sqrt{9w^2\cdot w}}{\sqrt{16}}=\dfrac{3w\sqrt{2}}{4}$

5. $\sqrt{-25}$; not a real number.

7. $\sqrt{\dfrac{p^5}{q^3}}=\dfrac{\sqrt{p^5}}{\sqrt{q^3}}$
$=\dfrac{p^2\sqrt{p}}{q\sqrt{q}}$
$=\dfrac{p^2\sqrt{p}}{q\sqrt{q}}\cdot\dfrac{\sqrt{q}}{\sqrt{q}}$
$=\dfrac{p^2\sqrt{pq}}{q^2}$

9. $\sqrt{\dfrac{49}{5t^3}}=\dfrac{\sqrt{49}}{\sqrt{t^2\cdot 5t}}$
$=\dfrac{7}{t\sqrt{5t}}\cdot\dfrac{\sqrt{5t}}{\sqrt{5t}}$
$=\dfrac{7\sqrt{5t}}{t\left(\sqrt{5t}\right)^2}$
$=\dfrac{7\sqrt{5t}}{5t^2}$

11. $\left(\sqrt{4x-6}\right)^2=4x-6$

13. $\left(\sqrt[3]{9p+7}\right)^3=9p+7$

15. $\left(\sqrt{w^2+2w-17}\right)^2=w^2+2w-17$

17. $\left(\sqrt{2x}\right)^2=2x$

19. $\left(\sqrt[4]{7r}\right)^4=7r$

21. $\sqrt{t}=7$
$\left(\sqrt{t}\right)^2=(7)^2$
$t=49$
Check: $\sqrt{49}=7$
$7=7$

23. $\sqrt{4x}=6$
$\left(\sqrt{4x}\right)^2=(6)^2$
$4x=36$
$x=9$
Check: $\sqrt{4\cdot 9}=6$
$\sqrt{36}=6$
$6=6$

25. $\sqrt{5y+1}=4$
$\left(\sqrt{5y+1}\right)^2=(4)^2$
$5y+1=16$
$5y=15$
$y=3$
Check: $\sqrt{5(3)+1}=4$
$\sqrt{16}=4$
$4=4$

27. $\sqrt[4]{2x+1}=2$
$\left(\sqrt[4]{2x+1}\right)^4=(2)^4$
$2x+1=16$
$2x=15$
$x=\dfrac{15}{2}$
Check: $\sqrt[4]{2\left(\dfrac{15}{2}\right)+1}=2$
$\sqrt[4]{16}=2$
$2=2$

29. $(2z-3)^{1/2}=9$

$$\sqrt{2z-3}=9$$
$$\left(\sqrt{2z-3}\right)^2=(9)^2$$
$$2z-3=81$$
$$2z=84$$
$$z=42$$

Check: $(2(42)-3)^{1/2}=9$
$$(81)^{1/2}=9$$
$$9=9$$

31. $\sqrt[3]{x-2}=3$

$$\left(\sqrt[3]{x-2}\right)^3=(3)^3$$
$$x-2=27$$
$$x=27$$

33. $(15-w)^{1/3}=-5$

$$\sqrt[3]{15-w}=-5$$
$$\left(\sqrt[3]{15-w}\right)^3=(-5)^3$$
$$15-w=-125$$
$$-w=-140$$
$$w=140$$

35. $\sqrt{x-16}=-3$

No solution; the even root of a number is never negative.

37. $\sqrt[3]{x+1}+3=-1$

$$\sqrt[3]{x+1}=-4$$
$$\left(\sqrt[3]{x+1}\right)^3=(-4)^3$$
$$x+1=-64$$
$$x=-65$$

39. $11=4\sqrt{3t}-5$

$$16=4\sqrt{3t}$$
$$4=\sqrt{3t}$$
$$(4)^2=\left(\sqrt{3t}\right)^2$$
$$16=3t$$
$$\frac{16}{3}=t$$

Check: $11=4\sqrt{3\left(\frac{16}{3}\right)}-5$
$$11=4\sqrt{16}-5$$
$$11=4\cdot 4-5$$
$$11=16-5$$
$$11=11$$

41. $\sqrt{6p-8}=p$

$$\left(\sqrt{6p-8}\right)^2=(p)^2$$
$$6p-8=p^2$$
$$0=p^2-6p+8$$
$$0=(p-4)(p-2)$$
$$p-4=0 \quad \text{or} \quad p-2=0$$
$$p=4 \quad \text{or} \quad p=2$$

Check:

$p=4$: $\sqrt{6\cdot 4-8}=4$
$$\sqrt{24-8}=4$$
$$\sqrt{16}=4$$
$$4=4$$

$p=2$: $\sqrt{6\cdot 2-8}=2$
$$\sqrt{12-8}=2$$
$$\sqrt{4}=2$$
$$2=2$$

43. $2x=\sqrt{4x+3}$

$$(2x)^2=\left(\sqrt{4x+3}\right)^2$$
$$4x^2=4x+3$$
$$4x^2-4x-3=0$$
$$(2x+1)(2x-3)=0$$
$$2x+1=0 \quad \text{or} \quad 2x-3=0$$
$$x=-\frac{1}{2} \quad \text{or} \quad x=\frac{3}{2}$$

Check:

$x=-\frac{1}{2}$: $2\left(-\frac{1}{2}\right)=\sqrt{4\left(-\frac{1}{2}\right)+3}$
$$-1=\sqrt{-2+3}$$
$$-1=\sqrt{1}$$
$$-1=\sqrt{1} \quad \text{False}$$

$x=\frac{3}{2}$: $2\left(\frac{3}{2}\right)=\sqrt{4\left(\frac{3}{2}\right)+3}$

$$3=\sqrt{6+3}$$
$$3=\sqrt{9}$$
$$3=3$$

Since $x=-\frac{1}{2}$ does not check, the solution is $x=\frac{3}{2}$.

45. $\sqrt[4]{h+4}=\sqrt[4]{2h-5}$

$$\left(\sqrt[4]{h+4}\right)^4=\left(\sqrt[4]{2h-5}\right)^4$$
$$h+4=2h-5$$
$$-h=-9$$
$$h=9$$

Check: $\sqrt[4]{9+4}=\sqrt[4]{2(9)-5}$

$$\sqrt[4]{13}=\sqrt[4]{18-5}$$
$$\sqrt[4]{13}=\sqrt[4]{13}$$

47. $\sqrt[3]{5a+3}=\sqrt[3]{a-13}$

$$\left(\sqrt[3]{5a+3}\right)^3=\left(\sqrt[3]{a-13}\right)^3$$
$$5a+3=a-13$$
$$4a=-16$$
$$a=-4$$

49. $\sqrt[4]{2x-5}=-1$

No solution; the even root of a number is never negative.

51. $r=\sqrt[3]{\frac{3V}{4\pi}}$

$$(r)^3=\left(\sqrt[3]{\frac{3V}{4\pi}}\right)^3$$
$$r^3=\frac{3V}{4\pi}$$
$$4\pi r^3=3V$$
$$\frac{4\pi r^3}{3}=V$$
$$V=\frac{4\pi r^3}{3}$$

53. $r=\pi\sqrt{r^2+h^2}$

$$\frac{r}{\pi}=\sqrt{r^2+h^2}$$
$$\left(\frac{r}{\pi}\right)^2=\left(\sqrt{r^2+h^2}\right)^2$$
$$\frac{r^2}{\pi^2}=r^2+h^2$$
$$\frac{r^2}{\pi^2}-r^2=h^2$$
$$\frac{r^2-\pi^2r^2}{\pi^2}=h^2$$
$$h^2=\frac{r^2-\pi^2r^2}{\pi^2}$$

55. $a^2+10a+25$

57. $25w^2-40w+16$

59. $5a-6\sqrt{5a}+9$

61. $\sqrt{a^2+2a+1}=a+5$

$$\left(\sqrt{a^2+2a+1}\right)^2=(a+5)^2$$
$$a^2+2a+1=a^2+10a+25$$
$$2a+1=10a+25$$
$$-8a=25$$
$$a=-3$$

Check: $\sqrt{(-3)^2+2(-3)+1}=-3+5$

$$\sqrt{9-6+1}=2$$
$$\sqrt{4}=2$$
$$2=2$$

63. $\sqrt{25w^2-2w-3}=5w-4$

$$\left(\sqrt{25w^2-2w-3}\right)^2=(5w-4)^2$$
$$25w^2-2w-3=25w^2-40w+16$$
$$-2w-3=-40w+16$$
$$38w=19$$
$$w=\frac{1}{2}$$

$$\text{Check: } \sqrt{25\left(\frac{1}{2}\right)^2 - 2\left(\frac{1}{2}\right) - 3} = 5\left(\frac{1}{2}\right) - 4$$
$$\sqrt{25\left(\frac{1}{4}\right) - 1 - 3} = \frac{5}{2} - 4$$
$$\sqrt{\frac{25}{4} - \frac{4}{4} - \frac{12}{4}} = \frac{5}{2} - \frac{8}{2}$$
$$\sqrt{\frac{9}{4}} = -\frac{3}{2}$$
$$\frac{3}{2} \neq -\frac{3}{2}$$

No solution

65.
$$\sqrt{9z^2 - z + 6} = 3z - 1$$
$$\left(\sqrt{9z^2 - z + 6}\right)^2 = (3z-1)^2$$
$$9z^2 - z + 6 = 9z^2 - 6z + 1$$
$$5z = -5$$
$$z = -1$$
$$\text{Check: } \sqrt{9(-1)^2 - (-1) + 6} = 3(-1) - 1$$
$$\sqrt{9(1) + 1 + 6} = -3 - 1$$
$$\sqrt{16} = -3 - 1$$
$$4 \neq -4$$

No solution

67.
$$\sqrt{5a-9} = \sqrt{5a} - 3$$
$$\left(\sqrt{5a-9}\right)^2 = \left(\sqrt{5a} - 3\right)^2$$
$$5a - 9 = 5a - 6\sqrt{5a} + 9$$
$$-18 = -6\sqrt{5a}$$
$$3 = \sqrt{5a}$$
$$3^2 = \left(\sqrt{5a}\right)^2$$
$$9 = 5a$$
$$\frac{9}{5} = a$$
$$\text{Check: } \sqrt{5\left(\frac{9}{5}\right) - 9} = \sqrt{5\left(\frac{9}{5}\right)} - 3$$
$$\sqrt{9-9} = \sqrt{9} - 3$$
$$0 = 3 - 3$$
$$0 = 0$$

69.
$$\sqrt{2h+5} - \sqrt{2h} = 1$$
$$\sqrt{2h+5} = 1 + \sqrt{2h}$$
$$\left(\sqrt{2h+5}\right)^2 = \left(1 + \sqrt{2h}\right)^2$$
$$2h + 5 = 1 + 2\sqrt{2h} + 2h$$
$$4 = 2\sqrt{2h}$$
$$2 = \sqrt{2h}$$
$$(2)^2 = \left(\sqrt{2h}\right)^2$$
$$4 = 2h$$
$$2 = h$$
$$\text{Check: } \sqrt{2(2)+5} - \sqrt{2(2)} = 1$$
$$\sqrt{9} - \sqrt{4} = 1$$
$$3 - 2 = 1$$

71.
$$\sqrt{t-9} = 3 + \sqrt{t}$$
$$\left(\sqrt{t-9}\right)^2 = \left(3 + \sqrt{t}\right)^2$$
$$t - 9 = 9 + 6\sqrt{t} + t$$
$$-18 = 6\sqrt{t}$$
$$-3 = \sqrt{t}$$
$$(-3)^2 = \left(\sqrt{t}\right)^2$$
$$9 = t$$
$$\text{Check: } \sqrt{9-9} = 3 + \sqrt{9}$$
$$0 = 3 + 3$$
$$0 \neq 6$$

No solution

73.
$$\sqrt{x^2+3} = 6 + x$$
$$\left(\sqrt{x^2+3}\right)^2 = (6+x)^2$$
$$x^2 + 3 = 36 + 12x + x^2$$
$$-33 = 12x$$
$$\frac{-33}{12} = x$$
$$\frac{-11}{4} = x$$

Check: $\sqrt{\left(\frac{-11}{4}\right)^2+3}=6+\left(-\frac{11}{4}\right)$

$\sqrt{\frac{121}{16}+\frac{48}{16}}=\frac{24}{4}-\frac{11}{4}$

$\sqrt{\frac{169}{16}}=\frac{13}{4}$

$\frac{13}{4}=\frac{13}{4}$

75. $\sqrt{3t-7}=2-\sqrt{3t+1}$

$\left(\sqrt{3t-7}\right)^2=\left(2-\sqrt{3t+1}\right)^2$

$3t-7=4-4\sqrt{3t+1}+3t+1$

$-12=-4\sqrt{3t+1}$

$3=\sqrt{3t+1}$

$3^2=\left(\sqrt{3t+1}\right)^2$

$9=3t+1$

$8=3t$

$\frac{8}{3}=t$

Check: $\sqrt{3\left(\frac{8}{3}\right)-7}=2-\sqrt{3\left(\frac{8}{3}\right)+1}$

$\sqrt{8-7}=2-\sqrt{8+1}$

$\sqrt{1}=2-\sqrt{9}$

$1=2-3$

$1\neq-1$

No solution

77. $\sqrt{8b-3}=4+\sqrt{8b+1}$

$\left(\sqrt{8b-3}\right)^2=\left(4+\sqrt{8b+1}\right)^2$

$8b-3=16+8\sqrt{8b+1}+8b+1$

$-20=8\sqrt{8b+1}$

$\frac{-5}{4}=\sqrt{8b+1}$

$\left(\frac{-5}{4}\right)^2=\left(\sqrt{8b+1}\right)^2$

$\frac{25}{16}=8b+1$

$\frac{9}{16}=8b$

$\frac{9}{128}=b$

Check: $\sqrt{8\left(\frac{9}{128}\right)-3}=4+\sqrt{8\left(\frac{9}{128}\right)+1}$

$\sqrt{\frac{9}{16}-\frac{48}{16}}=4+\sqrt{\frac{9}{16}+\frac{16}{16}}$

$\sqrt{-\frac{39}{16}}=4+\sqrt{\frac{25}{16}}$

$\sqrt{-\frac{39}{16}}=\frac{16}{4}+\frac{5}{4}$

$\sqrt{-\frac{39}{16}}\neq 6$

No solution

79. $\sqrt{p^2+35}=\sqrt{12p}$

$\left(\sqrt{p^2+35}\right)^2=\left(\sqrt{12p}\right)^2$

$p^2+35=12p$

$p^2-12p+35=0$

$(p-7)(p-5)=0$

$p-7=0$ or $p-5=0$

$p=7$ or $p=5$

Check:

$p=7$: $\sqrt{7^2+35}=\sqrt{12\cdot 7}$

$\sqrt{49+35}=\sqrt{84}$

$\sqrt{84}=\sqrt{84}$

$$p = 5: \sqrt{5^2 + 35} = \sqrt{12 \cdot 5}$$
$$\sqrt{25 + 35} = \sqrt{60}$$
$$\sqrt{60} = \sqrt{60}$$

81.
$$\sqrt{6m+7} = \sqrt{3m+3} + 1$$
$$\left(\sqrt{6m+7}\right)^2 = \left(\sqrt{3m+3} + 1\right)^2$$
$$6m + 7 = 3m + 3 + 2\sqrt{3m+3} + 1$$
$$6m + 7 = 3m + 4 + 2\sqrt{3m+3}$$
$$3m + 3 = 2\sqrt{3m+3}$$
$$\frac{3m+3}{2} = \sqrt{3m+3}$$
$$\left(\frac{3m+3}{2}\right)^2 = \left(\sqrt{3m+3}\right)^2$$
$$\frac{9m^2 + 18m + 9}{4} = 3m + 3$$
$$9m^2 + 18m + 9 = 4(3m + 3)$$
$$9m^2 + 18m + 9 = 12m + 12$$
$$9m^2 + 6m - 3 = 0$$
$$3(3m^2 + 2m - 1) = 0$$
$$3(3m - 1)(m + 1) = 0$$
$$(3m - 1)(m + 1) = 0$$
$$3m - 1 = 0 \quad \text{or} \quad m + 1 = 0$$
$$m = \frac{1}{3} \quad \text{or} \quad m = -1$$

Check:

$$m = \frac{1}{3}: \sqrt{6\left(\frac{1}{3}\right) + 7} = \sqrt{3\left(\frac{1}{3}\right) + 3} + 1$$
$$\sqrt{2 + 7} = \sqrt{1 + 3} + 1$$
$$\sqrt{9} = \sqrt{4} + 1$$
$$3 = 2 + 1$$
$$3 = 3$$

$$m = -1: \sqrt{6(-1) + 7} = \sqrt{3(-1) + 3} + 1$$
$$\sqrt{1} = \sqrt{0} + 1$$
$$1 = 0 + 1$$
$$1 = 1$$

83.
$$\sqrt{z+1} + \sqrt{2z+3} = 1$$
$$\sqrt{z+1} = -\sqrt{2z+3} + 1$$
$$\left(\sqrt{z+1}\right)^2 = \left(-\sqrt{2z+3} + 1\right)^2$$
$$z + 1 = 2z + 3 - 2\sqrt{2z+3} + 1$$
$$z + 1 = 2z + 4 - 2\sqrt{2z+3}$$
$$-z - 3 = -2\sqrt{2z+3}$$
$$\frac{-z-3}{-2} = \sqrt{2z+3}$$
$$\frac{z+3}{2} = \sqrt{2z+3}$$
$$\left(\frac{z+3}{2}\right)^2 = \left(\sqrt{2z+3}\right)^2$$
$$\frac{z^2 + 6z + 9}{4} = 2z + 3$$
$$z^2 + 6z + 9 = 4(2z + 3)$$
$$z^2 + 6z + 9 = 8z + 12$$
$$z^2 - 2z - 3 = 0$$
$$(z - 3)(z + 1) = 0$$
$$z - 3 = 0 \quad \text{or} \quad z + 1 = 0$$
$$z = 3 \quad \text{or} \quad z = -1$$

Check:

$$z = 3: \sqrt{3+1} + \sqrt{2 \cdot 3 + 3} = 1$$
$$\sqrt{4} + \sqrt{9} = 1$$
$$2 + 3 = 1$$
$$5 = 1 \quad \text{False}$$

$$z = -1: \sqrt{-1+1} + \sqrt{2(-1) + 3} = 1$$
$$\sqrt{0} + \sqrt{1} = 1$$
$$0 + 1 = 1$$
$$1 = 1$$

Since $z = 3$ does not check, the solution is $z = -1$.

85. $1+\sqrt{2t+3}+\sqrt{3t-5}=0$

$$1+\sqrt{2t+3}=-\sqrt{3t-5}$$
$$\left(1+\sqrt{2t+3}\right)^2=\left(-\sqrt{3t-5}\right)^2$$
$$1+2\sqrt{2t+3}+2t+3=3t-5$$
$$2\sqrt{2t+3}+2t+4=3t-5$$
$$2\sqrt{2t+3}=t-9$$
$$\left(2\sqrt{2t+3}\right)^2=(t-9)^2$$
$$2^2\left(\sqrt{2t+3}\right)^2=t^2-18t+81$$
$$4(2t+3)=t^2-18t+81$$
$$8t+12=t^2-18t+81$$
$$0=t^2-26t+69$$
$$0=(t-3)(t-23)$$

$t-3=0$ or $t-23=0$

$t=3$ or $t=23$

Check:

$t=3$: $1+\sqrt{2(3)+3}+\sqrt{3(3)-5}=0$

$$1+\sqrt{9}+\sqrt{4}=0$$
$$1+3+2=0$$
$$6=0 \quad \text{False}$$

$t=23$:

$$1+\sqrt{2(23)+3}+\sqrt{3(23)-5}=0$$
$$1+\sqrt{49}+\sqrt{64}=0$$
$$1+7+8=0$$
$$16=0 \quad \text{False}$$

Since neither value checks, there is no solution.

87. (a)

$$t(d)=\sqrt{\frac{d}{4.9}}$$
$$7.89=\sqrt{\frac{d}{4.9}}$$
$$(7.89)^2=\left(\sqrt{\frac{d}{4.9}}\right)^2$$
$$62.3\approx\frac{d}{4.9}$$
$$305.27\approx d$$

The height of the building is approximately 305 meters.

(b)

$$t(d)=\sqrt{\frac{d}{4.9}}$$
$$(9.69)^2=\left(\sqrt{\frac{d}{4.9}}\right)^2$$
$$93.9\approx\frac{d}{4.9}$$
$$460.11\approx d$$

The height of the building is approximately 460 meters.

89. (a)

$$C(x)=\sqrt{0.3x+1}$$
$$C(10)=\sqrt{0.3(10)+1}$$
$$C(10)=\sqrt{4}$$
$$C(10)=2$$

The airline's cost would be 2 million dollars.

(b) 320(10000) = 3.2 million

3.2 – 2.0 = 1.2 million dollar profit

(c)

$$4=\sqrt{0.3x+1}$$
$$4^2=\left(\sqrt{0.3x+1}\right)^2$$
$$16=0.3x+1$$
$$15=0.3x$$
$$50=x$$

Approximately 50,000 passengers.

91. (a) $\sqrt{x^2+4}=\sqrt{3^2+4}=\sqrt{13}$;

$x+2=3+2=5$

(b) These expressions are not equal.

93. (a)

$$t(x)=0.90\sqrt[5]{x^3}$$
$$4=0.90\sqrt[5]{x^3}$$
$$4.44\approx\sqrt[5]{x^3}$$
$$4.44^5\approx\left(\sqrt[5]{x^3}\right)^5$$
$$1725.5\approx x^3$$
$$\sqrt[3]{1725.5}\approx\sqrt[3]{x^3}$$
$$11.99\approx x$$

The weight of the turkey is approximately 12 lb.

(b) $t(x) = 0.90\sqrt[5]{x^3}$

$t(18) = 0.90\sqrt[5]{18^3}$

$t(18) = 0.90(5.66)$

$t(18) = 5.094$

An 18 lb turkey will take about 5.1 hours to cook.

95. $a^2 + b^2 = c^2$

$k^2 + 9^2 = c^2$

$k^2 + 81 = c^2$

$\sqrt{k^2 + 81} = \sqrt{c^2}$

$\sqrt{k^2 + 81} = c$

97. $a^2 + b^2 = c^2$

$h^2 + b^2 = 5^2$

$h^2 + b^2 = 25$

$b^2 = 25 - h^2$

$\sqrt{b^2} = \sqrt{25 - h^2}$

$b = \sqrt{25 - h^2}$

99. $a^2 + b^2 = c^2$

$a^2 + 14^2 = k^2$

$a^2 + 196 = k^2$

$a^2 = k^2 - 196$

$\sqrt{a^2} = \sqrt{k^2 - 196}$

$a = \sqrt{k^2 - 196}$

101. $\sqrt{4x - \sqrt{8x^2 + 1}} = 1$

$\left(\sqrt{4x - \sqrt{8x^2 + 1}}\right)^2 = (1)^2$

$4x - \sqrt{8x^2 + 1} = 1$

$-\sqrt{8x^2 + 1} = -4x + 1$

$\left(-\sqrt{8x^2 + 1}\right)^2 = (-4x + 1)^2$

$8x^2 + 1 = 16x^2 - 8x + 1$

$0 = 8x^2 - 8x$

$0 = 8x(x - 1)$

$8x = 0$ or $x - 1 = 0$

$x = 0$ or $x = 1$

Check:

$x = 0$: $\sqrt{4 \cdot 0 - \sqrt{8 \cdot 0^2 + 1}} = 1$

$\sqrt{0 - \sqrt{1}} = 1$

$\sqrt{0 - 1} = 1$

$\sqrt{-1} = 1$ False

$x = 1$: $\sqrt{4 \cdot 1 - \sqrt{8 \cdot 1^2 + 1}} = 1$

$\sqrt{4 - \sqrt{9}} = 1$

$\sqrt{4 - 3} = 1$

$\sqrt{1} = 1$

$1 = 1$

Since $x = 0$ does not check, the solution is $x = 1$.

103. $\sqrt[3]{q^3 + 9q^2 - 27} = q + 3$

$\left(\sqrt[3]{q^3 + 9q^2 - 27}\right)^3 = (q + 3)^3$

$q^3 + 9q^2 - 27 = q^3 + 9q^2 + 27q + 27$

$0 = 27q + 54$

$-54 = 27q$

$-2 = q$

105.

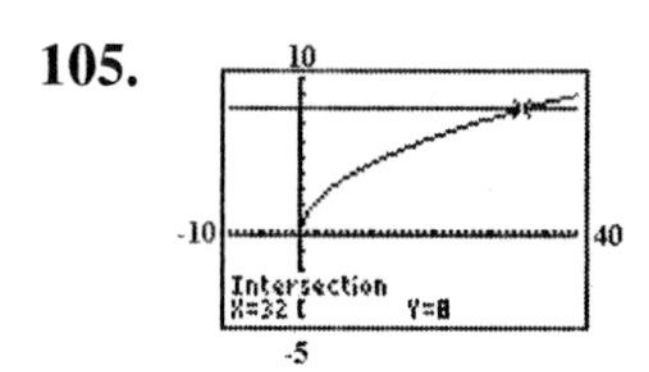

107.

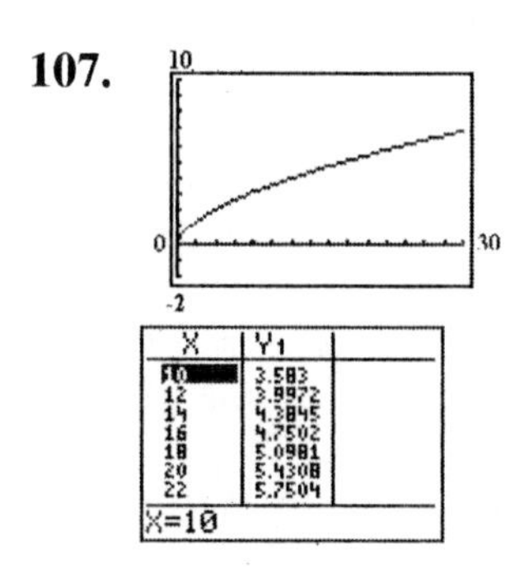

Section 10.8 Practice Exercises

1. $-2\sqrt{5}-3\sqrt{50}+\sqrt{125}$
$=-2\sqrt{5}-3\sqrt{25\cdot 2}+\sqrt{25\cdot 5}$
$=-2\sqrt{5}-15\sqrt{2}+5\sqrt{5}$
$=3\sqrt{5}-15\sqrt{2}$

3. $(3-\sqrt{x})(3+\sqrt{x})=9-x$

5. $\sqrt{5y-4}-2=4$
$\sqrt{5y-4}=6$
$\left(\sqrt{5y-4}\right)^2=6^2$
$5y-4=36$
$5y=40$
$y=8$
Check: $\sqrt{5(8)-4}-2=4$
$\sqrt{40-4}-2=4$
$\sqrt{36}-2=4$
$6-2=4$

7. $\sqrt[3]{3p+7}-\sqrt[3]{2p-1}=0$
$\sqrt[3]{3p+7}=\sqrt[3]{2p-1}$
$\left(\sqrt[3]{3p+7}\right)^3=\left(\sqrt[3]{2p-1}\right)^3$
$3p+7=2p-1$
$p=-8$

9. $\sqrt{36c+15}=6\sqrt{c}+1$
$\left(\sqrt{36c+15}\right)^2=\left(6\sqrt{c}+1\right)^2$
$36c+15=36c+12\sqrt{c}+1$
$14=12\sqrt{c}$
$(14)^2=\left(12\sqrt{c}\right)^2$
$196=144c$
$\frac{49}{36}=c$

Check: $\sqrt{36\left(\frac{49}{36}\right)+15}=6\sqrt{\left(\frac{49}{36}\right)}+1$
$\sqrt{49+15}=6\left(\frac{7}{6}\right)+1$
$\sqrt{64}=7+1$
$8=8$

11. $i=\sqrt{-1}$

13. $a-bi$

15. $\sqrt{-144}=12i$

17. $\sqrt{-3}=i\sqrt{3}$

19. $\sqrt{-20}=2i\sqrt{5}$

21. $3\sqrt{-18}+5\sqrt{-32}=3\sqrt{-1\cdot 9\cdot 2}+5\sqrt{-1\cdot 16\cdot 2}$
$=9i\sqrt{2}+20i\sqrt{2}$
$=29i\sqrt{2}$

23. $7\sqrt{-63}-4\sqrt{-28}=7\sqrt{-1\cdot 9\cdot 7}-4\sqrt{-1\cdot 4\cdot 7}$
$=21i\sqrt{7}-8i\sqrt{7}$
$=13i\sqrt{7}$

25. $\sqrt{-7}\sqrt{-7}=i\sqrt{7}\cdot i\sqrt{7}=i^2\sqrt{49}=-1(7)=-7$

27. $\sqrt{-9}\sqrt{-16}=3i\cdot 4i=12i^2=-12$

29. $\sqrt{-15}\sqrt{-6}=i\sqrt{15}\cdot i\sqrt{6}=i^2\sqrt{90}=-3\sqrt{10}$

31. $\frac{\sqrt{-50}}{\sqrt{-25}}=\frac{\sqrt{-1\cdot 25\cdot 2}}{5i}=\frac{5i\sqrt{2}}{5i}=\sqrt{2}$

33. $\frac{\sqrt{-90}}{\sqrt{10}}=\frac{\sqrt{-1\cdot 9\cdot 10}}{\sqrt{10}}=\frac{3i\sqrt{10}}{\sqrt{10}}=3i$

35. $\dfrac{2+\sqrt{-16}}{8}=\dfrac{2+i\sqrt{16}}{8}$
$=\dfrac{2+4i}{8}$
$=\dfrac{2(1+2i)}{8}$
$=\dfrac{1+2i}{4}$

37. $\dfrac{5-\sqrt{-75}}{10}=\dfrac{5-i\sqrt{75}}{10}$
$=\dfrac{5-i\sqrt{25\cdot 3}}{10}$
$=\dfrac{5-5i\sqrt{3}}{10}$
$=\dfrac{5\left(1-i\sqrt{3}\right)}{10}$
$=\dfrac{1-i\sqrt{3}}{2}$

39. $\dfrac{-6\pm\sqrt{-72}}{6}=\dfrac{-6\pm i\sqrt{72}}{6}$
$=\dfrac{-6\pm 6i\sqrt{2}}{6}$
$=\dfrac{6\left(-1\pm i\sqrt{2}\right)}{6}$
$=-1\pm i\sqrt{2}$

41. $\dfrac{-8\pm\sqrt{-48}}{4}=\dfrac{-8\pm i\sqrt{48}}{4}$
$=\dfrac{-8\pm 4i\sqrt{3}}{4}$
$=\dfrac{4\left(-2\pm i\sqrt{3}\right)}{4}$
$=-2\pm i\sqrt{3}$

43. $(2-i)+(5+7i)=(2+5)+(-1+7)i$
$=7+6i$

45. $\left(\dfrac{1}{2}+\dfrac{2}{3}i\right)-\left(\dfrac{1}{5}-\dfrac{5}{6}i\right)=\left(\dfrac{1}{2}-\dfrac{1}{5}\right)+\left(\dfrac{2}{3}+\dfrac{5}{6}\right)i$
$=\dfrac{3}{10}+\dfrac{9}{6}i$
$=\dfrac{3}{10}+\dfrac{3}{2}i$

47. $(1+3i)+(4-3i)=(1+4)+(3-3)i$
$=5+0i$

49. $(2+3i)-(1-4i)+(-2+3i)$
$=(2-1-2)+(3+4+3)i$
$=-1+10i$

51. $i^7=i^4\cdot i^3=1\cdot i^3=1\cdot -i=-i$

53. $i^{64}=\left(i^4\right)^{16}=1^{16}=1$

55. $i^{41}=i^{40}\cdot i=\left(i^4\right)^{10}\cdot i=1^{10}\cdot i=i$

57. $i^{52}=\left(i^4\right)^{13}=1$

59. $i^{23}=i^{20}\cdot i^3=\left(i^4\right)^5\cdot i^3=1^5\cdot i^3=i^3=-i$

61. $i^6=i^4\cdot i^2=1\cdot i^2=-1$

63. $(8i)(3i)=24i^2=-24=-24+0i$

65. $6i(1-3i)=6i-18i^2=18+6i$

67. $(2-10i)(3+2i)=6+4i-30i-20i^2$
$=6-26i+20$
$=26-26i$

69. $(5+2i)(5+2i)=-25-10i+10i+4i^2$
$=-25-4$
$=-29+0i$

71. $(4+5i)^2=16+40i+25i^2$
$=16+40i-25$
$=-9+40i$

73. $(2+i)(3-2i)(4+3i)$
$=(2+i)(12+9i-8i-6i^2)$
$=(2+i)(18+i)$
$=36+2i+18i+i^2$
$=35+20i$

75. Conjugate: $1-3i$
$(1+3i)(1-3i)=1-9i^2=1+9=10$

77. Conjugate: $4+3i$
$(4-3i)(4+3i)=16-9i^2=16+9=25$

79. $$\frac{2}{1+3i}\cdot\frac{1-3i}{1-3i}=\frac{2-6i}{1-9i^2}=\frac{2-6i}{1+9}=\frac{2-6i}{10}=\frac{1}{5}-\frac{3}{5}i$$

81. $$\frac{-i}{4-3i}\cdot\frac{4+3i}{4+3i}=\frac{-4i-3i^2}{16-9i^2}=\frac{-4i+3}{16+9}=\frac{3-4i}{25}=\frac{3}{25}-\frac{4}{25}i$$

83. $$\frac{5+2i}{5-2i}\cdot\frac{5+2i}{5+2i}=\frac{25+20i+4i^2}{25-4i^2}=\frac{25+20i-4}{25+4}=\frac{21+20i}{29}=\frac{21}{29}+\frac{20}{29}i$$

85. $$\frac{3}{2i}\cdot\frac{i}{i}=\frac{3i}{2i^2}=\frac{3i}{-2}=0-\frac{3}{2}i$$

87. $$\frac{3}{-i}\cdot\frac{i}{i}=\frac{3i}{-i^2}=\frac{3i}{1}=0+3i$$

89. $$7i^{-5}=\frac{7}{i^5}=\frac{7}{i^4\cdot i}=\frac{7}{i}\cdot\frac{i}{i}=\frac{7i}{i^2}=\frac{7i}{-1}=0-7i$$

91. $$12i^{-8}=\frac{12}{i^8}=\frac{12}{\left(i^4\right)^2}=12+0i$$

93. $$i^{-10}=\frac{1}{i^{10}}=\frac{1}{\left(i^4\right)^2\cdot i^2}=\frac{1}{i^2}=\frac{1}{-1}=-1+0i$$

Chapter 10 Review Exercises

1. **(a)** False, $\sqrt{0}=0$ is not positive.

(b) False, $\sqrt[3]{-8}=-2$.

3. **(a)** False

(b) True

5. $\sqrt[4]{625}=\sqrt[4]{5^4}=5$

7. $f(x)=\sqrt{x-1}$

(a) $f(10)=\sqrt{10-1}=\sqrt{9}=3$

(b) $f(1)=\sqrt{1-1}=\sqrt{0}=0$

(c) $f(8)=\sqrt{8-1}=\sqrt{7}$

(d) $x-1\geq 0$
$x\geq 1$
Domain: $[1,\infty)$

9. $$\frac{\sqrt[3]{2x}}{\sqrt[4]{2x}}+4$$

11. $a^2+b^2=c^2$
$15^2+b^2=17^2$
$225+b^2=289$
$b^2=64$
$b=8$
The length of the third side is 8 cm.

13. Yes, provided the expressions are well defined. For example:
$x^5 \cdot x^3 = x^8$ and $x^{1/5} \cdot x^{1/3} = x^{8/15}$.

15. Take the reciprocal of the base and change the exponent to positive.

17. $16^{-1/4} = \left(\frac{1}{16}\right)^{1/4} = \sqrt[4]{\frac{1}{16}} = \frac{1}{2}$

19. $\left(b^{1/2} \cdot b^{1/3}\right)^{12} = b^6 \cdot b^4 = b^{10}$

21. $\sqrt[4]{x^3} = x^{3/4}$

23. $10^{1/3} \approx 2.1544$

25. $147^{4/5} \approx 54.1819$

27. For a radical expression to be simplified the following conditions must be met:
(1) Factors of the radicand must have powers less than the index.
(2) There may be no fractions in the radicand.
(3) There may be no radical in the denominator of a fraction.

29. $\sqrt[4]{x^5yz^4} = \sqrt[4]{x^4z^4 \cdot xy} = xz\sqrt[4]{xy}$

31. $\sqrt[3]{\frac{-16x^7y^6}{z^9}} = \sqrt[3]{\frac{-8x^6y^6 \cdot 2x}{z^9}} = \frac{-2x^2y^2\sqrt[3]{2x}}{z^3}$

33. $\frac{1}{8}(5280) = 660; \ \frac{1}{2}(1.5) = 0.75$

$$a^2 + b^2 = c^2$$
$$(660)^2 + h^2 = (660.75)^2$$
$$435600 + h^2 = 436590.5625$$
$$h^2 = 990.5625$$
$$h \approx 31.47$$

The height of the bulge is approximately 31 feet.

35. Cannot be combined. The indices are different.

37. Can be combined.
$\sqrt[4]{3xy} + 2\sqrt[4]{3xy} = 3\sqrt[4]{3xy}$

39. $4\sqrt{7} - 2\sqrt{7} + 3\sqrt{7} = (4 - 2 + 3)\sqrt{7} = 5\sqrt{7}$

41.
$$\begin{aligned}\sqrt{50} + 7\sqrt{2} - \sqrt{8} &= 5\sqrt{2} + 7\sqrt{2} - 2\sqrt{2} \\ &= (5 + 7 - 2)\sqrt{2} \\ &= 10\sqrt{2}\end{aligned}$$

43. False; 5 and $3\sqrt{x}$ are not like radicals.

45. $a + b$ and $a - b$ are conjugates.

47. $\sqrt{3} \cdot \sqrt{12} = \sqrt{36} = 6$

49.
$$\begin{aligned}-2\sqrt{3}\left(\sqrt{3} - 3\sqrt{3}\right) &= -2\sqrt{9} + 6\sqrt{9} \\ &= (-2 \cdot 3) + (6 \cdot 3) \\ &= -6 + 18 \\ &= 12\end{aligned}$$

51.
$$\begin{aligned}&\left(\sqrt[3]{2x} - \sqrt[3]{4x}\right)^2 \\ &= \left(\sqrt[3]{2x}\right)^2 - 2\left(\sqrt[3]{2x}\sqrt[3]{4x}\right) + \left(\sqrt[3]{4x}\right)^2 \\ &= \sqrt[3]{4x^2} - 2\left(\sqrt[3]{8x^2}\right) + \sqrt[3]{16x^2} \\ &= \sqrt[3]{4x^2} - 4\sqrt[3]{x^2} + 2\sqrt[3]{2x^2}\end{aligned}$$

53.
$$\begin{aligned}\sqrt[3]{u} \cdot \sqrt{u^5} &= u^{1/3} \cdot u^{5/2} \\ &= u^{\frac{1}{3}+\frac{5}{2}} \\ &= u^{\frac{2}{6}+\frac{15}{6}} \\ &= u^{17/6} \\ &= \sqrt[6]{u^{17}} \\ &= \sqrt[6]{u^{12} \cdot u^5} \\ &= u^2\sqrt[6]{u^5}\end{aligned}$$

55. $\sqrt[3]{(a+b)} \cdot \sqrt[6]{(a+b)^5} = (a+b)^{1/3} \cdot (a+b)^{5/6}$
$= (a+b)^{\frac{1}{3}+\frac{5}{6}}$
$= (a+b)^{\frac{2}{6}+\frac{5}{6}}$
$= (a+b)^{7/6}$
$= \sqrt[6]{(a+b)^7}$
$= (a+b)\sqrt[6]{(a+b)}$

57. $\frac{2\sqrt{x}+\sqrt{5}}{\sqrt{5x}} \cdot \frac{\sqrt{5x}}{\sqrt{5x}} = \frac{2\sqrt{5x^2}+\sqrt{25x}}{\sqrt{25x^2}}$
$= \frac{2x\sqrt{5}+5\sqrt{x}}{5x}$

59. $\frac{2\sqrt{b}-1}{3\sqrt{b}-1} \cdot \frac{3\sqrt{b}+1}{3\sqrt{b}+1} = \frac{6b-\sqrt{b}-1}{9b-1}$

61. $\sqrt{2y} = 7$
$\left(\sqrt{2y}\right)^2 = 7^2$
$2y = 49$
$y = \frac{49}{2}$
Check: $\sqrt{2\left(\frac{49}{2}\right)} = 7$
$\sqrt{49} = 7$
$7 = 7$

63. $\sqrt[3]{2w-3} + 5 = 2$
$\sqrt[3]{2w-3} = -3$
$\left(\sqrt[3]{2w-3}\right)^3 = (-3)^3$
$2w-3 = -27$
$2w = -24$
$w = -12$

65. $\sqrt{t} + \sqrt{t-5} = 5$
$\sqrt{t-5} = 5-\sqrt{t}$
$\left(\sqrt{t-5}\right)^2 = \left(5-\sqrt{t}\right)^2$
$t-5 = 25-10\sqrt{t}+t$
$-30 = -10\sqrt{t}$
$3 = \sqrt{t}$
$3^2 = \left(\sqrt{t}\right)^2$
$9 = t$
Check: $\sqrt{9}+\sqrt{9-5} = 5$
$3+\sqrt{4} = 5$
$3+2 = 5$
$5 = 5$

67. $\sqrt{2m^2+4} - \sqrt{9m} = 0$
$\sqrt{2m^2+4} = \sqrt{9m}$
$\left(\sqrt{2m^2+4}\right)^2 = \left(\sqrt{9m}\right)^2$
$2m^2+4 = 9m$
$2m^2-9m+4 = 0$
$(2m-1)(m-4) = 0$

$2m-1 = 0$ or $m-4 = 0$
$m = \frac{1}{2}$ or $m = 4$
Check:
$m = \frac{1}{2}$: $\sqrt{2\left(\frac{1}{2}\right)^2+4} - \sqrt{9\left(\frac{1}{2}\right)} = 0$
$\sqrt{\frac{1}{2}+4} - \sqrt{\frac{9}{2}} = 0$
$\sqrt{\frac{9}{2}} - \sqrt{\frac{9}{2}} = 0$
$0 = 0$
$m = 4$: $\sqrt{2(4)^2+4} - \sqrt{9(4)} = 0$
$\sqrt{36} - \sqrt{36} = 0$
$0 = 0$

69. $v(d)=\sqrt{32d}$

(a) $v(20)=\sqrt{32(20)}\approx 25.3$

25.3 ft/sec; when the water depth is 20 ft, a wave travels about 25.3 ft/sec.

(b) $16=\sqrt{32d}$

$(16)^2=\left(\sqrt{32d}\right)^2$

$256=32d$

$8=d$

The depth is 8 ft when the wave travels 16 ft/sec.

71. $a+bi$ where a and b are real numbers and $i=\sqrt{-1}$.

73. For each case, simplify the expression by multiplying the numerator and denominator by the conjugate of the denominator.

75. $-\sqrt{-5}=-i\sqrt{5}$

77. $i^{38}=i^{36}\cdot i^2=(i^4)^9\cdot i^2=-1$

79. $i^{19}=i^{16}\cdot i^3=(i^4)^4\cdot i^3=-i$

81. $2i^{17}-3i^{23}+2i^{24}+4i^{34}$

$=2(i^{16}\cdot i)-3(i^{20}\cdot i^3)+2(i^4)^6+4(i^{32}\cdot i^2)$

$=2(i^4)^4\cdot i-3(i^4)^5\cdot i^3+2+4(i^4)^8\cdot i^2$

$=2i-3i^3+2+4i^2$

$=2i+3i+2-4$

$=-2+5i$

83. $(2i+4)(2i-4)=4i^2-16$

$=-4-16$

$=-20+0i$

85. $(5-i)^2=25-10i+i^2$

$=25-10i-1$

$=24-10i$

87. $\dfrac{-16-8i}{8}=-2-i$

real part: -2

imaginary part: -1

89. $\dfrac{2+i}{i^5}\cdot\dfrac{i^3}{i^3}=\dfrac{2i^3+i^4}{i^8}$

$=\dfrac{2i^3+i^4}{(i^4)^2}$

$=-2i+1$

$=1-2i$

91. $\dfrac{6\pm\sqrt{-144}}{3}=\dfrac{6\pm i\sqrt{144}}{3}$

$=\dfrac{6\pm 12i}{3}$

$=\dfrac{3(2\pm 4i)}{3}$

$=2\pm 4i$

Chapter 10 Test

1. (a) 6

(b) -6

3. (a) $\sqrt[3]{y^3}=y$

(b) $\sqrt[4]{y^4}=|y|$

5. $\sqrt{\dfrac{16}{9}}=\dfrac{4}{3}$

7. $\sqrt{a^4b^3c^5}=\sqrt{a^4b^2c^4\cdot bc}=a^2bc^2\sqrt{bc}$

9. $\sqrt{\dfrac{32w^6}{3w}}=\dfrac{\sqrt{32w^6}}{\sqrt{3w}}$

$=\dfrac{\sqrt{16w^6\cdot 2}}{\sqrt{3w}}$

$=\dfrac{4w^3\sqrt{2}}{\sqrt{3w}}$

$=\dfrac{4w^3\sqrt{2}}{\sqrt{3w}}\cdot\dfrac{\sqrt{3w}}{\sqrt{3w}}$

$=\dfrac{4w^3\sqrt{6w}}{3w}$

$=\dfrac{4w^2\sqrt{6w}}{3}$

11. $\dfrac{\sqrt[3]{10}}{\sqrt[4]{10}} = \dfrac{10^{1/3}}{10^{1/4}}$
$= 10^{\frac{1}{3}-\frac{1}{4}}$
$= 10^{\frac{4}{12}-\frac{3}{12}}$
$= 10^{1/12}$
$= \sqrt[12]{10}$

13. –0.3080

15. $\dfrac{t^{-1}\cdot t^{1/2}}{t^{1/4}} = \dfrac{t^{-1+\frac{1}{2}}}{t^{1/4}}$
$= \dfrac{t^{-1/2}}{t^{1/4}}$
$= t^{-\frac{1}{2}-\frac{1}{4}}$
$= t^{-\frac{2}{4}-\frac{1}{4}}$
$= t^{-3/4}$
$= \dfrac{1}{t^{3/4}}$

17. **(a)** $3\sqrt{x}\left(\sqrt{2}-\sqrt{5}\right) = 3\sqrt{2x} - 3\sqrt{5x}$

(b) $\left(\sqrt{2x}-3\right)^2 = 2x - 6\sqrt{2x} + 9$

19. **(a)** $\sqrt{-8} = i\sqrt{4\cdot 2} = 2i\sqrt{2}$

(b) $2\sqrt{-16} = 2i\sqrt{16} = 8i$

(c) $\dfrac{2\pm\sqrt{-8}}{4} = \dfrac{2\pm i\sqrt{8}}{4}$
$= \dfrac{2\pm 2i\sqrt{2}}{4}$
$= \dfrac{2\left(1\pm i\sqrt{2}\right)}{4}$
$= \dfrac{1\pm i\sqrt{2}}{2}$

21. $(4+i)(8+2i) = 4(8) + 4(2i) + 8i + i(2i)$
$= 32 + 8i + 8i + 2i^2$
$= 32 + 16i + 2i^2$
$= 32 + 16i + 2(-1)$
$= 32 + 16i - 2$
$= 30 + 16i$

23. $(10+3i)[(-5i+8)-(5-3i)]$
$= (10+3i)[(8-5)+(-5i+3i)]$
$= (10+3i)(3-2i)$
$= 30 - 20i + 9i - 6i^2$
$= 36 - 11i$

25. $r(V) = \sqrt[3]{\dfrac{3V}{4\pi}}$;

$r(10) = \sqrt[3]{\dfrac{3(10)}{4\pi}} = \sqrt[3]{\dfrac{30}{4\pi}} \approx 1.34$

The radius of a sphere of volume 10 cubic units is approximately 1.34 units.

27. $\sqrt[3]{2x+5} = -3$
$\left(\sqrt[3]{2x+5}\right)^3 = (-3)^3$
$2x + 5 = -27$
$2x = -32$
$x = -16$

29. $\sqrt{t+7} - \sqrt{2t-3} = 2$
$\sqrt{t+7} = \sqrt{2t-3} + 2$
$\left(\sqrt{t+7}\right)^2 = \left(\sqrt{2t-3}+2\right)^2$
$t + 7 = 2t - 3 + 4\sqrt{2t-3} + 4$
$t + 7 = 2t + 1 + 4\sqrt{2t-3}$
$-t + 6 = 4\sqrt{2t-3}$
$(-t+6)^2 = \left(4\sqrt{2t-3}\right)^2$
$t^2 - 12t + 36 = 16(2t-3)$
$t^2 - 12t + 36 = 32t - 48$
$t^2 - 44t + 84 = 0$
$(t-2)(t-42) = 0$
$t - 2 = 0$ or $t - 42 = 0$
$t = 2$ or $t = 42$

Check:

$t = 2$: $\sqrt{2+7} - \sqrt{2(2)-3} = 2$
$\sqrt{9} - \sqrt{1} = 2$
$3 - 1 = 2$
$2 = 2$

$t = 42$: $\sqrt{42+7} - \sqrt{2(42)-3} = 2$
$\sqrt{49} - \sqrt{81} = 2$
$7 - 9 = 2$
$-2 = 2$ False

Since $t = 42$ does not check, the solution is $t = 2$.

Cumulative Review Exercises Chapters 1–10

1. $6^2 - 2[5 - 8(3-1) + 4 \div 2]$
$= 36 - 2[5 - 8(2) + 2]$
$= 36 - 2[5 - 16 + 2]$
$= 36 - 2(-9)$
$= 36 + 18$
$= 54$

3. $9(2y+8) = 20 - (y+5)$
$18y + 72 = 20 - y - 5$
$18y + 72 = 15 - y$
$19y = -57$
$y = -3$

5. $2x + y = 9$
$y = -2x + 9$
$m = -2;$
$y - y_1 = m(x - x_1)$
$y + 1 = -2(x - 3)$
$y + 1 = -2x + 6$
$y = -2x + 5$

7. $2x - 3y = 0$
$-4x + 3y = -1$

Eliminate the y variable by adding the two equations together.

$2x - 3y = 0$
$-4x + 3y = -1$
$-2x = -1$
$x = \frac{1}{2}$

Substitute $x = \frac{1}{2}$ into either equation to solve for y.

$-4\left(\frac{1}{2}\right) + 3y = -1$
$-2 + 3y = -1$
$3y = 1$
$y = \frac{1}{3}$

The solution to the system is $\left(\frac{1}{2}, \frac{1}{3}\right)$.

9. $x = 6, y = 3, z = 8$

11. Not a function.

13. $\left(\frac{a^{3/2}b^{-1/4}c^{1/3}}{ab^{-5/4}c^{0}}\right)^{12} = \frac{a^{18}b^{-3}c^{4}}{a^{12}b^{-15}} = a^6b^{12}c^4$

15. $(2x+5)(x-3) = 2x^2 - 6x + 5x - 15$
$= 2x^2 - x - 15$

The degree of the product is 2nd degree.

17. $\frac{x^2 - x - 12}{x+3} = \frac{(x-4)(x+3)}{x+3} = x - 4$

19. $\sqrt[3]{\frac{54c^4}{cd^3}} = \sqrt[3]{\frac{27c^3 \cdot 2}{d^3}} = \frac{3c\sqrt[3]{2}}{d}$

21. $\frac{13i}{3+2i} \cdot \frac{3-2i}{3-2i} = \frac{39i - 26i^2}{9 - 4i^2}$
$= \frac{26 + 39i}{13}$
$= 2 + 3i$

23. (a) $y = \sqrt{2x+1}$

x-intercepts:

$0 = \sqrt{2x+1}$
$(0)^2 = \left(\sqrt{2x+1}\right)^2$
$0 = 2x + 1$
$-\frac{1}{2} = x$

The x-intercept is $\left(-\frac{1}{2}, 0\right)$.

y-intercept:

$$y=\sqrt{2\cdot 0+1}$$
$$y=\sqrt{1}$$
$$y=1$$

The y-intercept is (0, 1).

(b) $y = (x-3)(2x-5)$

x-intercepts:

$$0=(x-3)(2x-5)$$
$$x-3=0 \quad \text{or} \quad 2x-5=0$$
$$x=3 \quad \text{or} \quad x=\frac{5}{2}$$

The x-intercepts are (3, 0), $\left(\frac{5}{2}, 0\right)$.

y-intercept:

$$y=(0-3)(2\cdot 0-5)$$
$$y=(-3)(-5)$$
$$y=15$$

The y-intercept is (0, 15).

25. Let x represent the time required for both pumps working together to drain the pool.

	Work Rate	Time	Portion of Job Completed
Small Pump	$\frac{1}{20}$	x	$\frac{1}{20}x$
Large Pump	$\frac{1}{12}$	x	$\frac{1}{12}x$

$$\frac{1}{20}x+\frac{1}{12}x=1$$

The LCD is 60.

$$60\left(\frac{1}{20}x+\frac{1}{12}x\right)=60(1)$$
$$60\cdot\frac{1}{20}x+60\cdot\frac{1}{12}x=60$$
$$3x+5x=60$$
$$8x=60$$
$$x=7.5$$

It will take 7.5 hours.

27.
$$\frac{2t^2+9t+4}{t^2-16}\cdot\frac{t^2-4t}{2t^2-5t-3}$$
$$=\frac{(2t+1)(t+4)}{(t+4)(t-4)}\cdot\frac{t(t-4)}{(2t+1)(t-3)}$$
$$=\frac{t}{t-3}$$

29.
$$y^6-8=(y^2)^3-2^3$$
$$=(y^2-2)[(y^2)^2+(y^2)(2)+(2)^2]$$
$$=(y^2-2)(y^4+2y^2+4)$$

31.
$$|2x-7|+5\le 10$$
$$|2x-7|\le 5$$
$$-5\le 2x-7\le 5$$
$$2\le 2x\le 12$$
$$1\le x\le 6$$

Interval notation: [1, 6]

33. $x(x-2)\le 15$

Find the boundary points.

$$x(x-2)=15$$
$$x^2-2x=15$$
$$x^2-2x-15=0$$
$$(x-5)(x+3)=0$$
$$x-5=0 \quad \text{or} \quad x+3=0$$
$$x=5 \quad \text{or} \quad x=-3$$

Plot the boundary points on the number line and test a point from each region.

<u>Test $x=-4$</u>: $(-4)(-4-2)\le 15$
$$(-4)(-6)\le 15$$
$$24\le 15 \quad \text{False}$$

<u>Test $x=0$</u>: $0(0-2)\le 15$
$$0(-2)\le 15$$
$$0\le 15 \quad \text{True}$$

<u>Test $x=6$</u>: $6(6-2)\le 15$
$$6(4)\le 15$$
$$24\le 15 \quad \text{False}$$

Since the test point $x = 0$ is true, the solution is [–3, 5]. The endpoints are included in the solution because the symbol "≤" does include equality.

35. Let x represent the unit cost to develop small photos.
Let y represent the unit cost to develop midsize photos.

$$24x + 36y = 15.36$$
$$48x + 12y = 13.92$$

Multiply the first equation by –2, then add the equations.

$$\begin{aligned} -48x - 72y &= -30.72 \\ \underline{48x + 12y} &\underline{= 13.92} \\ -60y &= -16.8 \\ y &= 0.28 \end{aligned}$$

Substitute $y = 0.28$ into the first equation and solve for x.

$$\begin{aligned} 24x + 36(0.28) &= 15.36 \\ 24x + 10.08 &= 15.36 \\ 24x &= 5.28 \\ x &= 0.22 \end{aligned}$$

A small photo costs \$0.22 to develop, and a midsize photo costs \$0.28 to develop.

Chapter 11

Section 11.1 Practice Exercises

1. $x^2 = 100$
$x = \pm 10$

3. $a^2 = 5$
$a = \pm\sqrt{5}$

5. $v^2 + 11 = 0$
$v^2 = -11$
$v = \pm i\sqrt{11}$

7. $(p-5)^2 = 9$
$p - 5 = \pm 3$
$p = 5 \pm 3$
$p = 8$ or $p = 2$

9. $(x-2)^2 = 5$
$x - 2 = \pm\sqrt{5}$
$x = 2 \pm \sqrt{5}$

11. $(h-4)^2 = -8$
$h - 4 = \pm 2i\sqrt{2}$
$h = 4 \pm 2i\sqrt{2}$

13. $\left(a - \frac{1}{2}\right)^2 = \frac{3}{4}$
$a - \frac{1}{2} = \pm\frac{\sqrt{3}}{2}$
$a = \frac{1}{2} \pm \frac{\sqrt{3}}{2}$

15. $\left(x - \frac{3}{2}\right)^2 + \frac{7}{4} = 0$
$\left(x - \frac{3}{2}\right)^2 = -\frac{7}{4}$
$x - \frac{3}{2} = \pm\frac{i\sqrt{7}}{2}$
$x = \frac{3}{2} \pm \frac{i\sqrt{7}}{2}$

17. **(1)** Factoring and applying the zero-product rule:
$x^2 - 81 = 0$
$(x-9)(x+9) = 0$
$x = 9$ or $x = -9$

(2) Applying the square root property:
$x^2 - 81 = 0$
$x^2 = 81$
$x = \pm 9$

19. $x^2 + x^2 = 6^2$
$2x^2 = 36$
$x^2 = 18$
$x = \sqrt{18}$
$x = 4.2$ ft

21. $A = P(1+r)^t$

(a) $11664 = 10000(1+r)^2$
$1.1664 = (1+r)^2$
$\sqrt{1.1664} = 1 + 4$
$-1 \pm 1.08 = r$
$r \approx 0.08$ or $r \approx -2.08$
The interest rate is 8%.

(b) $7392.60 = 6000(1+r)^2$
$1.2321 = (1+r)^2$
$\sqrt{1.2321} = 1 + r$
$-1 \pm 1.11 \approx r$
$r \approx 0.11$ or $r \approx -2.11$
The interest rate is 11%.

(c) $6500 = 5000(1+r)^2$
$1.3 = (1+r)^2$
$\sqrt{1.3} = 1 + r$
$-1 \pm 1.1402 \approx r$
$r \approx 0.1402$ or $r \approx -2.1402$
The interest rate is 14.02%.

23. $x^2 - 6x + k;\ k = 9$
$x^2 - 6x + 9 = (x-3)^2$

25. $x^2 - 5y + k;\ k = \frac{25}{4}$

$$y^2 + 5y + \frac{25}{4} = \left(y + \frac{5}{2}\right)^2$$

27. $b^2 + \frac{2}{5}b + k;\ k = \frac{1}{25}$

$$b^2 + \frac{2}{5}b + \frac{1}{25} = \left(b + \frac{1}{5}\right)^2$$

29. **(1)** Write equation in the form $ax^2 + bx + c = 0$.

(2) Divide each term by a.

(3) Isolate the variable terms.

(4) Complete the square and a factor.

(5) Apply the square root property.

31.
$$\begin{aligned} t^2 + 8t + 15 &= 0 \\ t^2 + 8t &= -15 \\ t^2 + 8t + 16 &= -15 + 16 \\ (t+4)^2 &= 1 \\ t + 4 &= \pm 1 \\ t &= -4 \pm 1 \\ t &= -3,\ t = -5 \end{aligned}$$

33.
$$\begin{aligned} x^2 + 6x &= 16 \\ x^2 + 6x + 9 &= 16 + 9 \\ (x+3)^2 &= 25 \\ x + 3 &= \pm 5 \\ x &= -3 \pm 5 \\ x &= 2,\ x = -8 \end{aligned}$$

35.
$$\begin{aligned} p^2 + 4p + 6 &= 0 \\ p^2 + 4p &= -6 \\ p^2 + 4p + 4 &= -6 + 4 \\ (p+2)^2 &= -2 \\ p + 2 &= \pm i\sqrt{2} \\ p &= -2 \pm i\sqrt{2} \end{aligned}$$

37.
$$\begin{aligned} y^2 - 3y - 10 &= 0 \\ y^2 - 3y &= 10 \\ y^2 - 3y + \frac{9}{4} &= 10 + \frac{9}{4} \\ \left(y - \frac{3}{2}\right)^2 &= \frac{49}{4} \\ y - \frac{3}{2} &= \pm\frac{7}{2} \\ y &= \frac{3}{2} \pm \frac{7}{2} \\ y &= 5,\ y = -2 \end{aligned}$$

39.
$$\begin{aligned} 2a^2 + 4a + 5 &= 0 \\ a^2 + 2a + \frac{5}{2} &= 0 \\ a^2 + 2a &= -\frac{5}{2} \\ a^2 + 2a + 1 &= -\frac{5}{2} + 1 \\ (a+1)^2 &= -\frac{3}{2} \\ a + 1 &= \pm i\sqrt{\frac{3}{2}} \\ a + 1 &= \pm\frac{i\sqrt{6}}{2} \\ a &= -1 \pm \frac{i\sqrt{6}}{2} \end{aligned}$$

41.
$$\begin{aligned} 9x^2 - 36x + 40 &= 0 \\ x^2 - 4x + \frac{40}{9} &= 0 \\ x^2 - 4x &= -\frac{40}{9} \\ x^2 - 4x + 4 &= -\frac{40}{9} + 4 \\ (x-2)^2 &= -\frac{4}{9} \\ x - 2 &= \pm\frac{2}{3}i \\ x &= 2 \pm \frac{2}{3}i \end{aligned}$$

43.
$$p^2-\frac{2}{5}p=\frac{2}{25}$$
$$p^2-\frac{2}{5}p+\frac{1}{25}=\frac{2}{25}+\frac{1}{25}$$
$$p^2-\frac{2}{5}p+\frac{1}{25}=\frac{3}{25}$$
$$\left(p-\frac{1}{5}\right)^2=\frac{3}{25}$$
$$p-\frac{1}{5}=\pm\frac{\sqrt{3}}{5}$$
$$p=\frac{1}{5}\pm\frac{\sqrt{3}}{5}$$

45.
$$(2w+5)(w-1)=2$$
$$2w^2+3w-5=2$$
$$2w^2+3w-7=0$$
$$w^2+\frac{3}{2}w-\frac{7}{2}=0$$
$$w^2+\frac{3}{2}w=\frac{7}{2}$$
$$w^2+\frac{3}{2}w+\frac{9}{16}=\frac{7}{2}+\frac{9}{16}$$
$$w^2+\frac{3}{2}w+\frac{9}{16}=\frac{65}{16}$$
$$\left(w+\frac{3}{4}\right)^2=\frac{65}{16}$$
$$w+\frac{3}{4}=\pm\frac{\sqrt{65}}{4}$$
$$w=-\frac{3}{4}\pm\frac{\sqrt{65}}{4}$$

47.
$$n(n-4)=7$$
$$n^2-4n=7$$
$$n^2-4n+4=7+4$$
$$(n-2)^2=11$$
$$n-2=\pm\sqrt{11}$$
$$n=2\pm\sqrt{11}$$

49.
$$A=\pi r^2$$
$$\frac{A}{\pi}=r^2$$
$$\sqrt{\frac{A}{\pi}}=r \text{ or } \frac{\sqrt{A\pi}}{\pi}=r$$

51.
$$a^2+b^2+c^2=d^2$$
$$a^2=d^2-b^2-c^2$$
$$a=\sqrt{d^2-b^2-c^2}$$

53.
$$V=\frac{1}{3}\pi r^2h$$
$$3V=\pi r^2h$$
$$\frac{3V}{\pi h}=r^2$$
$$\sqrt{\frac{3V}{\pi h}}=r \text{ or } \frac{\sqrt{3V\pi h}}{\pi h}=r$$

55.
$$s=2\sqrt{x}$$
$$\frac{s}{2}=\sqrt{x}$$
$$\frac{s^2}{4}=x$$

57.
$$P(x)=-\frac{1}{8}x^2+5x$$

(a)
$$20=-\frac{1}{8}x^2+5x$$
$$-160=x^2-40x$$
$$-160+400=x^2-40x+400$$
$$240=(x-20)^2$$
$$\sqrt{240}=x-20$$
$$\pm15.5=x-20$$
$$20\pm15.5=x$$
$x\approx35.5$ thousand textbooks or
$x\approx4.5$ thousand textbooks.

(b) Profit increases to a point as more books are produced. Beyond that point, the market is "flooded" and profit decreases. Hence, there are two points at which the profit is $20,000. Producing 4.5 thousand books makes the same profit using fewer resources than producing 35.5 thousand books.

Section 11.2 Practice Exercises

1. $(x+5)^2 = 49$
$x+5 = \pm 7$
$x = -5 \pm 7$
$x = -12,\ x = 2$

3. $x^3 - 1 = (x-1)(x^2 + x + 1)$

5. $x^3 - 2x^2 - 9x + 18 = 0$
$(x^3 - 2x) - (9x - 18) = 0$
$x^2(x-2) - 9(x-2) = 0$
$(x-2)(x^2 - 9) = 0$
$(x-2)(x-3)(x+3) = 0$
$x = 2,\ x = 3,\ x = -3$

7. $9uv - 6u + 12v - 9 = 2u(4v-3) + 3(4v-3)$
$= (4v-3)(2u+3)$

9. $\dfrac{16-\sqrt{640}}{4} = \dfrac{16-8\sqrt{10}}{4} = 4 - 2\sqrt{10}$

11. $\dfrac{14-\sqrt{-147}}{7} = \dfrac{14-i\sqrt{147}}{7}$
$= \dfrac{14-7i\sqrt{3}}{7}$
$= 2 - i\sqrt{3}$

13. $x^2 + 2x + 1 = 0$
$a = 1,\ b = 2,\ c = 1$

15. $19m^2 - 8m + 0 = 0$
$a = 19,\ b = -8,\ c = 0$

17. $5p^2 + 0p - 21 = 0$
$a = 5,\ b = 0,\ c = -21$

19. $4n^2 - 8n - 5n^2 - 4 = 0$
$-n^2 - 3n - 4 = 0$
$n^2 + 3n + 4 = 0$
$a = 1,\ b = 3,\ c = 4$

21. $4 - 4(1)(1) = 0$
one rational solution

23. $64 - 4(19)(0) = 64$
two rational solutions

25. $0 - 4(5)(-21) = 420$
two irrational solutions

27. $9 - 4(1)(4) = -7$
two imaginary solutions

29. **(a)** Discriminant:
$b^2 - 4ac = (-3)^2 - 4(3)(-1) = 9 + 12 = 21$
Because the discriminant is positive, there are two real solutions.

(b) Because there are two real solutions in part (a), the related function has two x-intercepts.

31. **(a)** Discriminant:
$b^2 - 4ac = (4)^2 - 4(4)(1) = 16 - 16 = 0$
Because the discriminant is zero, there is one real solution.

(b) Because there is one real solution in part (a), the related function has one x-intercept.

33. **(a)** Discriminant:
$b^2 - 4ac = (-6)^2 - 4(2)(5)$
$= 36 - 40$
$= -4$
Because the discriminant is negative, there are two imaginary solutions.

(b) Because there are two imaginary solutions in part (a), the related function has no x-intercepts.

35. Factoring can be used as a method of solving a quadratic equation if the equation is factorable.

37. The quadratic formula can be used as a method for solving a quadratic equation for any equation written in the form $ax^2 + bx + c = 0$.

39. $a^2+11a-12=0$

$a=1, b=11, c=-12$

$$a=\frac{-11\pm\sqrt{121-4(1)(-12)}}{2}$$

$$a=\frac{-11\pm\sqrt{121+48}}{2}$$

$$a=\frac{-11\pm13}{2}$$

$$a=\frac{-11+13}{2}=\frac{2}{2}=1;$$

$$a=\frac{-11-13}{2}=\frac{-24}{2}=-12$$

41. $9y^2-2y+5=0$

$a=9, b=-2, c=5$

$$y=\frac{2\pm\sqrt{4-4(9)(5)}}{18}$$

$$y=\frac{2\pm\sqrt{4-180}}{18}$$

$$y=\frac{2\pm4i\sqrt{11}}{18}$$

$$y=\frac{1\pm2i\sqrt{11}}{9}$$

43. $12p^2-4p+5=0$

$a=12, b=-4, c=5$

$$p=\frac{4\pm\sqrt{16-4(12)(5)}}{24}$$

$$p=\frac{4\pm\sqrt{16-240}}{24}$$

$$p=\frac{4\pm\sqrt{-224}}{24}$$

$$p=\frac{4\pm4i\sqrt{14}}{24}$$

$$p=\frac{1\pm i\sqrt{14}}{6}$$

45. $z^2-2z-35=0$

$a=1, b=-2, c=-35$

$$z=\frac{2\pm\sqrt{4-4(1)(-35)}}{2}$$

$$z=\frac{2\pm\sqrt{4+140}}{2}$$

$$z=\frac{2\pm\sqrt{144}}{2}$$

$$z=\frac{2\pm12}{10}$$

$$z=\frac{2+12}{2}=\frac{14}{2}=7;$$

$$z=\frac{2-12}{2}=\frac{-10}{2}=-5$$

47. $a^2+3a-8=0$

$a=1, b=3, c=-8$

$$a=\frac{-3\pm\sqrt{9-4(1)(-8)}}{2}$$

$$a=\frac{-3\pm\sqrt{9+32}}{2}$$

$$a=\frac{-3\pm\sqrt{41}}{2}$$

49. $25x^2-20x+4=0$

$a=25, b=-20, c=4$

$$x=\frac{20\pm\sqrt{400-4(25)(4)}}{50}$$

$$x=\frac{20\pm\sqrt{400-400}}{50}$$

$$x=\frac{20}{50}=\frac{2}{5}$$

51. $w^2-6w+14=0$

$a=1, b=-6, c=14$

$$w=\frac{6\pm\sqrt{36-4(1)(14)}}{2}$$

$$w=\frac{6\pm\sqrt{36-56}}{2}$$

$$w=\frac{6\pm2i\sqrt{5}}{2}$$

$$w=3\pm i\sqrt{5}$$

Section 11.2 Practice Exercises

1. $(x+5)^2 = 49$
$x+5 = \pm 7$
$x = -5 \pm 7$
$x = -12, x = 2$

3. $x^3 - 1 = (x-1)(x^2 + x + 1)$

5. $x^3 - 2x^2 - 9x + 18 = 0$
$(x^3 - 2x) - (9x - 18) = 0$
$x^2(x-2) - 9(x-2) = 0$
$(x-2)(x^2 - 9) = 0$
$(x-2)(x-3)(x+3) = 0$
$x = 2, x = 3, x = -3$

7. $9uv - 6u + 12v - 9 = 2u(4v-3) + 3(4v-3)$
$= (4v-3)(2u+3)$

9. $\dfrac{16 - \sqrt{640}}{4} = \dfrac{16 - 8\sqrt{10}}{4} = 4 - 2\sqrt{10}$

11. $\dfrac{14 - \sqrt{-147}}{7} = \dfrac{14 - i\sqrt{147}}{7}$
$= \dfrac{14 - 7i\sqrt{3}}{7}$
$= 2 - i\sqrt{3}$

13. $x^2 + 2x + 1 = 0$
$a = 1, b = 2, c = 1$

15. $19m^2 - 8m + 0 = 0$
$a = 19, b = -8, c = 0$

17. $5p^2 + 0p - 21 = 0$
$a = 5, b = 0, c = -21$

19. $4n^2 - 8n - 5n^2 - 4 = 0$
$-n^2 - 3n - 4 = 0$
$n^2 + 3n + 4 = 0$
$a = 1, b = 3, c = 4$

21. $4 - 4(1)(1) = 0$
one rational solution

23. $64 - 4(19)(0) = 64$
two rational solutions

25. $0 - 4(5)(-21) = 420$
two irrational solutions

27. $9 - 4(1)(4) = -7$
two imaginary solutions

29. **(a)** Discriminant:
$b^2 - 4ac = (-3)^2 - 4(3)(-1) = 9 + 12 = 21$
Because the discriminant is positive, there are two real solutions.

(b) Because there are two real solutions in part (a), the related function has two x-intercepts.

31. **(a)** Discriminant:
$b^2 - 4ac = (4)^2 - 4(4)(1) = 16 - 16 = 0$
Because the discriminant is zero, there is one real solution.

(b) Because there is one real solution in part (a), the related function has one x-intercept.

33. **(a)** Discriminant:
$b^2 - 4ac = (-6)^2 - 4(2)(5)$
$= 36 - 40$
$= -4$
Because the discriminant is negative, there are two imaginary solutions.

(b) Because there are two imaginary solutions in part (a), the related function has no x-intercepts.

35. Factoring can be used as a method of solving a quadratic equation if the equation is factorable.

37. The quadratic formula can be used as a method for solving a quadratic equation for any equation written in the form $ax^2 + bx + c = 0$.

39. $a^2 + 11a - 12 = 0$

$a = 1, b = 11, c = -12$

$$a = \frac{-11 \pm \sqrt{121 - 4(1)(-12)}}{2}$$

$$a = \frac{-11 \pm \sqrt{121 + 48}}{2}$$

$$a = \frac{-11 \pm 13}{2}$$

$$a = \frac{-11 + 13}{2} = \frac{2}{2} = 1;$$

$$a = \frac{-11 - 13}{2} = \frac{-24}{2} = -12$$

41. $9y^2 - 2y + 5 = 0$

$a = 9, b = -2, c = 5$

$$y = \frac{2 \pm \sqrt{4 - 4(9)(5)}}{18}$$

$$y = \frac{2 \pm \sqrt{4 - 180}}{18}$$

$$y = \frac{2 \pm 4i\sqrt{11}}{18}$$

$$y = \frac{1 \pm 2i\sqrt{11}}{9}$$

43. $12p^2 - 4p + 5 = 0$

$a = 12, b = -4, c = 5$

$$p = \frac{4 \pm \sqrt{16 - 4(12)(5)}}{24}$$

$$p = \frac{4 \pm \sqrt{16 - 240}}{24}$$

$$p = \frac{4 \pm \sqrt{-224}}{24}$$

$$p = \frac{4 \pm 4i\sqrt{14}}{24}$$

$$p = \frac{1 \pm i\sqrt{14}}{6}$$

45. $z^2 - 2z - 35 = 0$

$a = 1, b = -2, c = -35$

$$z = \frac{2 \pm \sqrt{4 - 4(1)(-35)}}{2}$$

$$z = \frac{2 \pm \sqrt{4 + 140}}{2}$$

$$z = \frac{2 \pm \sqrt{144}}{2}$$

$$z = \frac{2 \pm 12}{10}$$

$$z = \frac{2 + 12}{2} = \frac{14}{2} = 7;$$

$$z = \frac{2 - 12}{2} = \frac{-10}{2} = -5$$

47. $a^2 + 3a - 8 = 0$

$a = 1, b = 3, c = -8$

$$a = \frac{-3 \pm \sqrt{9 - 4(1)(-8)}}{2}$$

$$a = \frac{-3 \pm \sqrt{9 + 32}}{2}$$

$$a = \frac{-3 \pm \sqrt{41}}{2}$$

49. $25x^2 - 20x + 4 = 0$

$a = 25, b = -20, c = 4$

$$x = \frac{20 \pm \sqrt{400 - 4(25)(4)}}{50}$$

$$x = \frac{20 \pm \sqrt{400 - 400}}{50}$$

$$x = \frac{20}{50} = \frac{2}{5}$$

51. $w^2 - 6w + 14 = 0$

$a = 1, b = -6, c = 14$

$$w = \frac{6 \pm \sqrt{36 - 4(1)(14)}}{2}$$

$$w = \frac{6 \pm \sqrt{36 - 56}}{2}$$

$$w = \frac{6 \pm 2i\sqrt{5}}{2}$$

$$w = 3 \pm i\sqrt{5}$$

53. $x^2 - x - 6 = 1$

$x^2 - x - 7 = 0$

$a = 1, b = -1, c = -7$

$x = \frac{1 \pm \sqrt{1 - 4(1)(-7)}}{2}$

$x = \frac{1 \pm \sqrt{1 + 28}}{2}$

$x = \frac{1 \pm \sqrt{29}}{2}$

55. $\frac{1}{2}y^2 + \frac{2}{3}y + \frac{2}{3} = 0$

$3y^2 + 4y + 4 = 0$

$a = 3, b = 4, c = 4$

$y = \frac{-4 \pm \sqrt{16 - 4(3)(4)}}{6}$

$y = \frac{-4 \pm \sqrt{16 - 48}}{6}$

$y = \frac{-4 \pm \sqrt{-32}}{6}$

$y = \frac{-4 \pm 4i\sqrt{2}}{6}$

$y = \frac{-2 \pm 2i\sqrt{2}}{3}$

57. $\frac{1}{5}h^2 + h + \frac{3}{5} = 0$

$h^2 + 5p + 3 = 0$

$a = 1, b = 5, c = 3$

$h = \frac{-5 \pm \sqrt{25 - 4(1)(3)}}{2}$

$h = \frac{-5 \pm \sqrt{25 - 12}}{2}$

$h = \frac{-5 \pm \sqrt{13}}{2}$

59. $0.01x^2 + 0.06x + 0.08 = 0$

$x^2 + 6x + 8 = 0$

$a = 1, b = 6, c = 8$

$x = \frac{-6 \pm \sqrt{36 - 4(1)(8)}}{2}$

$x = \frac{-6 \pm \sqrt{36 - 32}}{2}$

$x = \frac{-6 \pm \sqrt{4}}{2}$

$x = \frac{-6 \pm 2}{2}$

$x = \frac{-6 + 2}{2} = \frac{-4}{2} = -2;$

$x = \frac{-6 - 2}{2} = \frac{-8}{2} = -4$

61. $0.3t^2 + 0.7t - 0.5 = 0$

$3t^2 + 7t - 5 = 0$

$a = 3, b = 7, c = -5$

$t = \frac{-7 \pm \sqrt{49 - 4(3)(-5)}}{6}$

$t = \frac{-7 \pm \sqrt{49 + 60}}{6}$

$t = \frac{-7 \pm \sqrt{109}}{6}$

63. (a) $x^3 - 27 = (x - 3)(x^2 + 3x + 9)$

(b) $(x - 3)(x^2 + 3x + 9) = 0$

$x^2 + 3x + 9 = 0$ or $x - 3 = 0$

$x = 3$

$a = 1, b = 3, c = 9$

$x = \frac{-3 \pm \sqrt{9 - 4(1)(9)}}{2}$

$x = \frac{-3 \pm \sqrt{9 - 36}}{2}$

$x = \frac{-3 \pm \sqrt{-27}}{2}$

$x = \frac{-3 \pm 3i\sqrt{3}}{2}$

$x = 3,\ x = \frac{-3 \pm 3i\sqrt{3}}{2}$

65. **(a)** $3x^3 - 6x^2 + 6x = 3x(x^2 - 2x + 2)$

(b) $3x(x^2 - 2x + 2) = 0$

$3x = 0 \quad \text{or} \quad x^2 - 2x + 2 = 0$

$x = 0$

$a = 1, b = -2, c = 2$

$$x = \frac{2 \pm \sqrt{4 - 4(1)(2)}}{2}$$

$$x = \frac{2 \pm \sqrt{4 - 8}}{2}$$

$$x = \frac{2 \pm \sqrt{-4}}{2}$$

$$x = \frac{2 \pm 2i}{2}$$

$x = 1 \pm i$

$x = 0, x = 1 \pm i$

67. $V = lwh$

$s^2 = 27$

$s = \sqrt[3]{27}$

$s = 3$ ft

69. $a^2 + 3a + 4 = 0$

$a = 1, b = 3, c = 4$

$$a = \frac{-3 \pm \sqrt{9 - 4(1)(4)}}{2}$$

$$a = \frac{-3 \pm \sqrt{9 - 16}}{2}$$

$$a = \frac{-3 \pm \sqrt{-7}}{2}$$

$$a = \frac{-3 \pm i\sqrt{7}}{2}$$

71. $x^2 - 2 = 0$

$x^2 = 2$

$x = \pm\sqrt{2}$

73. $4y^2 + 8y - 5 = 0$

$(2y + 5)(2y - 1) = 0$

$2y + 5 = 0 \quad \text{or} \quad 2y - 1 = 0$

$2y = -5 \quad \text{or} \quad 2y = 1$

$y = -\frac{5}{2} \quad \text{or} \quad y = \frac{1}{2}$

75. $\left(x + \frac{1}{2}\right)^2 + 4 = 0$

$\left(x + \frac{1}{2}\right)^2 = -4$

$x + \frac{1}{2} = \pm 2i$

$x = -\frac{1}{2} \pm 2i$

77. $d(v) = \frac{v^2}{20} + v$

(a) $150 = \frac{v^2}{20} + v$

$3000 = v^2 + 20v$

$v^2 + 20v - 3000 = 0$

$$v = \frac{-20 \pm \sqrt{400 - 4(1)(-300)}}{2}$$

$$v = \frac{-20 \pm \sqrt{400 + 12000}}{2}$$

$$v = \frac{-20 \pm \sqrt{12400}}{2}$$

$$v = \frac{-20 \pm 111.36}{2}$$

$v \approx 45.68;\ v \approx -65.68$

$v = 46$ mph

(b) $100 = \frac{v^2}{20} + v$

$2000 = v^2 + 20v$

$v^2 + 20v - 2000 = 0$

$$v = \frac{-20 \pm \sqrt{400 - 4(1)(-2000)}}{2}$$

$$v = \frac{-20 \pm \sqrt{400 + 8000}}{2}$$

$$v = \frac{-20 \pm \sqrt{8400}}{2}$$

$$v \approx \frac{-20 \pm 91.65}{2}$$

$v \approx 35.83;\ v \approx -55.83$

$v \approx 36$ mph

79.
$$a^2 + b^2 = c^2$$
$$(b-2.1)^2 + b^2 = (10.2)^2$$
$$b^2 - 4.2b + 4.41 + b^2 = 104.04$$
$$2b^2 - 4.2b - 99.63 = 0$$
$$b = \frac{4.2 \pm \sqrt{17.64 - 4(2)(-99.63)}}{4}$$
$$b = \frac{4.2 \pm \sqrt{814.68}}{4}$$
$$b \approx 8.2;\ b - 2.1 \approx 6.1$$

–6.1 is not a possible answer because there cannot be a negative length. One leg is 8.2 meters. The other leg is 6.1 meters.

81. $N(t) = -1.43t^2 + 94.56t + 4825$

(a) $N(40) = -1.43(40)^2 + 94.56(40) + 4825$
$= 6319.4 \approx 6319$ thousand farms

(b) $N(t)$ gives approximate values.

(c) $5000 = -1.43t^2 + 94.56t + 4825$
$$1.43t^2 - 94.56t + 175 = 0$$
$$t = \frac{94.56 \pm \sqrt{94.56^2 - 4(1.43)(175)}}{2.86}$$
$$t \approx 64.22;\ t \approx 1.9$$

2 years after 1980, or 1892
64 years after 1890, or 1954

83.

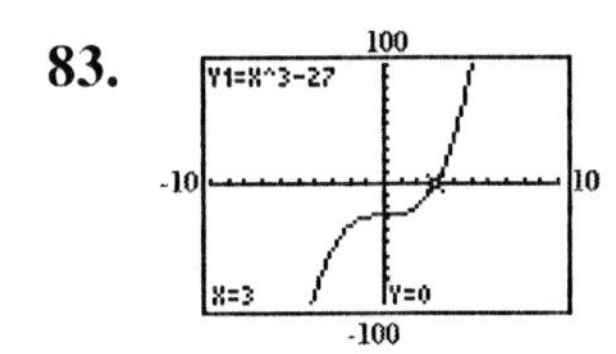

85.

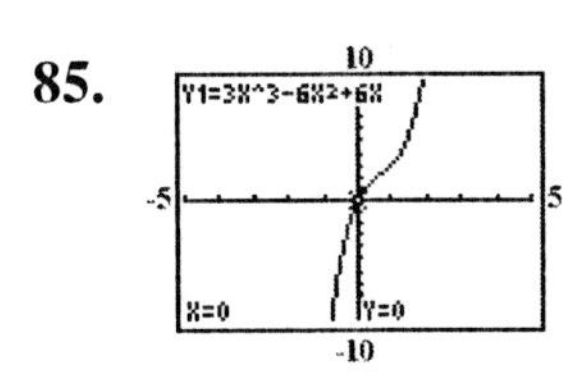

87. $P(t) = 1.12t^2 + 204.4t + 6697$

(a) $P(6) = 1.12(6)^2 + 204.4(6) + 6697$
$= 7963.72 \approx 7964$ thousand people

(b) $P(36) = 1.12(36)^2 + 204.4(36) + 6697$
$= 15506.92$
$\approx 15{,}507$ thousand people

(c) $10000 = 1.12t^2 + 204.4t + 6697$
$$1.12t^2 + 204.4t - 3303 = 0$$
$$t = \frac{-204.4 \pm \sqrt{204.4^2 - 4(1.12)(-3303)}}{2.24}$$
$$t \approx 14.93 \approx 15$$

15 years after 1974, or 1989

(d)

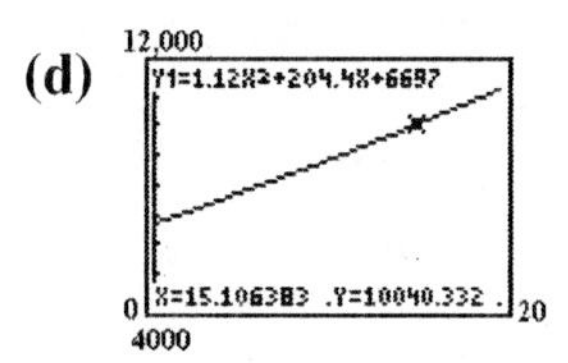

Section 11.3 Practice Exercises

1.
$$y + 6\sqrt{y} = 16$$
$$6\sqrt{y} = 16 - y$$
$$\left(6\sqrt{y}\right)^2 = (16-y)^2$$
$$36y = 256 - 32y + y^2$$
$$y^2 - 68y + 256 = 0$$
$$(y-4)(y-64) = 0$$
$$y = 4 \text{ or } y = 64$$

Check:
$$4 + 6\sqrt{4} = 16$$
$$4 + 6(2) = 16$$
$$4 + 12 = 16 \quad \text{True}$$

$$64 + 6\sqrt{64} = 16$$
$$64 + 6(8) = 16$$
$$64 + 48 = 16 \quad \text{False}$$

$y = 4$ is the only solution.

3.
$$2x + 3\sqrt{x} - 2 = 0$$
$$3\sqrt{x} = 2 - 2x$$
$$(3\sqrt{x})^2 = (2-2x)^2$$
$$9x = 4 - 8x + 4x^2$$
$$4x^2 - 17x + 4 = 0$$
$$(4x-1)(x-16) = 0$$

$x = \frac{1}{4}$ or $x = 16$

Check:

$$2\left(\frac{1}{4}\right) + 3\sqrt{\frac{1}{4}} - 2 = 0$$
$$\frac{1}{2} + 3\left(\frac{1}{2}\right) - 2 = 0$$
$$\frac{1}{2} + \frac{3}{2} - \frac{4}{2} = 0 \quad \text{True}$$

$$2(16) + 3\sqrt{16} - 2 = 0$$
$$32 + 3(4) - 2 = 0$$
$$32 + 12 - 2 = 0 \quad \text{False}$$

$x = \frac{1}{4}$ is the only solution.

5. $\sqrt{4b+1} - \sqrt{b-2} = 3$

$$\sqrt{4b+1} = 3 + \sqrt{b-1}$$
$$\left(\sqrt{4b+1}\right)^2 = \left(3 + \sqrt{b-2}\right)^2$$
$$4b + 1 = 9 + 6\sqrt{b-2} + b - 2$$
$$4b + 1 = 7 + b + 6\sqrt{b-2}$$
$$3b - 6 = 6\sqrt{b-2}$$
$$(3b-6)^2 = \left(6\sqrt{b-2}\right)^2$$
$$9b^2 - 36b + 36 = 36(b-2)$$
$$9b^2 - 36b + 36 = 36b - 72$$
$$9b^2 - 72b + 108 = 0$$
$$9(b^2 - 8b + 12) = 0$$
$$9(b-6)(b-2) = 0$$

$b = 6$ or $b = 2$

Check:

$$\sqrt{4(6)+1} - \sqrt{6-2} = 3$$
$$\sqrt{25} - \sqrt{4} = 3$$
$$5 - 2 = 3 \quad \text{True}$$

$$\sqrt{4(2)+1} - \sqrt{2-2} = 3$$
$$\sqrt{9} - \sqrt{0} = 3$$
$$3 = 3 \quad \text{True}$$

$b = 6$; $b = 2$

7. $\sqrt{w-6} + 3 = \sqrt{w+9}$

$$\left(\sqrt{w-6} + 3\right)^2 = \left(\sqrt{w+9}\right)^2$$
$$w - 6 + 6\sqrt{w-6} + 9 = w + 9$$
$$w + 3 + 6\sqrt{w-6} = w + 9$$
$$6\sqrt{w-6} = 6$$
$$\sqrt{w-6} = 1$$
$$\left(\sqrt{w-6}\right)^2 = (1)^2$$
$$w - 6 = 1$$
$$w = 7$$

Check:

$$\sqrt{7-6} + 3 = \sqrt{7+9}$$
$$1 + 3 = 4 \quad \text{True}$$

$w = 7$

9. $x^4 - 16 = 0$

$$(x^2 - 4)(x^2 + 4) = 0$$
$$x^2 - 4 = 0 \quad \text{or} \quad x^2 + 4 = 0$$
$$x^2 = 4 \quad \text{or} \quad x^2 = -4$$
$$x = \pm 2 \quad \text{or} \quad x = \pm 2i$$

11. $m^4 - 81 = 0$

$$(m^2 - 9)(m^2 + 9) = 0$$
$$m^2 - 9 = 0 \quad \text{or} \quad m^2 + 9 = 0$$
$$m^2 = 9 \quad \text{or} \quad m^2 = -9$$
$$m = \pm 3 \quad \text{or} \quad m = \pm 3i$$

13. $a^3 + 8 = 0$

$$(a+2)(a^2 - 2a + 4) = 0$$
$$a + 2 = 0 \quad \text{or} \quad a^2 - 2a + 4 = 0$$
$$a = -2 \quad \text{or} \quad a = \frac{2 \pm \sqrt{4 - 4(1)(4)}}{2}$$
$$a = -2 \quad \text{or} \quad a = 1 \pm i\sqrt{3}$$

15. $5p^3 - 5 = 0$

$$5(p^3 - 1) = 0$$
$$5(p-1)(p^2 + p + 1) = 0$$

$$p-1=0 \quad \text{or} \quad p^2+p+1=0$$
$$p=1 \quad \text{or} \quad p=\frac{-1\pm\sqrt{1-4(1)(1)}}{2}$$
$$p=1 \quad \text{or} \quad p=\frac{-1\pm i\sqrt{3}}{2}$$

17. (a) $u^2+10u+24=0$
$(u+6)(u+4)=0$
$u+6=0 \quad \text{or} \quad u+4=0$
$u=-6 \quad \text{or} \quad u=-4$

(b) $(y^2+5y)^2+10(y^2+5y)+24=0$

Substitute $u=y^2+5y$.

$u^2+10u+24=0$
$(u+6)(u+4)=0$
$u+6=0 \quad \text{or} \quad u+4=0$
$u=-6 \quad \text{or} \quad u=-4$

Substitute $u=-6$.

$y^2+5y=-6$
$y^2+5y+6=0$
$(y+3)(y+2)=0$
$y+3=0 \quad \text{or} \quad y+2=0$
$y=-3 \quad \text{or} \quad y=-2$

Substitute $u=-4$.

$y^2+5y=-4$
$y^2+5y+4=0$
$(y+4)(y+1)=0$
$y+4=0 \quad \text{or} \quad y+1=0$
$y=-4 \quad \text{or} \quad y=-1;$
$y=-3, y=-2, y=-4, y=-1$

19. (a) $u^2-2u-24=0$
$(u-6)(u+4)=0$
$u-6=0 \quad \text{or} \quad u+4=0$
$u=6 \quad \text{or} \quad u=-4$

(b) $(x^2-5x)^2-2(x^2-5x)-24=0$

Substitute $u=x^2-5x$.

$u^2-2u-24=0$
$(u-6)(u+4)=0$
$u-6=0 \quad \text{or} \quad u+4=0$
$u=6 \quad \text{or} \quad u=-4$

Substitute $u=6$.

$x^2-5x=6$
$x^2-5x-6=0$
$(x-6)(x+1)=0$
$x-6=0 \quad \text{or} \quad x+1=0$
$x=6 \quad \text{or} \quad x=-1$

$x^2-5x=-4$
$x^2-5x+4=0$
$(x-4)(x-1)=0$
$x-4=0 \quad \text{or} \quad x-1=0$
$x=4 \quad \text{or} \quad x=1;$
$x=6, x=-1, x=4, x=1$

21. $(4x+5)^2+3(4x+5)+2=0$

Substitute $u=4x+5$.

$u^2+3u+2=0$
$(u+2)(u+1)=0$
$u+2=0 \quad \text{or} \quad u+1=0$
$u=-2 \quad \text{or} \quad u=-1$

Substitute $u=-2$.

$4x+5=-2$
$4x=-7$
$x=\frac{-7}{4}$

Substitute $u=-1$.

$4x+5=-1$
$4x=-6$
$x=\frac{-3}{2};$

$x=-\frac{7}{4}, x=-\frac{3}{2}$

23. $16\left(\frac{x+6}{4}\right)^2+8\left(\frac{x+6}{4}\right)+1=0$

Substitute $u=\frac{x+6}{4}$.

$$16u^2+8u+1=0$$
$$(4u+1)(4u+1)=0$$
$$4u+1=0$$
$$4u=-1$$
$$u=-\frac{1}{4}$$

Substitute $u=-\frac{1}{4}$.

$$\frac{x+6}{4}=\frac{-1}{4}$$
$$x+6=-1$$
$$x=-7$$

25. $(x^2-2x)^2+2(x^2-2x)=3$

Substitute $u=x^2-2x$.

$$u^2+2u=3$$
$$u^2+2u-3=0$$
$$(u+3)(u-1)=0$$
$$u+3=0 \quad \text{or} \quad u-1=0$$
$$u=-3 \quad \text{or} \quad u=1$$

Substitute $u=-3$.

$$x^2-2x=-3$$
$$x^2-2x+3=0$$
$$x=\frac{2\pm\sqrt{4-4(1)(3)}}{2}$$
$$x=\frac{2\pm 2i\sqrt{2}}{2}=1\pm i\sqrt{2}$$

Substitute $u=1$.

$$x^2-2x=1$$
$$x^2-2x-1=0$$
$$x=\frac{2\pm\sqrt{4-4(1)(-1)}}{2}$$
$$x=\frac{2\pm 2\sqrt{2}}{2}=1\pm\sqrt{2}$$
$$x=1\pm i\sqrt{2},\ x=1\pm\sqrt{2}$$

27. $x^4-13x^2+36=0$

Substitute $u=x^2$.

$$u^2-13u+36=0$$
$$(u-9)(u-4)=0$$
$$u-9=0 \quad \text{or} \quad u-4=0$$
$$u=9 \quad \text{or} \quad u=4$$

Substitute $u=9$.

$$x^2=9$$
$$x=\pm 3$$

Substitute $u=4$.

$$x^2=4$$
$$x=\pm 2;$$

$x=3, x=-3, x=2, x=-2$

29. $x^6-9x^3+8=0$

Substitute $u=x^3$.

$$u^2-9u+8=0$$
$$(u-8)(u-1)=0$$
$$u-8=0 \quad \text{or} \quad u-1=0$$
$$u=8 \quad \text{or} \quad u=1$$

Substitute $u=8$.

$$x^3=8$$
$$x^3-8=0$$
$$(x-2)(x^2+2x+4)=0$$
$$x-2=0 \quad \text{or} \quad x^2+2x+4=0$$
$$x=2 \quad \text{or} \quad x=\frac{-2\pm\sqrt{4-4(1)(4)}}{2}$$
$$x=\frac{-2\pm 2i\sqrt{3}}{2}$$
$$x=\frac{-1\pm i\sqrt{3}}{2}$$

Substitute $u=1$.

$$x^3=1$$
$$x^3-1=0$$
$$(x-1)(x^2+x+1)=0$$
$$x-1=0 \quad \text{or} \quad x^2+x+1=0$$
$$x=1 \quad \text{or} \quad x=\frac{-1\pm\sqrt{1-4(1)(1)}}{2}$$
$$x=\frac{-1\pm i\sqrt{3}}{2}$$

$x=2, x=1,\ x=-1\pm i\sqrt{3},\ x=\frac{-1\pm i\sqrt{3}}{2}$

31. $m^{2/3} - m^{1/3} - 6 = 0$

$(m^{1/3})^2 - m^{1/3} - 6 = 0$

Substitute $u = m^{1/3}$.

$u^2 - u - 6 = 0$

$(u-3)(u+2) = 0$

$u - 3 = 0$ or $u + 2 = 0$

$u = 3$ or $u = -2$

Substitute $u = 3$.

$m^{1/3} = 3$

$\sqrt[3]{m} = 3$

$m = 27$

Substitute $u = -2$.

$m^{1/3} = -2$

$\sqrt[3]{m} = -2$

$m = -8;$

$m = 27, m = -8$

33. $2t^{2/5} + 7t^{1/5} + 3 = 0$

$2(t^{1/5})^2 + 7t^{1/5} + 3 = 0$

Substitute $u = t^{1/5}$.

$2u^2 + 7u + 3 = 0$

$(2u+1)(u+3) = 0$

$2u + 1 = 0$ or $u + 3 = 0$

$2u = -1$ or $u = -3$

$u = -\frac{1}{2}$

Substitute $u = -\frac{1}{2}$.

$t^{1/5} = -\frac{1}{2}$

$\sqrt[5]{t} = -\frac{1}{2}$

$t = -\frac{1}{32}$

Substitute $u = -3$.

$t^{1/5} = -3$

$\sqrt[5]{t} = -3$

$t = -243;$

$t = -\frac{1}{32}, t = -243$

35. $x^2 - 4 = 0$

(a) $(x-2)(x+2) = 0$

$x - 2 = 0$ or $x + 2 = 0$

$x = 2$ or $x = -2$

(b) $x^2 = 4$

$x = \pm 2$

(c) $x = \frac{0 \pm \sqrt{0 - 4(1)(-4)}}{2}$

$= \frac{\pm\sqrt{16}}{2}$

$= \frac{\pm 4}{2}$

$= \pm 2$

37. $a^3 + 16a - a^2 - 16 = 0$

$(a^3 + 16a) - (a^2 + 16) = 0$

$a(a^2 + 16) - (a^2 + 16) = 0$

$(a^2 + 16)(a - 1) = 0$

$a^2 + 16 = 0$ or $a - 1 = 0$

$a^2 = -16$ or $a = 1$

$a = \pm 4i;$

$a = \pm 4i, a = 1$

39. $x^3 + 5x - 4x^2 - 20 = 0$

$(x^3 + 5x) - (4x^2 + 20) = 0$

$x(x^2 + 5) - 4(x^2 + 5) = 0$

$(x^2 + 5)(x - 4) = 0$

$x^2 + 5 = 0$ or $x - 4 = 0$

$x^2 = -5$ or $x = 4$

$x = \pm i\sqrt{5};$

$x = \pm i\sqrt{5}, x = 4$

41. (a) $x^4 + 4x^2 + 4 = 0$

$(x^2 + 2)(x^2 + 2) = 0$

$x^2 = -2$

$x = \pm i\sqrt{2}$

(b) 2 imaginary solutions, 0 real solutions

(c) No x-intercepts

(d)

43. (a)
$$x^4 - x^3 - 6x^2 = 0$$
$$x^2(x^2 - x - 6) = 0$$
$$x^2(x-3)(x+2) = 0$$
$$x^2 = 0 \quad \text{or} \quad x-3=0 \quad \text{or} \quad x+2=0$$
$$x = 0 \quad \text{or} \quad x = 3 \quad \text{or} \quad x = -2;$$
$$x = 0, x = 3, x = -2$$

(b) 3 real solutions, 0 imaginary solutions

(c) 3 x-intercepts

(d)

Section 11.4 Practice Exercises

1. The value of k shifts the graph of $y = x^2$ vertically.

3.

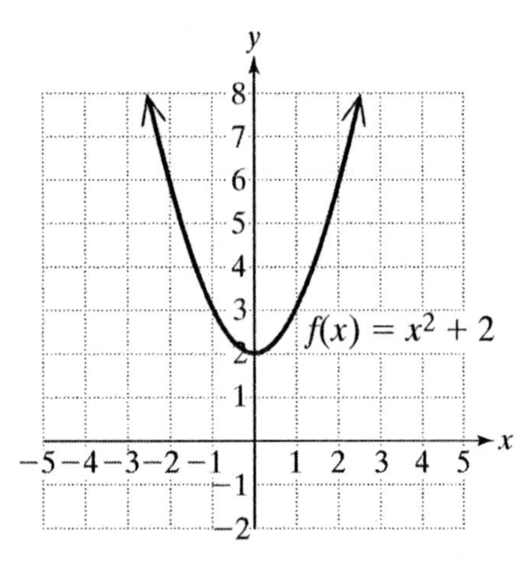

5.

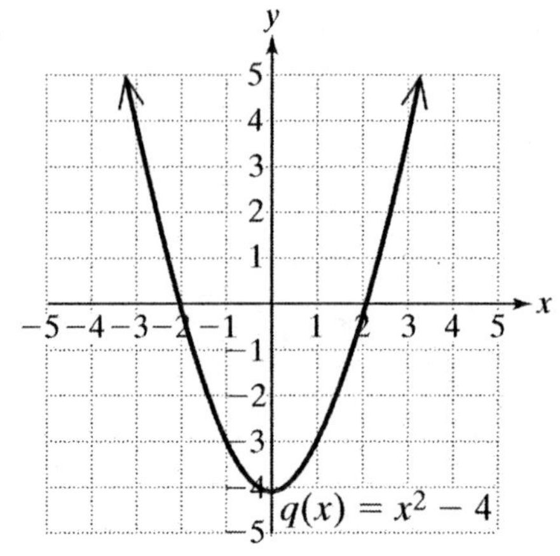

7.

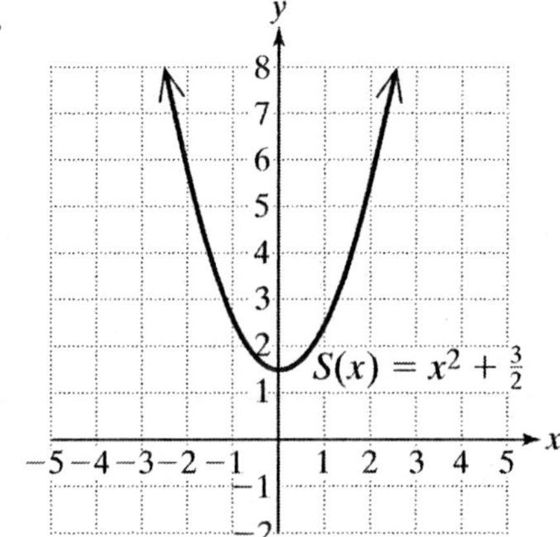

9.

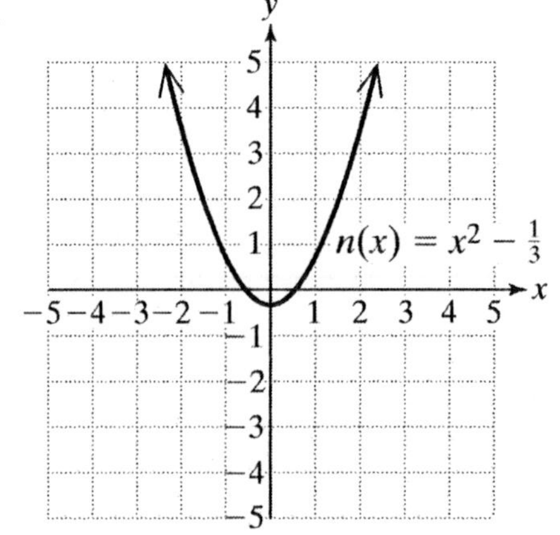

11.

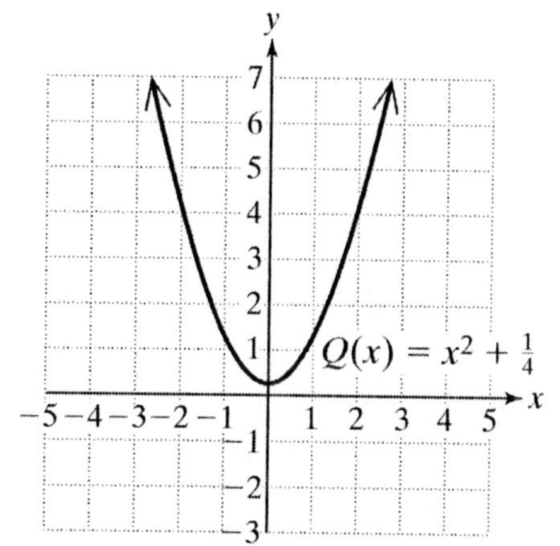

13.

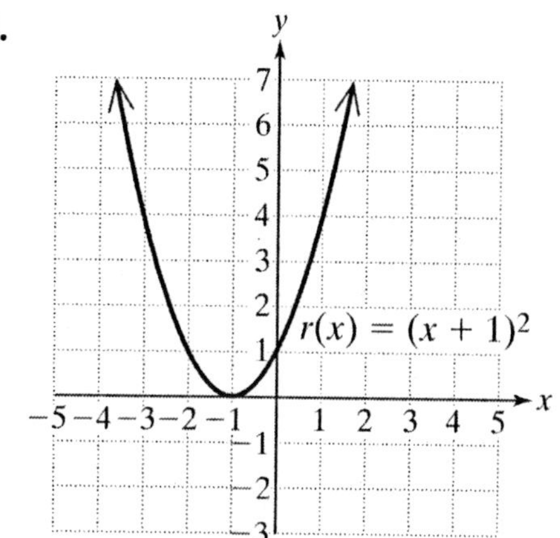

15.

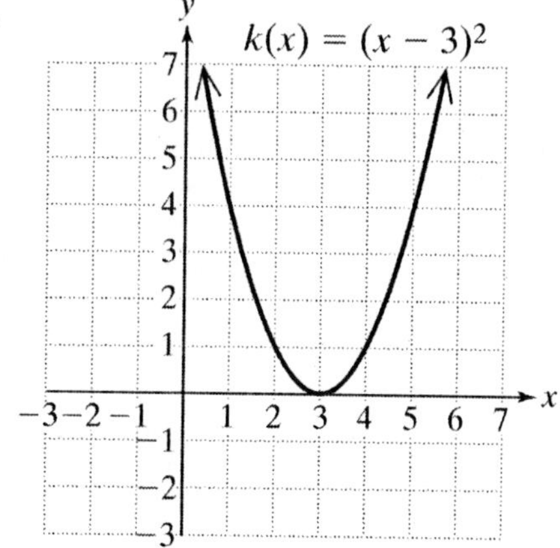

17.

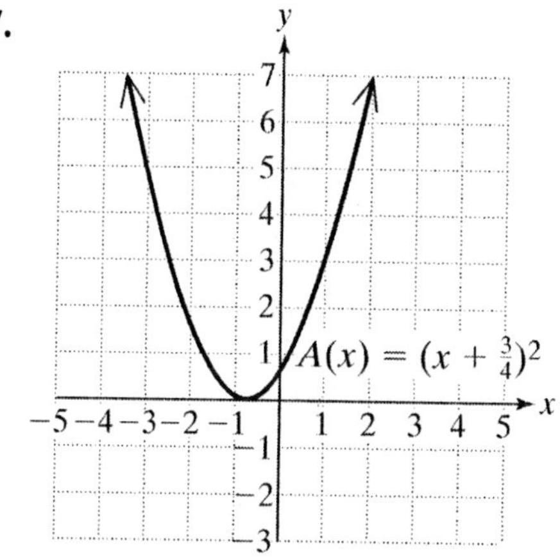

19.

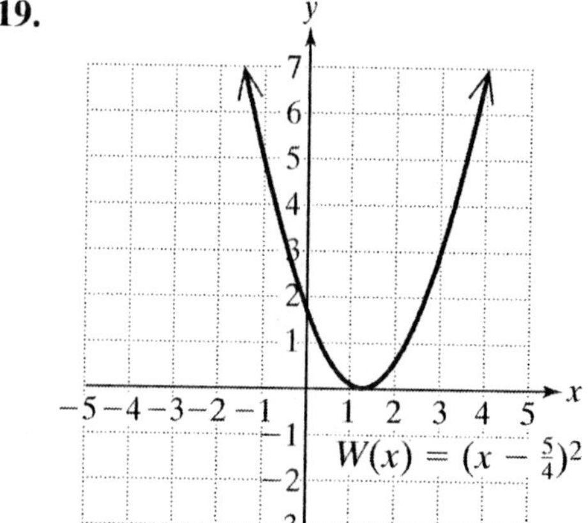

21.

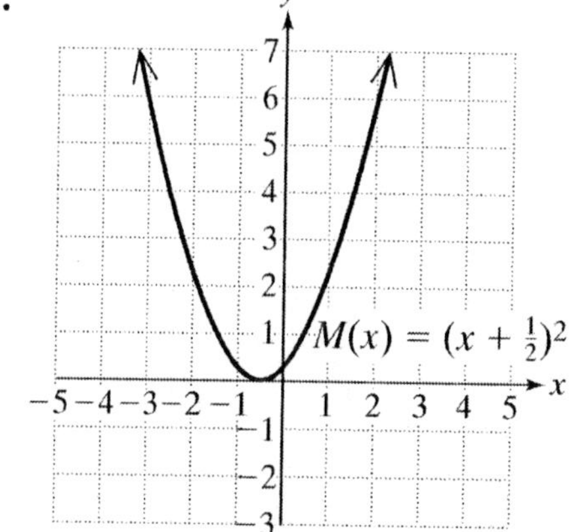

23. The value of a vertically stretches or shrinks the graph of $y = x^2$.

25. d

27. g

29. a

31. b

33.

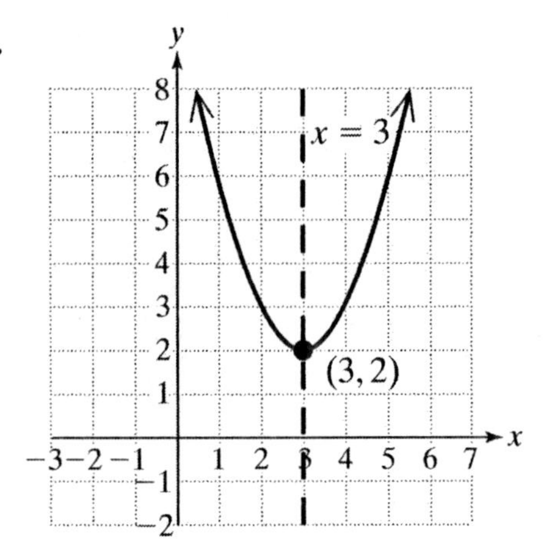

35.

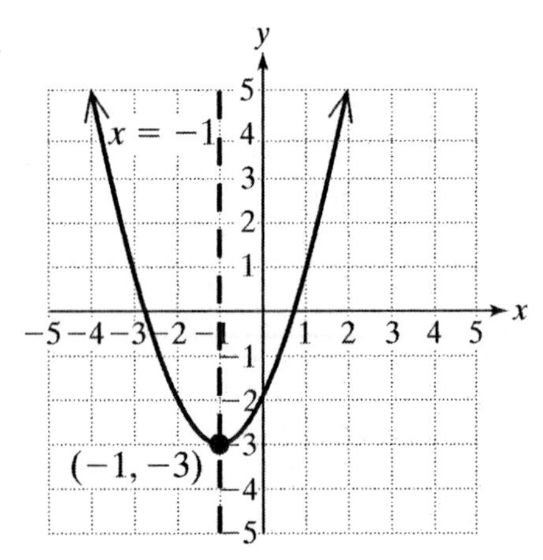

37.

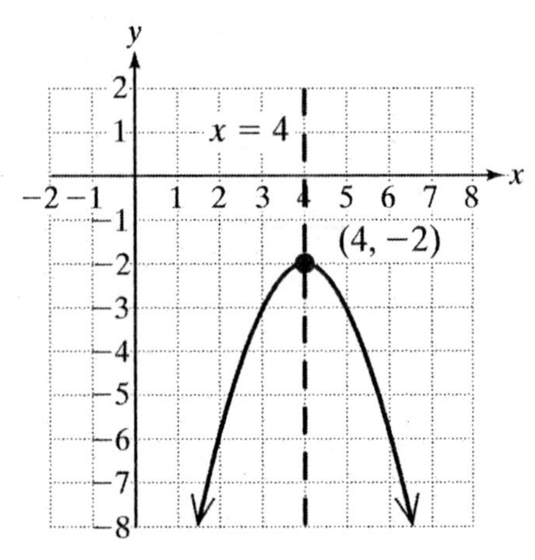

39.

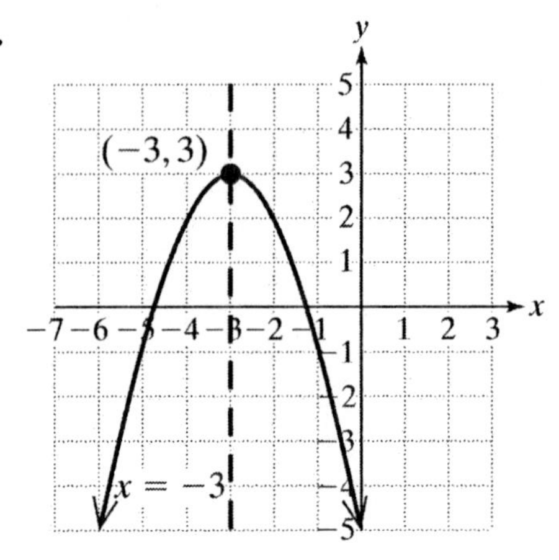

41.

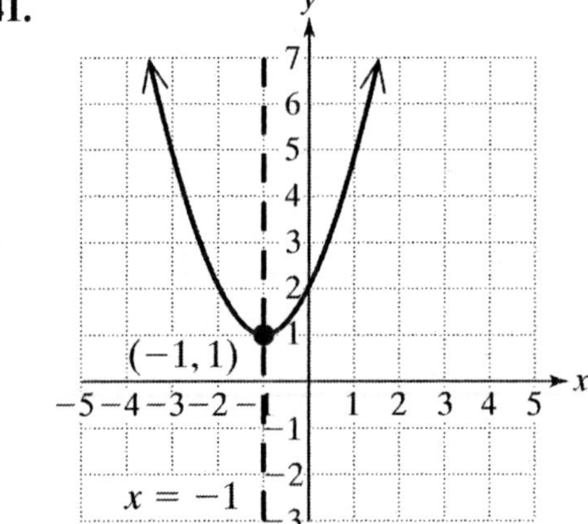

43. Vertex $(6, -9)$ is a minimum point with minimum value of -9.

45. Vertex $(2, 5)$ is a maximum point with maximum value 5.

47. Vertex $(-8, -3)$ is a minimum point with minimum value -3.

49. Vertex $\left(-\frac{3}{4}, \frac{21}{4}\right)$ is a maximum point with maximum value $\frac{21}{4}$.

51. Vertex $\left(7, -\frac{3}{2}\right)$ is a minimum point with minimum value $-\frac{3}{2}$.

53. True, since the parabola opens down.

55. False, since the minimum value corresponds to the y-value of 8.

57. **(a)** (60, 30)

(b) 30 feet

(c) $H(0) = (0-60)2+30$
$= 3600+30$
$= 40+30$
$= 70$ feet

59.

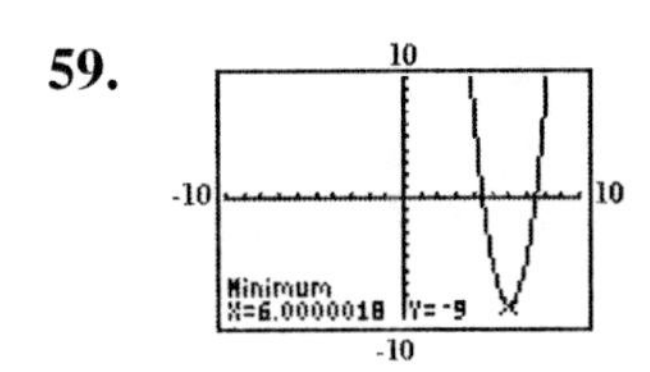

61.

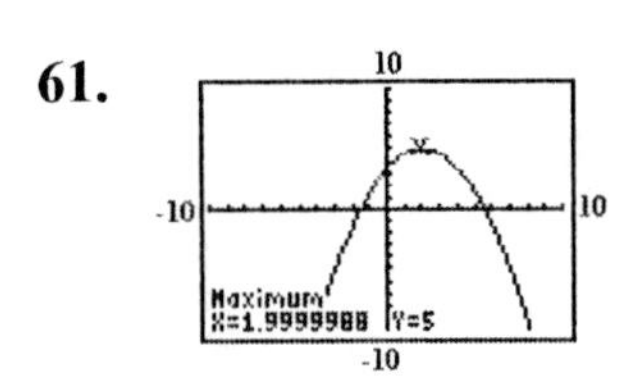

63.

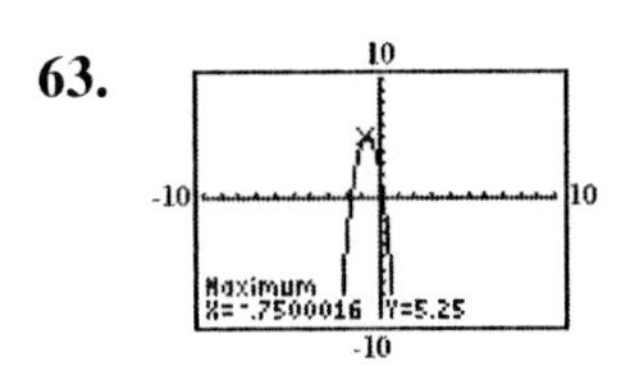

Section 11.5 Practice Exercises

1. $f(x)$ is like the graph of $y = x^2$ but opening downward and stretched vertically by a factor of 2.

3. $Q(x)$ is the graph of $y = x^2$ shifted down $\frac{8}{3}$ units.

5. $s(x)$ is the graph of $y = x^2$ shifted to the right 4 units.

7. $k = \left[\frac{1}{2}(-8)\right]^2 = 16$

9. $k = \left[\frac{1}{2}(7)\right]^2 = \frac{49}{4}$

11. $k = \left[\frac{1}{2}\left(\frac{2}{9}\right)\right]^2 = \frac{1}{81}$

13. $k = \left[\frac{1}{2}\left(\frac{-1}{3}\right)\right]^2 = \frac{1}{36}$

15. $g(x) = (x^2 - 8x) + 5$
$= (x^2 - 8x + 16) + 5 - 16$
$= (x-4)^2 - 11$
Vertex: (4, –11)

17. $n(x) = 2(x^2 + 6x) + 13$
$= 2(x^2 + 6x + 9) + 13 - 18$
$= 2(x+3)^2 - 5$
Vertex: (–3, –5)

19. $p(x) = -3(x^2 - 2x) - 5$
$= -3(x^2 - 2x + 1) - 5 + 3$
$= -3(x-1)^2 - 2$
Vertex: (1, –2)

21. $k(x) = (x^2 + 7x) - 10$
$= \left(x^2 + 7x + \frac{49}{4}\right) - 10 - \frac{49}{4}$
$= \left(x + \frac{7}{2}\right)^2 - \frac{89}{4}$
Vertex: $\left(-\frac{7}{2}, -\frac{89}{4}\right)$

23. $f(x) = (x^2 + 8x) - 1$
$= (x^2 + 8x + 16) - 1 - 16$
$= (x+4)^2 - 17$
Vertex: (–4, –17)

25. $h = \frac{-(-4)}{2(1)} = 2;$

$Q(2) = (2)^2 - 4(2) + 7 = 3$
Vertex: (2, 3)

27. $h=\dfrac{-(-6)}{2(-3)}=-1;$

$r(-1)=-3(-1)^2-6(-1)-5=-2$

Vertex: $(-1,-2)$

29. $h=\dfrac{-(8)}{2(1)}=-4;$

$N(-4)=(-4)^2+8(-4)+1=-15$

Vertex: $(-4,-15)$

31. $h=\dfrac{-(1)}{2\left(\dfrac{1}{2}\right)}=-1;$

$m(-1)=\dfrac{1}{2}(-1)^2+(-1)+\dfrac{5}{2}=2$

Vertex: $(-1, 2)$

33. $h=\dfrac{-(2)}{2(-1)}=1;$

$k(1)=-(1)^2+2(1)+2=3$

Vertex: $(1, 3)$

35. (a) $h=\dfrac{-(9)}{2(1)}=-\dfrac{9}{2};$

$y=\left(-\dfrac{9}{2}\right)^2+9\left(-\dfrac{9}{2}\right)+8=-\dfrac{49}{4}$

Vertex: $\left(-\dfrac{9}{2},-\dfrac{49}{4}\right)$

(b) $y=(0)^2+9(0)+8=8;\ (0, 8)$

(c) $0=(x+8)(x+1)$

$x+8=0$ or $x+1=0$; $(-8, 0)$, $(-1, 0)$

(d)

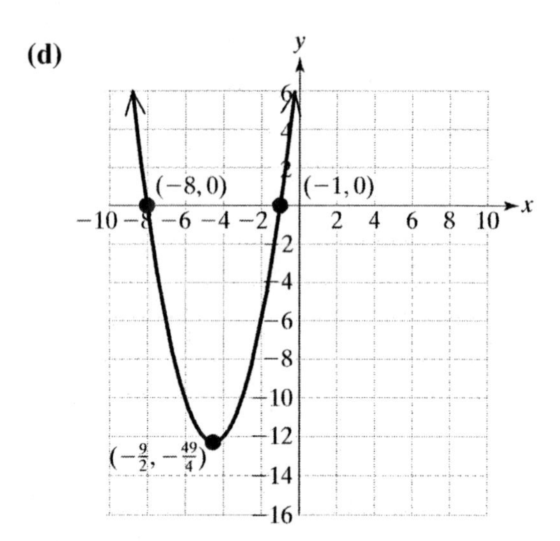

37. (a) $h=\dfrac{-(-2)}{2(2)}=\dfrac{1}{2};$

$y=2\left(\dfrac{1}{2}\right)^2-2\left(\dfrac{1}{2}\right)+4=\dfrac{7}{2}$

Vertex: $\left(\dfrac{1}{2},\dfrac{7}{2}\right)$

(b) $y=2(0)^2-2(0)+4=4;\ (0, 4)$

(c) $0=2(x^2-x+2)$

$x^2-x+2=0$ has complex solutions; no x-intercepts

(d)

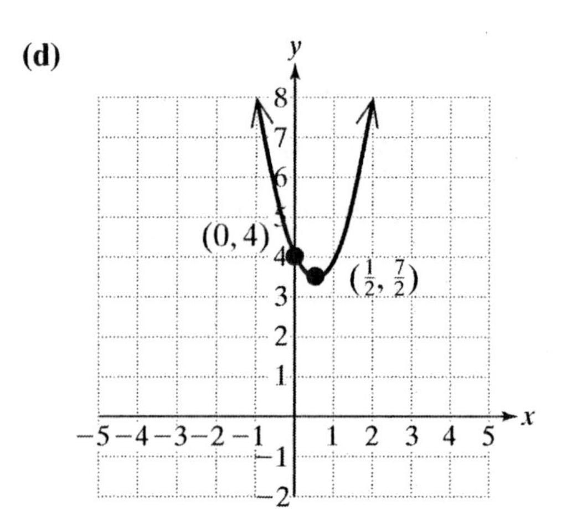

39. **(a)** $h = \frac{-(3)}{2(-1)} = \frac{3}{2};$

$y = -\left(\frac{3}{2}\right)^2 + 3\left(\frac{3}{2}\right) - \frac{9}{4} = 0$

Vertex: $\left(\frac{3}{2}, 0\right)$

(b) $y = -(0)^2 + 3(0) - \frac{9}{4} = -\frac{9}{4};\ \left(0, -\frac{9}{4}\right)$

(c) $0 = -\left(x^2 - 3x + \frac{9}{4}\right)$

$0 = -\left(x - \frac{3}{2}\right)\left(x - \frac{3}{2}\right)$

$x - \frac{3}{2} = 0;\ \left(\frac{3}{2}, 0\right)$

(d)

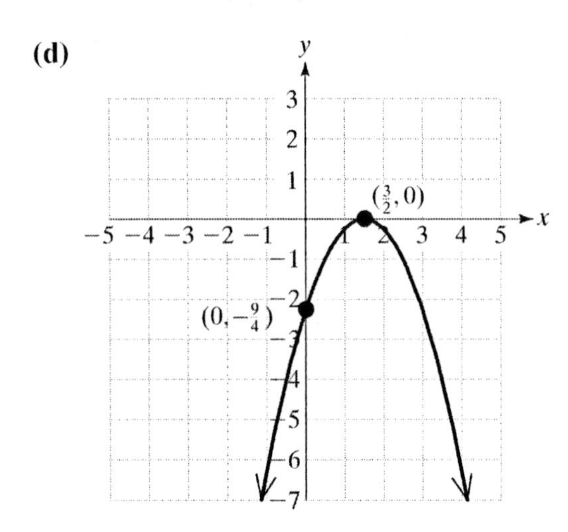

41. **(a)** $P(28) = -0.857(28)^2 + 56.1(28) - 880$

$P(28) = 18.9$ thousand miles

(b) $h = \frac{-(56.1)}{2(-0.857)} \approx 32.7$ psi

43. **(a)** $P(100) = -\frac{1}{50}(100)^2 + 12(100) - 550$

$= \$450$

(b) $P(150) = -\frac{1}{50}(150)^2 + 12(150) - 550$

$= \$800$

(c) $0 = -\frac{1}{50}x^2 + 12x - 550$

$0 = x^2 - 600x + 27500$

$0 = (x - 50)(x - 550)$

$x - 50 = 0$ or $x - 550 = 0$

$x = 50$ books $\quad x = 550$ books

(d) $h = \frac{-(12)}{2\left(-\frac{1}{50}\right)} = 300;$

$P(200) = -\frac{1}{50}(300)^2 + 12(300) - 550$

$= \$1250$

Vertex: (300, 1250)

(e)

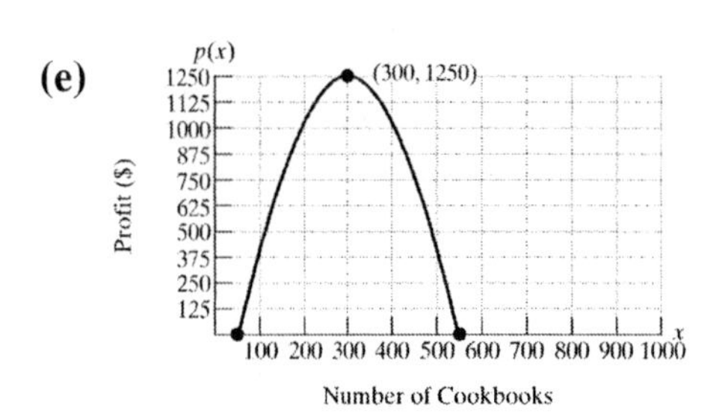

(f) 300 books produced will yield a maximum profit of \$1250.

45. **(a)** The perimeter of a rectangle is calculated using $P = 2l + 2w$. In this case, one dimension is covered by the barn leaving $P = 2x + y$. Since the perimeter is given to be 200 feet, $2x + y = 200$.

(b) $2x + y = 200$

$y = 200 - 2x$

Since $A = xy$ then $A = x(200 - 2x)$.

(c) $A = -2x^2 + 200x$

$x = \frac{-(200)}{2(-2)} = 50$

$y = 200 - 2(50) = 100$

The maximum area will be obtained with the dimensions 50 ft by 100 ft.

47.

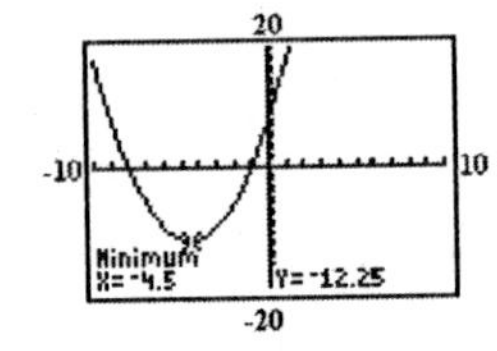

49.

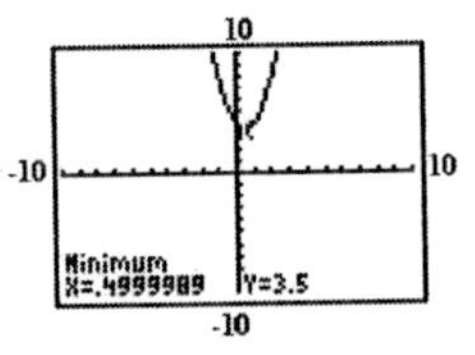

51.

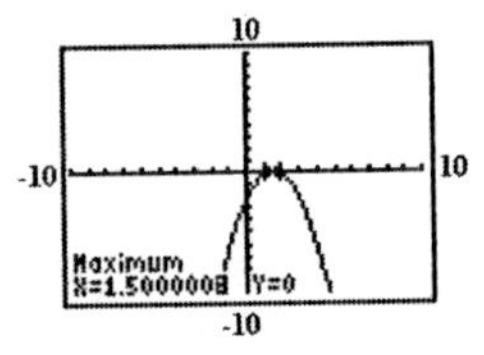

Chapter 11 Review Exercises

1. $x^2 = 5$

$x = \pm\sqrt{5}$

3. $a^2 = -81$

$a = \pm 9i$

5. $(x-2)^2 = 72$

$x - 2 = \pm 6\sqrt{2}$

$x = 2 \pm 6\sqrt{2}$

7. $(3y-1)^2 = 3$

$3y - 1 = \pm\sqrt{3}$

$3y = 1 \pm \sqrt{3}$

$y = \dfrac{1 \pm \sqrt{3}}{3}$

9. $a^2 + b^2 = c^2$

$(5)^2 + b^2 = (10)^2$

$25 + b^2 = 100$

$b^2 = 75$

$b = \pm 5\sqrt{3}$

$b \approx 8.7$ inches

11. $s^2 = 150$

$s \approx 12.2$ inches

13. $k = \dfrac{81}{4}$; $x^2 - 9x + \dfrac{81}{4} = \left(x - \dfrac{9}{2}\right)^2$

15. $k = \dfrac{1}{25}$; $z^2 - \dfrac{2}{5}z + \dfrac{1}{25} = \left(z - \dfrac{1}{5}\right)^2$

17. $4y^2 - 12y + 13 = 0$

$y^2 - 3y + \dfrac{13}{4} = 0$

$y^2 - 3y = -\dfrac{13}{4}$

$y^2 - 3y + \dfrac{9}{4} = -\dfrac{13}{4} + \dfrac{9}{4}$

$\left(y - \dfrac{3}{2}\right)^2 = -1$

$y - \dfrac{3}{2} = \pm i$

$y = \dfrac{3}{2} \pm i$

19. $b^2 + \dfrac{7}{2}b = 2$

$b^2 + \dfrac{7}{2}b + \dfrac{49}{16} = 2 + \dfrac{49}{16}$

$\left(b + \dfrac{7}{4}\right)^2 = \dfrac{81}{16}$

$b + \dfrac{7}{4} = \pm\dfrac{9}{4}$

$b = -\dfrac{7}{4} \pm \dfrac{9}{4}$

$b = \dfrac{1}{2}$; $b = -4$

21. $-t^2 + 8t - 25 = 0$

$t^2 - 8t = -25$

$t^2 - 8t + 16 = -25 + 16$

$(t-4)^2 = -9$

$t - 4 = \pm 3i$

$t = 4 \pm 3i$

23. $x^2 - 5x + 6 = 0$

$b^2 - 4ac = 25 - 4(1)(6) = 1$

2 rational solutions

25. $z^2 - 17z + 23 = 0$
$b^2 - 4ac = 289 - 4(1)(23) = 197$
2 irrational solutions

27. $25b^2 + 10b + 1 = 0$
$b^2 - 4ac = 100 - 4(25)(1) = 0$
1 rational solution

29. $y^2 - 4y + 1 = 0$

$$y = \frac{4 \pm \sqrt{16 - 4(1)(1)}}{2}$$
$$y = \frac{4 \pm \sqrt{12}}{2}$$
$$y = \frac{4 \pm 2\sqrt{3}}{2}$$
$$y = 2 \pm \sqrt{3}$$

31. $6a^2 - 7a - 10 = 0$

$$a = \frac{7 \pm \sqrt{49 - 4(6)(-10)}}{12}$$
$$a = \frac{7 \pm \sqrt{289}}{12}$$
$$a = \frac{7 \pm 17}{12}$$
$$a = -\frac{5}{6};\ a = 2$$

33. $b^2 - \frac{3}{5}b - \frac{4}{25} = 0$

$$25b^2 - 15b - 4 = 0$$
$$b = \frac{15 \pm \sqrt{225 - 4(25)(-4)}}{100}$$
$$b = \frac{15 \pm \sqrt{625}}{100}$$
$$b = \frac{15 \pm 25}{100}$$
$$b = \frac{4}{5},\ b = -\frac{1}{5}$$

35. $-x^2 + 4x + 32 = 0$

$$x = \frac{-4 \pm \sqrt{16 - 4(-1)(32)}}{-2}$$
$$x = \frac{-4 \pm \sqrt{144}}{-2}$$
$$x = \frac{-4 \pm 12}{-2}$$
$$x = 8;\ x = -4$$

37. $5x^2 - 20 = 0$

$$x = \frac{0 \pm \sqrt{0 - 4(5)(-20)}}{10}$$
$$x = \frac{\sqrt{400}}{10}$$
$$x = \frac{\pm 20}{10}$$
$$x = 2;\ x = -2$$

39. $D(s) = \frac{1}{10}s^2 - 3s + 22$

(a) $D(150) = \frac{1}{10}(150)^2 - 3(150) + 22$
$= 1822$ ft

(b) $1000 = \frac{1}{10}s^2 - 3s + 22$

$$0 = \frac{1}{10}s^2 - 3s - 978$$
$$0 = s^2 - 30s - 9780$$
$$s = \frac{30 \pm \sqrt{900 - 4(1)(-9780)}}{2}$$
$$s = 115 \text{ ft/sec}$$

(c) $\frac{115 \text{ ft}}{\text{sec}} \cdot \frac{3600 \text{ sec}}{\text{hr}} \cdot \frac{\text{mile}}{5280 \text{ ft}} = 78 \text{ mph}$

41. $x - 4\sqrt{x} - 21 = 0$
Let $u = \sqrt{x}$ and $u^2 = x$.

$$u^2 - 4u - 21 = 0$$
$$(u - 7)(u + 3) = 0$$
$$u - 7 = 0 \quad \text{or} \quad u + 3 = 0$$
$$u = 7 \quad \text{or} \quad u = -3$$

Substitute $u = -3$.

$\sqrt{x} = -3$ not possible

Substitute $u = 7$.

$\sqrt{x} = 7$

$x = 49$

43. $y^4 - 11y^2 + 18 = 0$

$(y^2 - 9)(y^2 - 2) = 0$

$y^2 = 9$ or $y^2 = 2$

$y = \pm 3$ or $y = \pm\sqrt{2}$

45. $t^{2/5} + t^{1/5} - 6 = 0$

Let $u = t^{1/5}$ and $u^2 = t^{2/5}$.

$u^2 + u - 6 = 0$

$(u + 3)(u - 2) = 0$

$u + 3 = 0$ or $u - 2 = 0$

$u = -3$ or $u = 2$

Substitute $u = -3$.

$t^{1/5} = -3$

$t = -243$

Substitute $u = 2$.

$t^{1/5} = 2$

$t = 32$

47. $\sqrt{4a - 3} - \sqrt{8a + 1} = -2$

$\sqrt{4a - 3} = -2 + \sqrt{8a + 1}$

$4a - 3 = \left(-2 + \sqrt{8a + 1}\right)^2$

$4a - 3 = 4 - 4\sqrt{8a + 1} + 8a + 1$

$-4a - 8 = -4\sqrt{8a + 1}$

$a + 2 = \sqrt{8a + 1}$

$a^2 + 4a + 4 = 8a + 1$

$a^2 - 4a + 3 = 0$

$(a - 3)(a - 1) = 0$

$a = 3$ or $a = 1$

Check:

$\sqrt{4(3) - 3} - \sqrt{8(3) + 1} = -2$

$\sqrt{12 - 3} - \sqrt{24 + 1} = -2$

$\sqrt{9} - \sqrt{25} = -2$

$3 - 5 = -2;$

$\sqrt{4(1) - 3} - \sqrt{8(1) + 1} = -2$

$\sqrt{4 - 3} - \sqrt{8 + 1} = -2$

$\sqrt{1} - \sqrt{9} = -2$

$1 - 3 = -2$

49.

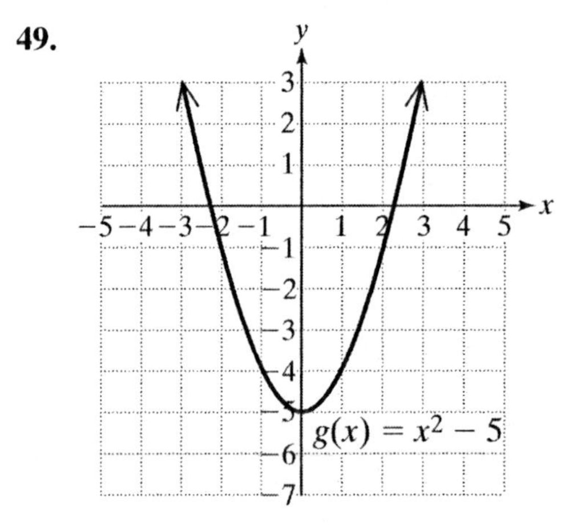

51.

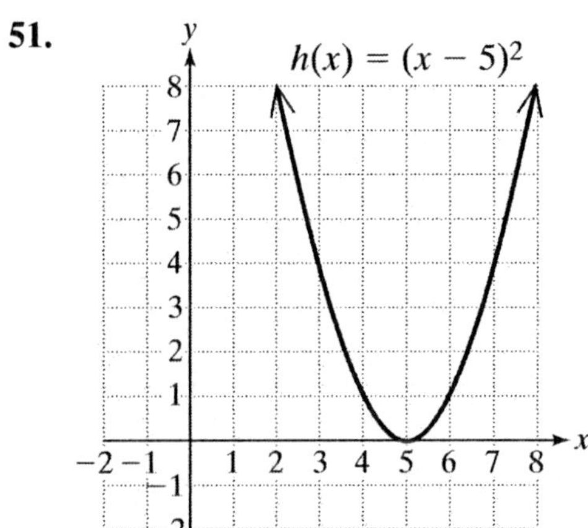

53. 

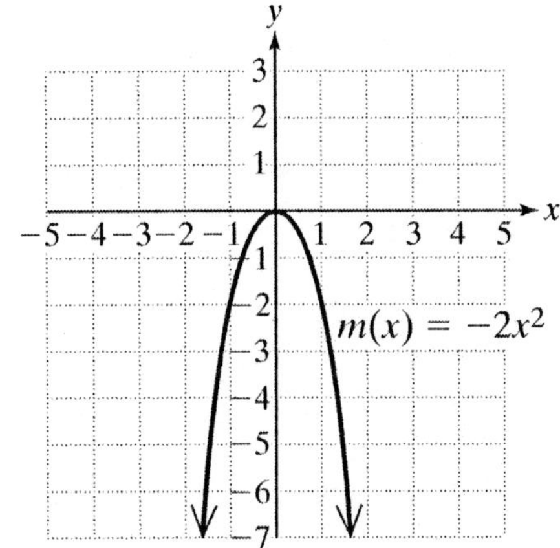

55. 

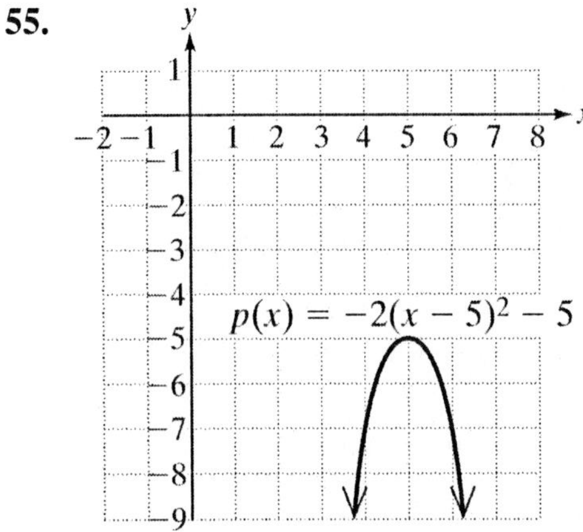

57. Vertex: $\left(4, \dfrac{5}{3}\right)$

Since $a = \dfrac{1}{3} > 0,$ the parabola opens up making the vertex a minimum point. The minimum value is $\dfrac{5}{3}$.

59. $x = -\dfrac{2}{11}$

61.
$$\begin{aligned} z(x) &= (x^2 - 6x) + 7 \\ &= (x^2 - 6x + 9) + 7 - 9 \\ &= (x-3)^2 - 2 \end{aligned}$$
Vertex: $(3, -2)$

63.
$$\begin{aligned} p(x) &= -5(x^2 + 2x) - 13 \\ &= -5(x^2 + 2x + 1) - 13 + 5 \\ &= -5(x+1)^2 - 8 \end{aligned}$$
Vertex: $(-1, -8)$

65. $h = \dfrac{-(4)}{2(-2)} = 1;$

$f(1) = -2(1)^2 + 4(1) - 17 = -15$

Vertex: $(1, -15)$

67. $h = \dfrac{-(-3)}{2(3)} = \dfrac{1}{2};$

$$m\left(\frac{1}{2}\right) = 3\left(\frac{1}{2}\right)^2 - 3\left(\frac{1}{2}\right) + 11 = \frac{41}{4}$$

Vertex: $\left(\dfrac{1}{2}, \dfrac{41}{4}\right)$

69. (a)
$$\begin{aligned} y(12) &= -\frac{1}{1152}(12)^2 + \frac{1}{12}(12) \\ &= 0.875 \text{ grams} \end{aligned}$$
$$\begin{aligned} y(36) &= -\frac{1}{1152}(36)^2 + \frac{1}{12}(36) \\ &= 1.875 \text{ grams} \end{aligned}$$
$$\begin{aligned} y(48) &= -\frac{1}{1152}(48)^2 + \frac{1}{12}(48) \\ &= 2 \text{ grams} \end{aligned}$$
$$\begin{aligned} y(60) &= -\frac{1}{1152}(60)^2 + \frac{1}{12}(60) \\ &= 1.875 \text{ grams} \end{aligned}$$

(b) Find the vertex.

$$h = \frac{-\frac{1}{12}}{2\left(-\frac{1}{1152}\right)} = \frac{\frac{1}{12}}{\frac{1}{576}} = \frac{1}{12} \cdot \frac{576}{1} = 48$$

From part (a), $y(48) = 2$. The vertex is $(2, 48)$. The maximum yield of toxin is 2 grams after 48 hours.

Chapter 11 Test

1. $(x+3)^2 = 25$
$$x+3 = \pm\sqrt{25}$$
$$x+3 = \pm 5$$
$$x = -3 \pm 5$$
$$x = -3+5 \quad \text{or} \quad x = -3-5$$
$$x = 2 \quad \text{or} \quad x = -8$$

3. $(m+1)^2 = -1$
$$m+1 = \pm\sqrt{-1}$$
$$m+1 = \pm i$$
$$m = -1 \pm i$$

5. $2x^2 + 12x - 36 = 0$
$$\frac{2x^2}{2} + \frac{12x}{2} - \frac{36}{2} = 0$$
$$x^2 + 6x - 18 = 0$$
$$x^2 + 6x = 18$$
$$x^2 + 6x + 9 = 18 + 9$$
$$(x+3)^2 = 27$$
$$x+3 = \pm\sqrt{27}$$
$$x+3 = \pm 3\sqrt{3}$$
$$x = -3 \pm 3\sqrt{3}$$

7. **(a)** $x^2 - 3x + 12 = 0$

(b) $a = 1, b = -3, c = 12$

(c) $b^2 - 4ac = 9 - 4(1)(12) = -39$

(d) 2 imaginary solutions

9. $3x^2 - 4x + 1 = 0$

$a = 3, b = -4, c = 1$
$$x = \frac{-b \pm \sqrt{b^2 - 4ac}}{2a}$$
$$= \frac{-(-4) \pm \sqrt{(-4)^2 - 4(3)(1)}}{2(3)}$$
$$= \frac{4 \pm \sqrt{4}}{6}$$
$$= \frac{4 \pm 2}{6}$$
$$x = \frac{4+2}{6} \quad \text{or} \quad x = \frac{4-2}{6}$$
$$x = \frac{6}{6} \quad \text{or} \quad x = \frac{2}{6}$$
$$x = 1 \quad \text{or} \quad x = \frac{1}{3}$$

11. $14 = \frac{1}{2}(2h-3)(h)$
$$28 = 2h^2 - 3h$$
$$0 = 2h^2 - 3h - 28$$
$$h = \frac{3 \pm \sqrt{9 - 4(2)(-28)}}{4}$$
$$h = \frac{3 \pm \sqrt{233}}{4}$$
$$h = 4.6$$

Height is 4.6 feet.
Base is $2h - 3 = 6.2$ feet.

13. $x - \sqrt{x} - 6 = 0$
$$x - 6 = \sqrt{x}$$
$$(x-6)^2 = \left(\sqrt{x}\right)^2$$
$$x^2 - 12x + 36 = x$$
$$x^2 - 13x + 36 = 0$$
$$(x-9)(x-4) = 0$$
$$x - 9 = 0 \quad \text{or} \quad x - 4 = 0$$
$$x = 9 \quad \text{or} \quad x = 4$$

Check: $x = 9$
$$9 - \sqrt{9} - 6 = 0$$
$$9 - 3 - 6 = 0$$
$$0 = 0 \quad \text{True}$$

Check: $x = 4$
$$4 - \sqrt{4} - 6 = 0$$
$$4 - 2 - 6 = 0$$
$$-4 = 0 \quad \text{False}$$

Since $x = 4$ does not check, the solution is $x = 9$.

15. $f(x) = x^2 - 6x + 8$

y-intercept:
$$y = 0^2 - 6(0) + 8 = 8$$
$(0, 8)$

x-intercept:
$0 = x^2 - 6x + 8$
$0 = (x-4)(x-2)$
$x = 4$ or $x = 2$
$(4, 0)$; $(2, 0)$
Graph c

17. $p(x) = -2x^2 - 8x - 6$
y-intercept:
$y = -2(0)^2 - 8(0) - 6 = -6$
$(0, -6)$
x-intercept
$0 = -2x^2 - 8x - 6$
$0 = -2(x^2 + 4x + 3)$
$0 = -2(x+1)(x+3)$
$x = -1, x = -3$
$(-1, 0)$; $(-3, 0)$
Graph d

19. $0 = -\dfrac{x^2}{256} + x$
$0 = -x^2 + 256x$
$0 = -x(x - 256)$
$x = 0, x = 256$
The rocket will hit the ground 256 feet from the launch pad.

21. $y = x^2 - 2$ is the graph of $y = x^2$ shifted down 2 units.

23. $y = -4x^2$ is the graph of $y = 4x^2$ opening downward instead of upward.

25. (a) $g(x) = 2(x^2 - 10x) + 51$
$= 2(x^2 - 10x + 25) + 51 - 50$
$= 2(x-5)^2 + 1$
Vertex: $(5, 1)$

(b) $h = \dfrac{-(-20)}{2(2)} = 5$

$g(5) = 2(5)^2 - 20(5) + 51 = 1$
Vertex: $(5, 1)$

Cumulative Review Exercises, Chapters 1–11

1. (a) $A \cup B = \{2, 4, 6, 8, 10, 12, 16\}$

(b) $A \cap B = \{2, 8\}$

3. $4^0 - \left(\dfrac{1}{2}\right)^{-3} - 81^{1/2} = 1 - 2^3 - \sqrt{81}$
$= 1 - 8 - 9$
$= -16$

5. (a) $x^2 + 2x^2 - 9x - 18 = x^2(x+2) - 9(x+2)$
$= (x+2)(x^2 - 9)$
$= (x+2)(x-3)(x+3)$

(b)
$$\begin{array}{r} x^2 + 5x + 6 \\ x-3 \overline{\smash{)}\, x^3 + 2x^2 - 9x - 18} \\ \underline{-(x^3 - 3x^2)} \\ 5x^2 - 9x \\ \underline{-(5x^2 - 15x)} \\ 6x - 18 \\ \underline{-(6x - 18)} \\ 0 \end{array}$$

7. $\dfrac{4}{\sqrt{2x}} \cdot \dfrac{\sqrt{2x}}{\sqrt{2x}} = \dfrac{4\sqrt{2x}}{2x} = \dfrac{2\sqrt{2x}}{x}$

9. $\begin{cases} \dfrac{1}{9}x - \dfrac{1}{3}y = -\dfrac{13}{9} \\ x - \dfrac{1}{2}y = \dfrac{9}{2} \end{cases}$

Multiply the first equation by 9 and the second equation by 2.
$x - 3y = -13$
$2x - y = 9$
Multiply the first equation by -2.
$-2x + 6y = 26$
$\underline{2x - y = 9}$
$5y = 35$
$y = 7$
Substitute $y = 7$ into either original equation.

$$x - \frac{1}{2}(7) = \frac{9}{2}$$
$$x - \frac{7}{2} = \frac{9}{2}$$
$$x = \frac{16}{2}$$
$$x = 8$$

The solution to the system is (8, 7).

11. $(x-3)^2 + 16 = 0$
$$(x-3)^2 = -16$$
$$x - 3 = \pm 4i$$
$$x = 3 \pm 4i$$

13. Half of 10 squared, 25.

15. $3x - 5y = 10$

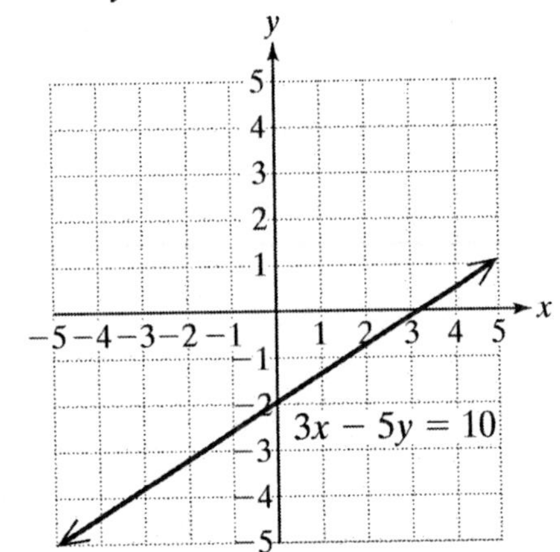

17. (a)

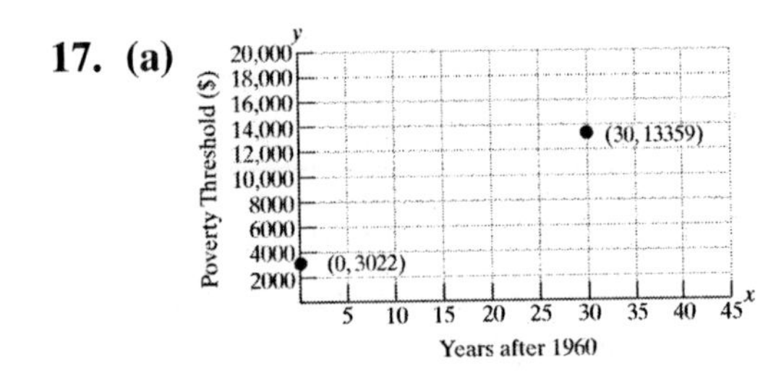

(b) $m = \frac{y_2 - y_1}{x_2 - x_1} = \frac{13359 - 3022}{30 - 0} = 345$

(c) $y = 345x + 3022$
When $x = 20$,
$y = 345(20) + 3022 = \$9922$.
When $x = 20$, $y = \$9922$.

19. The domain element "3" has more than one corresponding range element.

21. $y = k\frac{x}{z};$
$$15 = k\frac{50}{10}$$
$$15 = 5k$$
$$3 = k;$$
$$y = 3\left(\frac{65}{5}\right)$$
$$y = \frac{65}{5}$$
$$y = 39$$

23. $g(x) = \sqrt{2-x}$

(a) $g(-7) = \sqrt{2-(-7)} = \sqrt{9} = 3$

(b) $g(0) = \sqrt{2-0} = \sqrt{2}$

(c) $g(3) = \sqrt{2-3} = \sqrt{-1} = i$ (not a real number)

25. (a) $(-\infty, 2]$

(b) $(-\infty, 1) \cup \{2\} \cup [3, 4]$

(c) $f(-2) = 2$

(d) $f(1) = 3$

(e) $f(0) = 2$

(f) $f(x) = 0$ when $x = -3$

27. $2x - 3 \le x + 5$ or $2x + 1 \ge -3$
$x - 3 \le 5$ or $2x \ge -4$
$x \le 8$ or $x \ge -2$

Interval notation: $(-\infty, \infty)$

29. $2x^2 + 8x - 10 \ge 0$

Find the boundary points.
$$2x^2 + 8x - 10 = 0$$
$$2(x^2 + 4x - 5) = 0$$
$$2(x+5)(x-1) = 0$$
$x + 5 = 0$ or $x - 1 = 0$
$x = -5$ or $x = 1$

Plot the boundary points on the number line

and test a point from each region.

Test $x = -6$: $2(-6)^2 + 8(-6) - 10 \geq 0$

$14 \geq 0$ True

Test $x = 0$: $2(0)^2 + 8(0) - 10 \geq 0$

$-10 \geq 0$ False

Test $x = 2$: $2(2)^2 + 8(2) - 10 \geq 0$

$14 \geq 0$ True

Since the test points $x = -6$ and $x = 2$ are true, the solution is $(-\infty, -5] \cup [1, \infty)$. The endpoints are included in the solution since the symbol "≥" includes equality.

31. $3x - 2y < 6$

Graph the related equation

$3x - 2y = 6$

using the intercepts.

x-intercept: $3x - 2(0) = 6$

$3x = 6$

$x = 2$

The x-intercept is (2, 0).

y-intercept: $3(0) - 2y = 6$

$-2y = 6$

$y = -3$

The y-intercept is (0, –3).

Since the symbol "<" does not include equality, use a dashed boundary line. Test points on each side of the line to decide whether to shade above or below the line.

Test Point Above: (0, 0)

$3(0) - 2(0) < 6$

$0 < 6$ True

Test Point Below: (0, –4)

$3(0) - 2(-4) < 6$

$8 < 6$ False

Shade the region above the boundary.

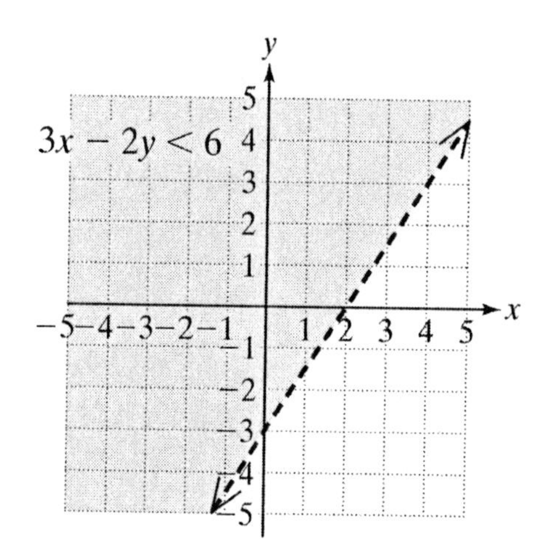

33. $\dfrac{1}{p} + \dfrac{1}{q} = \dfrac{1}{f}$

LCD = pqf

$$pqf\left(\frac{1}{p}\right) + pqf\left(\frac{1}{q}\right) = pqf\left(\frac{1}{f}\right)$$

$$qf + pf = pq$$

$$f(q+p) = pq$$

$$f = \frac{pq}{q+p}$$

35. $\dfrac{y - \frac{4}{y-3}}{y-4}$

LCD = $y - 3$

$$\frac{(y-3)\left(y - \frac{4}{y-3}\right)}{(y-3)(y-4)} = \frac{y(y-3) - 4}{(y-3)(y-4)}$$

$$= \frac{y^2 - 3y - 4}{(y-3)(y-4)}$$

$$= \frac{(y-4)(y+1)}{(y-3)(y-4)}$$

$$= \frac{y+1}{y-3}$$

37. $x^2 - 16x + 2 = 0$

$x^2 - 16x = -2$

Add $\left[\frac{1}{2}(-16)\right]^2 = 64$ to both sides.

$x^2 - 16x + 64 = -2 + 64$

$(x-8)^2 = 62$

$x - 8 = \pm\sqrt{62}$

$x = 8 \pm \sqrt{62}$

Chapter 12

Section 12.1 Practice Exercises

1. $(f+g)(x) = f(x)+g(x)$
$= (x+4)+(2x^2+4x)$
$= 2x^2+5x+4$
Domain: $(-\infty, \infty)$

3. $(g-f)(x) = g(x)-f(x)$
$= (2x^2+4x)-(x+4)$
$= 2x^2+4x-x-4$
$= 2x^2+3x-4$
Domain: $(-\infty, \infty)$

5. $(f \cdot h)(x) = f(x)\cdot h(x)$
$= (x+4)\sqrt{x-4}$
Domain: $x-1 \geq 1$
$x \geq 1$
$[1, \infty)$

7. $(g \cdot f)(x) = g(x)\cdot f(x)$
$= (2x^2+4x)(x+4)$
$= 2x^3+8x^2+4x^2+16x$
$= 2x^3+12x^2+16x$
Domain: $(-\infty, \infty)$

9. $\left(\frac{h}{f}\right)(x) = \frac{h(x)}{f(x)} = \frac{\sqrt{x-1}}{x+4}$
Domain: $x \geq 1$ and $x \neq -4$
$[1, \infty)$

11. $\left(\frac{f}{g}\right)(x) = \frac{f(x)}{g(x)} = \frac{x+4}{2x^2+4x}$
Domain: The denominator is zero when
$2x^2+4x=0$
$2x(x+2)=0$
$2x=0$ or $x+2=0$
$x=0$ or $x=-2$
We exclude these values from the domain.
$(-\infty, -2) \cup (-2, 0) \cup (0, \infty)$

13. $(f \circ g)(x) = f(g(x))$
$= f(2x^2+4x)$
$= (2x^2+4x)+4$
$= 2x^2+4x+4$
Domain: $(-\infty, \infty)$

15. $(f \circ k)(x) = f(k(x))$
$= f\left(\frac{1}{x}\right)$
$= \left(\frac{1}{x}\right)+4$
$= \frac{1}{x}+4$
Domain: $(-\infty, 0) \cup (0, \infty)$

17. $(k \circ h)(x) = k(h(x))$
$= k\left(\sqrt{x-1}\right)$
$= \frac{1}{\sqrt{x-1}}$
Domain: $(1, \infty)$

19. No

21. $(m \cdot r)(0) = m(0)\cdot r(0)$
$= 0^3 \cdot \sqrt{0+4}$
$= 0 \cdot \sqrt{4}$
$= 0 \cdot 2$
$= 0$

23. $(m+r)(-4) = m(-4)+r(-4)$
$= (-4)^3+\sqrt{(-4)+4}$
$= -64+\sqrt{0}$
$= -64+0$
$= -64$

25. $(r \circ n)(3) = r(n(3))$
$n(3) = (3)-3 = 0$
$r(n(3)) = r(0) = \sqrt{0+4} = \sqrt{4} = 2$

27. $(p \circ m)(-1) = p(m(-1))$
$m(-1) = (-1)^3 = -1$
$p(m(-1)) = p(-1) = \frac{1}{(-1)+2} = \frac{1}{1} = 1$

29. $(m \circ p)(2) = m(p(2))$

$p(2) = \frac{1}{(2)+2} = \frac{1}{4}$

$m(p(2)) = m\left(\frac{1}{4}\right) = \left(\frac{1}{4}\right)^3 = \frac{1}{64}$

31. $(r+p)(-3) = r(-3) + p(-3)$

$= \sqrt{(-3)+4} + \frac{1}{(-3)+2}$

$= \sqrt{1} + \frac{1}{-1}$

$= 1 - 1$

$= 0$

33. $(m \circ p)(-2) = m(p(-2))$

$p(-2) = \frac{1}{(-2)+2} = \frac{1}{0}$

Undefined

35. $(f+g)(2) = f(2) + g(2) = 2 + (-2) = 0$

37. $(f \cdot g)(-1) = f(-1) \cdot g(-1) = 1 \cdot 1 = 1$

39. $\left(\frac{g}{f}\right)(0) = \frac{g(0)}{f(0)} = \frac{0}{2} = 0$

41. $\left(\frac{f}{g}\right)(0) = \frac{f(0)}{g(0)} = \frac{2}{0}$ Undefined

43. $(g \circ f)(-2) = g(f(-1)) = g(1) = -1$

45. $(f \circ g)(-4) = f(g(-4)) = f(2) = 2$

47. $(g \circ g)(2) = g(g(2)) = g(-2) = 2$

49. (a) $P(x) = 5.98x - (2.2x + 1) = 3.78x - 1$

(b) $P(50) = 3.78(50) - 1 = \$188$

51. (a) $F(t) = 0.925t + 13.083 - 0.725t - 8.683$

$F(t) = 0.2t + 4.4$

$F(t)$ represents the outstanding child support (in billions) according to the year t.

(b) $F(0) = 0.2(0) + 4.4 = 4.4$

$F(2) = 0.2(2) + 4.4 = 4.8$

$F(4) = 0.2(4) + 4.4 = 5.2$

$F(0) = 4.4$ means in 1985, 4.4 billion dollars of child support was not paid.
$F(2) = 4.8$ means in 1987, 4.8 billion dollars of child support was not paid.
$F(4) = 5.2$ means in 1989, 5.2 billion dollars of child support was not paid.

53. (a) $(D \circ r)(t) = D(r(t))$

$= D(80t)$

$= 7(80t)$

$= 560t$

This function represents the total distance Joe travels as a function of time.

(b) $(D \circ r)(10) = 560(10) = 5600$ feet

Section 12.2 Practice Exercises

1. yes

3. no

5. yes

7. yes

9. no

11. yes

13. $g^{-1} = \{(5, 3), (1, 8), (9, -3), (2, 0)\}$

15. $r^{-1} = \{(3, a), (6, b), (9, c)\}$

17. The function is not one-to-one. Colorado and Connecticut each have 6 representatives.

19. $y = x + 4$

To find the inverse, interchange x and y.

$x = y + 4$

$x - 4 = y$

$y = x - 4$

$h^{-1}(x) = x - 4$

21. $y = \frac{1}{3}x - 2$

To find the inverse, interchange x and y.

$$x = \frac{1}{3}y - 2$$
$$x + 2 = \frac{1}{3}y$$
$$3(x+2) = 3\left(\frac{1}{3}y\right)$$
$$3(x+2) = y$$
$$y = 3(x+2)$$
$$m^{-1}(x) = 3(x+2)$$

23. $y = -x + 10$

To find the inverse, interchange x and y.

$$x = -y + 10$$
$$x - 10 = -y$$
$$\frac{x-10}{-1} = \frac{-y}{-1}$$
$$-x + 10 = y$$
$$p^{-1}(x) = -x + 10$$

25. $y = x^3$

To find the inverse, interchange x and y.

$$x = y^3$$
$$\sqrt[3]{x} = y$$
$$f^{-1}(x) = \sqrt[3]{x}$$

27. $y = \sqrt[3]{2x-1}$

To find the inverse, interchange x and y.

$$x = \sqrt[3]{2y-1}$$
$$(x)^3 = \left(\sqrt[3]{2y-1}\right)^3$$
$$x^3 = 2y - 1$$
$$x^3 + 1 = 2y$$
$$\frac{x^3+1}{2} = y$$
$$f^{-1}(x) = \frac{x^3+1}{2}$$

29. **(a)** $f(4) = 0.3048(4) = 1.2192$ m;
$f(50) = 0.3048(50) = 15.24$ m

(b) $y = 0.3048x$

To find the inverse, interchange x and y.

$$x = 0.3048y$$
$$\frac{1}{0.3048}x = y$$
$$f^{-1}(x) = \frac{1}{0.3048}x$$

(c) $f^{-1}(1500) = \frac{1}{0.3048}(1500) \approx 4921.3$ ft

31. False, $x = 2$ is not a function.

33. True, any function of the form $f(x) = mx + b$ $(m \neq 0)$ has an inverse.

35. False, $k(1) = 1$ and $k(-1) = 1$.

37. True

39. $(b, 0)$

41. For example: $f(x) = x$

43. **(a)** Domain $\{x|x \leq 0\}$, Range $\{y|y \geq -4\}$

(b) Domain $\{x|x \geq -4\}$, Range $\{y|y \leq 0\}$

45. **(a)** $\{x|-2 \leq x \leq 0\}$

(b) $\{y|0 \leq y \leq 2\}$

(c) $\{x|0 \leq x \leq 2\}$

(d) $\{y|-2 \leq y \leq 0\}$

(e)

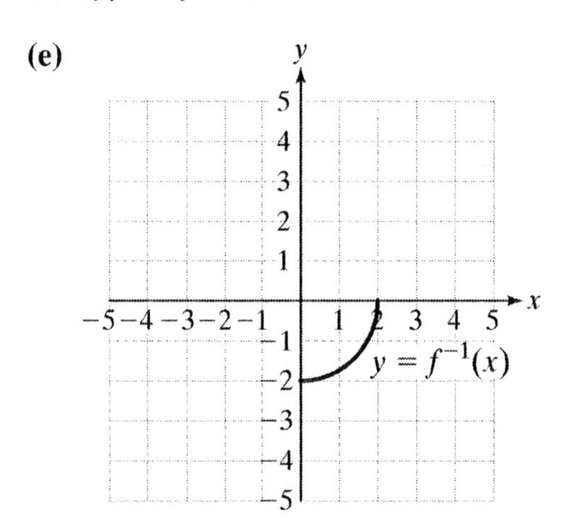

47. **(a)** $\{x|2 \leq x \leq 5\}$

(b) $\{y|0 \le y \le 3\}$

(c) $\{x|0 \le x \le 3\}$

(d) $\{y|2 \le y \le 5\}$

(e)

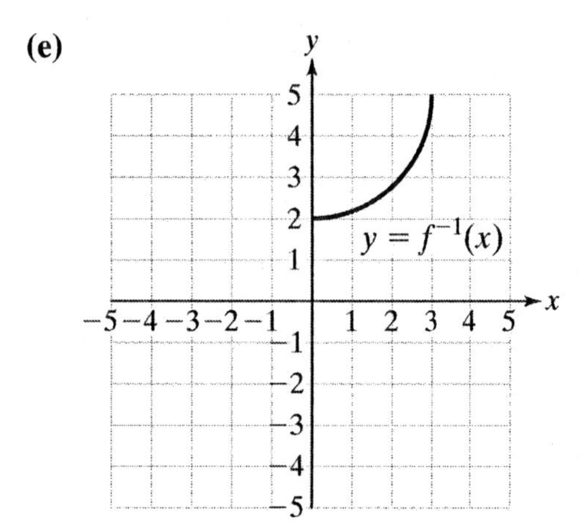

49. $y = \dfrac{3-x}{x+3}$

To find the inverse, interchange x and y.

$$x = \frac{3-y}{y+3}$$
$$x(y+3) = 3-y$$
$$xy + 3x = 3-y$$
$$xy + y = 3-3x$$
$$y(x+1) = 3-3x$$
$$y = \frac{3-3x}{x+1}$$
$$p^{-1}(x) = \frac{3-3x}{x+1},\ x \ne -1$$

51. $y = \dfrac{4}{x+2}$

To inverse, interchange x and y.

$$x = \frac{4}{y+2}$$
$$x(y+2) = 4$$
$$xy + 2x = 4$$
$$xy = 4-2x$$
$$y = \frac{4-2x}{x}$$
$$w^{-1}(x) = \frac{4-2x}{x},\ x \ne 0$$

53. $y = x^2 - 1;\ x \ge 0$

To find the inverse, interchange x and y.

$$x = y^2 - 1$$
$$x + 1 = y^2$$
$$\sqrt{x+1} = y$$
$$m^{-1}(x) = \sqrt{x+1}$$

55. $y = x^2 - 1;\ x \le 0$

To find the inverse, interchange x and y.

$$x = y^2 - 1$$
$$x + 1 = y^2$$
$$-\sqrt{x+1} = y$$
$$g^{-1}(x) = -\sqrt{x+1}$$

57. $v(x) = \sqrt{x+16}$

To find the inverse, interchange x and y.

$$x = \sqrt{y+16}$$
$$x^2 = y + 16$$
$$x^2 - 16 = y$$
$$v^{-1}(x) = x^2 - 16,\ x \ge 0$$

59. $u(x) = -\sqrt{x+16}$

To find the inverse, interchange x and y.

$$x = -\sqrt{y+16}$$
$$x^2 = y + 16$$
$$x^2 - 16 = y$$
$$u^{-1}(x) = x^2 - 16,\ x \le 0$$

61. $k(x) = x^3 - 4;\ k^{-1}(x) = \sqrt[3]{x+4}$

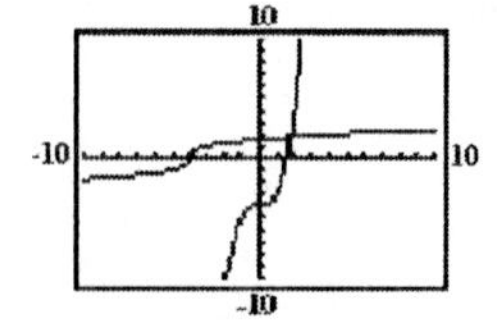

63. $m(x) = 3x - 4;\ m^{-1}(x) = \dfrac{x+4}{3}$

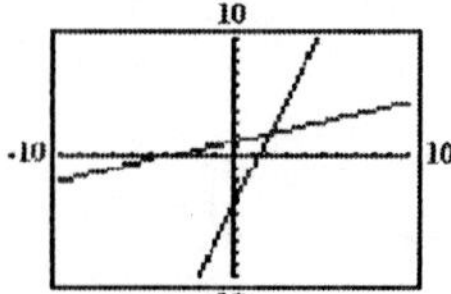

Section 12.3 Practice Exercises

1. 25

3. $\frac{1}{10^3} = \frac{1}{1000}$

5. $\sqrt{36} = 6$

7. $\left(\sqrt[4]{16}\right)^3 = 2^3 = 8$

9. 5.8731

11. 1385.4557

13. 0.0063

15. 0.8950

17. (a) $3^x = 3^2$
$x = 2$

(b) $3^x = 3^3$
$x = 3$

(c) between 2 and 3, closer to 2

19. (a) $2^x = 2^4$
$x = 4$

(b) $2^x = 2^5$
$x = 5$

(c) between 4 and 5, closer to 5

21. $f(0) = \left(\frac{1}{5}\right)^0 = 1;$

$f(1) = \left(\frac{1}{5}\right)^1 = \frac{1}{5};$

$f(2) = \left(\frac{1}{5}\right)^2 = \frac{1}{25};$

$f(-1) = \left(\frac{1}{5}\right)^{-1} = 5;$

$f(-2) = \left(\frac{1}{5}\right)^{-2} = 5^2 = 25$

23. $h(0) = (\pi)^2 = 1;$

$h(1) = (\pi)^1 \approx 3.14;$

$h(-1) = (\pi)^{-1} \approx 0.32;$

$h\left(\sqrt{2}\right) = (\pi)^{\sqrt{2}} \approx 5.05;$

$h(\pi) = (\pi)^\pi \approx 36.46$

25. $r(0) = (3)^{0+2} = (3)^2 = 9;$

$r(1) = (3)^{1+2} = (3)^3 = 27;$

$r(2) = (3)^{2+2} = (3)^4 = 81;$

$r(-1) = (3)^{-1+2} = (3)^1 = 3;$

$r(-2) = (3)^{-2+2} = (3)^0 = 1;$

$r(-3) = (3)^{-3+2} = (3)^{-1} = \frac{1}{3}$

27. If $b > 1$, the graph is increasing. If $0 < b < 1$, the graph is decreasing.

29.

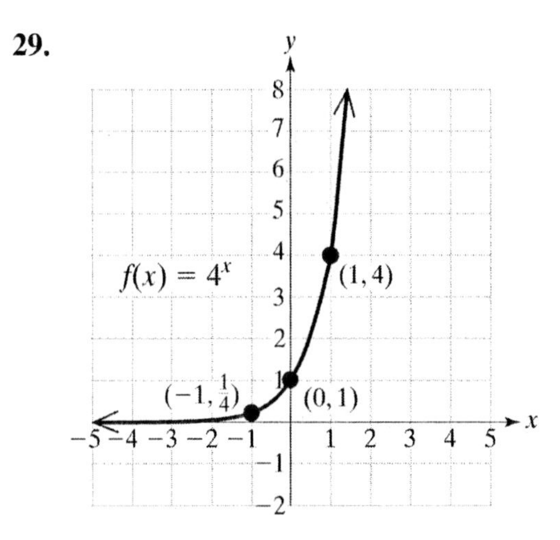

31.

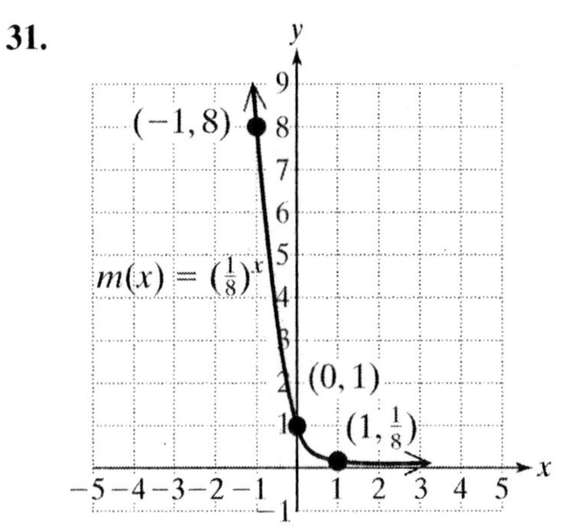

33.

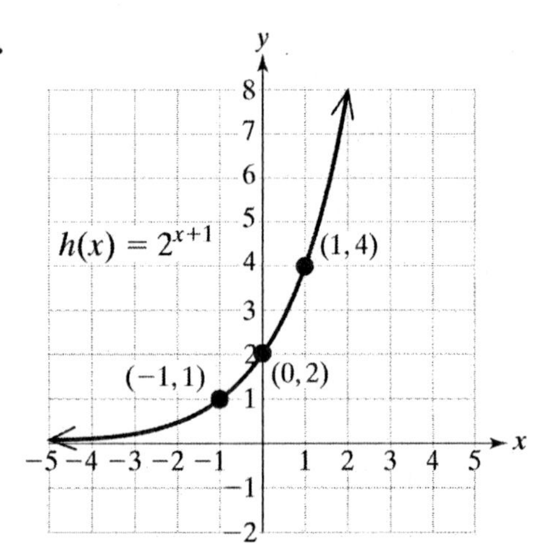

35.

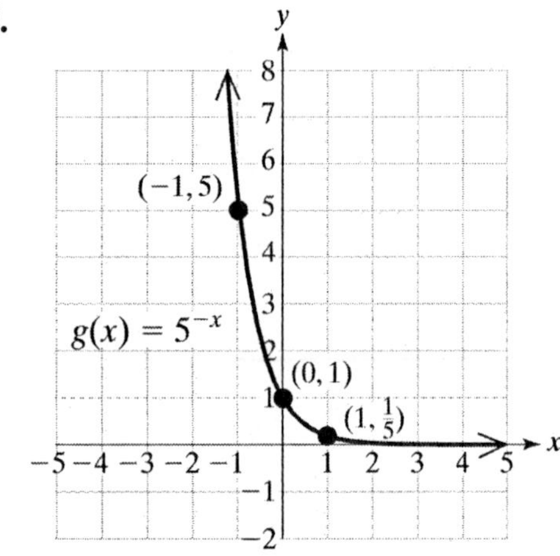

37. **(a)** $A(5) = 1000(2)^{5/7} = \$1640.67$

(b) $A(10) = 1000(2)^{10/7} = \$2691.80$

(c) $A(0) = 1000(2)^{0/7} = \$1000$
After 0 years, only the initial amount will be in the account.
$A(7) = 1000(2)^{7/7} = \$2000$
The amount invested will double in 7 years.

39. **(a)** $I(t) = 3{,}600{,}000(1.0036)^t$

(b) $S(t) = 3{,}500{,}000(1.012)^t$

(c) $I(20) = 3{,}600{,}000(1.0036)^{20} = 3{,}900{,}000;$

$S(20) = 3{,}500{,}000(1.012)^{20} = 4{,}400{,}000;$

$I(40) = 3{,}600{,}000(1.0036)^{40} = 4{,}200{,}000;$

$S(40) = 3{,}500{,}000(1.012)^{40} = 5{,}600{,}000;$

$I(60) = 3{,}600{,}000(1.0036)^{60} = 4{,}500{,}000;$

$S(60) = 3{,}500{,}000(1.012)^{60} = 7{,}200{,}000$

(d) Since Singapore has a higher growth rate, the population of Singapore will eventually overtake the population of Ireland.

(e) The population density (number of people per square mile) is more than 100 times as large for Singapore as for Ireland.

41.

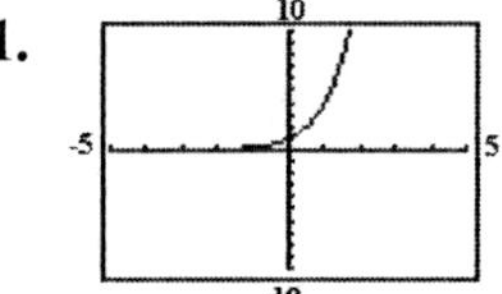

43.

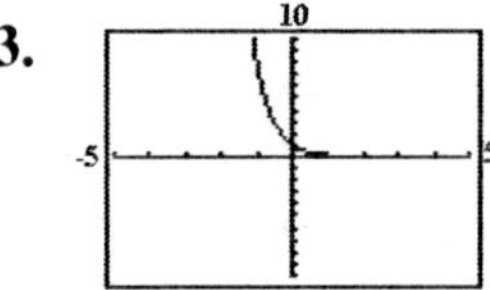

45.

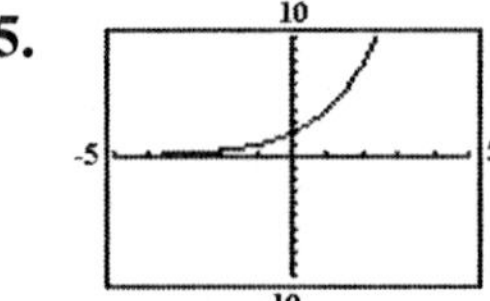

47.

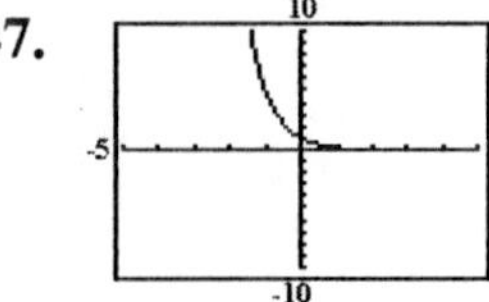

49. (a)

(b) (7, \$2000), (14, \$4000), (21, \$8000)
7 years, 14 years, 21 years

Section 12.4 Practice Exercises

1. ii

3. (a) $g(-2) = 3^{-2} = \dfrac{1}{3^2} = \dfrac{1}{9};$

$g(-1) = 3^{-1} = \dfrac{1}{3};$

$g(0) = 3^0 = 1;$

$g(1) = 3^1 = 3;$

$g(2) = 3^2 = 9$

(b)

5. (a) $s(-2) = \left(\dfrac{2}{5}\right)^{-2} = \dfrac{2^{-2}}{5^{-2}} = \dfrac{5^2}{2^2} = \dfrac{25}{4};$

$s(-1) = \left(\dfrac{2}{5}\right)^{-1} = \dfrac{2^{-1}}{5^{-1}} = \dfrac{5}{2};$

$s(0) = \left(\dfrac{2}{5}\right)^{0} = 1;$

$s(1) = \left(\dfrac{2}{5}\right)^{1} = \dfrac{2}{5};$

$s(2) = \left(\dfrac{2}{5}\right)^{2} = \dfrac{2^2}{5^2} = \dfrac{4}{25}$

(b)

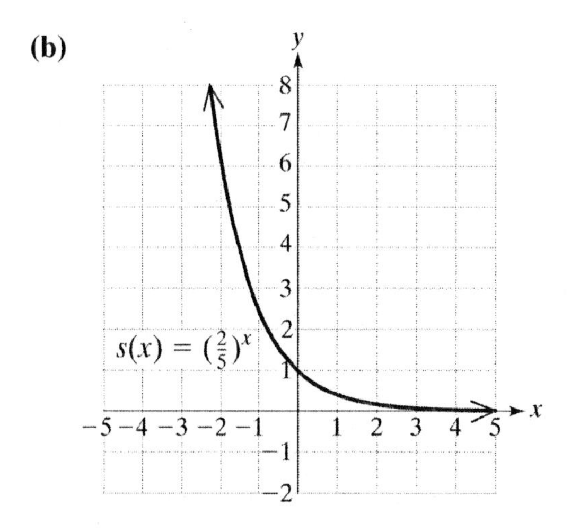

7. $b^y = x$

9. $\log_{10}(1000) = 3$

11. $\log_8(2) = \dfrac{1}{3}$

13. $\log_8\left(\dfrac{1}{64}\right) = -2$

15. $\log_b(x) = y$

17. $\log_e(x) = y$

19. $\log_H(q) = m$

21. $125^{2/3} = 25$

23. $25^{-1/2} = \frac{1}{5}$

25. $2^7 = 128$

27. $b^y = 82$

29. $2^x = 7$

31. $\left(\frac{1}{2}\right)^6 = x$

33. $3^x = 27$
$3^x = 3^3$
$x = 3$

35. $2^x = \frac{1}{16}$
$2^x = 2^{-4}$
$x = -4$

37. $8^x = 2$
$(2^3)^x = 2$
$2^{3x} = 2^1$
$3x = 1$
$x = \frac{1}{3}$

39. $8^x = 8$
$8^x = 8^1$
$x = 1$

41. $9^x = 9^3$
$x = 3$

43. $7^x = 1$
$7^x = 7^0$
$x = 0$

45. $10^x = 100$
$10^x = 10^2$
$x = 2$

47. $10^x = 10000$
$10^x = 10^4$
$x = 4$

49. $10^x = 0.1$
$10^x = 10^{-1}$
$x = -1$

51. $10^x = 0.001$
$10^x = 10^{-3}$
$x = -3$

53. 0.7782

55. 0.4971

57. −1.5051

59. −2.2676

61. 5.5315

63. −7.4202

65. **(a)** slightly less than 2

(b) slightly more than 1

(c) $\log 93 \approx 1.9685$; $\log 12 \approx 1.0792$

67. **(a)** $f\left(\frac{1}{64}\right) = \log_4\left(\frac{1}{64}\right)$

Thus, $4^x = \frac{1}{64}$
$4^x = 4^{-3}$
$x = -3;$

$f\left(\frac{1}{16}\right) = \log_4\left(\frac{1}{16}\right)$

Thus, $4^x = \frac{1}{16}$
$4^x = 4^{-2}$
$x = -2;$

$f\left(\frac{1}{4}\right) = \log_4\left(\frac{1}{4}\right)$

Thus, $4^x = \frac{1}{4}$

$4^x = 4^{-1}$

$x = -1;$

$f(1) = \log_4(1)$

Thus, $4^x = 1$

$4^x = 4^0$

$x = 0;$

$f(4) = \log_4(4)$

Thus, $4^x = 4$

$4^x = 4^1$

$x = 1;$

$f(16) = \log_4(16)$

Thus, $4^x = 16$

$4^x = 4^2$

$x = 2;$

$f(64) = \log_4(64)$

Thus, $4^x = 64$

$4^x = 4^3$

$x = 3$

(b)

$f(x) = \log_4 x$

69. $3^y = x$

x	y
$\frac{1}{9}$	-2
$\frac{1}{3}$	-1
1	0
3	1
9	2

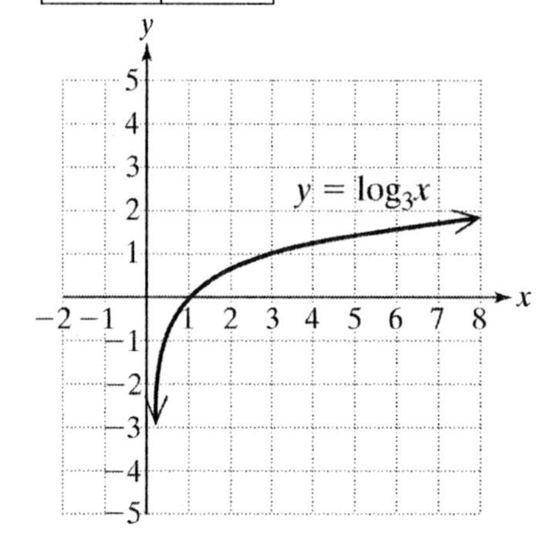

71. $\left(\frac{1}{2}\right)^y = x$

x	y
4	-2
2	-1
1	0
$\frac{1}{2}$	1
$\frac{1}{4}$	2

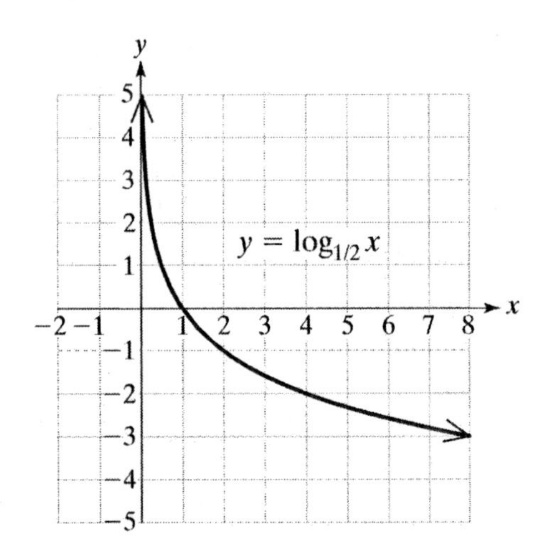

73. $\{x|x > 0\}$

75. $\{x|x > 0\}$

77. $x - 5 > 0$
$x > 5$
Interval notation: $(5, \infty)$

79. $x + 1.2 > 0$
$x > -1.2$
Interval notation: $(-1.2\ \infty)$

81. $x^2 > 0$
all real numbers except 0
Interval notation: $(-\infty, 0) \cup (0, \infty)$

83. **(a)**

t (months)	0	1	2	6	12	24
$S_1(t)$	91	82.0	76.7	65.6	57.6	49.1
$S_2(t)$	88	83.5	80.8	75.3	71.3	67.0

(b) Group 1: $t(0) = 91$
Group 2: $t(0) = 88$

(c) Method II

85. **(a)** $\text{pH} = -\log[\text{H}^+] = -\log(0.0002) \approx 3.7$

(b) $\text{pH} = -\log[\text{H}^+] = -\log(1.0 \times 10^{-11}) = 11$

87. Domain: $(-2, \infty)$
Asymptote: $x = -2$

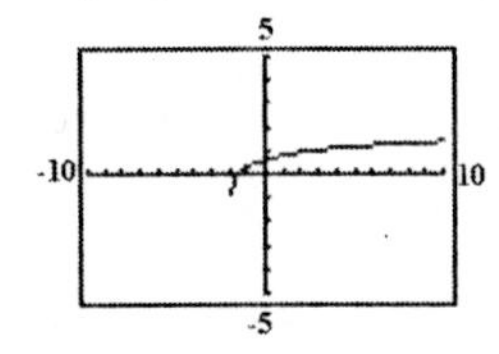

89. Domain: $(-8, \infty)$
Asymptote: $x = -8$

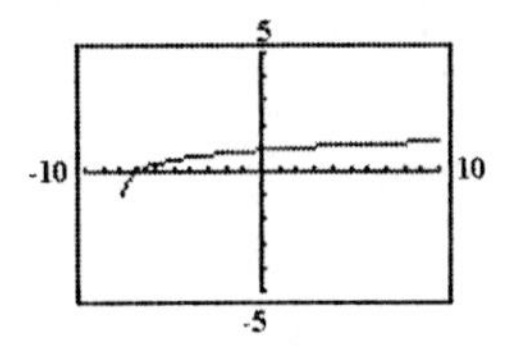

91. Domain: $(-\infty, 3)$
Asymptote: $x = 3$

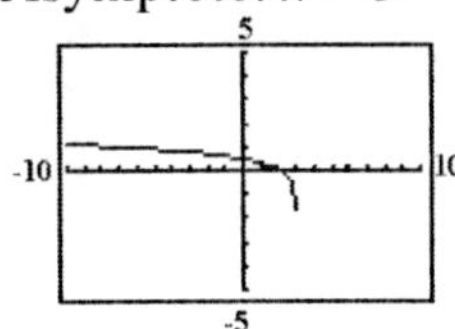

Section 12.5 Practice Exercises

1. $\frac{1}{8^2} = \frac{1}{64}$

3. $2^x = 32$
$2^x = 2^5$
$x = 5$

5. 2.9707

7. 1.4314

9. d

11. b

13. For example: $\log_{10} 1 = 0$

15. For example: $\log_4 4^2 = 2$

17. $3^x = 3$
$3^x = 3^1$
$x = 1$

19. $5^x = 5^4$
$x = 4$

21. 11

23. 3

25. $3^x = 1$
$3^x = 3^0$
$x = 0$

27. 9

29. Expressions a and c are equivalent.

31. Expressions a and c are equivalent.

33. $\log_3 x - \log_3 5$

35. $\log 2 + \log x$

37. $4\log_{10} x$

39. $\log_4 a + \log_4 b - \log_4 c$

41. $\frac{1}{2}\log_b x + \log_b y - (3\log_b z + \log_b w)$
$= \frac{1}{2}\log_b x + \log_b y - 3\log_b z - \log_b w$

43. log(CABIN)

45. $\log_3 x^2 - \log_3 y^3 + \log_3 z = \log_3\left(\frac{x^2 z}{y^3}\right)$

47. $\log_b x - \log_b x^3 + \log_b x^4 = \log_b\left(\frac{x}{x^3} \cdot x^4\right)$
$= \log_b(x^2)$

49. $\log_8 a^5 - \log_8 1 + \log_8 8$
$\log_8 a^5 - 0 + 1$
$\log_8 a^5 + 1$ or $\log_8(8a^5)$

51. **(a)** $B = 10(\log I - \log I_0)$
$B = 10\log I - 10\log I_0$

(b) $B = 10\log I - 10\log(10^{-16})$
$B = 10\log I + 160\log 10$
$B = 10\log I + 160$

53. (a) M
$= 4.71 + 2.5\log(3.9\times10^{26}) - 2.5\log(L)$
$= 4.71 + 2.5[\log(3.9\times10^{26}) - \log(L)]$
$= 4.71 + 2.5\log\left(\dfrac{3.9\times10^{26}}{L}\right)$

(b) $M = 4.71 + 2.5\log\left(\dfrac{3.9\times10^{26}}{3.9\times10^{26}}\right)$
$= 4.71 + 2.5\log(1)$
$= 4.71 + 0$
$= 4.71$

(c) $M = 4.71 + 2.5\log\left(\dfrac{3.9\times10^{26}}{8.2\times10^{27}}\right)$
$\approx 4.71 + 2.5\log(0.048)$
$\approx 4.71 + 2.5(-1.32)$
≈ 1.4

55. (a) Domain: $(-\infty, 0) \cup (0, \infty)$

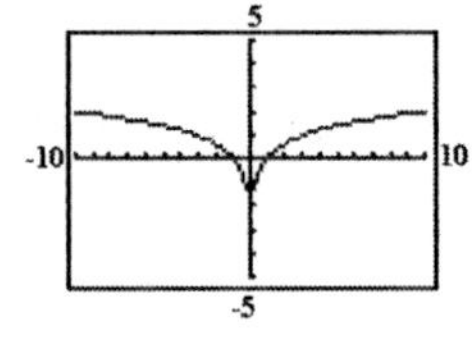

(b) Domain: $(0, \infty)$

5
-10
10
-5

(c) They are equivalent for all x in the intersection of their domains, $(0, \infty)$.

Section 12.6 Practice Exercises

1.

x	$f(x)$
−3	$\frac{8}{27}$
−2	$\frac{4}{9}$
−1	$\frac{2}{3}$
0	1
1	$\frac{3}{2}$
2	$\frac{9}{4}$
3	$\frac{27}{8}$

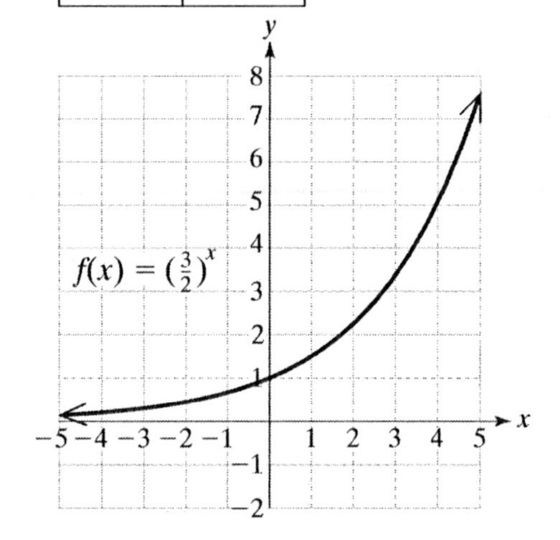

3.

x	$q(x)$
−0.75	−0.60
−0.50	0.30
−0.25	0.12
0	0
1	0.30
2	0.48

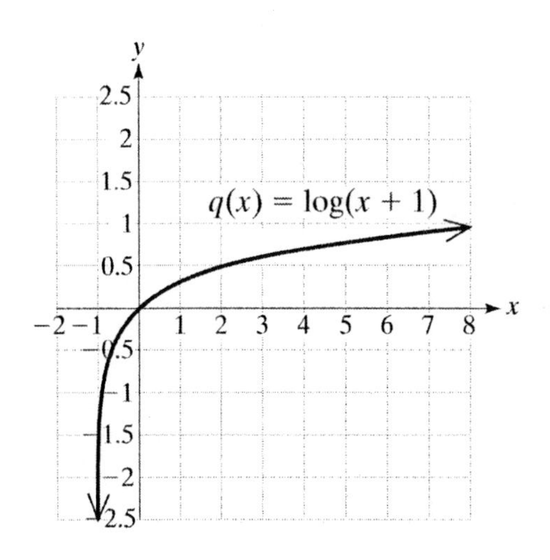

5. $\ln a^4 + \ln\sqrt{b} - \ln c$

   $4\ln a + \frac{1}{2}\ln b - \ln c$

7. $\frac{1}{5}\ln\left(\frac{ab}{c^2}\right)$

   $\frac{1}{5}\ln ab - \frac{1}{5}\ln c^2$

   $\frac{1}{5}\ln a + \frac{1}{5}\ln b - \frac{2}{5}\ln c$

9. $\ln a^2 - \ln b - \ln\sqrt[3]{c}$

   $= \ln a^2 - \left(\ln b + \ln\sqrt[3]{c}\right)$

   $= \ln\left(\frac{a^2}{b\sqrt[3]{c}}\right)$

11. $\ln x^4 - \ln y^3 - \ln z = \ln x^4 - (\ln y^3 + \ln z)$

    $= \ln\left(\frac{x^4}{y^3 z}\right)$

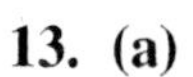

13. (a)

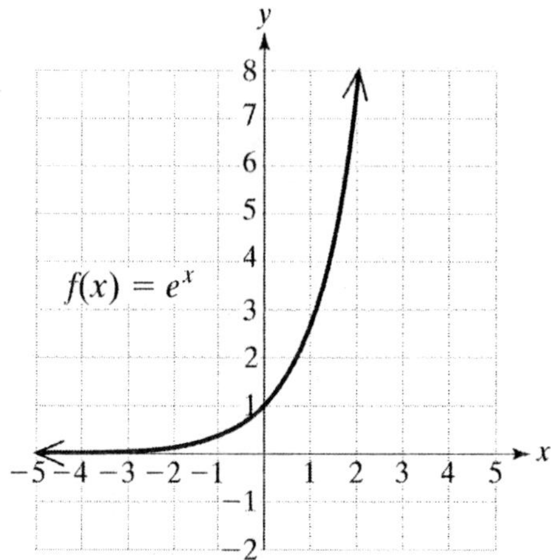

(b) Domain: $(-\infty, \infty)$
Range: $(0, \infty)$

(c)

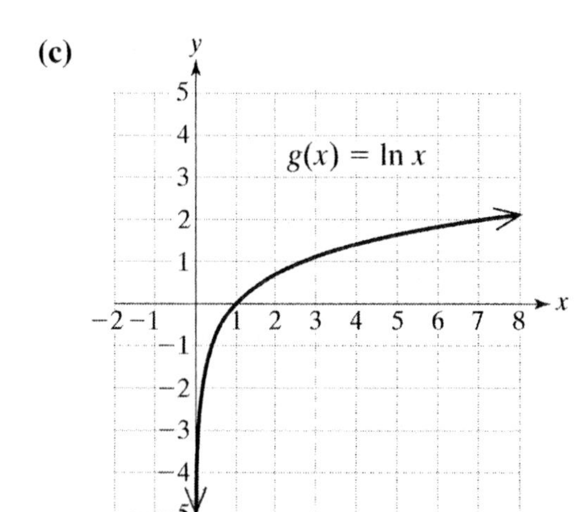

(d) Domain: $(0, \infty)$
Range: $(-\infty, \infty)$

15. Domain: $(-\infty, \infty)$
    Range: $(0, \infty)$

x	y
–4	0.05
–3	0.14
–2	0.37
–1	1
0	2.72
1	7.39
2	20.09

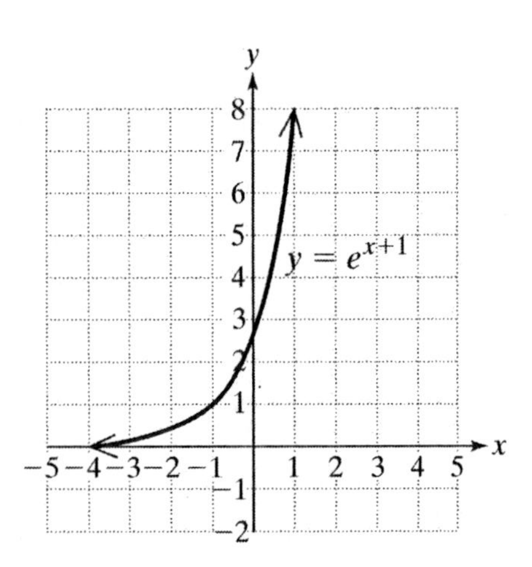

17. Domain: $(2, \infty)$
Range: $(-\infty, \infty)$

x	y
2.25	−1.39
2.50	−0.69
2.75	−0.29
3	0
4	0.69
5	1.10
6	1.39

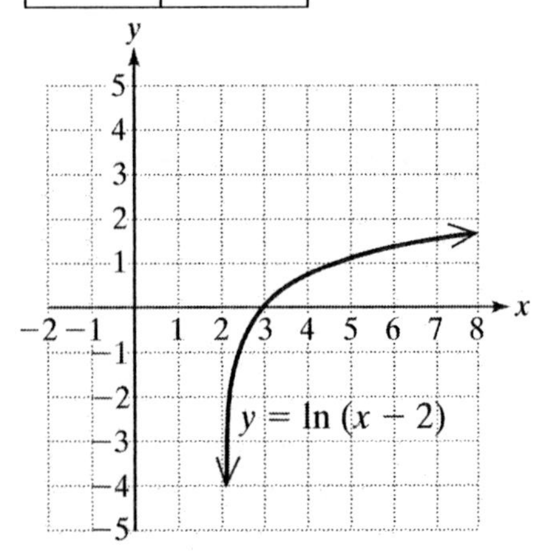

19. **(a)** 2.9570

(b) 2.9570

(c) They are the same.

21. $\dfrac{\log 7}{\log 2} \approx 2.8074$

23. $\dfrac{\log 24}{\log 8} \approx 1.5283$

25. $\dfrac{\log 0.012}{\log 8} \approx -2.1269$

27. $\dfrac{\log 1}{\log 9} = 0$

29. $\dfrac{\log\left(\frac{1}{100}\right)}{\log 4} \approx -3.3219$

31. $\dfrac{\log 0.0006}{\log 7} \approx -3.8124$

33. **(a)** between 2 and 3

(b) slightly less than 3

(c) $\log_3 15 \approx 2.4650$, $\log_3 25 \approx 2.9299$

35. **(a)** slightly more than 1

(b) slightly less than 2

(c) $\log_6 10 \approx 1.2851$, $\log_6 30 \approx 1.8982$

37. **(a)** $D(0) = 91 + 160 \ln(0 + 1) = 91$ deaths

(b) $D(4) = 91 + 160 \ln(4 + 1) \approx 349$
Sept. 5, 349 deaths;
$D(9) = 91 + 160 \ln(9 + 1) \approx 459$
Sept. 10, 459 deaths;
$D(19) = 91 + 160 \ln(19 + 1) \approx 570$
Sept. 20, 570 deaths

39. **(a)** $t = \dfrac{\ln(2)}{0.055} \approx 12.6$ years

(b) $t = \dfrac{\ln(2)}{0.08} \approx 8.7$ years

(c) $t = 2(8.7) = 17.4$ years

41. **(a)** $A(8) = 5000\left(1+\frac{0.045}{4}\right)^{4(8)} = \$7152.26$

(b) $A(8) = 5000\left(1+\frac{0.055}{4}\right)^{4(8)} = \$7740.30$

(c) $A(8) = 5000\left(1+\frac{0.07}{4}\right)^{4(8)} = \$8711.07$

(d) $A(8) = 5000\left(1+\frac{0.09}{4}\right)^{4(8)} = \$10,190.52$

An investment grows more rapidly at higher interest rates.

43. **(a)** $A(8) = 15000\left(1+\frac{0.05}{1}\right)^{1(8)}$
$= \$22,161.83$

(b) $A(8) = 15000\left(1+\frac{0.05}{4}\right)^{4(8)}$
$= \$22,321.96$

(c) $A(8) = 15000\left(1+\frac{0.05}{12}\right)^{12(8)}$
$= \$22,358.78$

(d) $A(8) = 15000\left(1+\frac{0.05}{365}\right)^{365(8)}$
$= \$22,376.76$

(e) $A(8) = 15000(e)^{0.05(8)} = \$22,377.37$

More money is earned at a greater number of compound periods per year.

45. **(a)** $A(5) = 10000(e)^{0.06(5)} = \$13,498.59$

(b) $A(10) = 10000(e)^{0.06(10)} = \$18,221.19$

(c) $A(15) = 10000(e)^{0.06(15)} = \$24,596.03$

(d) $A(20) = 10000(e)^{0.06(20)} = \$33,201.17$

(e) $A(30) = 10000(e)^{0.06(30)} = \$60,496.47$

More money is earned over a longer period of time.

47. **(a)–(b)**

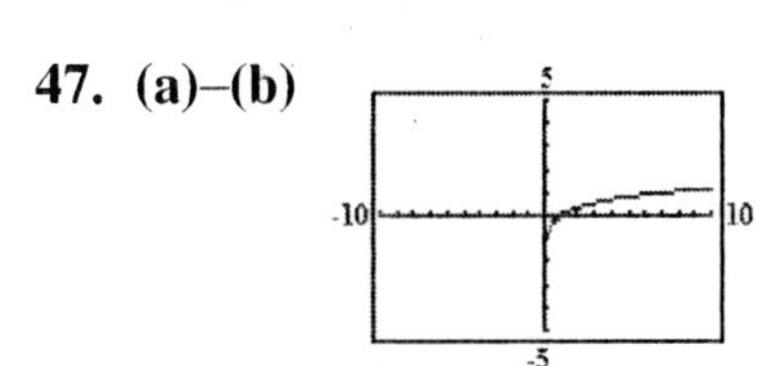

(c) They appear to be the same.

49. **(a)–(b)**

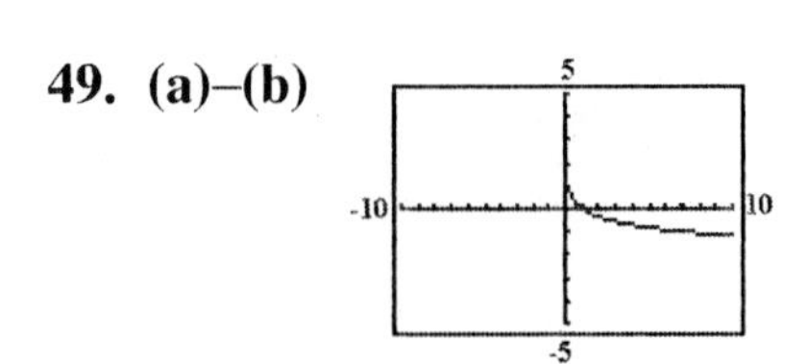

(c) They appear to be the same.

51.

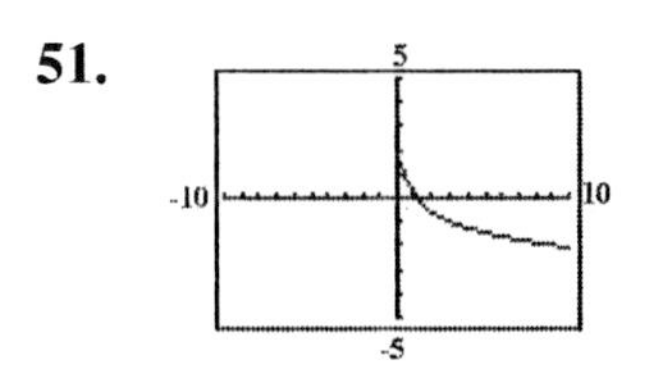

53.

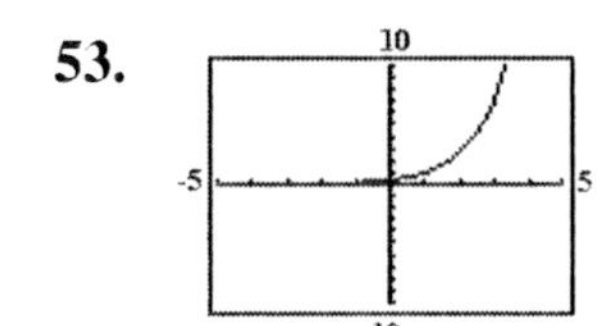

55.

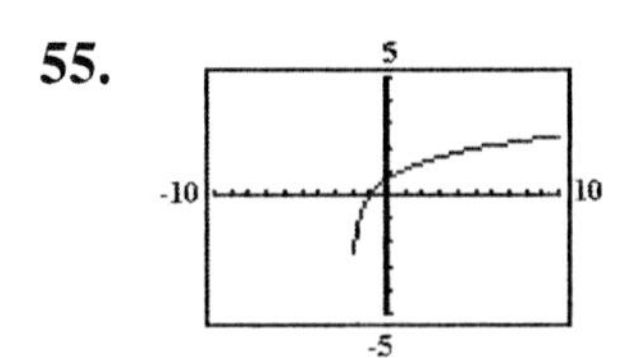

Section 12.7 Practice Exercises

1. **(a)**

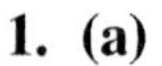

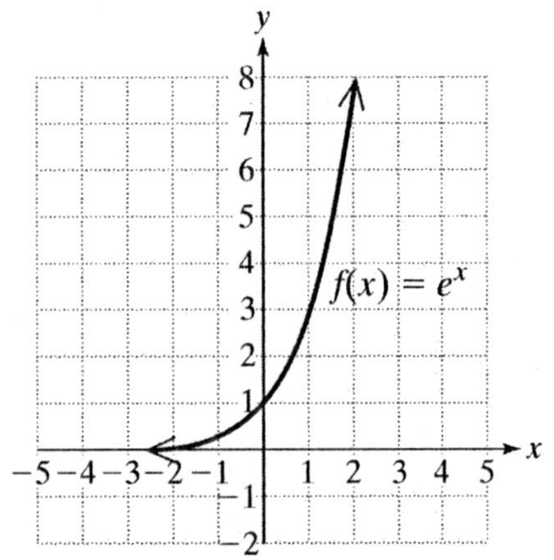

(b) Domain: $(-\infty, \infty)$
Range: $(0, \infty)$

3. **(a)**

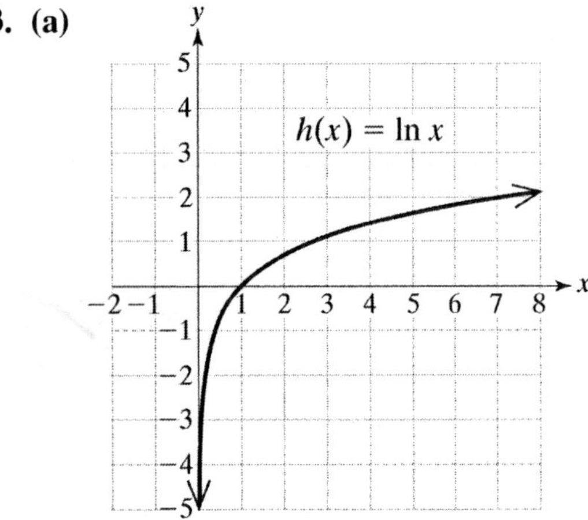

(b) $x = 0$

(c) Domain: $(0, \infty)$
Range: $(-\infty, \infty)$

5. $\log_b[(x-1)(x+2)]$

7. $\log_b\left(\dfrac{x}{1-x}\right)$

9. $x-5=0$
$x=5$
Domain: $(5, \infty)$

11. $x+2=0$
$x=-2$
Domain: $(-2, \infty)$

13. $2x+1=0$
$2x=-1$
$x=-\dfrac{1}{2}$
Domain: $\left(-\dfrac{1}{2}, \infty\right)$

15. $5^x = 5^4$
$x=4$

17. $2^{-x} = 2^6$
$-x=6$
$x=-6$

19. $(6^2)^x = 6^1$
$6^{2x} = 6^1$
$2x=1$
$x=\dfrac{1}{2}$

21. $4^{2x-1} = 4^3$
$2x-1=3$
$2x=4$
$x=2$

23. $(3^4)^{3x-4} = 3^{-5}$
$3^{12x-16} = 3^{-5}$
$12x-16=-5$
$12x=11$
$x=\dfrac{11}{12}$

25. $\ln 8^a = \ln 21$
$a\ln 8 = \ln 21$
$a = \dfrac{\ln 21}{\ln 8} \approx 1.464$

27. $\ln e^x = \ln 8.1254$
$x \ln e = \ln 8.1254$
$x = \ln 8.1254 \approx 2.095$

29. $\ln 10^t = \ln 0.0138$
$t \ln e = \ln 0.0138$
$t = \ln 0.0138 \approx -1.860$

31. $\ln e^{0.07h} = \ln 15$
$0.07h \ln e = \ln 15$
$0.07h = \ln 15$
$h = \dfrac{\ln 15}{0.07} \approx 38.686$

33. $\ln 32e^{0.04m} = \ln 128$
$\ln 32 + \ln e^{0.04m} = \ln 128$
$\ln 32 + 0.04m \ln e = \ln 128$
$0.04m = \ln 128 - \ln 32$
$0.04m = \ln \dfrac{128}{32}$
$0.04m = \ln 4$
$m = \dfrac{\ln 4}{0.04} \approx 34.657$

35. $\ln 3^{x+1} = \ln 5^x$
$(x+1)\ln 3 = x \ln 5$
$x \ln 3 + \ln 3 = x \ln 5$
$x \ln 3 - x \ln 5 = -\ln 3$
$x(\ln 3 - \ln 5) = -\ln 3$
$x = \dfrac{-\ln 3}{\ln 3 - \ln 5} \approx 2.151$

37. $10000 = 5000e^{0.07t}$
$2 = e^{0.07t}$
$\ln 2 = \ln e^{0.07t}$
$\ln 2 = 0.07t \ln e$
$\ln 2 = 0.07t$
$\dfrac{\ln 2}{0.07} = t$
Approximately 9.9 years

39. (a) $A(5) = 10(0.5)^{5/14}$
$A(5) \approx 7.8$ grams

(b) $4 = 10(0.5)^{t/14}$
$\dfrac{4}{10} = (0.5)^{t/14}$
$\ln 0.4 = \ln(0.5)^{t/14}$
$\ln 0.4 = \dfrac{t}{14}\ln(0.5)$
$\dfrac{\ln 0.4}{\ln 0.5} = \dfrac{t}{14}$
$14\left(\dfrac{\ln 0.4}{\ln 0.5}\right) = t$
$t \approx 18.5$ days

41. (a) $2002 - 1998 = 4$
$P(4) = 1237(1.0095)^4$
≈ 1285 million people

(b) $2012 - 1998 = 14$
$P(14) = 1237(1.0095)^{14}$
≈ 1412 million people

(c) $2000 = 1237(1.0095)^t$
$\dfrac{2000}{1237} = 1.0095^t$
$\ln\left(\dfrac{2000}{1237}\right) = \ln 1.0095^t$
$\ln\left(\dfrac{2000}{1237}\right) = t \ln 1.0095$
$\dfrac{\ln\left(\frac{2000}{1237}\right)}{\ln 1.0095} = t$
$50.8 \approx t$
1998 + 51 = the year 2049

43. (a) $A(0) = 500e^{0.0277(0)}$
$= 500e^0$
$= 500(1)$
$= 500$ bacteria

(b) $A(10) = 500e^{0.0277(10)}$
$= 500e^{0.277}$
≈ 660 bacteria

(c)
$$1000 = 500e^{0.0277t}$$
$$\frac{1000}{500} = e^{0.0277t}$$
$$2 = e^{0.0277t}$$
$$\ln(2) = \ln(e^{0.0277t})$$
$$\ln 2 = 0.0277t$$
$$\frac{\ln 2}{0.0277} = t$$
$$t \approx 25 \text{ minutes}$$

45.
$$1000000 = 10000e^{0.12(t)}$$
$$100 = e^{0.12(t)}$$
$$\ln 100 = \ln e^{0.12(t)}$$
$$\ln 100 = 0.12t \ln e$$
$$\ln 100 = 0.12t$$
$$\frac{\ln 100}{0.12} = t$$
$$t \approx 38.4 \text{ years}$$

47.
$$3^2 = x$$
$$x = 9$$

49. $10^{42} = p$

51.
$$e^{0.08} = x$$
$$x \approx 1.083$$

53.
$$x^2 = 25$$
$$x^2 = 5^2$$
The bases must be equal since the exponents are equal. Thus, $x = 5$.

55.
$$b^4 = 10000$$
$$b^4 = 10^4$$
The bases must be equal since the exponents are equal. Thus, $b = 10$.

57.
$$y^{1/2} = 5$$
$$y = 25$$

59.
$$4^3 = c + 5$$
$$64 = c + 5$$
$$59 = c$$

61.
$$5^1 = 4y + 1$$
$$5 = 4y + 1$$
$$4 = 4y$$
$$1 = y$$

63.
$$\log_3 k(2k+3) = 2$$
$$3^2 = k(2k+3)$$
$$9 = 2k^2 + 3k$$
$$2k^2 + 3k - 9 = 0$$
$$(2k-3)(k+3) = 0$$
$$2k - 3 = 0 \quad \text{or} \quad k + 3 = 0$$
$$2k = 3 \quad \text{or} \quad k \neq -3$$
$$k = \frac{3}{2}$$

65.
$$x + 2 = 3x - 6$$
$$-2x = -8$$
$$x = 4$$

67.
$$\log_5 \frac{3t+2}{t} = \log_5 4$$
$$\frac{3t+2}{t} = 4$$
$$3t + 2 = 4t$$
$$-t + 2 = 0$$
$$-t = -2$$
$$t = 2$$

69.
$$\log_{10} 4m = \log_{10} 2(m-3)$$
$$4m = 2m - 6$$
$$2m = -6$$
$$m \neq -3$$
No solution

71.
$$(\log_2 x)^2 - 12\log_2 x + 32 = 0$$
$$(\log_2 x - 8)(\log_2 x - 4) = 0$$
$$\log_2 x - 8 = 0 \quad \text{or} \quad \log_2 x - 4 = 0$$
$$\log_2 x = 8 \quad \text{or} \quad \log_2 x = 4$$
$$2^8 = x \quad \text{or} \quad 2^4 = x$$
$$256 = x \quad \text{or} \quad 16 = x$$

73. (a) $P(43) = 2e^{-0.0079(43)} \approx 1.42 \text{ kg}$

(b) No, since 1.42 kg < 1.5 kg.

75. **(a)** **(i)** $R = \log\left(\dfrac{10^t I_0}{I_0}\right) = \log(10^5) = 5$

(ii) $R = \log\left(\dfrac{10^6 I_0}{I_0}\right) = \log(10^6) = 6$

(b) $7.1 = \log\left(\dfrac{I}{I_0}\right)$

$\dfrac{I}{I_0} = 10^{7.1}$ (exponential form)

$I = 10^{7.1} I_0$

$10^{7.1}$ times ($\approx$ 12,590,000) more intense

77.

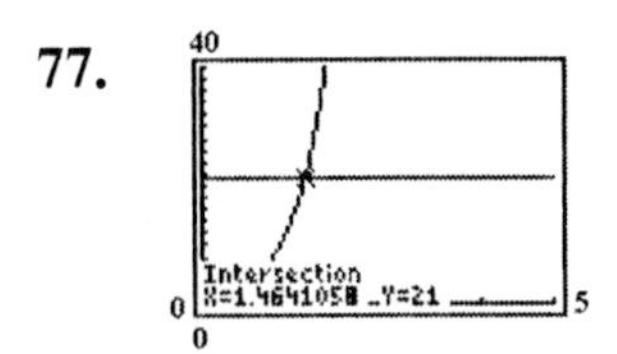

79.

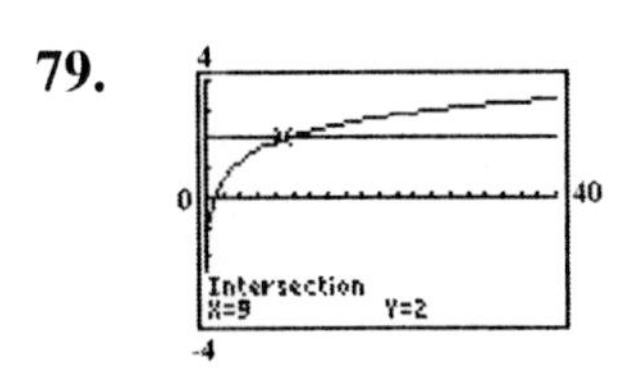

Chapter 12 Review Exercises

1. $(f - g)(x) = f(x) - g(x)$

$= (x - 7) - (-2x^3 - 8x)$

$= x - 7 + 2x^3 + 8x$

$= 2x^3 + 9x - 7$

Domain: $(-\infty, \infty)$

3. $(f \cdot n)(x) = f(x) \cdot n(x)$

$= (x - 7) \cdot \dfrac{1}{x - 2}$

$= \dfrac{x - 7}{x - 2}$

Domain: $(-\infty, 2) \cup (2, \infty)$

5. $\left(\dfrac{f}{g}\right)(x) = \dfrac{f(x)}{g(x)} = \dfrac{x - 7}{-2x^3 - 8x}$

Domain: The denominator is zero when

$-2x^3 - 8x = 0$

$2x(x^2 + 4) = 0$

$-2x = 0$ or $x^2 + 4 = 0$

$x = 0$

(The quantity $x^2 + 4$ is never 0.) We exclude this value from the domain.

$(-\infty, 0) \cup (0, \infty)$

7. $(m \circ f)(x) = m(f(x)) = m(x - 7) = \sqrt{x - 7}$

Domain: $[7, \infty)$

9. $(m \circ g)(-2) = m(g(-2))$

$g(-2) = -2(-2)^3 - 8(-2)$

$= -2(-8) + 16$

$= 16 + 16$

$= 32$

$m(g(-2)) = m(32) = \sqrt{32} = 4\sqrt{2}$

11. $(f \circ g)(4) = f(g(4))$

$g(4) = -2(4)^3 - 8(4)$

$= -2(64) - 32$

$= -128 - 32$

$= -160$

$f(g(4)) = f(-160) = -160 - 7 = -167$

13. **(a)** $(g \circ f)(x) = g(f(x))$

$= g(2x + 1)$

$= (2x + 1)^2$

$= 4x^2 + 4x + 1$

(b) $(f \circ g)(x) = f(g(x))$

$= f(x^2)$

$= 2(x)^2 + 1$

$= 2x^2 + 1$

(c) No, $f \circ g \neq g \circ f$.

15. $(f \cdot g)(-2) = f(-2) \cdot g(-2)$
$= (-1)(3)$
$= -3$

17. $(f - g)(2) = f(2) - g(2) = 2 - 3 = -1$

19. $(f \circ g)(4) = f(g(4)) = f(1) = 1$

21. yes

23. no

25. $\{(5, 3), (9, 2), (-1, 0), (1, 4)\}$

27. $y = 3 - 4x$
To find the inverse, interchange x and y.
$x = 3 - 4y$
$x - 3 = -4y$
$-\frac{1}{4}(x-3) = -\frac{1}{4}(-4y)$
$-\frac{1}{4}x + \frac{3}{4} = y$
$p^{-1}(x) = -\frac{1}{4}x + \frac{3}{4}$

29. $y = \sqrt[5]{x} + 3$
To find the inverse, interchange x and y.
$x = \sqrt[5]{y} + 3$
$x - 3 = \sqrt[5]{y}$
$(x-3)^5 = y$
$g^{-1}(x) = (x-3)^5$

31. $y = \frac{x-2}{x+2}$
To find the inverse, interchange x and y.
$x = \frac{y-2}{y+2}$
$x(y+2) = y - 2$
$xy + 2x = y - 2$
$xy - y = -2x - 2$
$y(x-1) = -2x - 2$
$y = \frac{-2x-2}{x-1}$
$m^{-1}(x) = \frac{-2x-2}{x-1}$

33. $(f \circ g)(x) = 5\left(\frac{1}{5}x + \frac{2}{5}\right) - 2 = x + 2 - 2 = x;$
$(g \circ f)(x) = \frac{1}{5}(5x-2) + \frac{2}{5} = x - \frac{2}{5} + \frac{2}{5} = x$

35. **(a)** Domain: $x \geq -1$; Range: $y \geq 0$

(b) Domain: $x \geq 0$; Range: $y \geq -1$

37. 1024

39. 2

41. 8.825

43. 1.627

45.

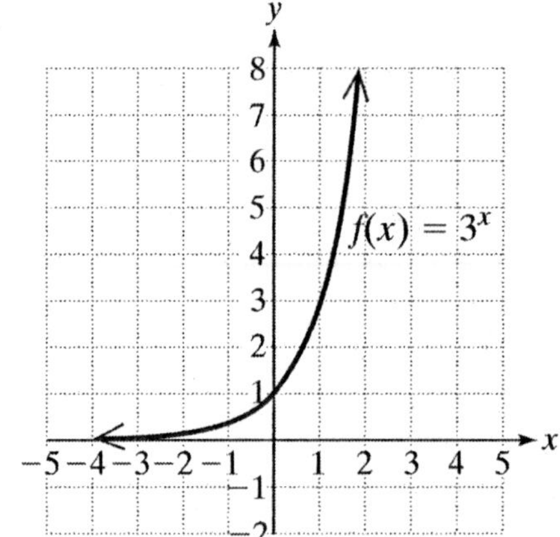

47.

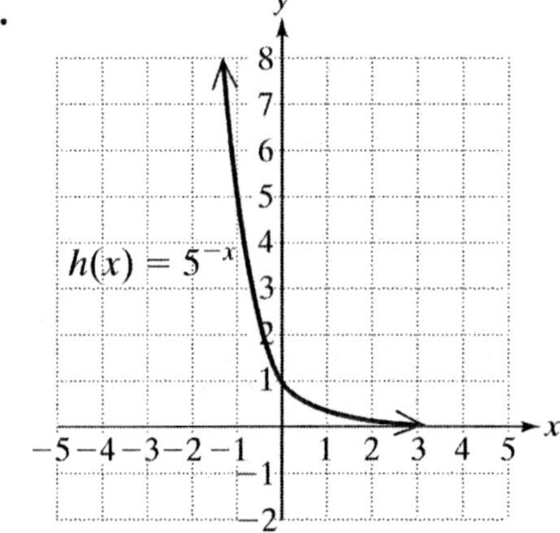

49. **(a)** horizontal

(b) $y = 0$

51. $3^x = \dfrac{1}{27}$

$3^x = 3^{-3}$

$x = -3$

53. 1

55. $2^x = 16$

$2^x = 2^4$

$x = 4$

57. $10^x = 100{,}000$

$10^x = 10^5$

$x = 5$

59.

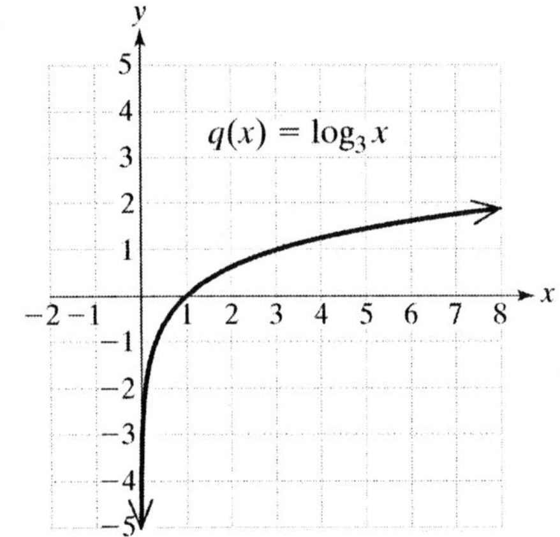

61. **(a)** vertical

(b) $x = 0$

63. 1

65. $\left(\dfrac{1}{2}\right)^x = 1$

$\left(\dfrac{1}{2}\right)^x = \left(\dfrac{1}{2}\right)^0$

$x = 0$

67. **(a)** $\log_b x + \log_b y$

(b) $\log_b\left(\dfrac{x}{y}\right)$

(c) $p\log_b x$

69. $\log_b 2^2 = 2\log_b 2 \approx 2(0.693) = 1.386$

71. $\log_b 5^2 = 2\log_b 5 \approx 2(1.609) = 3.218$

73. $\log_b 2(3)(5) = \log_b 2 + \log_b 3 + \log_b 5$
$\approx 0.693 + 1.099 + 1.609$
$= 3.401$

75. $\log_b 5^2 - \log_b 3 = 2\log_b 5 - \log_b 3$
$\approx 2(1.609) - 1.099$
$= 2.119$

77. $12\log_b 2 \approx 12(0.693) = 8.316$

79. $\log_3 a^{1/2} + \log_3 b^{1/2} - \log_3 c^2 - \log_3 d^4$

$= \log_3\left(\dfrac{\sqrt{ab}}{c^2 d^4}\right)$

81. $\log_4 18^{-1} + \log_4 6 + \log_4 3 - \log_4 1$

$= \log_4\left(\dfrac{18^{-1}\cdot 6\cdot 3}{1}\right)$

$= \log_4\left(\dfrac{6\cdot 3}{18}\right)$

$= \log_4 1$

$= 0$

83. 14.0940

85. 57.2795

87. –2.1972

89. –4.1227

91. –2.2366

93. 1.3029

95. $\dfrac{\log 80}{\log 9} \approx 1.9943$

97. $\dfrac{\log 100}{\log\left(\frac{1}{3}\right)} \approx -4.1918$

99. $\frac{\log(0.0062)}{\log 4} \approx -3.6668$

101. **(a)** $S(0) = 75e^{-0.5(0)} + 20 = 95$

The student's score is 95 at the end of the course.

(b) $S(6) = 75e^{-0.5(6)} + 20 \approx 23.7$

The student's score is 23.7 after 6 months.

(c) $S(12) = 75e^{-0.5(12)} + 20 \approx 20.2$

The student's score is 20.2 after 1 year.

(d) Yes, the limiting value is 20.

103. $(-\infty, \infty)$

105. $(-\infty, \infty)$

107. $x + 5 > 0$

$x > -5$

$(-5, \infty)$

109. $3x - 4 > 0$

$3x > 4$

$x > \frac{4}{3}$

$\left(\frac{4}{3}, \infty\right)$

111. $5^3 = x$

$x = 125$

113. $6^3 = y$

$y = 216$

115. $10^3 = 2w - 1$

$1000 = 2w - 1$

$1001 = 2w$

$\frac{1001}{2} = w$

$500.5 = w$

117. $\log p + \log(p-3) = 1$

$\log p(p-3) = 1$

$10^1 = p(p-3)$

$10 = p^2 - 3p$

$p^2 - 3p - 10 = 0$

$(p-5)(p+2) = 0$

$p - 5 = 0$ or $p + 2 = 0$

$p = 5$ or $p \neq -2$

$p = 5$

119. $(2^2)^{3x+5} = 2^4$

$2^{6x+10} = 2^4$

$6x + 10 = 4$

$6x = -6$

$x = -1$

121. $\ln 4^a = \ln 21$

$a \ln 4 = \ln 21$

$a = \frac{\ln 21}{\ln 4} \approx 2.1962$

123. $\ln e^{-x} = \ln 0.1$

$-x \ln e = \ln 0.1$

$-x = \ln 0.1$

$x = -\ln 0.1 \approx 2.3026$

125. $\log 10^{2n} = \log 1512$

$2n \log 10 = \log 1512$

$2n = \log 1512$

$n = \frac{\log 1512}{2} \approx 1.5898$

127. **(a)** $A(7) = 2e^{-0.0862(7)} \approx 1.09$ micrograms

(b) $A(30) = 2e^{-0.0862(30)}$

≈ 0.15 micrograms

(c)
$$\begin{aligned}0.5 &= 2e^{-0.0862(t)}\\ 0.25 &= e^{-0.0862(t)}\\ \ln 0.25 &= \ln(e^{-0.0862(t)})\\ \ln 0.25 &= -0.0862t \ln e\\ \ln 0.25 &= -0.0862t\\ \frac{\ln 0.25}{-0.0862} &= t\\ 16.08 \text{ days} &\approx t\end{aligned}$$

129. (a) $V(0) = 15000e^{-0.15(0)} = \$15{,}000$

The initial value of the car is $15,000.

(b) $V(10) = 15000e^{-0.15(10)} \approx \$3347$

The value of the car after 10 years is $3347.

(c)
$$\begin{aligned}5000 &= 15000e^{-0.15(t)}\\ \frac{1}{3} &= e^{-0.15(t)}\\ \ln\left(\frac{1}{3}\right) &= \ln(e^{-0.15(t)})\\ \ln\left(\frac{1}{3}\right) &= -0.15t \ln e\\ \ln\left(\frac{1}{3}\right) &= -0.15t\\ \frac{\ln\left(\frac{1}{3}\right)}{-0.15} &= t\\ 7.3 \text{ years} &\approx t\end{aligned}$$

(d) Yes, the limiting value is 0.

Chapter 12 Test

1. $\left(\frac{f}{g}\right)(x) = \frac{f(x)}{g(x)} = \frac{x-4}{\sqrt{x+2}}$

3.
$$\begin{aligned}(g \circ f)(x) &= g(f(x))\\ &= g(x-4)\\ &= \sqrt{(x-4)+2}\\ &= \sqrt{x-2}\end{aligned}$$

5.
$$\begin{aligned}(f-g)(7) &= f(7) - g(7)\\ &= (7-4) - \sqrt{7+2}\\ &= 3 - \sqrt{9}\\ &= 3-3\\ &= 0\end{aligned}$$

7. $(h \circ g)(14) = h(g(14))$

$g(14) = \sqrt{14+2} = \sqrt{16} = 4$

$h(g(14)) = h(4) = \frac{1}{4}$

9. $\left(\frac{g}{f}\right)(x) = \frac{g(x)}{f(x)} = \frac{\sqrt{x+2}}{x-4}$

Domain: $x + 2 \geq 0$ and $x - 4 \neq 0$

$x \geq -2$ and $x \neq 4$

$[-2, 4) \cup (4, \infty)$

11. b

13. $y = (x-1)^2$

To find the inverse, interchange x and y.

$$\begin{aligned}x &= (y-1)^2\\ \sqrt{x} &= y - 1\\ \sqrt{x} + 1 &= y\\ g^{-1}(x) &= \sqrt{x} + 1,\ x \geq 0\end{aligned}$$

15. (a) 4.6416

(b) 32.2693

(c) 687.2913

17. (a) $\log_{16} 8 = \frac{3}{4}$

(b) $x^5 = 31$

19. $\log_b n = \frac{\log_a n}{\log_a b}$

21. (a) $-(\log_3 3 - \log_3 9x)$
$= -\log_3 3 + \log_3 9x$
$= -1 + \log_3 9 + \log_3 x$
$= -1 + \log_3 3^2 + \log_3 x$
$= -1 + 2\log_3 3 + \log_3 x$
$= -1 + 2(1) + \log_3 x$
$= 1 + \log_3 x$

(b) $\log_{10} 10^{-5} = -5\log_{10} 10 = -5$

23. (a) 1.6487

(b) 0.0498

(c) −1.0986

(d) 1

25. (a) $p(4) = 92 - 20\ln(4+1) \approx 59.8$
59.8% of the material is retained after 4 months.

(b) $p(12) = 92 - 20\ln(12+1) \approx 40.7$
40.7% of the material is retained after 1 year.

(c) $p(0) = 92 - 20\ln(0+1) \approx 92$
92% of the material is retained at the end of the course.

27. (a) $P(0) = \dfrac{1{,}500{,}000}{1+5000e^{-0.8(0)}} \approx 300$
There are 300 bacteria initially.

(b) $P(6) = \dfrac{1{,}500{,}000}{1+5000e^{-0.8(6)}}$
$\approx 35{,}588$ bacteria

(c) $P(12) = \dfrac{1{,}500{,}000}{1+5000e^{-0.8(12)}}$
$\approx 1{,}120{,}537$ bacteria

(d) $P(18) = \dfrac{1{,}500{,}000}{1+5000e^{-0.8(18)}}$
$\approx 1{,}495{,}831$ bacteria

(e) Yes, the limiting amount appears to be 1,500,000.

29. $\left(\dfrac{1}{2}\right)^{-5} = x$
$32 = x$

31. $3^{x+4} = 3^{-3}$
$x + 4 = -3$
$x = -7$

33. $\ln e^{2.4x} = \ln 250$
$2.4x \ln e = \ln 250$
$x = \dfrac{\ln 250}{2.4}$
$x \approx 2.301$

35. (a) $A(5) = 2000e^{0.075(5)} \approx \$2909.98$

(b) $2P = P \cdot e^{0.075(t)}$
$2 = e^{0.075(t)}$
$\ln 2 = \ln e^{0.075(t)}$
$\ln 2 = 0.075(t)\ln e$
$\ln 2 = 0.075(t)$
$\dfrac{\ln 2}{0.075} = t$
$t \approx 9.24$ years to double

Cumulative Review Exercises, Chapters 1–12

1. $\dfrac{8-4\cdot 4+15\div 5}{|4|} = \dfrac{8-16+3}{4} = -\dfrac{5}{4}$

3. $t-2 \overline{)\, t^4 + 0t^3 - 13t^2 + 0t + 36}$ with quotient $t^3 + 2t^2 - 9t - 18$

$$\begin{array}{l} t^4 + 0t^3 - 13t^2 + 0t + 36 \\ -(t^4 - 2t^3) \\ \quad 2t^3 - 13t^2 \\ \quad -(2t^3 - 4t^2) \\ \qquad -9t^2 + 0t \\ \qquad -(-9t^2 + 18t) \\ \qquad\quad -18t + 36 \\ \qquad\quad -(-18t + 36) \\ \qquad\qquad 0 \end{array}$$

Quotient: $t^3 + 2t^2 - 9t - 18$; remainder: 0

5. $\dfrac{4}{\sqrt[3]{40}} \cdot \dfrac{\sqrt[3]{25}}{\sqrt[3]{25}} = \dfrac{4\sqrt[3]{25}}{\sqrt[3]{1000}} = \dfrac{4\sqrt[3]{25}}{10} = \dfrac{2\sqrt[3]{25}}{5}$

7. $\dfrac{2^{\frac{2}{5}-\frac{-8}{5}} d^{\frac{1}{5}-\frac{1}{10}}}{c^{\frac{3}{4}-\frac{-1}{4}}} = \dfrac{2^2 d^{1/10}}{c^1} = \dfrac{4d^{1/10}}{c}$

9. $\dfrac{4-3i}{2+5i} \cdot \dfrac{2-5i}{2-5i} = \dfrac{8-20i-6i+15i^2}{4-25i^2}$

$= \dfrac{8-26i-15}{4+25}$

$= \dfrac{-7-26i}{29}$

$= -\dfrac{7}{29} - \dfrac{26}{29}i$

11. $\dfrac{22}{33} = \dfrac{2}{3}$

13.

	20% alcohol	100% alcohol	50% alcohol
Number of liters	8	x	y
Number of liters of alcohol	0.20(8)	$1x$	$0.50(y)$

(Amount of 20% alcohol)
+ (Amount of pure alcohol)
= (Amount of 50% alcohol)
$\Rightarrow 8 + x = y$
(Amount of alcohol in 20%)
+ (Amount of alcohol in pure alcohol)
= (Amount of alcohol in resulting solution)
$\Rightarrow 0.20(8) + 1x = 0.50y$

$8 + x = y$
$0.20(8) + 1x = 0.50y$

Since the first equation is solved for y, substitute this expression into the second equation.

$0.20(8) + 1x = 0.50(8 + x)$
$1.6 + x = 4 + 0.5x$
$0.5x = 2.4$
$x = 4.8$ liters

Use 4.8 liters of pure alcohol.

15. $\left[\begin{array}{cc|c} 5 & 10 & 25 \\ -2 & 6 & -20 \end{array}\right]$

$\frac{1}{5}R_1 \Rightarrow R_1 \left[\begin{array}{cc|c} 1 & 2 & 5 \\ -2 & 6 & -20 \end{array}\right]$

$\frac{1}{2}R_2 \Rightarrow R_2 \left[\begin{array}{cc|c} 1 & 2 & 5 \\ -1 & 3 & -10 \end{array}\right]$

$R_1 + R_2 \Rightarrow R_2 \left[\begin{array}{cc|c} 1 & 2 & 5 \\ 0 & 5 & -5 \end{array}\right]$

$\frac{1}{5}R_2 \Rightarrow R_2 \left[\begin{array}{cc|c} 1 & 2 & 5 \\ 0 & 1 & -1 \end{array}\right]$

$-2R_2 + R_1 \Rightarrow R_1 \left[\begin{array}{cc|c} 1 & 0 & 7 \\ 0 & 1 & -1 \end{array}\right]$

The solution to the system is $(7, -1)$.

17. $ax - bx = c + d$
$x(a - b) = c + d$
$x = \dfrac{c+d}{a-b}$

19. $1 - kT = \left(\dfrac{V_0}{V}\right)^2$

$-kT = -1 + \left(\dfrac{V_0}{V}\right)^2$

$T = \dfrac{-1 + \left(\frac{V_0}{V}\right)^2}{-k}$

$T = \dfrac{1 - \left(\frac{V_0}{V}\right)^2}{k}$ or $T = \dfrac{V_2 - V_0^2}{kV^2}$

21. **(a)** $(f \cdot g)(t) = 6(-5t) = -30t$

(b) $(g \circ h)(t) = -5(2t^2) = -10t^2$

(c) $(h - g)(t) = 2t^2 - (-5t) = 2t^2 + 5t$

23. **(a)** $x = 2$

(b) $y = 6$

(c) Find the slope of $2x + y = 4$
(related equation $y = -2x + 4$)
Slope -2

Slope of perpendicular line is $\frac{1}{2}$

Point-slope form:

$$y-6=\frac{1}{2}(x-2)$$
$$y-6=\frac{1}{2}x-1$$
$$y=\frac{1}{2}x+5$$

25. Let x represent the measure of smallest angle.
Let $2x$ represent the measure of largest angle.
Let $x + 20$ represent the measure of middle angle.

$$x+2x+x+20=80$$
$$4x+20=180$$
$$4x=160$$
$$x=40$$

$2x = 80$
$x + 20 = 60$
The angles measure 40°, 80°, and 60°.

27. **(a)** vi **(b)** i **(c)** v **(d)** x **(e)** ii **(f)** ix **(g)** iv **(h)** viii **(i)** vii **(j)** iii

29. $y=5x-\frac{2}{3}$

To find the inverse, interchange x and y.

$$x=5y-\frac{2}{3}$$
$$x+\frac{2}{3}=5y$$
$$\frac{1}{5}\left(x+\frac{2}{3}\right)=\frac{1}{5}(5y)$$
$$\frac{1}{5}x+\frac{2}{15}=y$$
$$f^{-1}(x)=\frac{1}{5}x+\frac{2}{15}$$

31. $$\frac{5(x-2)}{(x-2)(x-2)}\div\frac{5(x^2-25)}{3(5-x)}\cdot\frac{(x+5)(x^2-5x+25)}{3(2x+1)}=\frac{5(x-2)}{(x-2)(x-2)}\cdot\frac{3(5-x)}{5(x+5)(x-5)}\cdot\frac{(x+5)(x^2-5x+25)}{3(2x+1)}$$
$$=\frac{-(x^2-5x+25)}{(x-2)(2x+1)}$$
$$=\frac{-x^2+5x-25}{(x-2)(2x+1)}$$

33. **(a)** $x-4 \neq 0$ and $x+2 \neq 0$

$x \neq 4$ and $x \neq -2$

(b) $2(x+2)=5(x-4)$

$2x+4=5x-20$

$-3x=-24$

$x=8$

(c) Boundary numbers: –2, 4, 8

Test Point –3: $\frac{2}{-3-4} \geq \frac{5}{-3+2}$

$\frac{2}{-7} \geq -5$ True

Test Point 0: $\frac{2}{0-4} \geq \frac{5}{0+2}$

$-\frac{1}{2} \geq \frac{5}{2}$ False

Test Point 5: $\frac{2}{5-4} \geq \frac{5}{5+2}$

$2 \geq \frac{5}{7}$ True

Test Point 9: $\frac{2}{9-4} \geq \frac{5}{9+2}$

$\frac{2}{5} \geq \frac{5}{11}$ False

Solution: $(-\infty, -2) \cap (4, 8]$

35. $\left(\sqrt{-x}\right)^2 = (x+6)^2$

$-x = x^2+12x+36$

$0 = x^2+13x+36$

$0=(x+9)(x+4)$

$x+9=0$ or $x+4=0$

$x=-9$ or $x=-4$

Check: $\sqrt{-(-9)} = -9+6$

$3=-3$ False

Check: $\sqrt{-(-4)} = -4+6$

$2=2$ True

The solution is $x=-4$.

37. **(a)** $P(6)=4{,}000{,}000\left(\frac{1}{2}\right)^{6/6} = 2{,}000{,}000;$

$P(12)=4{,}000{,}000\left(\frac{1}{2}\right)^{12/6} = 1{,}000{,}000;$

$P(18)=4{,}000{,}000\left(\frac{1}{2}\right)^{18/6} = 500{,}000;$

$P(24)=4{,}000{,}000\left(\frac{1}{2}\right)^{6/6} = 250{,}000;$

$P(30)=4{,}000{,}000\left(\frac{1}{2}\right)^{30/6} = 125{,}000$

(b)

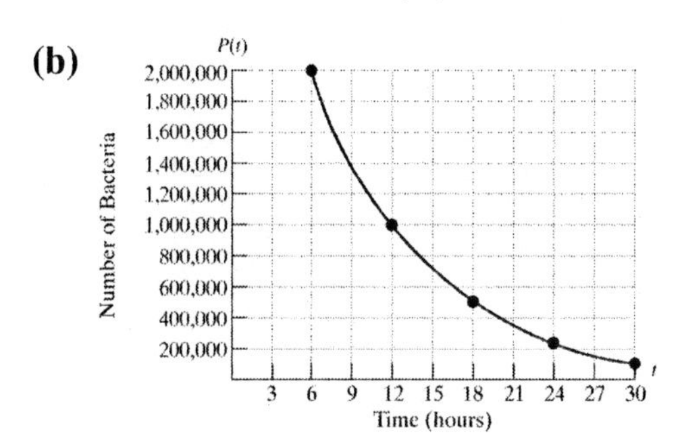

(c) $15{,}625 = 4{,}000{,}000\left(\frac{1}{2}\right)^{t/6}$

$\frac{1}{256} = \left(\frac{1}{2}\right)^{t/6}$

$\left(\frac{1}{2}\right)^{8} = \left(\frac{1}{2}\right)^{t/6}$

$8 = \frac{t}{6}$

48 hours $= t$

39. **(a)** 217.0723

(b) 23.1407

(c) 0.1768

(d) 3.7293

(e) –0.4005

(f) 2.6047

41. $\ln e^x = \ln 100$

$x \ln e = \ln 100$

$x = \ln 100 \approx 4.6052$

43. $\log\sqrt{z} - \log x^2 - \log y^3 = \log\left(\frac{\sqrt{z}}{x^2y^3}\right)$

Chapter 13

Section 13.1 Practice Exercises

1. $d=\sqrt{(-2-3)^2+(7-(-9))^2}=\sqrt{281}$

3. $d=\sqrt{(0-(-3))^2+(5-8)^2}=\sqrt{18}=3\sqrt{2}$

5. $d=\sqrt{\left(\frac{2}{3}-\left(-\frac{5}{6}\right)\right)^2+\left(\frac{1}{5}-\frac{3}{10}\right)^2}$

$=\sqrt{\frac{226}{100}}$

$=\frac{\sqrt{226}}{10}$

7. $d=\sqrt{(4-4)^2+(13-(-6))^2}=\sqrt{361}=19$

9. $d=\sqrt{(8-(-2))^2+(-6-(-6))^2}=\sqrt{100}=10$

11. $d=\sqrt{\left(3\sqrt{5}-\left(-\sqrt{5}\right)\right)^2+\left(2\sqrt{7}-\left(-3\sqrt{7}\right)\right)^2}$

$=\sqrt{255}$

13. Subtract 5 and -7, $5-(-7)=12$.

15.
$$10=\sqrt{(4-(-4))^2+(7-y)^2}$$
$$10=\sqrt{64+49-14y+y^2}$$
$$10=\sqrt{113-14y+y^2}$$
$$100=113-14y+y^2$$
$$0=y^2-14y+13$$
$$0=(y-13)(y-1)$$
$$y-13=0 \quad \text{or} \quad y-1=0$$
$$y=13 \quad \text{or} \quad y=1$$

17.
$$5=\sqrt{(x-4)^2+(2-(-1))^2}$$
$$5=\sqrt{x^2-8x+16+9}$$
$$5=\sqrt{x^2-8x+25}$$
$$25=x^2-8x+25$$
$$0=x^2-8x$$
$$0=x(x-8)$$
$$x=0 \quad \text{or} \quad x-8=0$$
$$x=0 \quad \text{or} \quad x=8$$

19. $d_1=\sqrt{(-3-(-2))^2+(2-(-4))^2}=\sqrt{37}$;

$d_2=\sqrt{(-2-3)^2+(-4-3)^2}=\sqrt{74}$;

$d_3=\sqrt{(-3-3)^2+(2-3)^2}=\sqrt{37}$

These three points are vertices of a right triangle since $d_1^2+d_3^2=d_2^2$.

21. $d_1=\sqrt{(-3-4)^2+(-2-(-3))^2}=\sqrt{50}$;

$d_2=\sqrt{(4-1)^2+(-3-5)^2}=\sqrt{73}$;

$d_3=\sqrt{(-3-1)^2+(-2-5)^2}=\sqrt{65}$

These three points are not vertices of a right triangle since $d_1^2+d_3^2\neq d_2^2$.

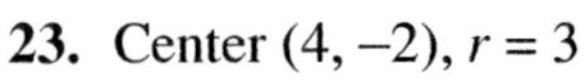

23. Center $(4,-2)$, $r=3$

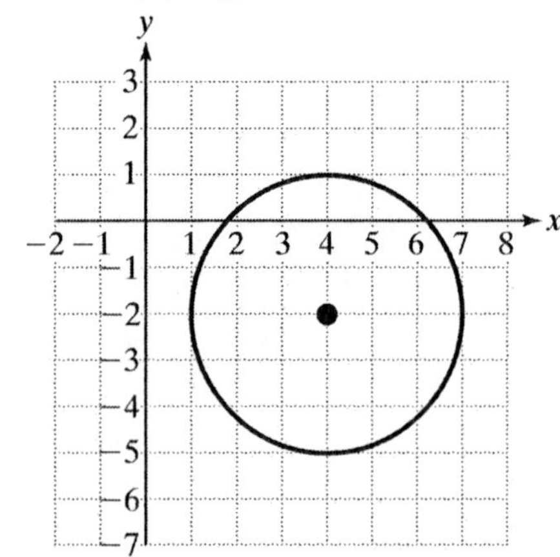

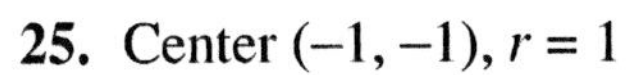
25. Center $(-1, -1)$, $r = 1$

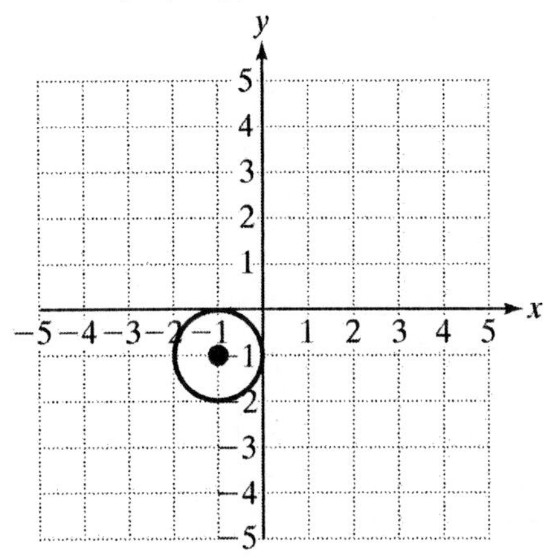

27. Center $(0, 5)$, $r = 5$

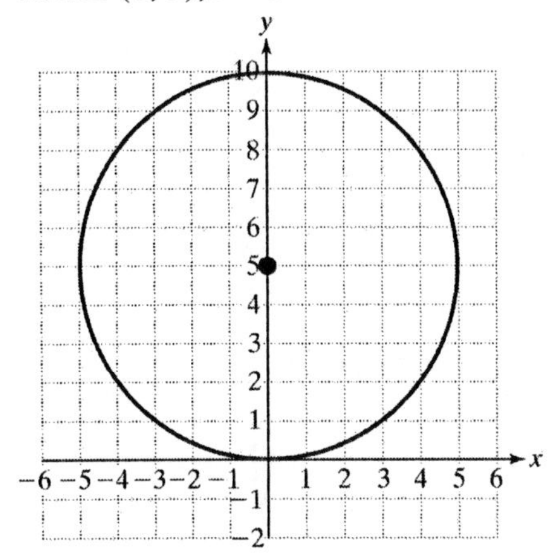

29. Center $(3, 0)$, $r = 2\sqrt{2}$

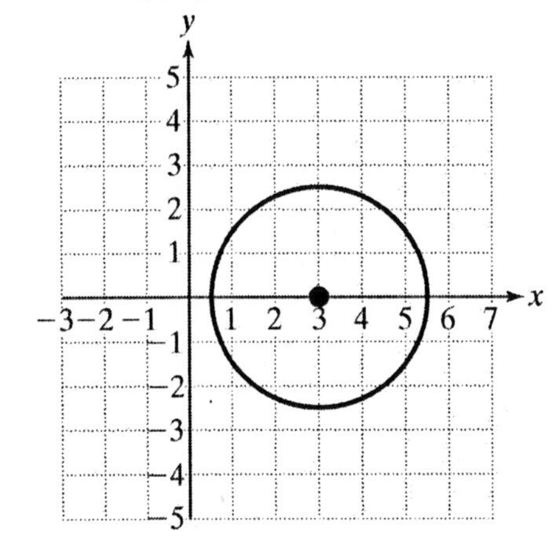

31. Center $(0, 0)$, $r = \sqrt{6}$

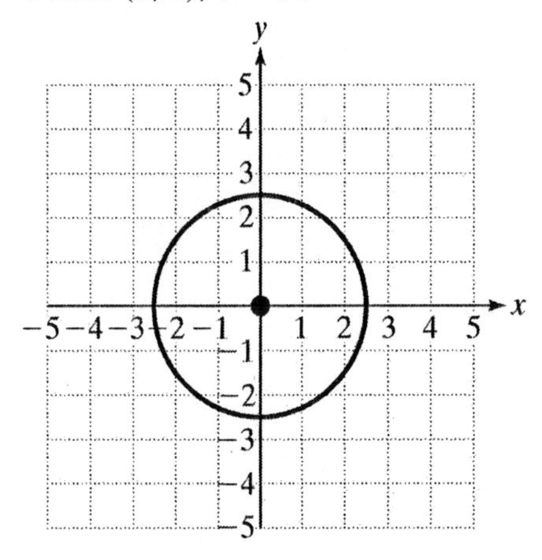

33. Center $\left(-\frac{4}{5}, 0\right)$, $r = \frac{8}{5}$

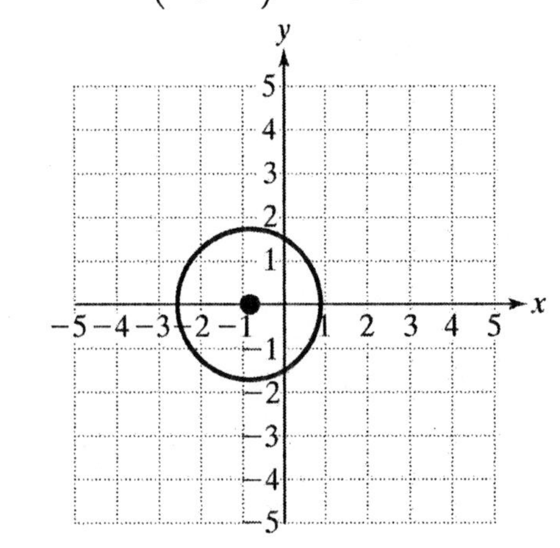

35.
$$x^2 - 2x + y^2 - 6y = 26$$
$$x^2 - 2x + 1 + y^2 - 6y + 9 = 26 + 1 + 9$$
$$(x-1)^2 + (y-3)^2 = 36$$

Center $(1, 3)$, $r = 6$

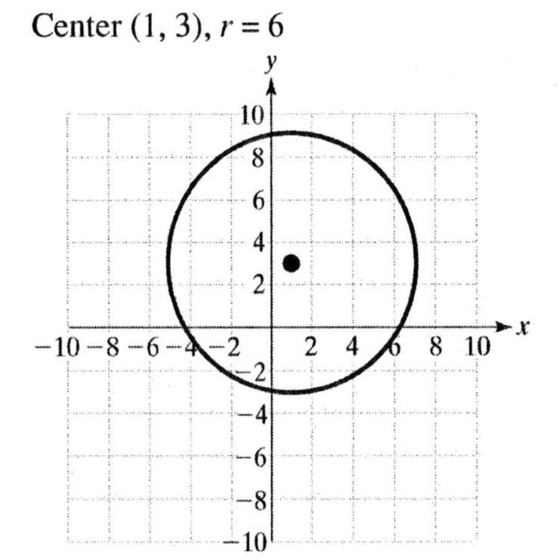

37.
$$x^2 + y^2 + 6y = -\frac{65}{9}$$
$$x^2 + y^2 + 6y + 9 = -\frac{65}{9} + 9$$
$$x^2 + (y+3)^2 = \frac{16}{9}$$

Center $(0, -3)$, $r = \frac{4}{3}$

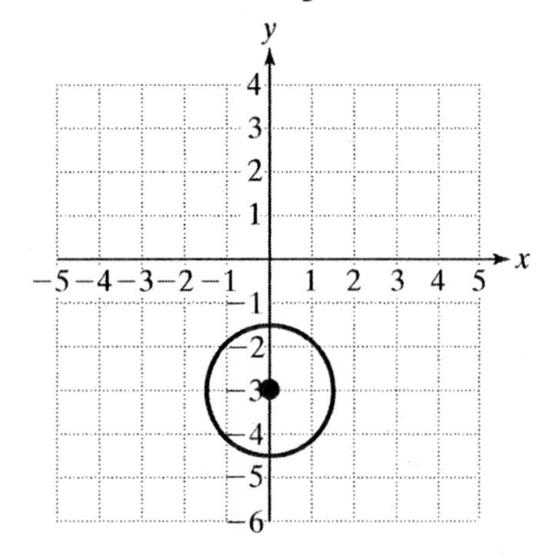

39. $x^2 + y^2 = 4$

41. $x^2 + (y-2)^2 = 4$

43. $(x+2)^2 + (y-2)^2 = 9$

45. $x^2 + y^2 = 49$

47. $\frac{1}{2}(12) = 6;\ (x+3)^2 + (y+4)^2 = 36$

49. $\left(\frac{x_1 + x_2}{2}, \frac{y_1 + y_2}{2}\right) = \left(\frac{-2+4}{2}, \frac{1+3}{2}\right) = (1, 2)$

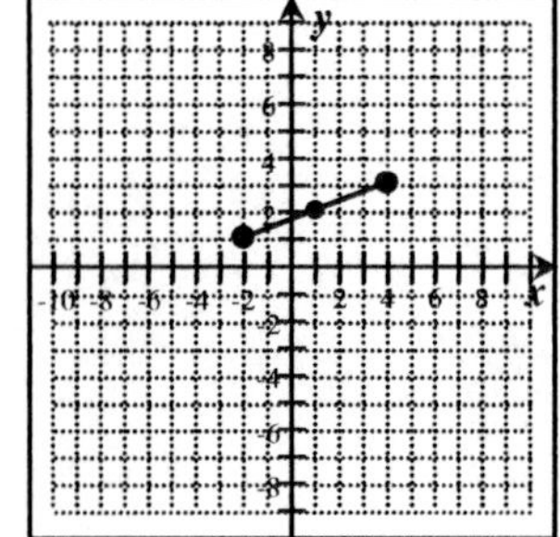

51. $\left(\frac{x_1 + x_2}{2}, \frac{y_1 + y_2}{2}\right) = \left(\frac{-4+2}{2}, \frac{-2+2}{2}\right)$
$= (-1, 0)$

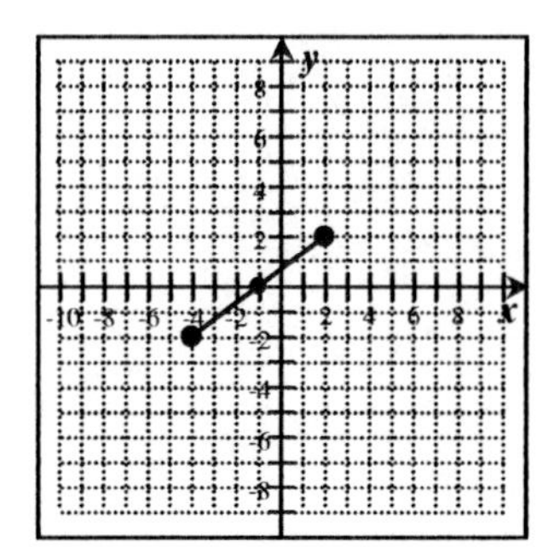

53. $\left(\frac{x_1 + x_2}{2}, \frac{y_1 + y_2}{2}\right) = \left(\frac{4+(-6)}{2}, \frac{0+12}{2}\right)$
$= (-1, 6)$

55. $\left(\frac{x_1 + x_2}{2}, \frac{y_1 + y_2}{2}\right) = \left(\frac{-3+3}{2}, \frac{8+(-2)}{2}\right)$
$= (0, 3)$

57. $\left(\frac{x_1 + x_2}{2}, \frac{y_1 + y_2}{2}\right) = \left(\frac{5+(-6)}{2}, \frac{2+1}{2}\right)$
$= \left(-\frac{1}{2}, \frac{3}{2}\right)$

59. $\left(\frac{x_1 + x_2}{2}, \frac{y_1 + y_2}{2}\right)$
$= \left(\frac{-2.4+1.6}{2}, \frac{-3.1+1.1}{2}\right)$
$= (-0.4, -1)$

61. $(x_1, y_1) = (30, 20)$
$(x_2, y_2) = (50, -5)$
$\left(\frac{x_1 + x_2}{2}, \frac{y_1 + y_2}{2}\right) = \left(\frac{30+50}{2}, \frac{20+(-5)}{2}\right)$
$= (40, 7.5)$
They should meet 40 miles east and 7.5 miles north of the warehouse.

63.

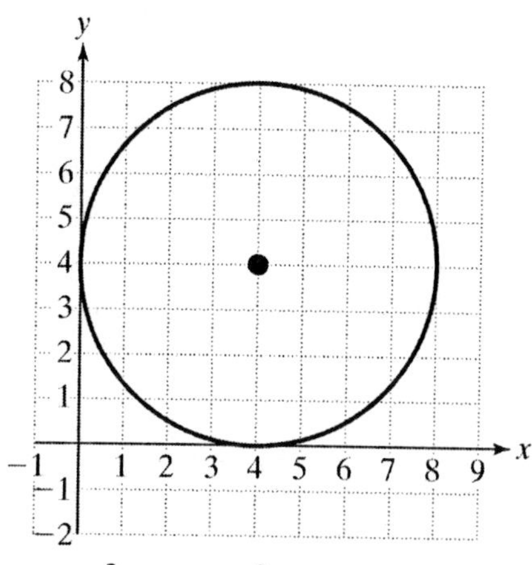

$(x-4)^2+(y-4)^2=16$

65. $(x-1)^2+(y-1)^2=r^2$

$(-4-1)^2+(3-1)^2=r^2$

$25+4=r^2$

$29=r^2$

Thus, the equation is

$(x-1)^2+(y-1)^2=29.$

67. Center $(4, -2)$; $r = 3$

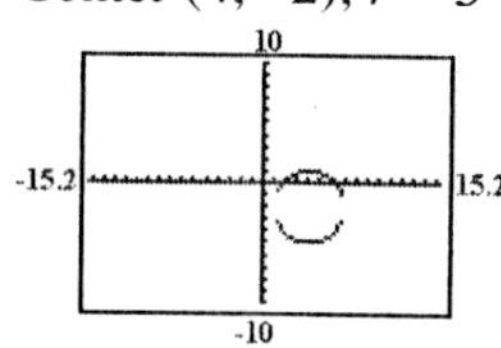

69. Center $(0, 5)$; $r = 5$

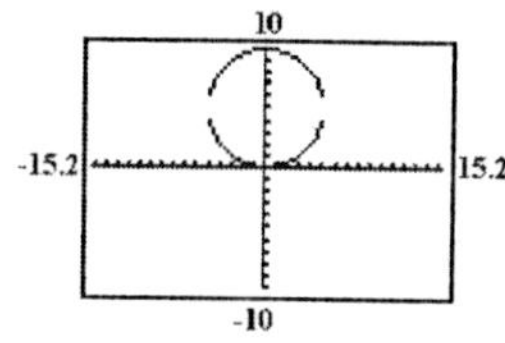

71. Center $(0, 0)$; $r = 2.4$

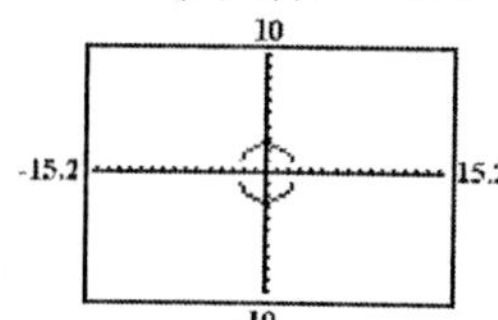

Section 13.2 Practice Exercises

1. For a parabola whose equation is written in the form $y=a(x-h)^2+k$, if $a>0$ the parabola opens upward, if $a<0$ the parabola opens downward. For a parabola written in the form $x=a(y-k)^2+h$, if $a>0$ the parabola opens right, if $a<0$ the parabola opens left.

3. Vertical axis of symmetry; opens upward.

5. Vertical axis of symmetry; opens downward.

7. Horizontal axis of symmetry; opens right.

9. Horizontal axis of symmetry; opens left.

11. Vertical axis of symmetry; opens downward.

13. The focus is $(h, k+p)$ where $p=\frac{1}{4a}$.

15. The directrix is the line $x=h-p$ where $p=\frac{1}{4a}$.

17. True

19. False

21. True

23. $y=\frac{1}{2}(x-0)^2+0$

$h=0, k=0, a=\frac{1}{2}$

Vertex: $(0, 0)$

Focus: $p=\frac{1}{4a}=\frac{1}{4\left(\frac{1}{2}\right)}=\frac{1}{2}$

$\left(0, 0+\frac{1}{2}\right)=\left(0, \frac{1}{2}\right)$

Directrix: $y=k-p$

$y=0-\frac{1}{2}=-\frac{1}{2}$

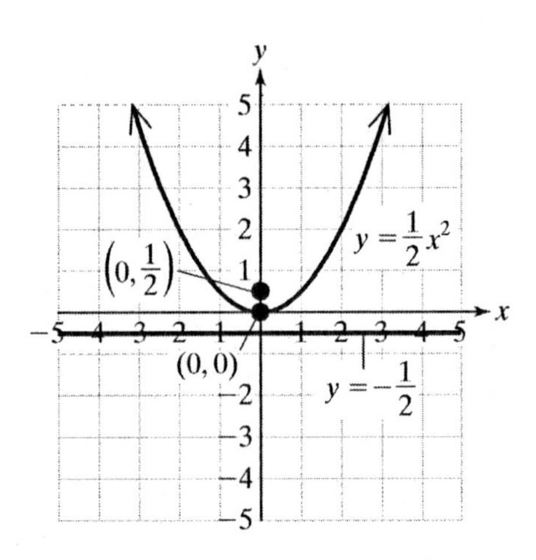

25. $y = -4(x-0)^2 + 0$

$h = 0, k = 0, a = -4$

Vertex: (0, 0)

Focus: $p = \frac{1}{4a} = \frac{1}{4(-4)} = -\frac{1}{16}$

$$\left(0, 0 + \left(-\frac{1}{16}\right)\right) = \left(0, -\frac{1}{16}\right)$$

Directrix: $y = k - p$

$$y = 0 - \left(-\frac{1}{16}\right) = \frac{1}{16}$$

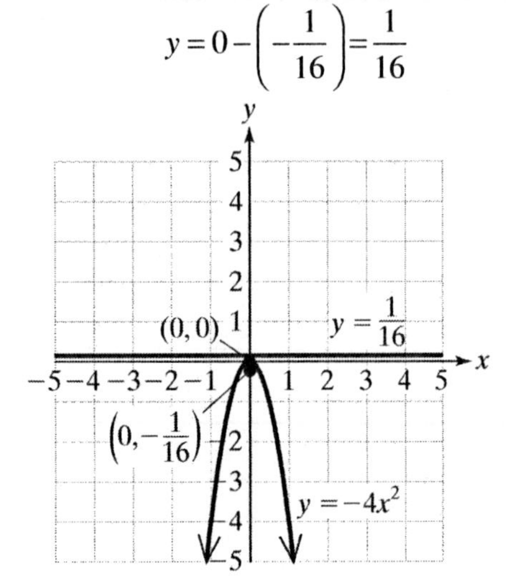

27. $y = (x-3)^2 + 0$

$h = 3, k = 0, a = 1$

Vertex: (3, 0)

Focus: $p = \frac{1}{4a} = \frac{1}{4(1)} = \frac{1}{4}$

$$\left(3, 0 + \frac{1}{4}\right) = \left(3, \frac{1}{4}\right)$$

Directrix: $y = k - p$

$$y = 0 - \frac{1}{4} = -\frac{1}{4}$$

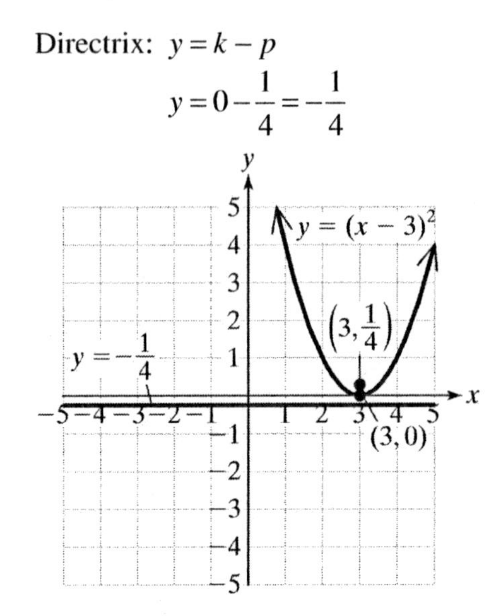

29. $x = (y-0)^2 + 0$

$h = 0, k = 0, a = 1$

Vertex: (0, 0)

Focus: $p = \frac{1}{4a} = \frac{1}{4(1)} = \frac{1}{4}$

$$\left(0 + \frac{1}{4}, 0\right) = \left(\frac{1}{4}, 0\right)$$

Directrix: $x = h - p$

$$x = 0 - \frac{1}{4} = -\frac{1}{4}$$

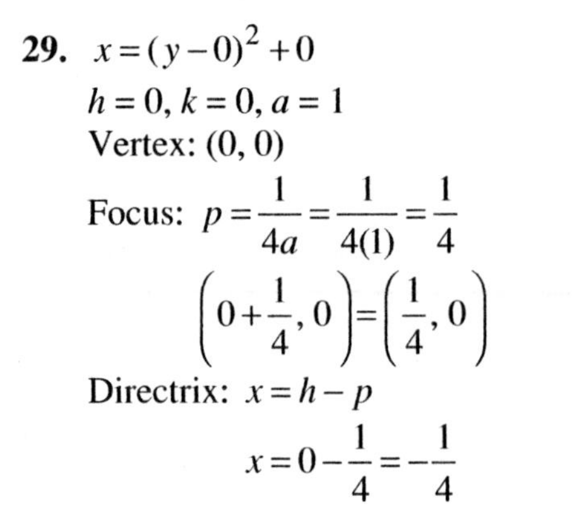

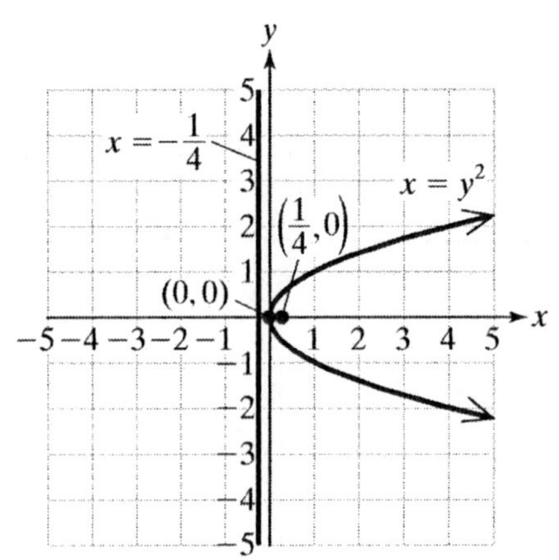

31. $x = -2(y+1)^2 + 0$

$h = 0, k = -1, a = -2$

Vertex: $(0, -1)$

Focus: $p = \dfrac{1}{4a} = \dfrac{1}{4(-2)} = -\dfrac{1}{8}$

$$\left(0 + \left(-\frac{1}{8}\right), -1\right) = \left(-\frac{1}{8}, -1\right)$$

Directrix: $x = h - p$

$$x = 0 - \left(-\frac{1}{8}\right) = \frac{1}{8}$$

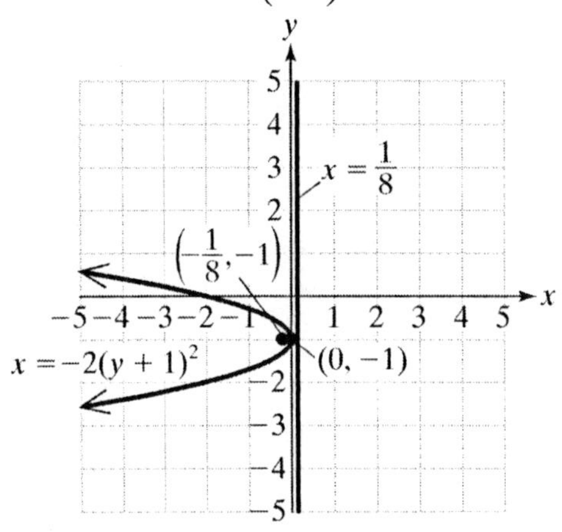

33. $y = \dfrac{1}{8}(x+2)^2 - 5$

$h = -2, k = -5, a = \dfrac{1}{8}$

Vertex: $(-2, -5)$

Focus: $p = \dfrac{1}{4a} = \dfrac{1}{4\left(\frac{1}{8}\right)} = \dfrac{1}{\frac{1}{2}} = 2$

$$(-2, -5+2) = (-2, -3)$$

Directrix: $y = k - p$

$$y = -5 - 2 = -7$$

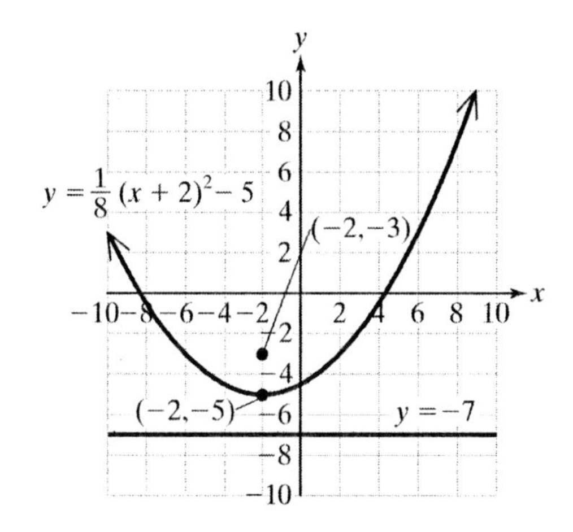

35. $x = -\dfrac{1}{12}(y-3)^2 - 3$

$h = -3, k = 3, a = -\dfrac{1}{12}$

Vertex: $(-3, 3)$

Focus: $p = \dfrac{1}{4a} = \dfrac{1}{4\left(-\frac{1}{12}\right)} = \dfrac{1}{-\frac{1}{3}} = -3$

$$(-3 + (-3), 3) = (-6, 3)$$

Directrix: $x = h - p$

$$x = -3 - (-3) = 0$$

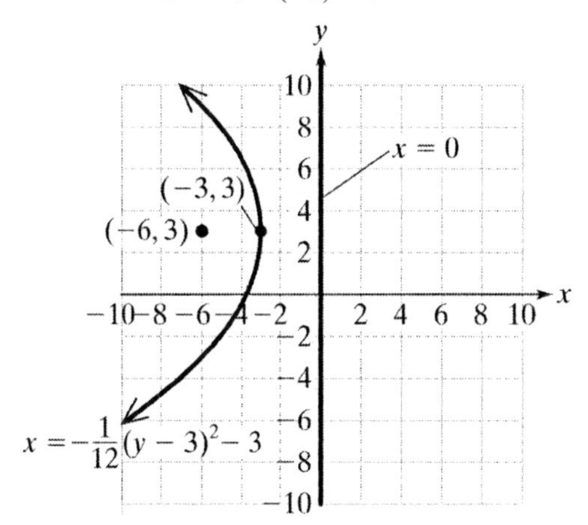

37. $k = \left[\dfrac{1}{2}(-4)\right]^2 = (-2)^2 = 4$

39. $k = \left[\dfrac{1}{2}(2)\right]^2 = (1)^2 = 1$

41. $k = \left[\dfrac{1}{2}(1)\right]^2 = \left(\dfrac{1}{2}\right)^2 = \dfrac{1}{4}$

43. $k=\left[\frac{1}{2}(-5)\right]^2=\left(\frac{-5}{2}\right)^2=\frac{25}{4}$

45. $y=(x-h)^2+k$
$y=(x^2-4x)+3$
$y=(x^2-4x+4-4)+3$
$y=(x^2-4x+4)-4+3$
$y=(x-2)^2-1$
$a=1, h=2, k=-1$
Vertex: $(2,-1)$
Focus: $\left(2,-1+\frac{1}{4}\right)=\left(2,-\frac{3}{4}\right)$
Directrix: $y=-1-\frac{1}{4}=-\frac{5}{4}$

47. $x=(y-k)^2+h$
$x=(y^2+2y)+6$
$x=(y^2+2y+1-1)+6$
$x=(y^2+2y+1)-1+6$
$x=(y+1)^2+5$
$a=1, h=5, k=-1$
Vertex: $(5,-1)$
Focus: $\left(5+\frac{1}{4},-1\right)=\left(\frac{21}{4},-1\right)$
Directrix: $x=5-\frac{1}{4}=\frac{19}{4}$

49. $y=(x-h)^2+k$
$y=-2(x^2-4x)$
$y=-2(x^2-4x+4-4)$
$y=-2(x^2-4x+4)+8$
$y=-2(x-2)^2+8$
$a=-2, h=2, k=8$
Vertex: $(2,8)$
Focus: $\left(2,8+\left(-\frac{1}{8}\right)\right)=\left(2,\frac{63}{8}\right)$
Directrix: $y=8-\left(-\frac{1}{8}\right)=\frac{65}{8}$

51. $y=(x-h)^2+k$
$y=(x^2-3x)+2$
$y=\left(x^2-3x+\frac{9}{4}-\frac{9}{4}\right)+2$
$y=\left(x^2-3x+\frac{9}{4}\right)-\frac{9}{4}+2$
$y=\left(x-\frac{3}{2}\right)^2-\frac{1}{4}$
$a=1,\ h=\frac{3}{2},\ k=-\frac{1}{4}$
Vertex: $\left(\frac{3}{2},-\frac{1}{4}\right)$
Focus: $\left(\frac{3}{2},-\frac{1}{4}+\frac{1}{4}\right)=\left(\frac{3}{2},0\right)$
Directrix: $y=-\frac{1}{4}-\frac{1}{4}=-\frac{2}{4}=-\frac{1}{2}$

53. $x=(y-k)^2+h$
$x=-2(y^2-8y)+1$
$x=-2(y^2-8y+16-16)+1$
$x=-2(y^2-8y+16)+32+1$
$x=-2(y-4)^2+33$
$a=-2, h=33, k=4$
Vertex: $(33,4)$
Focus: $\left(33+\left(-\frac{1}{8}\right),4\right)=\left(\frac{263}{8},4\right)$
Directrix: $x=33-\left(-\frac{1}{8}\right)=\frac{265}{8}$

55. $y=\frac{1}{50}x^2$
$a=\frac{1}{50},\ h=0, k=0$
Focus: $\left(0,0+\frac{1}{\frac{4}{50}}\right)=\left(0,0+\frac{1}{\frac{2}{25}}\right)$
$=\left(0,0+\frac{25}{2}\right)$
$=\left(0,\frac{25}{2}\right)$

57. Opens downward.

59. Opens right.

61. Opens downward.

63. Opens left.

65. Opens right.

67. Opens upward.

Section 13.3 Practice Exercises

1.

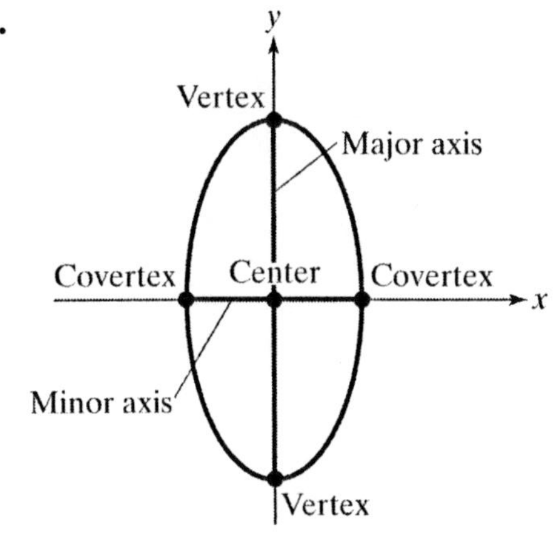

3. Center: (0, 0); vertical

5. Center: (2, 5); horizontal

7. Divide both sides by 10.

$$\frac{(x+2)^2}{2}+\frac{(y+4)^2}{10}=1$$

Center: (–2, –4); vertical

9. Center: (–4, 1); horizontal

11. Center: (0, 0)
Vertices: (0, 3) and (0, –3)
Co-vertices: (2, 0) and (–2, 0)

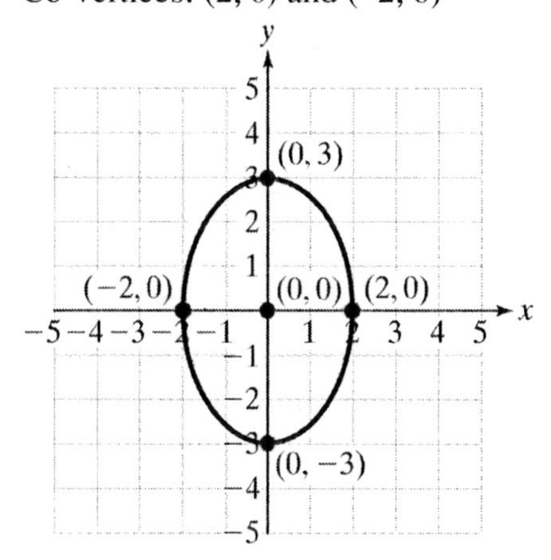

13. Center: (0, 0)
Vertices: (5, 0) and (–5, 0)
Co-vertices: (0, 4) and (0, 4)

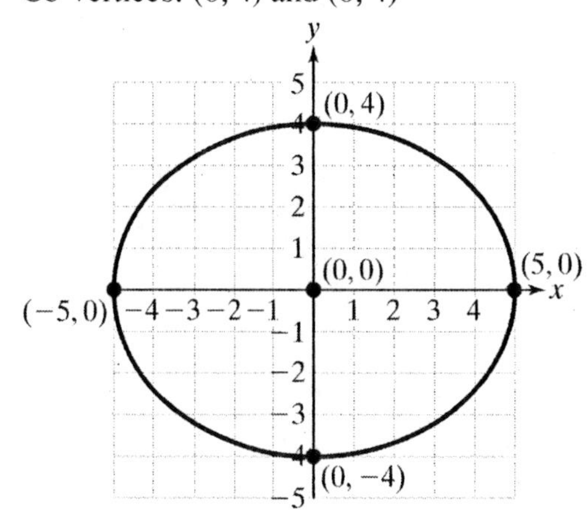

15. Center: (0, 2)
Vertices: (4, 2) and (–4, 2)
Co-vertices: (0, 0) and (0, 4)

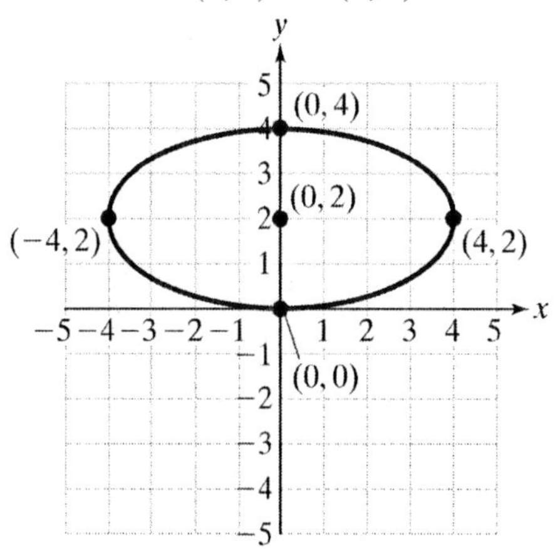

17. Center: (–1, –3)
Vertices: (–1, 3) and (–1, –9)
Co-vertices: (–2, –3) and (0, –3)

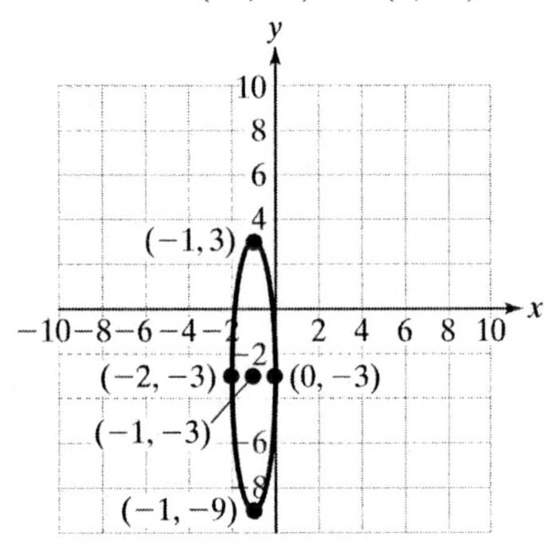

19. Center: (3, –1)
Vertices: (10, –1) and (–4, –1)
Co-vertices: (3, 4) and (3, –6)

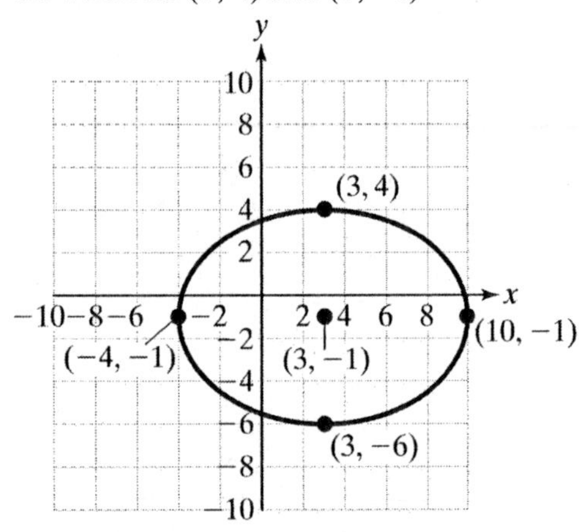

21. The length of the string is constant and the two tacks are fixed points. The sum of the distances from each tack to a point on the curve is constant, therefore an elliptical curve is traced.

23. $\frac{(x-h)^2}{a^2}+\frac{(y-h)^2}{b^2}=1$

$$\frac{x^2}{200^2}+\frac{y^2}{100^2}=1$$
$$\frac{50^2}{40000}+\frac{y^2}{10000}=1$$
$$\frac{2500}{40000}+\frac{y^2}{10000}=1$$
$$\frac{2500}{40000}+\frac{y^2}{10000}=1$$
$$\frac{y^2}{10000}=1-\frac{2500}{40000}$$
$$\frac{y^2}{10000}=\frac{37500}{40000}$$
$$y^2=\frac{375000000}{40000}$$
$$y^2=9375$$
$$y=97 \text{ feet}$$

25.

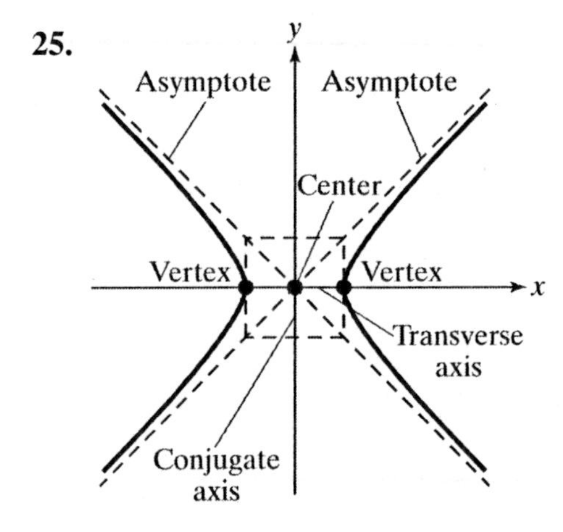

27. Center: (0, 0); horizontal

29. Center: (0, 0); vertical

31. Divide both sides by 24.
$$\frac{(x+1)^2}{24}-\frac{(y+8)^2}{4}=1$$
Center: (–1, –8); horizontal

33. Center: (–1, –4); vertical

35. Center: (0, 0)
Vertices: (3, 0) and (–3, 0)

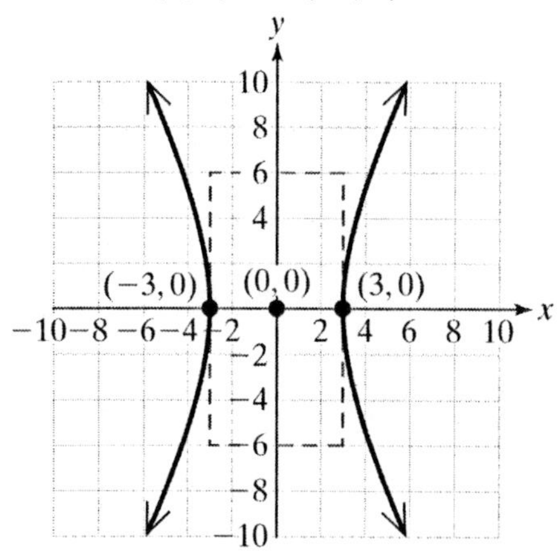

37. Center: (0, 0)
Vertices: (0, 3) and (0, –3)

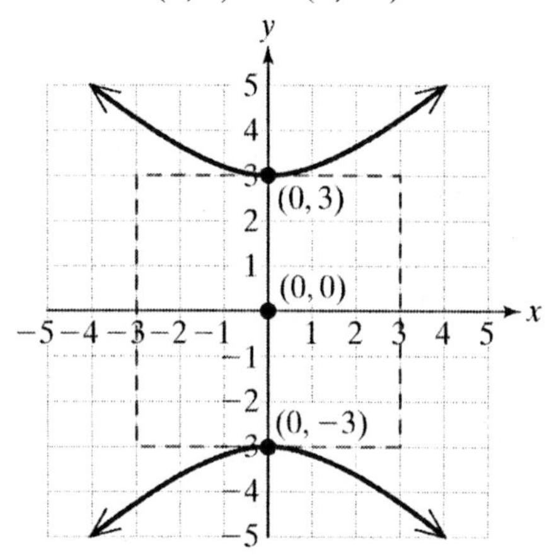

39. Center: (2, 2)
Vertices: (–4, 2) and (8, 2)

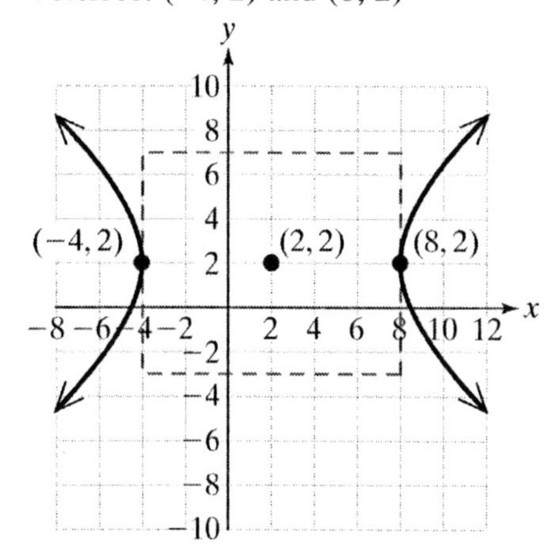

41. Center: (–4, 0)
Vertices: (–4, –2) and (–4, 2)

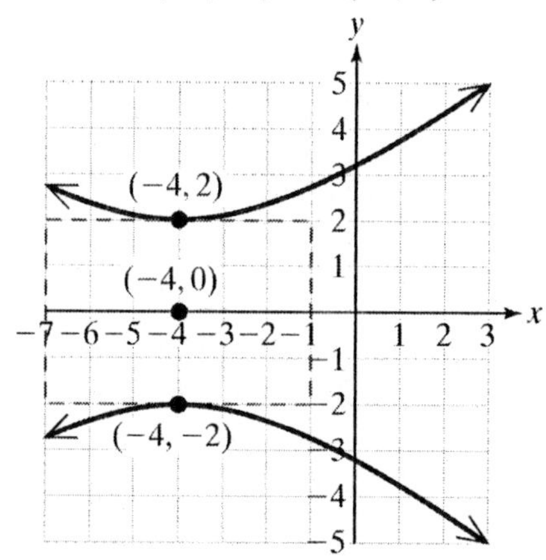

43. Center: (–6, –2)
Vertices: (–11, –2) and (–1, –2)

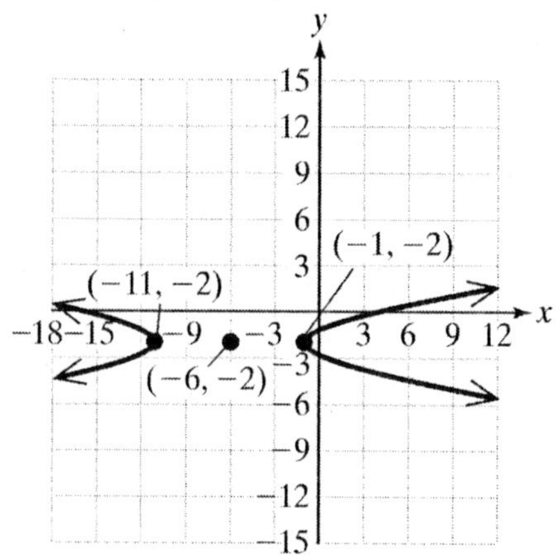

45. One line passes through the diagonal containing the points (–4, –4) and (4, 4). The other line passes through the diagonal containing the points (–4, 4) and (4, –4).

$$m = \frac{4+4}{4+4} = 1 \qquad m = \frac{-4-4}{4+4} = -1$$

$$y - y_1 = m(x - x_1) \qquad y - y_1 = m(x - x_1)$$
$$y + 4 = 1(x + 4) \qquad y - 4 = -1(x + 4)$$
$$y = x + 4 - 4 \qquad y = -x - 4 + 4$$
$$y = x \qquad y = -x$$

47. One line passes through the diagonal containing the points (–4, –7) and (8, –1). The other line passes through the diagonal containing the points (–4, –1) and (8, –7).

$m = \frac{-1+7}{8+4} = \frac{1}{2}$

$y - y_1 = m(x - x_1)$

$y + 7 = \frac{1}{2}(x+4)$

$y = \frac{1}{2}x + 2 - 7$

$y = \frac{1}{2}x - 5$

$m = \frac{-7+1}{8+4} = -\frac{1}{2}$

$y - y_1 = m(x - x_1)$

$y + 1 = -\frac{1}{2}(x+4)$

$y = -\frac{1}{2}x - 2 - 1$

$y = -\frac{1}{2}x - 3$

49. An ellipse can be written in the form

$\frac{(x-h)^2}{a^2} + \frac{(y-k)^2}{b^2} = 1$ or

$\frac{(x-h)^2}{b^2} + \frac{(y-k)^2}{a^2} = 1$, whereas a hyperbola can be written in the form

$\frac{(x-h)^2}{a^2} - \frac{(y-k)^2}{b^2} = 1$ or

$\frac{(y-k)^2}{a^2} - \frac{(x-h)^2}{b^2} = 1.$

51. Ellipse

53. Hyperbola

$\frac{y^2}{5} - \frac{(x+1)^2}{5} = 1$

55. Ellipse

$\frac{x^2}{12} + \frac{y^2}{9} = 1$

57. Hyperbola

$\frac{y^2}{1} - \frac{x^2}{1} = 1$

59. Find c when $a = 5$ and $b = 3$; vertical axis

$c^2 = a^2 - b^2$

$c^2 = 5^2 - 3^2$

$c^2 = 25 - 9$

$c^2 = 16$

$c = \pm 4$

The foci are (0, 4) and (0, –4).

61. Find c when $a = 10$ and $b = 6$; horizontal axis

$c^2 = a^2 - b^2$

$c^2 = 10^2 - 6^2$

$c^2 = 100 - 36$

$c^2 = 64$

$c = \pm 8$

The foci are (8, 0) and (–8, 0).

63. Find c when $a = 4$ and $b = 3$; horizontal transverse axis

$c^2 = a^2 + b^2$

$c^2 = 4^2 + 3^2$

$c^2 = 16 + 9$

$c^2 = 25$

$c = \pm 5$

The foci are (5, 0) and (–5, 0).

65. Find c when $a = 12$ and $b = 5$; vertical transverse axis

$c^2 = a^2 + b^2$

$c^2 = 12^2 + 5^2$

$c^2 = 144 + 25$

$c^2 = 169$

$c = \pm 13$

The foci are (0, 13) and (0, –13).

Section 13.4 Practice Exercises

1. $d = \sqrt{(x_1 - x_2)^2 + (y_1 - y_2)^2}$

3. The set of all points 2 units from the point (–1, 1) is a circle with $r = 2$.

equation $(x+1)^2 + (y-1)^2 = 4$.

5. Circle

7. Parabola

9. Hyperbola

11. Ellipse

13. none, one or two

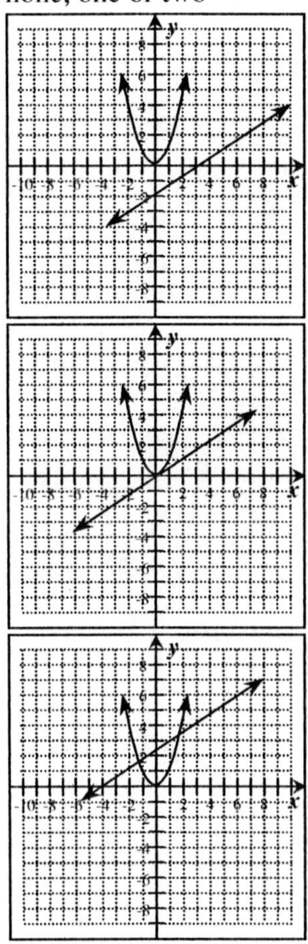

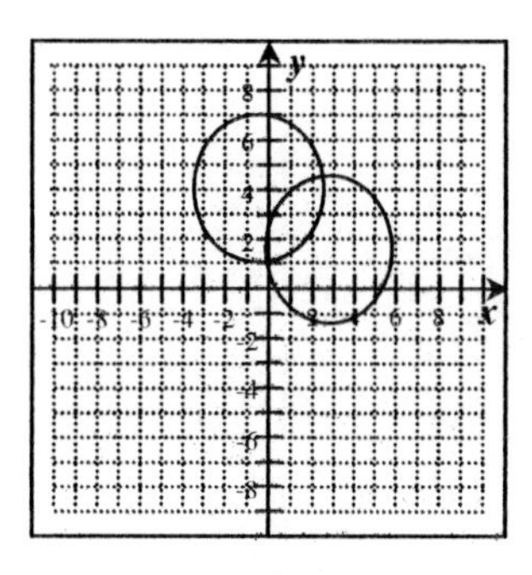

15. none, one or two

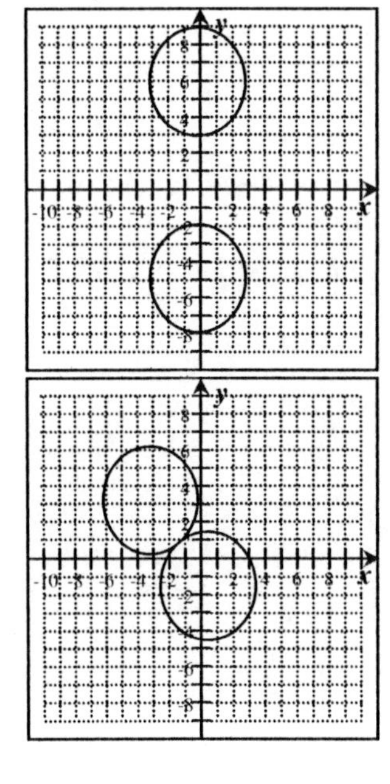

17. none, one, two, three, or four

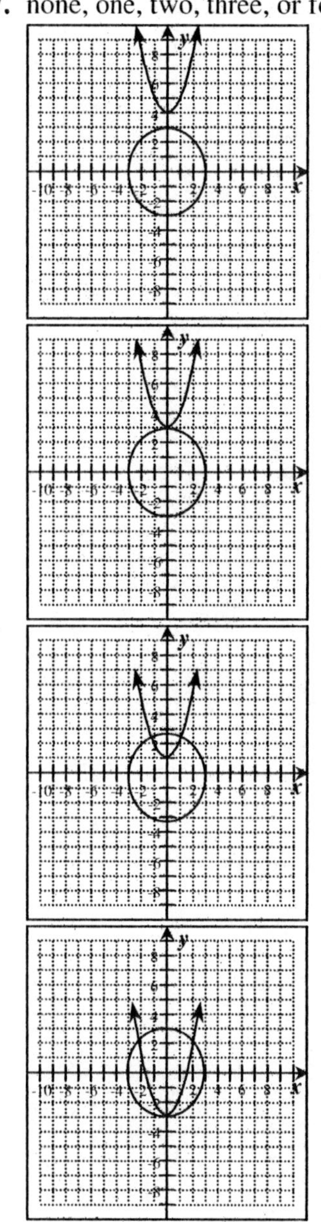

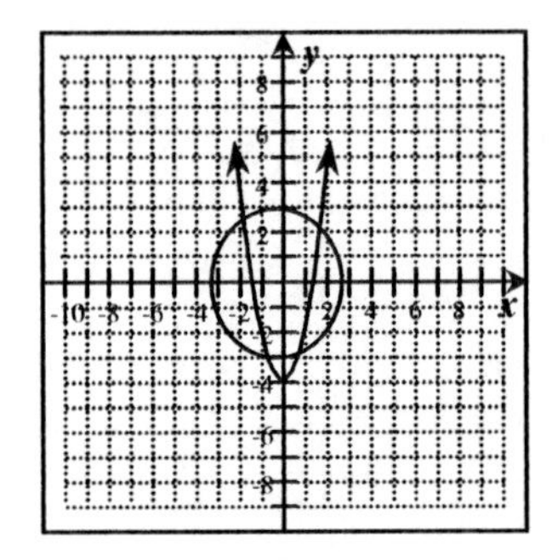

19. none, one, two, three or four

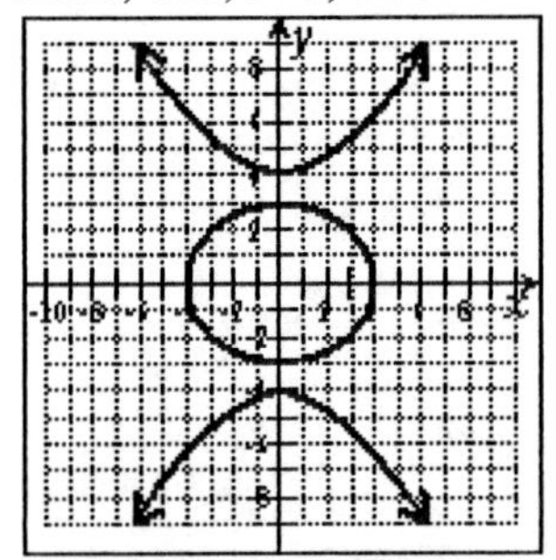

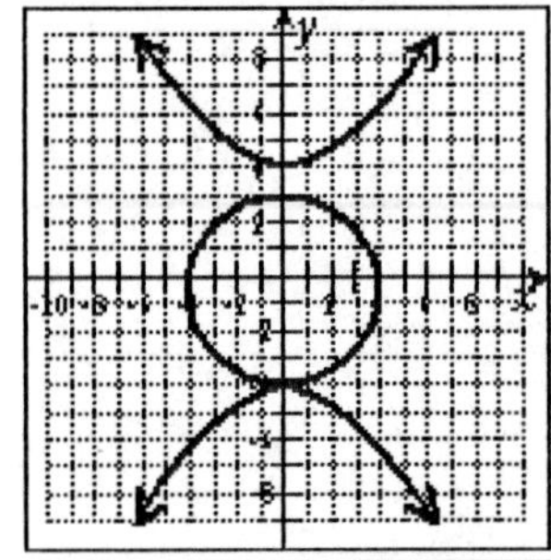

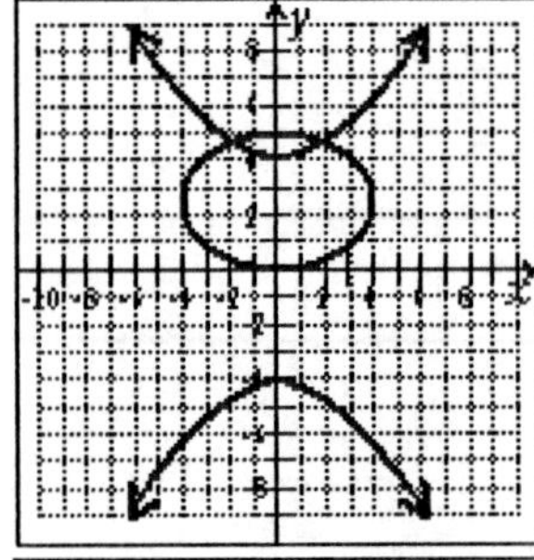

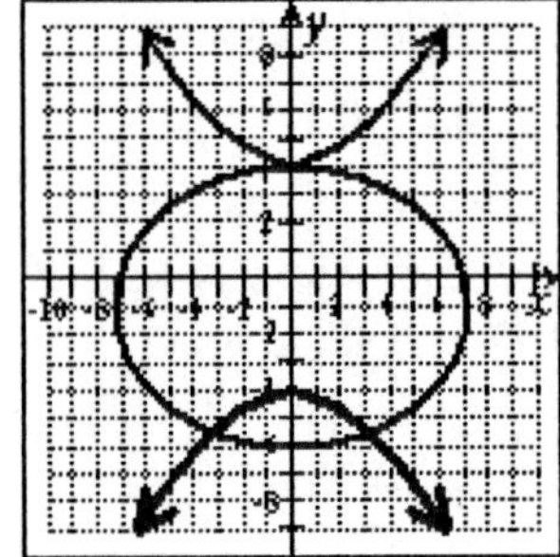

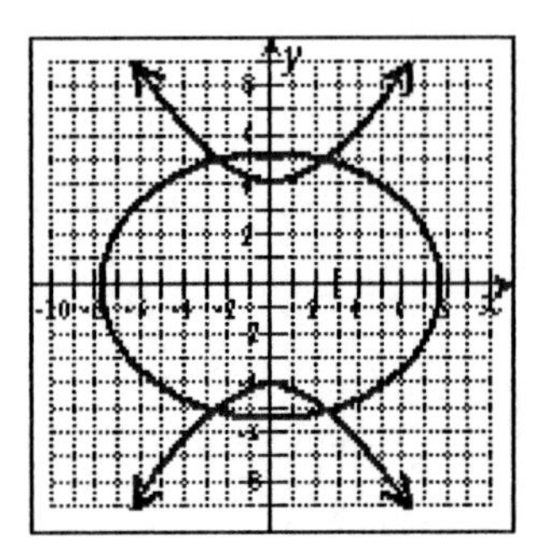

21.

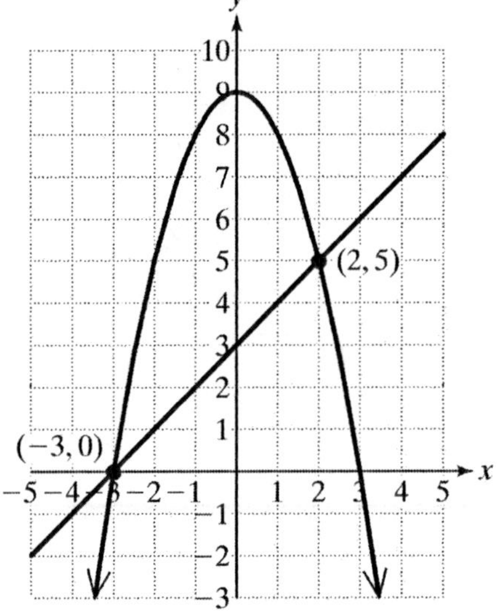

Substitute $x + 3$ for y in the second equation:

$$x^2 + (x+3) = 9$$
$$x^2 + x - 6 = 0$$
$$(x+3)(x-2) = 0$$
$$x+3=0 \quad \text{or} \quad x=2=0$$
$$x=-3 \quad \text{or} \quad x=2$$

Substituting $x = -3$ in $y = x + 3$ gives $y = 0$.
Substituting $x = 2$ in $y = x + 3$ gives $y = 5$.
The points of intersections are (–3, 0) and (2, 5).

23.

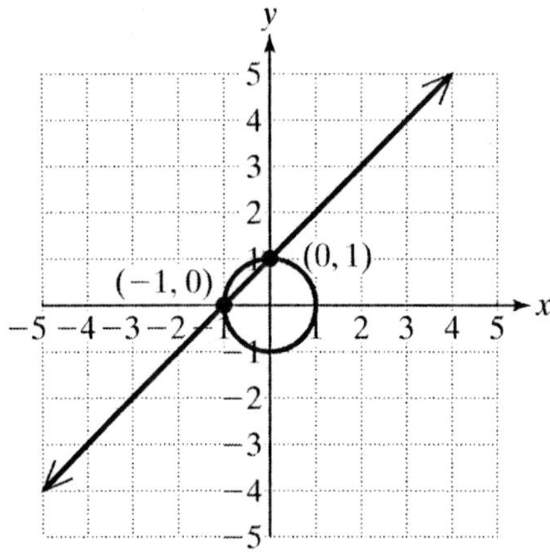

Substitute $x + 1$ for y in the first equation:

$$x^2 + (x+1)^2 = 1$$
$$x^2 + x^2 + 2x + 1 = 1$$
$$2x^2 + 2x = 0$$
$$2x(x+1) = 0$$

$2x = 0$ or $x + 1 = 0$

$x = 0$ or $x = -1$

Substituting $x = 0$ in $y = x + 1$ gives $y = 1$.
Substituting $x = -1$ in $y = x + 1$ gives $y = 0$.
The points of intersection are: (0, 1) and (−1, 0).

25.

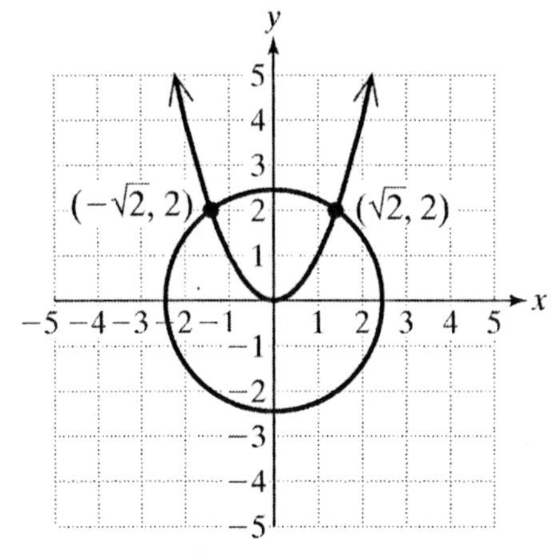

Substitute x^2 for y in the first equation:

$$x^2 + (x^2)^2 = 6$$
$$x^2 + x^4 - 6 = 0$$
$$x^4 + x^2 - 6 = 0$$
$$(x^2 + 3)(x^2 - 2) = 0$$

$x^2 + 3 = 0$ or $x^2 - 2 = 0$

$x^2 \neq -3$ or $x^2 = 2$

$x = \pm\sqrt{2}$

Substituting $x = \sqrt{2}$ in $y = x^2$ gives $y = 2$.
Substituting $x = -\sqrt{2}$ in $y = x^2$ gives $y = 2$.
The points of intersection are: $\left(\sqrt{2}, 2\right)$ and $\left(-\sqrt{2}, 2\right)$

27. Substitute $\sqrt{x}$ for y in the first equation:

$$x^2 + \left(\sqrt{x}\right)^2 = 20$$
$$x^2 + x - 20 = 0$$
$$(x+5)(x-4) = 0$$

$x + 5 = 0$ or $x - 4 = 0$

$x = -5$ or $x = 4$

Substituting $x = -5$ in $y = \sqrt{x}$ is not a real number.
Substituting $x = 4$ in $y = \sqrt{x}$ is not a real number.
Substituting $x = 4$ in $y = \sqrt{x}$ gives $y = 2$.
The point of intersection is: (4, 2).

29. Substitute $-\sqrt{x}$ for y in the first equation:

$$-\sqrt{x} = x^2$$

Since x^2 must always be non-negative, the only possible x-value is 0.
The point of intersection is: (0, 0).

31. Substitute $(x-3)^2$ for y in the first equation:

$$(x-3)^2 = x^2$$
$$x^2 - 6x + 9 = x^2$$
$$-6x + 9 = 0$$
$$-6x = -9$$
$$x = \frac{3}{2}$$

Substituting $x = \frac{3}{2}$ in $y = (x-3)^2$ gives $y = \frac{9}{4}$. The point of intersection is: $\left(\frac{3}{2}, \frac{9}{4}\right)$.

33. Substitute $4x$ for y in the first equation:

$$4x = x^2 + 6x$$
$$0 = x^2 + 2x$$
$$0 = x(x + 2)$$

$x = 0$ or $x = -2$

Substituting $x = 0$ in $y = 4x$ gives:

$y = 4(0) = 0$

Substituting $x = -2$ in $y = 4x$ gives:

$y = 4(-2) = -8$

The points of intersection are (0, 0) and (−2, −8).

35. Adding the two equations gives:

$$2x^2 = 18$$
$$x^2 = 9$$
$$x = \pm 3$$

Substituting $x = 3$ in $x^2 + y^2 = 13$ gives:

$$9 + y^2 = 13$$
$$y^2 = 4$$
$$y = \pm 2$$

Substituting $x = -3$ in $x^2 + y^2 = 13$ gives:

$$9 + y^2 = 13$$
$$y^2 = 4$$
$$y = \pm 2$$

The points of intersection are: (3, 2), (3, −2), (−3, 2), and (−3, −2).

37. Multiplying equation 2 by −9 gives the system:

$$\begin{cases} 9x^2 + 4y^2 = 36 \\ -9x^2 - 9y^2 = -81 \end{cases}$$

Adding the two equations gives:

$$-5y^2 = -45$$
$$y^2 = 9$$
$$y = \pm 3$$

Substituting $y = 3$ in $x^2 + y^2 = 9$ gives:

$$x^2 + 9 = 9$$
$$x^2 = 0$$
$$x = 0$$

Substituting $y = -3$ in $x^2 + y^2 = 9$ gives:

$$x^2 + 9 = 9$$
$$x^2 = 0$$
$$x = 0$$

The points of intersection are: (0, 3) and (0, −3).

39. Multiplying equation 1 by 3 and equation 2 by 4 gives the system:

$$\begin{cases} 9x^2 + 12y^2 = 48 \\ 8x^2 - 12y^2 = 20 \end{cases}$$

Adding the two equations gives:

$$17x^2 = 68$$
$$x^2 = 4$$
$$x = \pm 2$$

Substituting $x = 2$ in $3x^2 + 4y^2 = 16$ gives:

$$12 + 4y^2 = 16$$
$$4y^2 = 4$$
$$y = \pm 1$$

Substituting $x = -2$ in $3x^2 + 4y^2 = 16$ gives:

$$12 + 4y^2 = 16$$
$$4y^2 = 4$$
$$y = \pm 1$$

The points of intersection are: (2, 1), (2, −1), (−2, 1) and (−2, −1).

41. Adding the two equations gives:

$$2y = 0$$
$$y = 0$$

Substituting $y = 0$ in $y = x^2 - 2$ gives:

$$0 = x^2 - 2$$
$$x^2 = 2$$
$$x = \pm\sqrt{2}$$

The points of intersection are: $\left(\sqrt{2}, 0\right)$ and $\left(-\sqrt{2}, 0\right)$.

43. Multiplying equation 1 by −36 and equation 2 by 4 gives the system:

$$\begin{cases} -9x^2 - 4y^2 = -36 \\ 4x^2 + 4y^2 = 16 \end{cases}$$

Adding the two equations gives:

$$\begin{aligned} -5x^2 &= -20 \\ x^2 &= 4 \\ x &= \pm 2 \end{aligned}$$

Substituting $x = 2$ in $x^2 + y^2 = 4$ gives:

$$\begin{aligned} 4 + y^2 &= 4 \\ y^2 &= 0 \\ y &= 0 \end{aligned}$$

Substituting $x = -2$ in $x^2 + y^2 = 4$ gives:

$$\begin{aligned} 4 + y^2 &= 4 \\ y^2 &= 0 \\ y &= 0 \end{aligned}$$

The points of intersection are: (2, 0) and (–2, 0).

45. Multiplying equation 2 by –72 gives the system:

$$\begin{cases} x^2 + 6y^2 = 9 \\ -8x^2 - 6y^2 = -72 \end{cases}$$

Adding the two equations gives:

$$\begin{aligned} -7x^2 &= -63 \\ x^2 &= 9 \\ x &= \pm 3 \end{aligned}$$

Substituting $x = 3$ in $x^2 + 6y^2 = 9$ gives:

$$\begin{aligned} (3)^2 + 6y^2 &= 9 \\ 9 + 6y^2 &= 9 \\ 6y^2 &= 0 \\ y^2 &= 0 \\ y &= 0 \end{aligned}$$

Substituting $x = -3$ in $x^2 + 6y^2 = 9$ gives:

$$\begin{aligned} (-3)^2 + 6y^2 &= 9 \\ 9 + 6y^2 &= 9 \\ 6y^2 &= 0 \\ y^2 &= 0 \\ y &= 0 \end{aligned}$$

The points of intersection are (3, 0) and (–3, 0).

47. Multiplying the first equation by –1 gives the system:

$$\begin{cases} -x^2 + xy = 4 \\ 2x^2 - xy = 12 \end{cases}$$

Adding the two equations gives:

$$\begin{aligned} x^2 &= 16 \\ x &= \pm 4 \end{aligned}$$

Substituting $x = 4$ in $x^2 - xy = -4$ gives:

$$\begin{aligned} (4)^2 - (4)y &= -4 \\ 16 - 4y &= -4 \\ 16 - 4y &= -4 \\ -4y &= -20 \\ y &= 5 \end{aligned}$$

Substituting $x = -4$ in $x^2 - xy = -4$ gives:

$$\begin{aligned} (-4)^2 - (-4)y &= -4 \\ 16 + 4y &= -4 \\ 4y &= -20 \\ y &= -5 \end{aligned}$$

The points of intersection are (4, 5) and (–4, –5).

49. Let x represent the first number.
Let y represent the second number.

$$\begin{cases} x + y = 7 \\ x^2 + y^2 = 25 \end{cases}$$

Solving for y in equation 1 gives $y = -x + 7$. Substituting this expression in equation 2 gives:

$$\begin{aligned} x^2 + (-x+7)^2 &= 25 \\ x^2 + x^2 - 14x + 49 &= 25 \\ 2x^2 - 14x + 24 &= 0 \\ 2(x^2 - 7x + 12) &= 0 \\ 2(x-4)(x-3) &= 0 \end{aligned}$$

$2 \neq 0$ or $x - 4 = 0$ or $x - 3 = 0$
$x = 4$ or $x = 3$

If $x = 4$, $y = -4 + 7 = 3$.
If $x = 3$, $y = -3 + 7 = 4$.

51. Let x represent the first number.
Let y represent the second number.

$$\begin{cases} x^2 + y^2 = 32 \\ x^2 - y^2 = 18 \end{cases}$$

Adding the two equations gives:

$2x^2 = 50$

$x^2 = 25$

$x = \pm 5$

Substituting $x = 5$ in $x^2 + y^2 = 32$ gives:

$x^2 + 25 = 32$

$x^2 = 7$

$x = \pm\sqrt{7}$

Substituting $x = -5$ in $x^2 + y^2 = 32$ gives:

$25 + y^2 = 32$

$y^2 = 7$

$y = \pm\sqrt{7}$

The numbers are: 5 and $\sqrt{7}$, 5 and $-\sqrt{7}$, -5 and $\sqrt{7}$, -5 and $-\sqrt{7}$.

53.

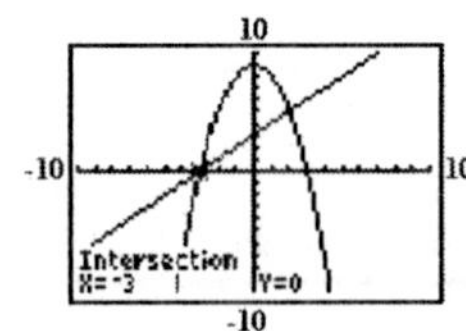

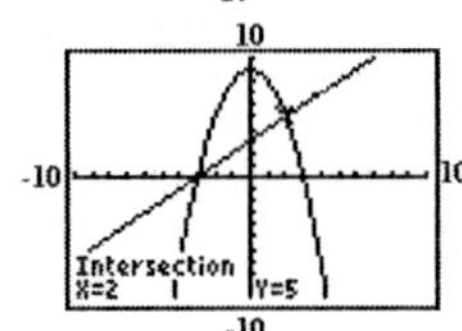

55.

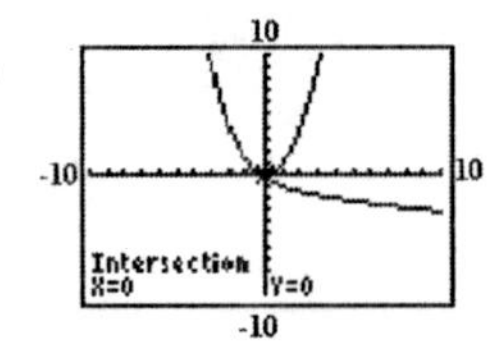

57. No solution

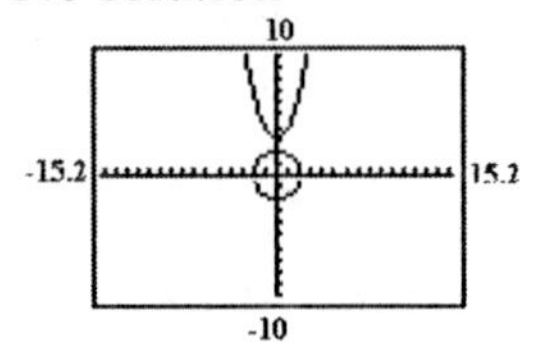

Section 13.5 Practice Exercises

1. g

3. h

5. f

7. k

9. a

11. d

13. $4x^2 - 2x + 1 + y^2 < 3$

$4(4)^2 - 2(4) + 1 + (-2)^2 < 3$

$64 - 8 + 1 + 4 < 3$

$61 < 3$ False

15. $y < x^2$

$-2 < (1)^2$

$-2 < 1$ True

$y > x^2 - 4$

$-2 > (1)^2 - 4$

$-2 > -3$ True

Since the point satisfied both inequalities, the answer is "True."

17. (a) The related equation $x^2 + y^2 = 9$ is a circle of radius 3 centered at the origin. Graph the related equation using a solid curve since the symbol "≤" includes equality.

Test Point Inside: (0, 0)

$x^2 + y^2 \le 9$

$(0)^2 + (0)^2 \le 9$

$0 \le 9$ False

Test Point Outside: (4, 0)

$x^2 + y^2 \le 9$

$(4)^2 + (0)^2 \le 9$

$16 \le 9$ False

Since the inequality is true at the test point (0, 0), shade the region inside the circle.

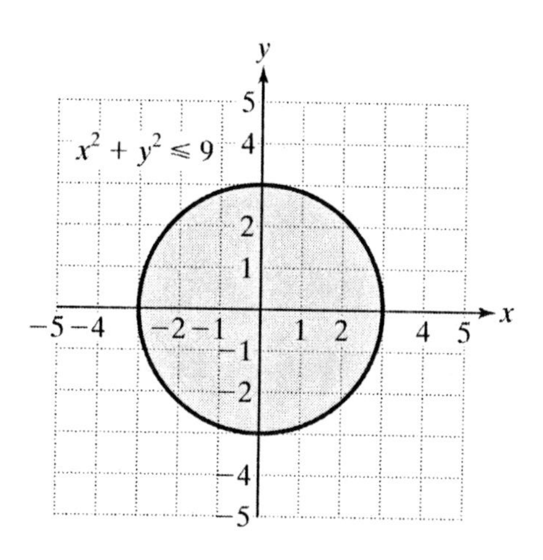

(b) The set of points on the outside the circle $x^2 + y^2 = 9$.

(c) The set of points on the circle $x^2 + y^2 = 9$.

19. (a) The related equation $y = x^2 + 1$ is a parabola which opens upward, with vertex (0, 1). Graph the related equation using a solid curve since the symbol "≥" includes equality.

Test Point Above: (0, 2)

$y \geq x^2 + 1$

$2 \geq (0)^2 + 1$

$2 \geq 1$ True

Test Point Below: (0, 0)

$y \geq x^2 + 1$

$0 \geq (0)^2 + 1$

$0 \geq 1$ False

Since the inequality is true at the test point (0, 2), shade the region above the parabola.

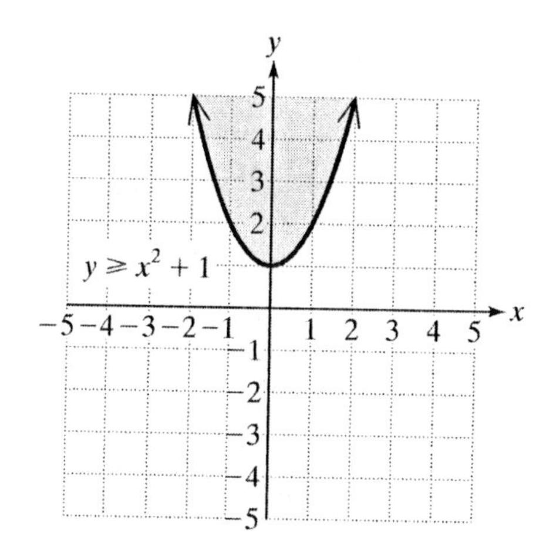

(b) The parabola $y = x^2 + 1$ would be drawn as a dashed curve.

21. The related equation $2x + y = 1$, which we can rewrite as $y = -2x + 1$, is a line with y-intercept (0, 1) and slope –2. Graph the related equation using a solid line since the symbol "≥" includes equality.

Test Point Above: (0, 2)

$2x + y \geq 1$

$2(0) + (2) \geq 1$

$2 \geq 1$ True

Test Point Below: (0, 0)

$2x + y \geq 1$

$2(0) + (0) \geq 1$

$0 \geq 1$ False

Since the inequality is true at the test point (0, 2), shade the region above the line.

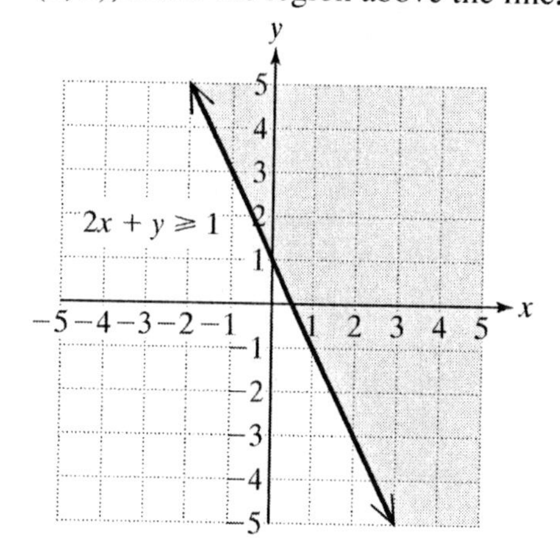

23. The related equation $x = y^2$ is a parabola which opens to the right. Graph the related equation using a solid curve since the symbol "≤" includes equality.

Test Point "Right": (1, 0)

$x \le y^2$

$1 \le 0^2$

$1 \le 0$ False

Test Point "Left" : (–1, 0)

$x \le y^2$

$-1 \le 0^2$

$-1 \le 0$ True

Since the inequality is true at the test point (–1, 0), shade the region left of the parabola.

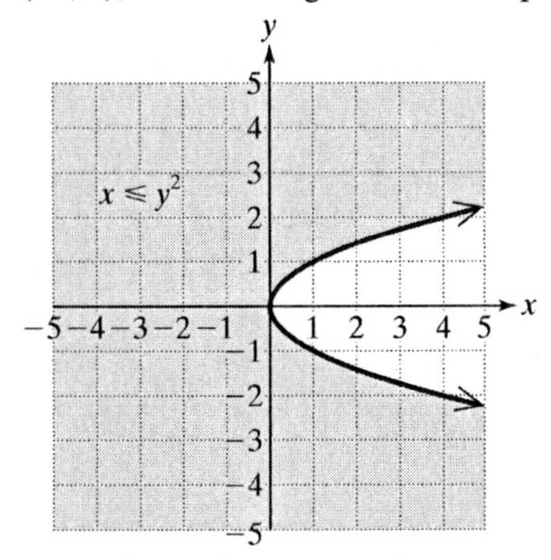

25. The related equation $(x-1)^2 + (y+2)^2 = 9$ is a circle of radius 3, centered at (1, –2). Graph the related equation using a dashed curve since the symbol ">" does not include equality.

Test Point Inside: (1, –2)

$(x-1)^2 + (y+2)^2 > 9$

$(1-1)^2 + (-2+2)^2 > 9$

$0^2 + 0^2 > 9$

$0 > 9$ False

Test Point Outside: (5, –2)

$(x-1)^2 + (y+2)^2 > 9$

$(5-1)^2 + (-2+2)^2 > 9$

$4^2 + 0^2 > 9$

$16 > 9$ True

Since the inequality is true at the test point (5, –2), shade the region outside the circle.

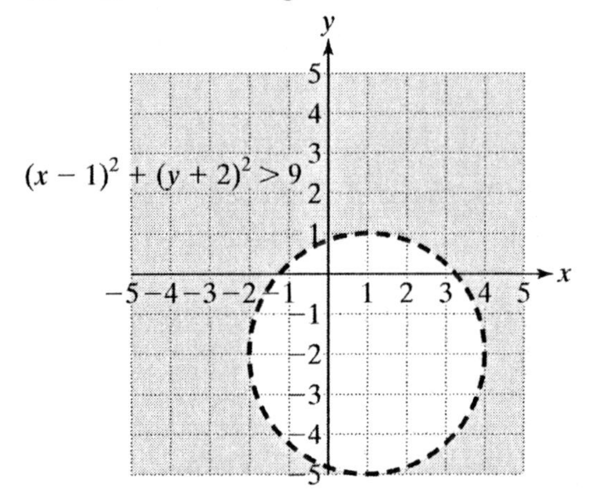

27. The related equation $x^2 + y^2 + 2x - 24 = 0$ is a circle. Rewrite it in standard form.

$x^2 + 2x + y^2 = 24$

$x^2 + 2x + 1 + y^2 = 24 + 1$

$(x+1)^2 + y^2 = 25$

This is a circle of radius 5, centered at (–1, 0). Graph the related equation using a dashed curve since the symbol "<" does not include equality.

Test Point Inside: (–1, 0)

$x^2 + y^2 + 2x - 24 < 0$

$(-1)^2 + (0)^2 + 2(-1) - 24 < 0$

$1 + 0 - 2 - 24 < 0$

$-25 < 0$ True

Test Point Outside: (5, 0)

$x^2 + y^2 + 2x - 24 < 0$

$(5)^2 + (0)^2 + 2(5) - 24 < 0$

$25 + 0 + 10 - 24 < 0$

$11 < 0$ False

Since the inequality is true at the test point (−1, 0), shade the region inside the circle.

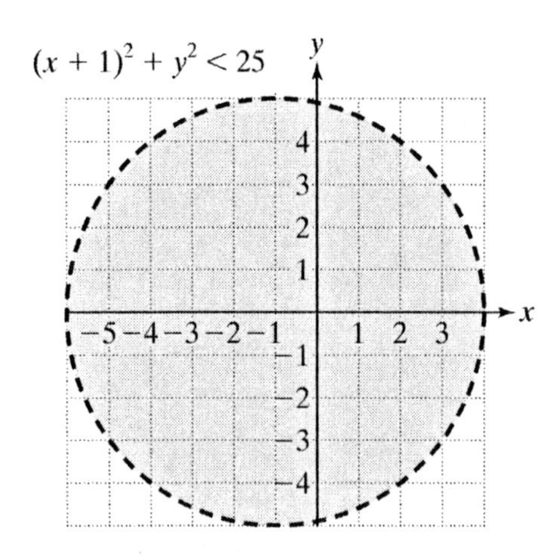

29. The related equation $9x^2-y^2=9$ is a hyperbola. Rewrite it in standard form.

$$\frac{9x^2}{9}-\frac{y^2}{9}=\frac{9}{9}$$

$$x^2-\frac{y^2}{9}=1$$

This is a hyperbola with horizontal transverse axis, centered at the origin. Graph the related equation using a dashed curve since the symbol ">" does not include equality. Select a test point from each of the three regions.

Test Point: (−2, 0)

$$9x^2-y^2>9$$
$$9(-2)^2-(0)^2>9$$
$$9(4)+0>9$$
$$36>9 \quad \text{True}$$

Test Point: (0, 0)

$$9x^2-y^2>9$$
$$9(0)^2-(0)^2>9$$
$$9(0)+0>9$$
$$0>9 \quad \text{False}$$

Test Point: (2, 0)

$$9x^2-y^2>9$$
$$9(2)^2-(0)^2>9$$
$$9(4)+0>9$$
$$36>9 \quad \text{True}$$

Since the inequality is true at the test points (−2, 0) and (2, 0), shade the regions to the left of the left branch of the hyperbola and to the right of the right branch of the hyperbola.

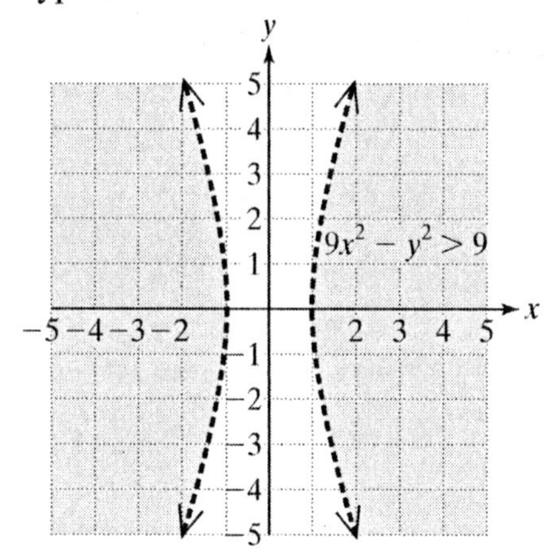

31. The related equation $x^2+16y^2=16$ is an ellipse. Rewrite it in standard form.

$$\frac{x^2}{16}+\frac{16y^2}{16}=\frac{16}{16}$$

$$\frac{x^2}{16}+y^2=1$$

This is an ellipse with horizontal major axis, centered at the origin. Graph the related equation using a solid curve since the symbol "≤" includes equality.

Test Point Inside: (0, 0)

$$x^2+16y^2\le 16$$
$$(0)^2+16(0)^2\le 16$$
$$0+0\le 16$$
$$0\le 16 \quad \text{True}$$

Test Point Outside: (0, 2)

$$x^2+16y^2\le 16$$
$$(0)^2+16(2)^2\le 16$$
$$0+16(4)\le 16$$
$$64\le 16 \quad \text{False}$$

Since the inequality is true at the test point (0, 0), shade the region inside the ellipse.

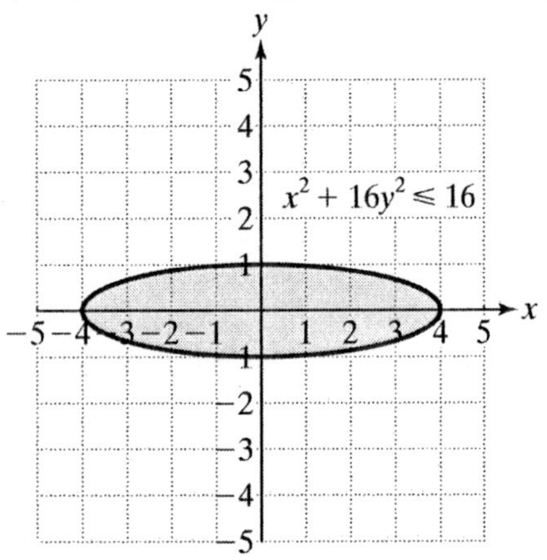

33. Graph related equation, $y = \ln x$, using a solid curve since the symbol "≤" includes equality.

Test Point Above: (1, 1)

$y \leq \ln x$

$1 \leq \ln 1$

$1 \leq 0$ False

Test Point Below: (2, 0)

$y \leq \ln x$

$0 \leq \ln 2$

$0 \leq 0.6931$ True

Since the inequality is true at the test point (2, 0) shade the region below the logarithmic curve.

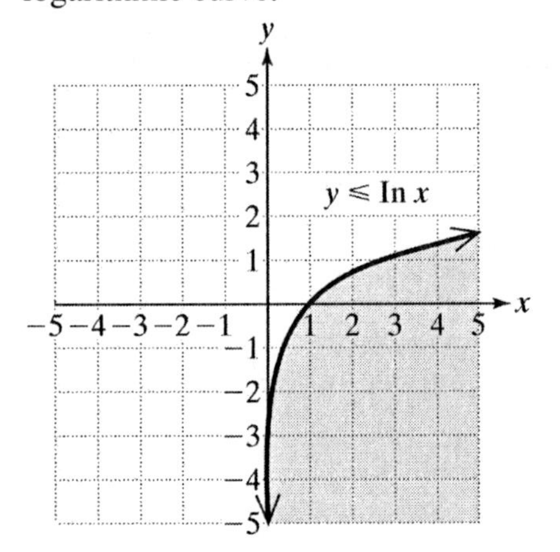

35. Graph related equation, $y = 5^x$, using a dashed curve since the symbol ">" does not include equality.

Test Point Above: (0, 2)

$y > 5^x$

$2 > 5^0$

$2 > 1$ True

Test Point Below: (0, 0)

$y > 5^x$

$0 > 5^0$

$0 > 1$ False

Since the inequality is true at the test point (0, 2), shade the region above the exponential curve.

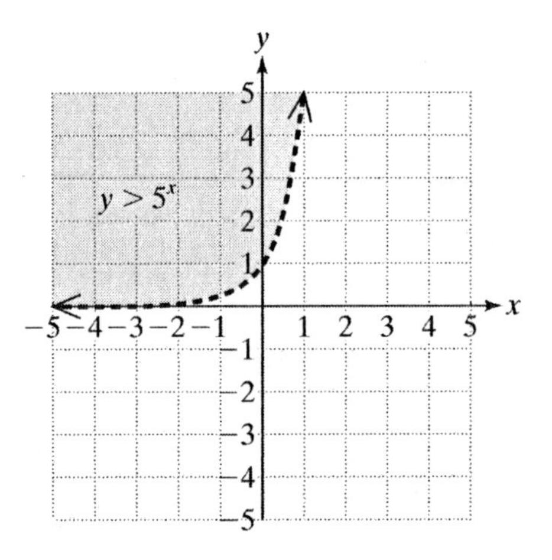

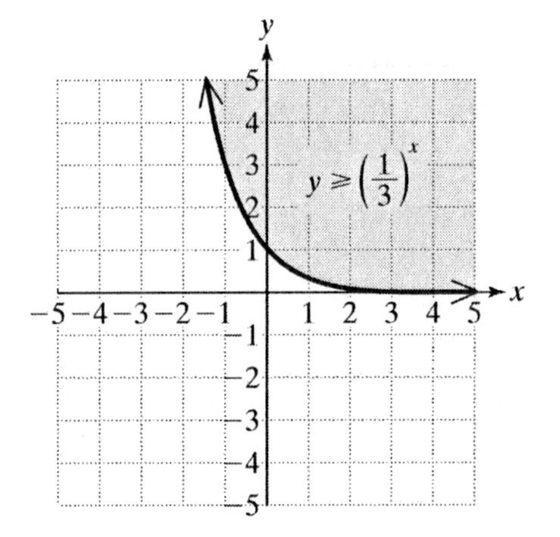

37. The solution to $y \le \sqrt{x}$ is the set of points on and below the curve $y = \sqrt{x}$.

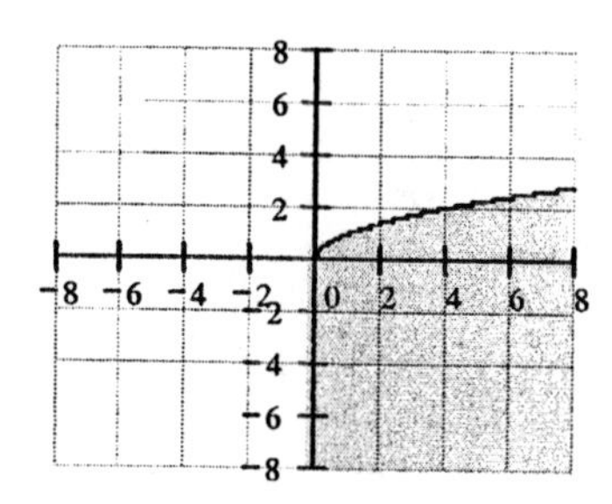

The solution to $x \ge 1$ is the set of points on and to the right of the line $x = 1$.

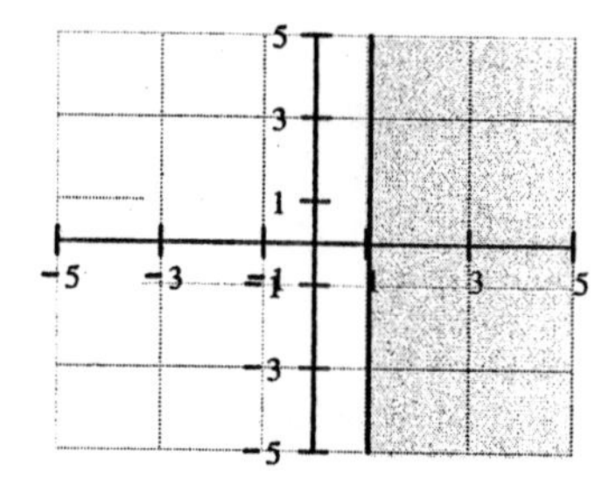

The solution to the system of inequalities is the intersection of the solution sets of the individual inequalities.

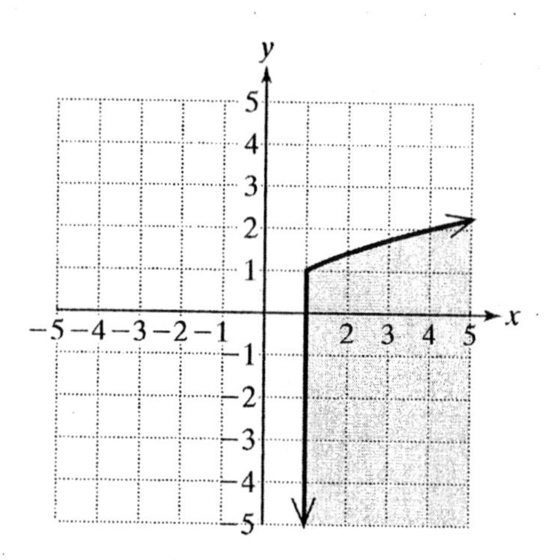

39. The solution to $\dfrac{x^2}{36} + \dfrac{y^2}{25} < 1$ is the set of points inside the ellipse $\dfrac{x^2}{36} + \dfrac{y^2}{25} = 1$.

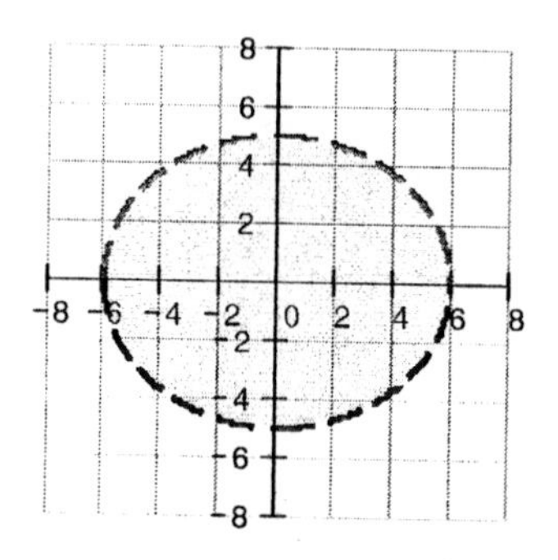

The solution to $x^2 + y^2 \ge 4$ is the set of points on and outside the circle $x^2 + y^2 = 4$.

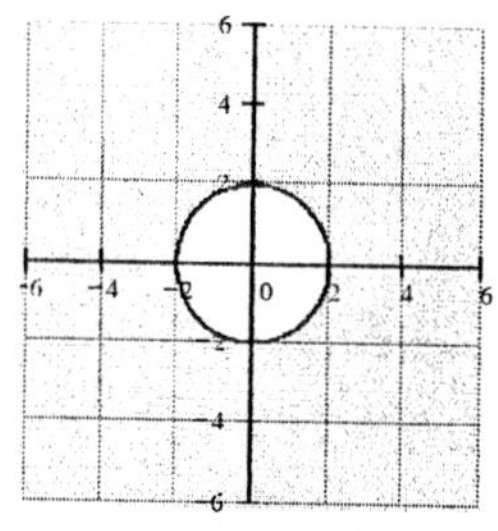

The solution to the system of inequalities is the intersection of the solution sets of the individual inequalities.

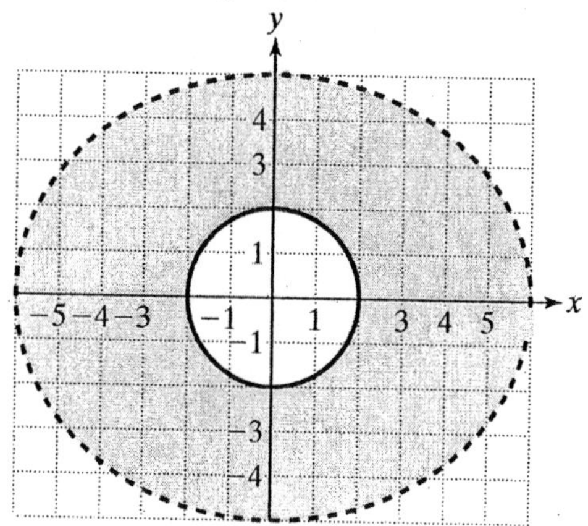

41. The solution to $y < x^2$ is the set of points below the parabola $y = x^2$.

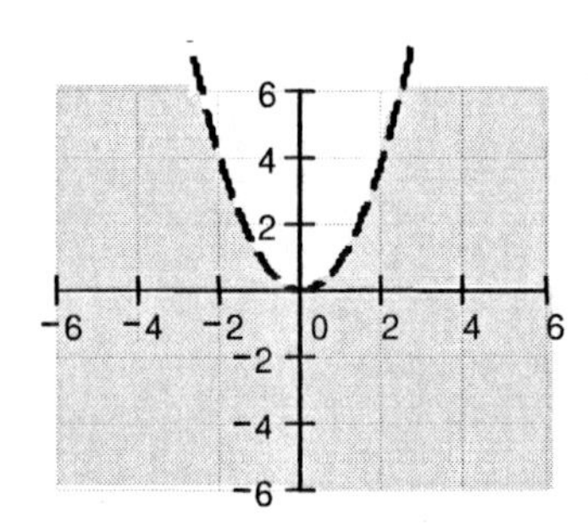

The solution to $y > x^2 - 4$ is the set of points above the parabola $y = x^2$.

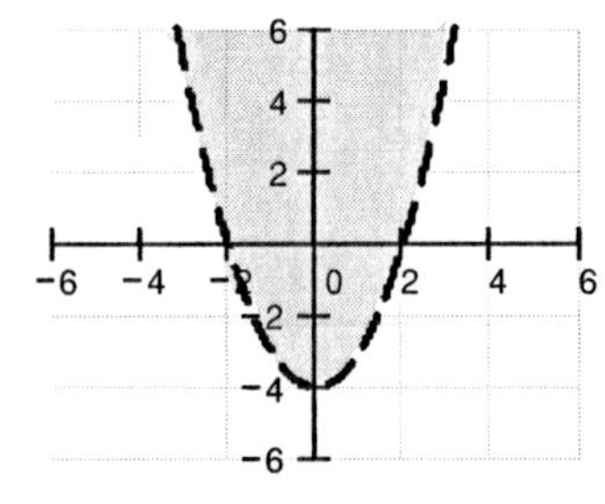

The solution to the system of inequalities is the intersection of the solution sets of the individual inequalities.

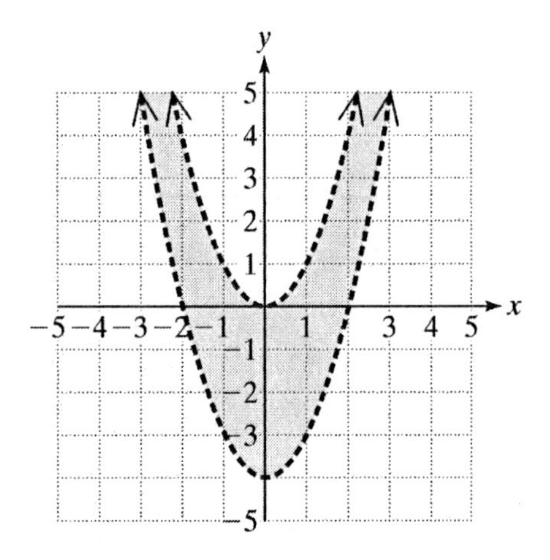

43. The solution to $y \le \dfrac{1}{x}$ is the set of points on and between the two branches of the curve $y = \dfrac{1}{x}$.

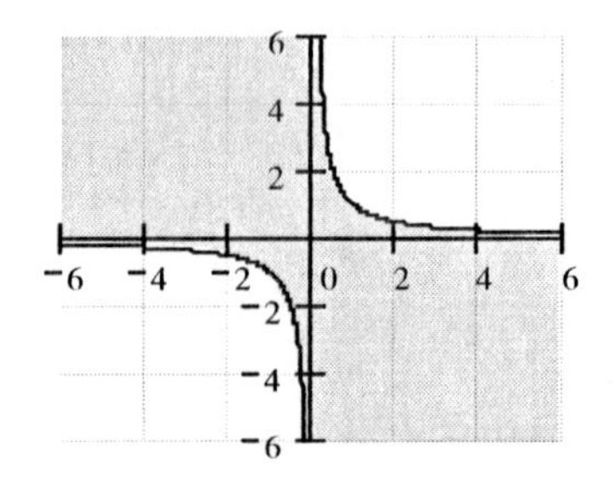

The solution to $y \ge 0$ is the set of points on and above the line $y = 0$.

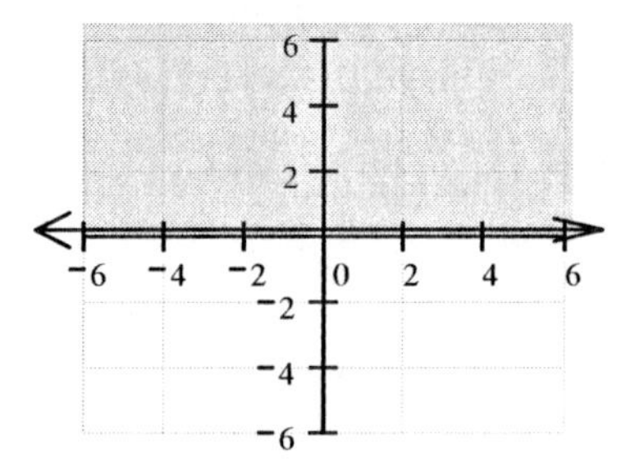

The solution to $y \le x$ is the set of points on and below the line $y = x$.

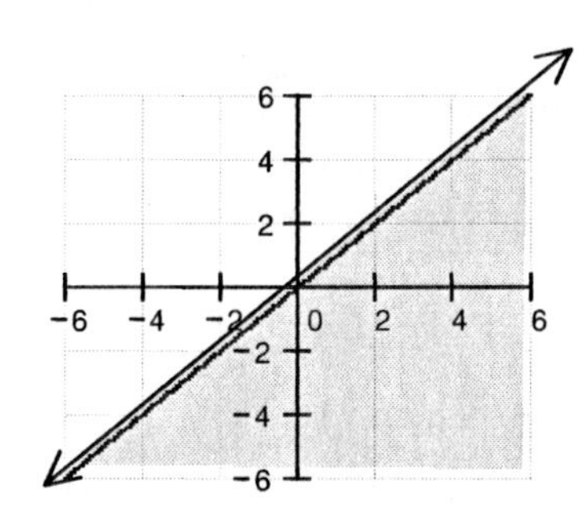

The solution to the system of inequalities is the intersection of the solution sets of the individual inequalities.

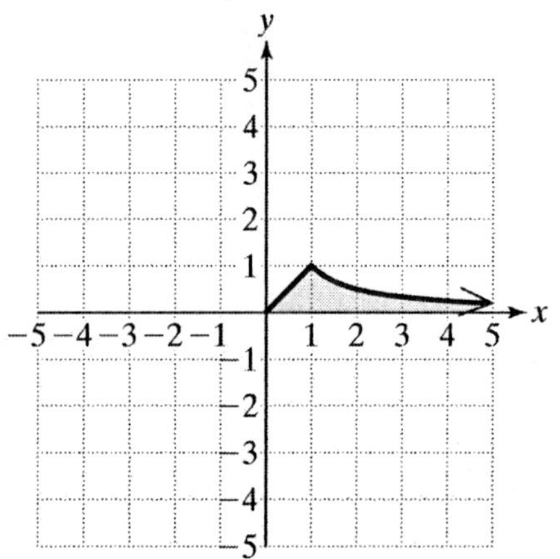

45. The solution to $x^2 + y^2 \geq 25$ is the set of points on and outside the circle $x^2 + y^2 = 25$.

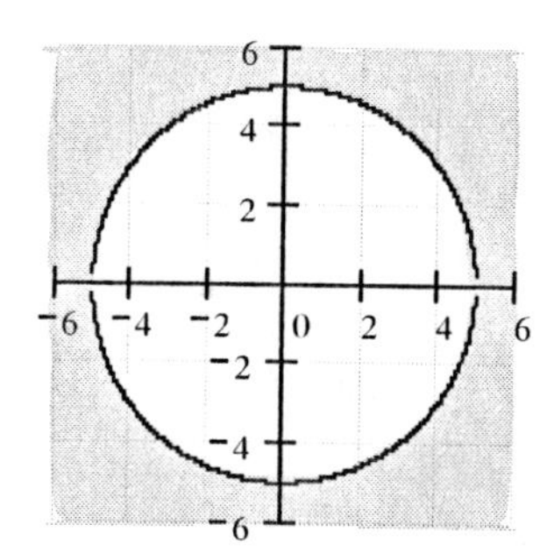

The solution to $x^2 + y^2 \leq 9$ is the set of points on and inside the circle $x^2 + y^2 = 9$.

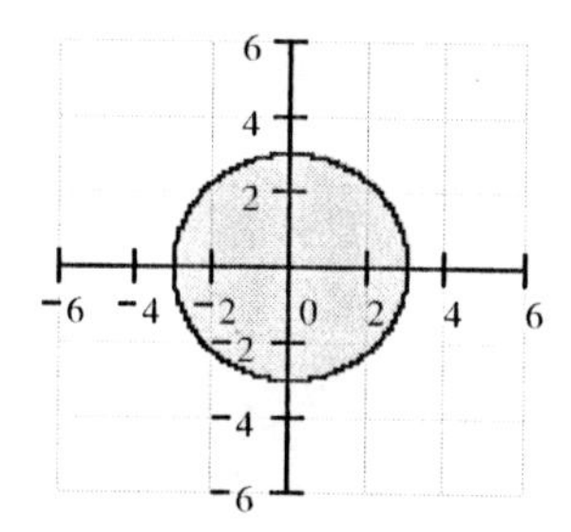

The solution to the system of inequalities is the intersection of the solution sets of the individual inequalities. Since the solution sets do not intersect, there is no solution.

47. The solution to $x < -(y-1)^2 + 3$ is the set of points to the left of the parabola $x = -(y-1)^2 + 3$.

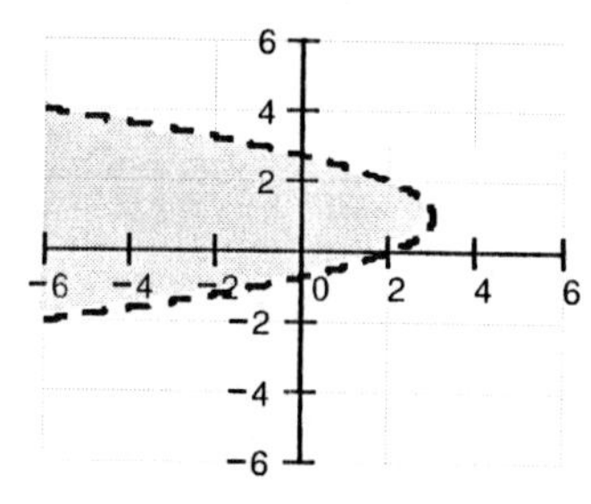

The solution to $x + y > 2$ is the set of points above the line $x + y = 2$.

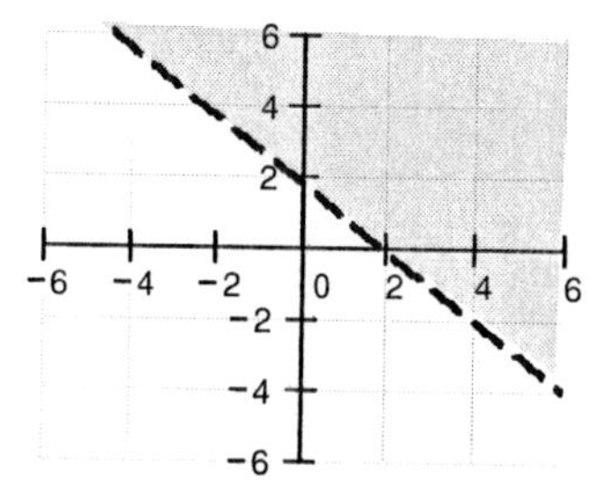

The solution to the system of inequalities is the intersection of the solution sets of the individual inequalities.

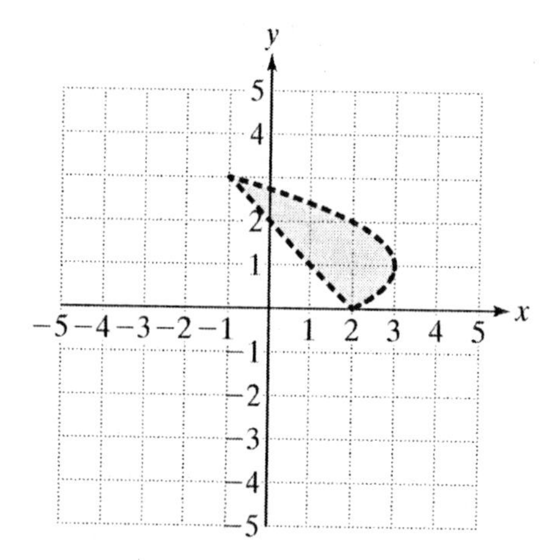

49. The solution to $x^2 + y^2 \leq 25$ is the set of points on and inside the circle $x^2 + y^2 = 25$.

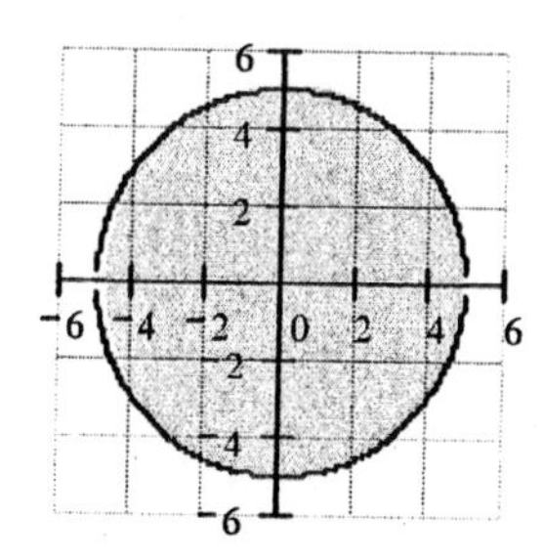

The solution to $x \leq \frac{4}{3}x$ is the set of points on and below the line $y = \frac{4}{3}x$.

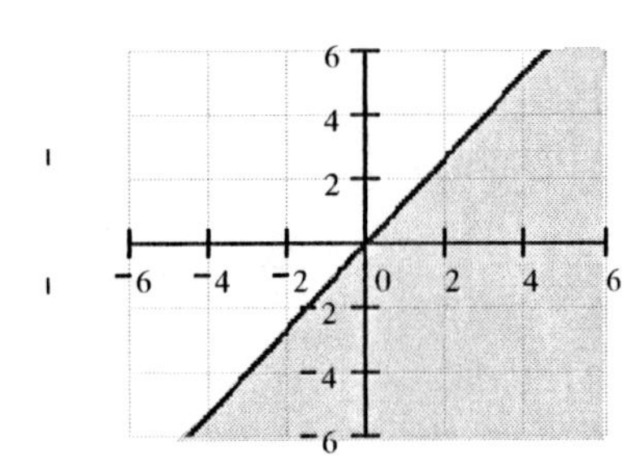

The solution to $y \geq -\frac{4}{3}x$ is the set of points on and above the line $y = -\frac{4}{3}x$.

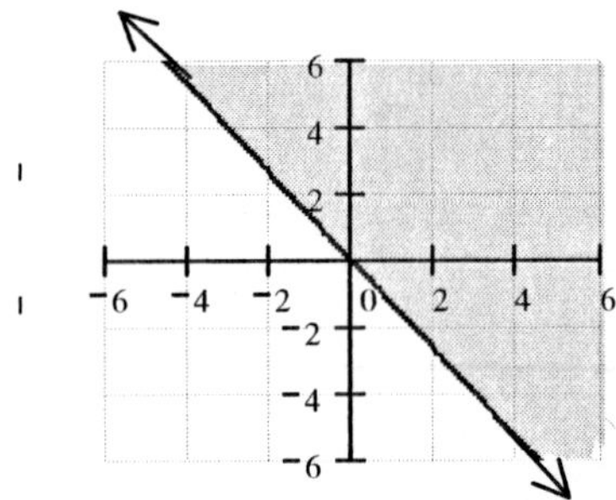

The solution to the system of inequalities is the intersection of the solution sets of the individual inequalities.

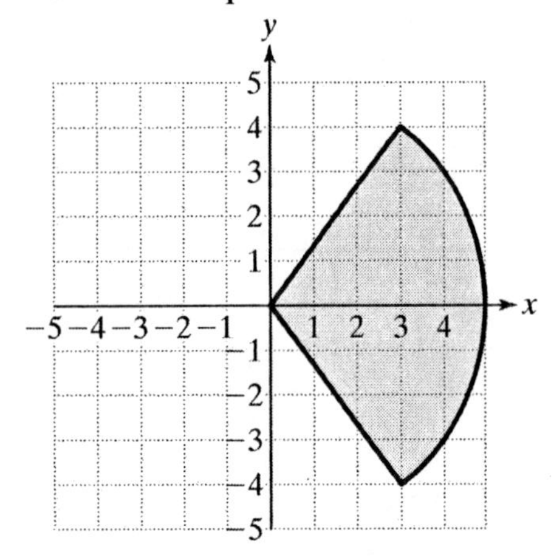

51. The solution to $x^2 + y^2 \leq 36$ is the set of points on and inside the circle $x^2 + y^2 = 36$.

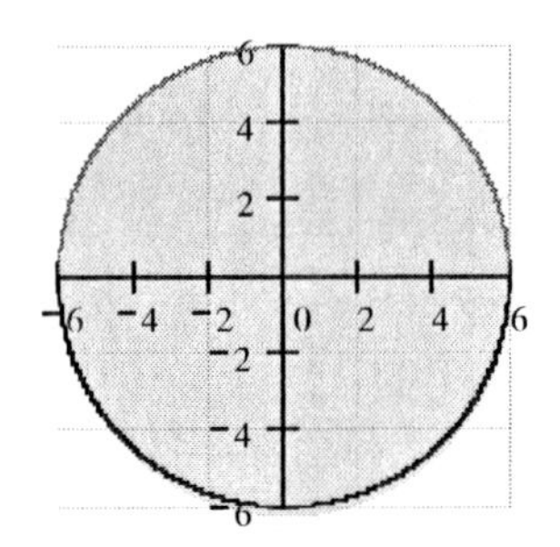

The solution to $x + y \geq 0$ is the set of points on and above the line $x + y = 0$.

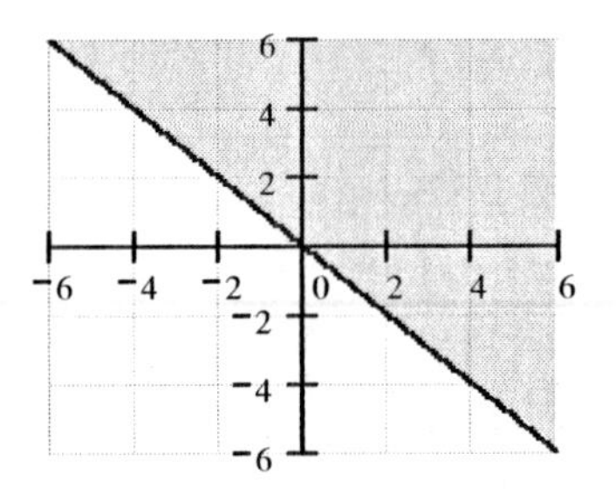

(a) The solution to two inequalities joined by the word "and" is the intersection of the solution sets of the two inequalities.

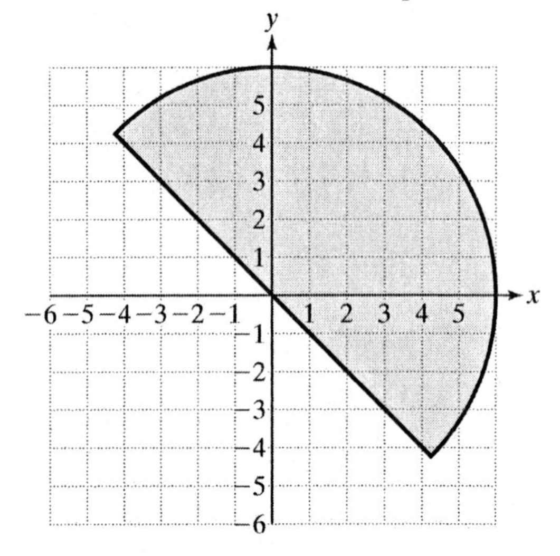

(b) The solution to two inequalities joined by the word "or" is the union of the solution sets of the two inequalities.

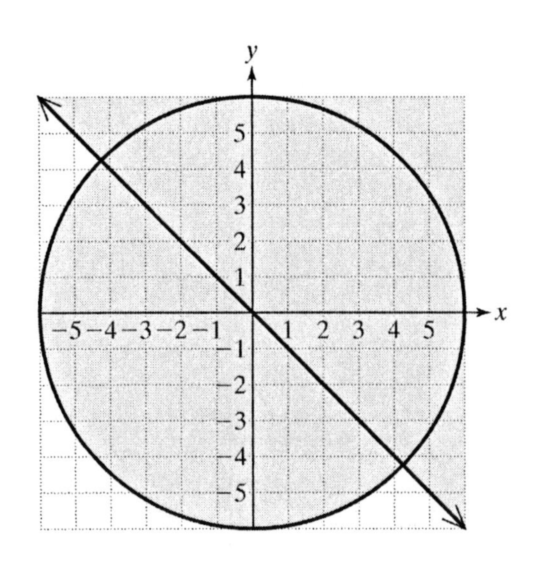

53. The solution to $y+1 \geq x^2$ is the set of points on and above the parabola $y+1=x^2$, which we can write as $y=x^2-1$.

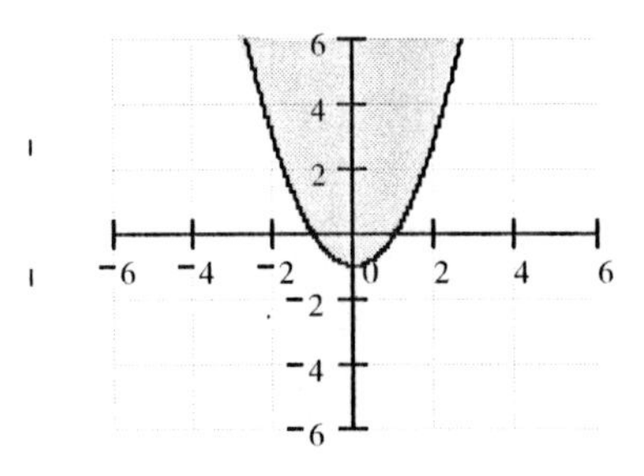

The solution to $y+1 \leq -x^2$ is the set of points on and below the parabola $y+1=-x^2$, which we can write as $y=-x^2-1$.

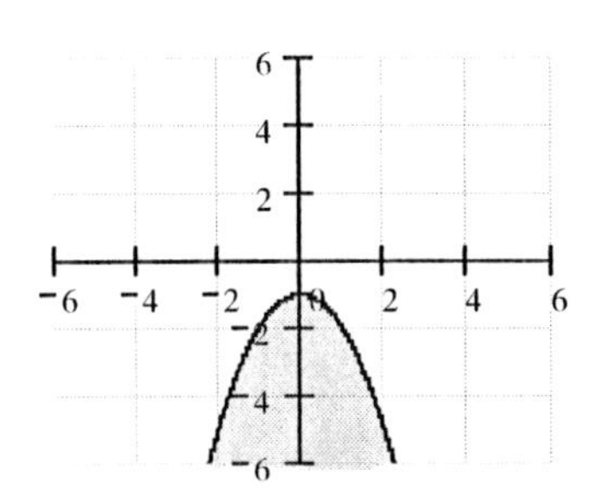

(a) The solution to two inequalities joined by the word "and" is the intersection of the solution sets of the two inequalities. In this case, the intersection is the point (0, −1).

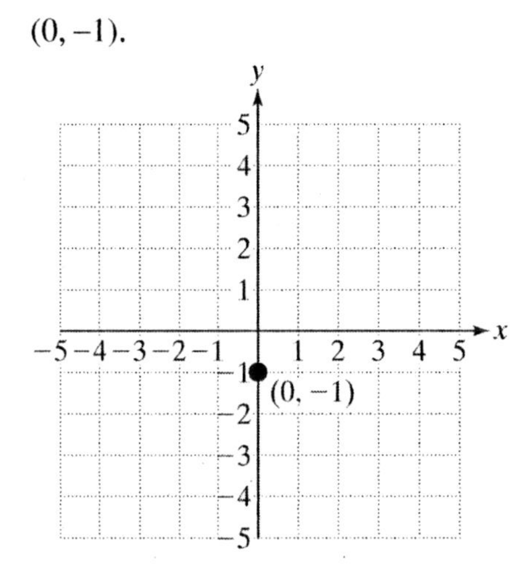

(b) The solution to two inequalities joined by the word "or" is the union of the solution sets of the two inequalities.

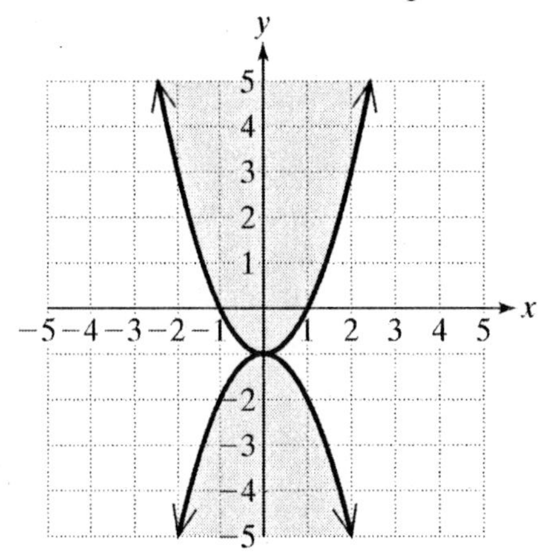

Chapter 13 Review Exercises

1. $d=\sqrt{(-6-0)^2+(3-1)^2}=\sqrt{40}=2\sqrt{10}$

3.
$$5=\sqrt{(x-2)^2+(5-9)^2}$$
$$5=\sqrt{x^2-4x+4+16}$$
$$5=\sqrt{x^2-4x+20}$$
$$25=x^2-4x+20$$
$$0=x^2-4x-5$$
$$0=(x+1)(x-5)$$
$$x+1=0 \quad \text{or} \quad x-5=0$$
$$x=-1 \quad \text{or} \quad x=5$$

5. $d_1 = \sqrt{(-2-1)^2 + (-3-3)^2} = \sqrt{45} = 3\sqrt{5};$
$d_2 = \sqrt{(1-5)^2 + (3-11)^2} = \sqrt{80} = 4\sqrt{5};$
$d_3 = \sqrt{(-2-5)^2 + (-3-11)^2} = \sqrt{245} = 7\sqrt{5};$
These three points are collinear since $d_1 + d_2 = d_3$.

7. Center $(12, 3)$, $r = 4$

9. Center $(-3, -8)$, $r = 2\sqrt{5}$

11. **(a)** $x^2 + y^2 = 64$

(b) $(x-8)^2 + (y-8)^2 = 64$

13.
$$x^2 + 4x + y^2 + 16y = -60$$
$$x^2 + 4x + 4 + y^2 + 16y + 64 = -60 + 4 + 64$$
$$(x+2)^2 + (y+8)^2 = 8$$

15.
$$x^2 - 6x + y^2 - \frac{2}{3}y = -\frac{1}{9}$$
$$x^2 - 6x + 9 + y^2 - \frac{2}{3}y + \frac{1}{9} = -\frac{1}{9} + 9 + \frac{1}{9}$$
$$(x-3)^2 + \left(y - \frac{1}{3}\right)^2 = 9$$

17. $\frac{1}{2}(6) = 3$
$x^2 + (y-2)^2 = 9$

19. $\left(\frac{x_1 + x_2}{2}, \frac{y_1 + y_2}{2}\right)$
$= \left(\frac{1.2 + (-4.1)}{2}, \frac{-3.7 + (-8.3)}{2}\right)$
$= (-1.45, -6)$

21. Horizontal axis of symmetry; parabola opens right

23. Vertical axis of symmetry; parabola opens upward

25. Vertex: $(-2, 0)$
$p = \frac{1}{4(1)} = \frac{1}{4}$

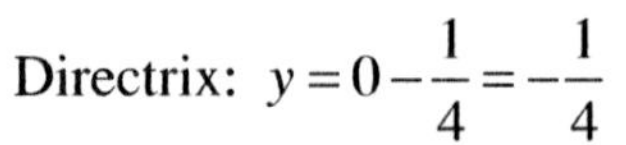

Focus: $\left(-2, 0 + \frac{1}{4}\right) = \left(-2, \frac{1}{4}\right)$

Directrix: $y = 0 - \frac{1}{4} = -\frac{1}{4}$

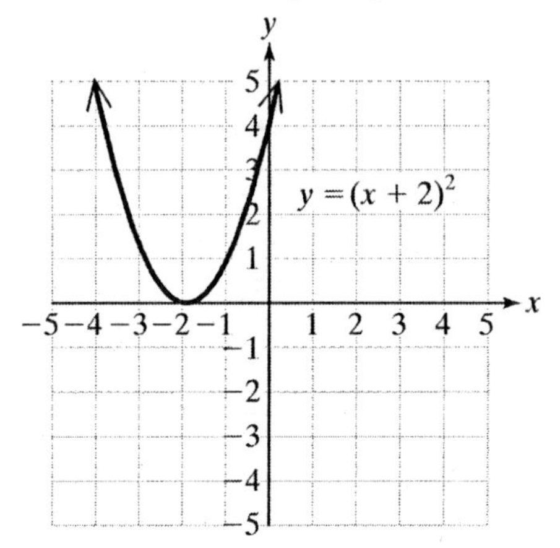

27. Vertex: $(-1, 0)$
$p = \frac{1}{4(2)} = \frac{1}{8}$

Focus: $\left(-1 + \frac{1}{8}, 0\right) = \left(-\frac{7}{8}, 0\right)$

Directrix: $x = -1 - \frac{1}{8} = -\frac{9}{8}$

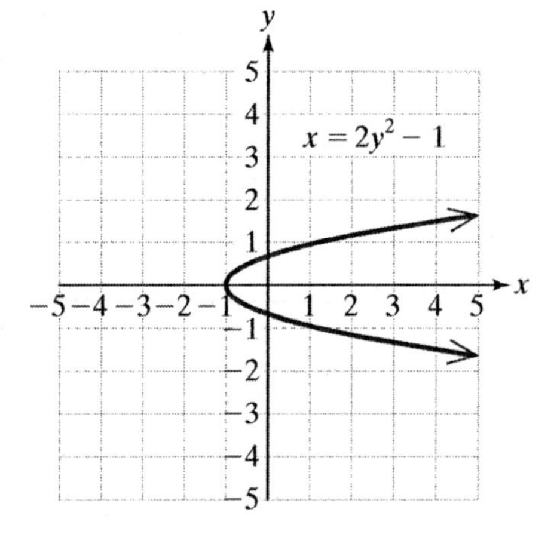

29.
$$x = (y^2 + 4y) + 2$$
$$x = (y^2 + 4y + 4 - 4) + 2$$
$$x = (y^2 + 4y + 4) - 4 + 2$$
$$x = (y+2)^2 - 2$$
Vertex: $(-2, -2)$
$p = \frac{1}{4(1)} = \frac{1}{4}$

Focus: $\left(-2+\frac{1}{4}, -2\right) = \left(-\frac{7}{4}, -2\right)$

Directrix: $x = -2 - \frac{1}{4} = -\frac{9}{4}$

31. $y = -2(x^2 + x)$

$y = -2\left(x^2 + x + \frac{1}{4} - \frac{1}{4}\right)$

$y = -2\left(x^2 + x + \frac{1}{4}\right) + \frac{1}{2}$

$y = -2\left(x + \frac{1}{2}\right)^2 + \frac{1}{2}$

Vertex: $\left(-\frac{1}{2}, \frac{1}{2}\right)$

$p = \frac{1}{4(-2)} = -\frac{1}{8}$

Focus: $\left(-\frac{1}{2}, \frac{1}{2} - \frac{1}{8}\right) = \left(-\frac{1}{2}, \frac{3}{8}\right)$

Directrix: $y = \frac{1}{2} + \frac{1}{8} = \frac{5}{8}$

33. Center: (0, 0); major axis is vertical

35. Center: (1, 0); major axis is horizontal

37.

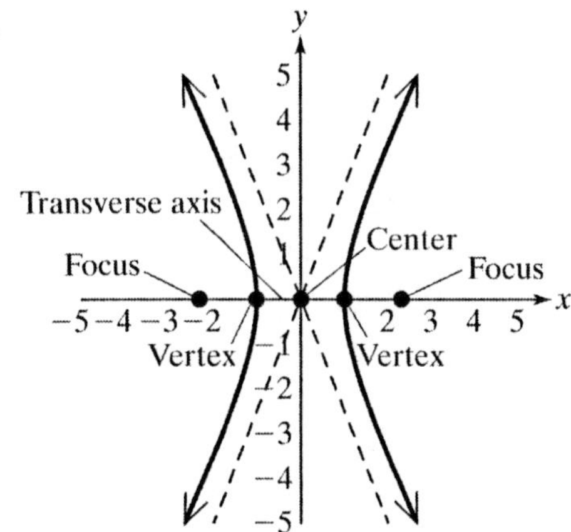

39. Center: (0, 0); transverse axis is vertical

41. Center: (–1, 0); transverse axis is horizontal

43. Ellipse

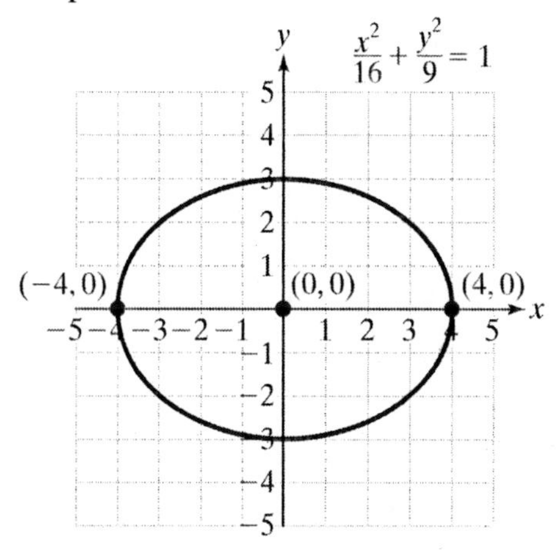

45. Hyperbola

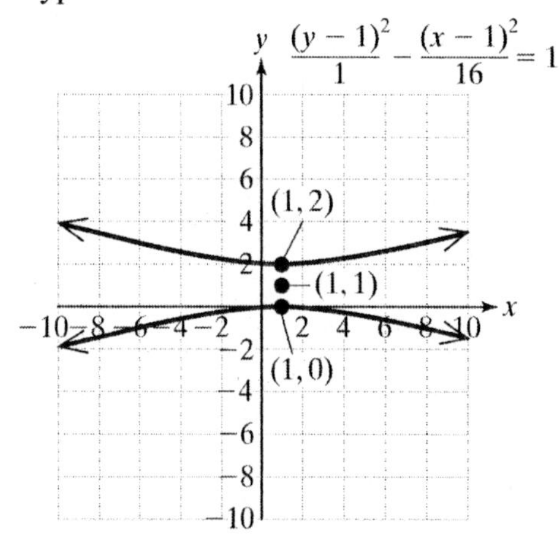

47. (a) $4x + 2y = 10$ is a line, $y = x^2 - 10$ is a parabola.

(b)

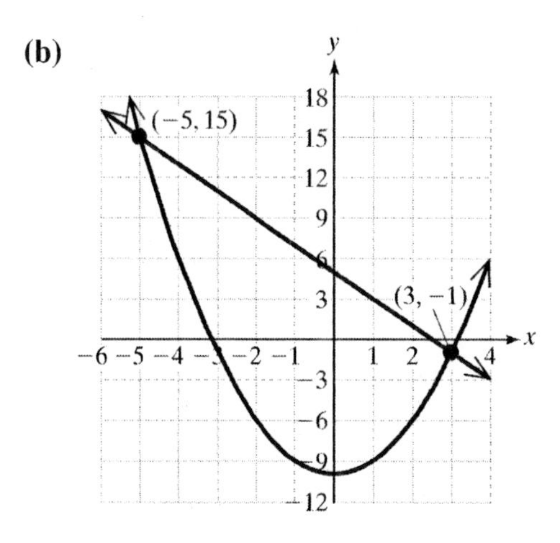

(c) Substitute x^2-10 for y in the first equation:

$$4x+2(x^2-10)=10$$
$$4x+2x^2-20=10$$
$$2x^2+4x-30=0$$
$$2(x^2+2x-15)=0$$
$$2(x+5)(x-3)=0$$
$$x+5=0 \quad \text{or} \quad x-3=0$$
$$x=-5 \quad \text{or} \quad x=3$$

Substituting $x=-5$ in $y=x^2-10$ gives

$y=(-5)^2-10=25-10=15.$

Substituting $x=3$ in $y=x^2-10$ gives

$y=(3)^2-10=9-10=-1.$

The points of intersection are: $(-5, 15)$ and $(3, -1)$.

49. (a) $x^2+y^2=16$ is a circle, $x-2y=8$ is a line.

(b)

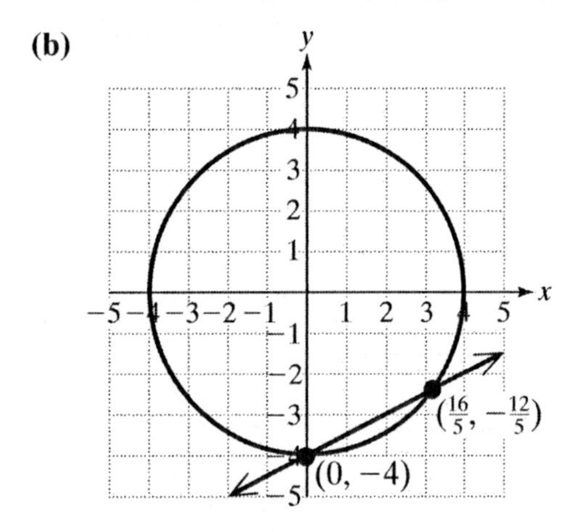

(c) Solve $x-2y=8$ for y. Substitute $\frac{1}{2}x-4$ for y in the first equation:

$$x^2+\left(\frac{1}{2}x-4\right)^2=9$$
$$x^2+\frac{1}{4}x^2-4x+16=16$$
$$\frac{5}{4}x^2-4x=0$$
$$5x^2-16x=0$$
$$x(5x-16)=0$$
$$x=0 \quad \text{or} \quad 5x-16=0$$
$$x=0 \quad \text{or} \quad x=\frac{16}{5}$$

Substituting $x=0$ in $y=\frac{1}{2}x-4$ gives

$y=\frac{1}{2}(0)-4=0-4=-4.$

Substituting $x=\frac{16}{5}$ in $y=\frac{1}{2}x-4$ gives

$y=\frac{1}{2}\left(\frac{16}{5}\right)-4=\frac{8}{5}-4=-\frac{12}{5}.$

The points of intersection are: $(0, -4)$ and $\left(\frac{16}{5}, -\frac{12}{5}\right)$.

51. Solve the second equation for y. Substitute $x+4$ for y in the first equation:

$$x^2+4(x+4)^2=29$$
$$x^2+4(x^2+8x+16)=29$$
$$x^2+4x^2+32x+64=29$$
$$5x^2+32x+35=0$$
$$(5x+7)(x+5)=0$$
$$5x+7=0 \quad \text{or} \quad x+5=0$$
$$x=-\frac{7}{5} \quad \text{or} \quad x=-5$$

Substituting $x=-\frac{7}{5}$ in $y=x+4$ gives

$y=-\frac{7}{5}+4=\frac{13}{5}.$

Substituting $x=-5$ in $y=x+4$ gives

$y=-5+4=-1.$

The points of intersection are: $\left(-\frac{7}{5}, \frac{13}{5}\right)$ and $(-5, -1)$.

53. Substitute x^2 for y in the second equation:

$$6x^2-(x^2)^2=8$$
$$6x^2-x^4-8=0$$
$$x^4-6x^2+8=0$$
$$(x^2-4)(x^2-2)=0$$
$$x^2-4=0 \quad \text{or} \quad x^2-2=0$$
$$x^2=4 \quad \text{or} \quad x^2=2$$
$$x=\pm 2 \quad \text{or} \quad x=\pm\sqrt{2}$$

Substituting $x = 2$ in $y=x^2$ gives $y = 4$.

Substituting $x = -2$ in $y=x^2$ gives $y = 4$.

Substituting $x=\sqrt{2}$ in $y=x^2$ gives $y = 2$.

Substituting $x=-\sqrt{2}$ in $y=x^2$ gives $y = 2$.

The points of intersection are: (2, 4), (–2, 4), $\left(\sqrt{2},\,2\right)$ and $\left(-\sqrt{2},\,2\right)$

55. Adding the two equations gives:

$$2x^2=72$$
$$x^2=36$$
$$x=\pm 6$$

Substituting $x = 6$ in $x^2+y^2=61$ gives:

$$36+y^2=61$$
$$y^2=25$$
$$y=\pm 5$$

Substituting $x = 6$ in $x^2+y^2=61$ gives:

$$36+y^2=61$$
$$y^2=25$$
$$y=\pm 5$$

The points of intersection are: (6, 5), (6, –5), (–6, 5), and (–6, –5).

57. The related equation $x - 2y = -2$, which we can rewrite as $y=\frac{1}{2}x+1$, is a line with y-intercept (0, 1) and slope $\frac{1}{2}$. Graph the related equation using a solid line since the symbol "≥" includes equality.

Test Point Above: (0, 2)

$$x-2y\ge -2$$
$$(0)-2(2)\ge -2$$
$$-4\ge -2 \quad \text{False}$$

Test Point Below: (0, 0)

$$x-2y\ge -2$$
$$(0)-2(0)\ge -2$$
$$0\ge -2 \quad \text{True}$$

Since the inequality is true at the test point (0, 0), shade the region below the line.

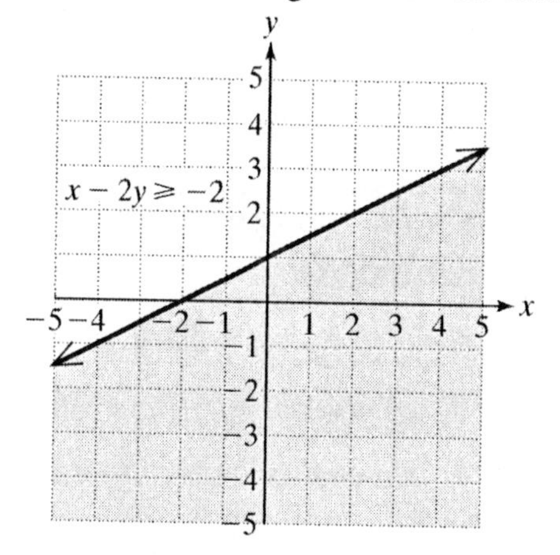

59. The related equation $\frac{x^2}{25}+\frac{y^2}{4}=1$ is an ellipse with horizontal major axis, centered at the origin. Graph the related equation using a dashed curve since the symbol ">" does not include equality.

Test Point Inside: (0, 0)

$$\frac{x^2}{25}+\frac{y^2}{4}>1$$
$$\frac{(0)^2}{25}+\frac{(0)^2}{4}>1$$
$$0+0>1$$
$$0>1 \quad \text{False}$$

Test Point Outside: (0, 3)

$$\frac{x^2}{25}+\frac{y^2}{4}>1$$
$$\frac{(0)^2}{25}+\frac{(3)^2}{4}>1$$
$$0+\frac{9}{4}>1$$
$$\frac{9}{4}>1 \quad \text{True}$$

Since the inequality is true at the test point (0, 3), shade the region outside the ellipse.

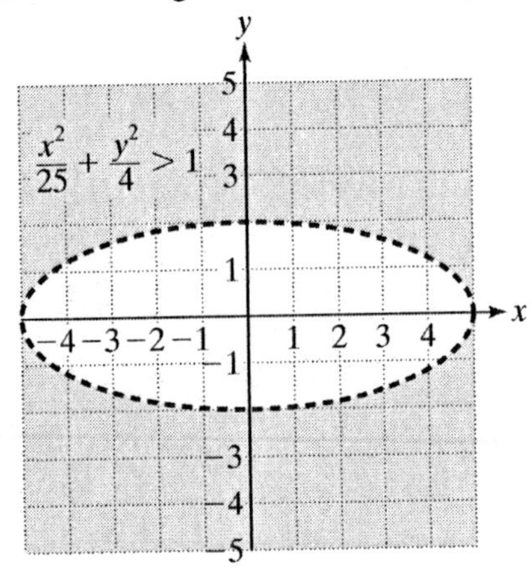

61. The related equation $(x+2)^2+(y+1)^2=4$ is a circle of radius 2, centered at (–2, –1). Graph the related equation using a solid curve since the symbol "≤" includes equality.

Test Point Inside: (–2, –1)

$$(x+2)^2+(y+1)^2\le 4$$
$$(-2+2)^2+(-1+1)^2\le 4$$
$$0^2+0^2\le 4$$
$$0\le 4 \quad \text{True}$$

Test Point Outside: (–2, 2)

$$(x+2)^2+(y+1)^2\le 4$$
$$(-2+2)^2+(2+1)^2\le 4$$
$$0^2+3^2\le 4$$
$$9\le 4 \quad \text{False}$$

Since the inequality is true at the test point (–2, –1), shade the region inside the circle.

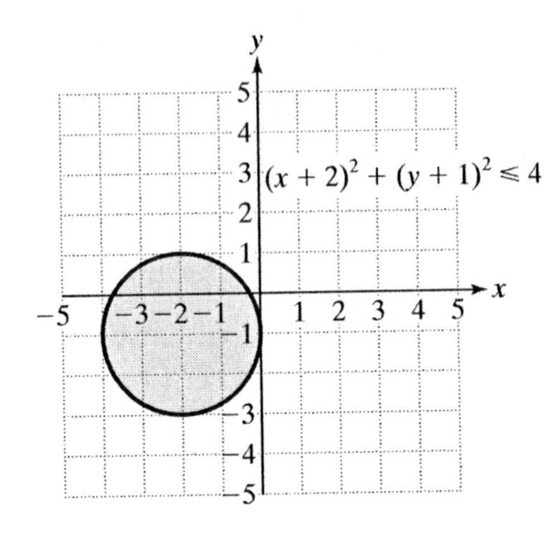

63. The related equation $y=x^2-1$ is a parabola which opens upward, with vertex (0, –1). Graph the related equation using a dashed curve since the symbol ">" does not include equality.

Test Point Above: (0, 0)

$$y>x^2-1$$
$$0>(0)^2-1$$
$$0>-1 \quad \text{True}$$

Test Point Below: (0, –2)

$$y>x^2-1$$
$$-2>(0)^2-1$$
$$-2>-1 \quad \text{False}$$

Since the inequality is true at the test point (0, 0), shade the region above the parabola.

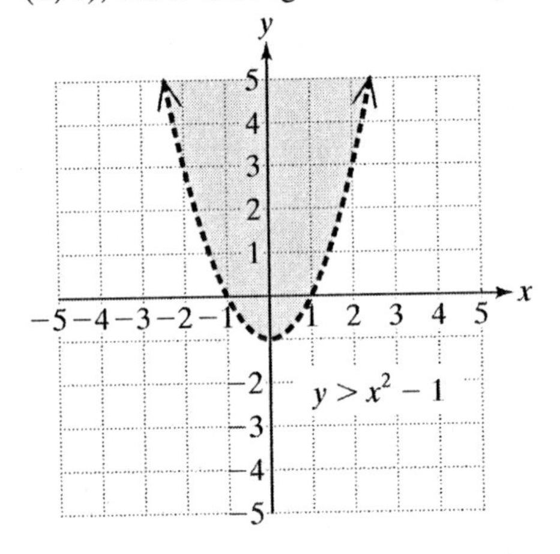

65. The solution to $y \le 3^x$ is the set of points on and below the curve $y = 3^x$.

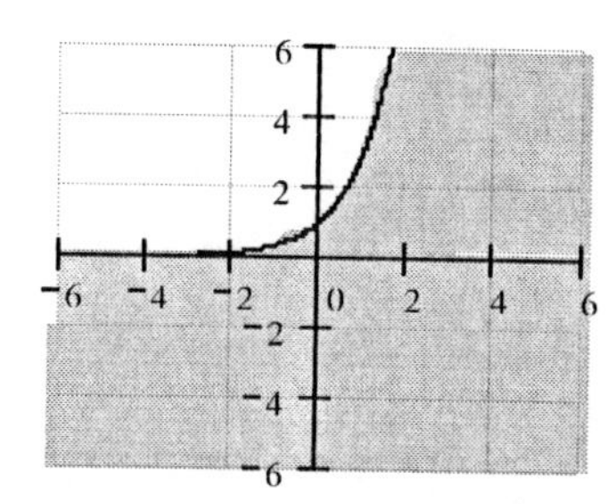

The solution to $x^2 + y^2 \le 9$ is the set of points on and inside the circle $x^2 + y^2 = 9$.

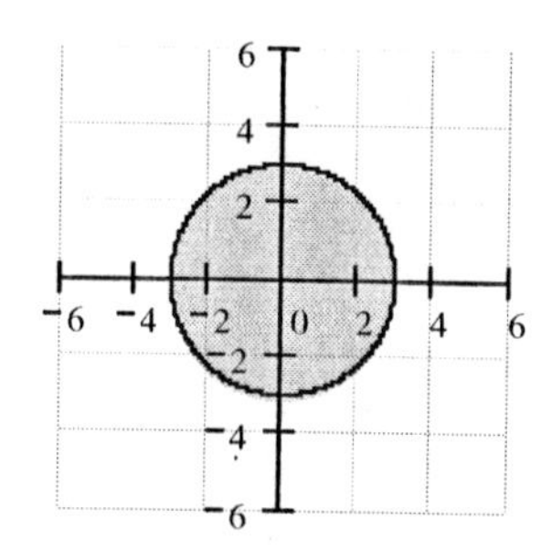

The solution to the system of inequalities is the intersection of the solution sets of the individual inequalities.

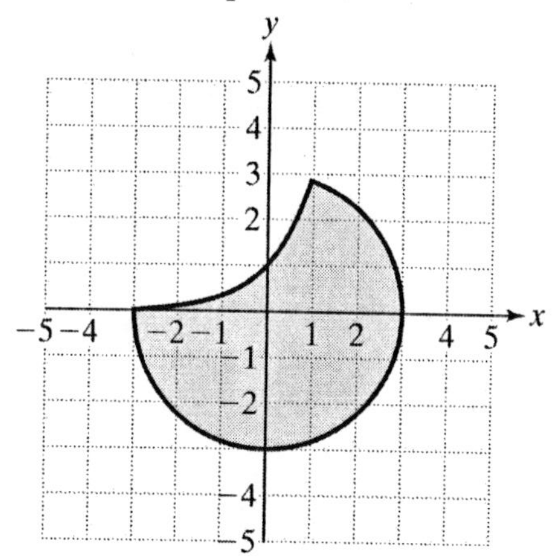

67. The solution to $\frac{y^2}{9} - x^2 \ge 1$ is the set of points on, above the upper branch of, and below the lower branch of the hyperbola $\frac{y^2}{9} - x^2 = 1$.

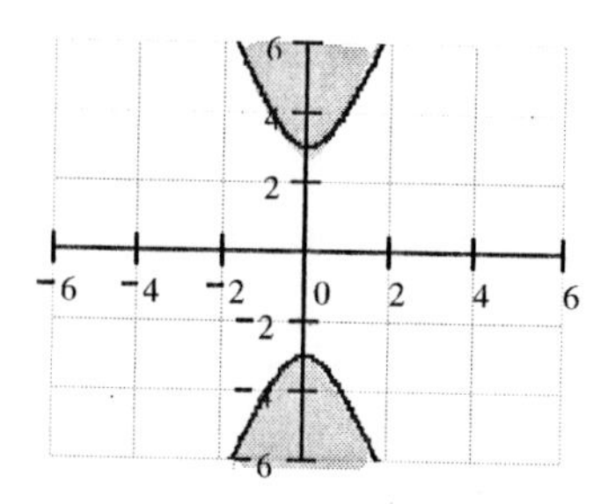

The solution to $x^2 + (y-2)^2 \le 4$ is the set of points on and inside the circle $x^2 + (y-2)^2 = 4$.

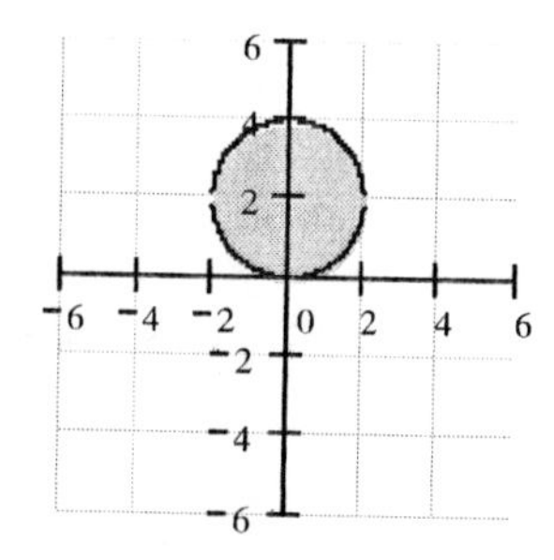

The solution to the system of inequalities is the intersection of the solution sets of the individual inequalities.

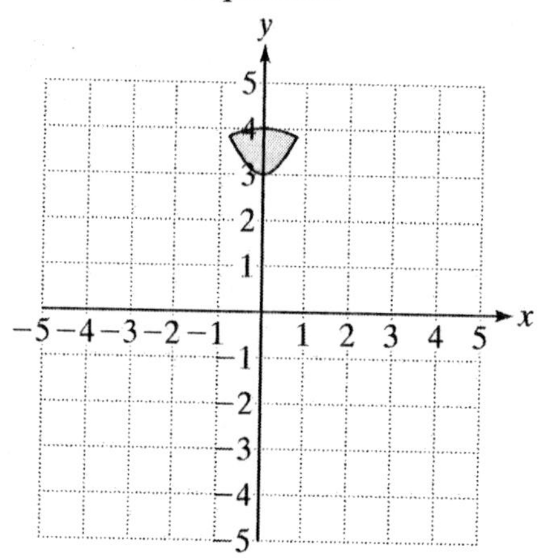

Chapter 13 Test

1. Vertex: (3, –6)

$$p = \frac{1}{4\left(-\frac{1}{4}\right)} = 1$$

Focus: (3 – 1, –6) = (2, –6)
Directrix: x = 3 + 1 = 4

3. $d = \sqrt{(5-(-2))^2 + (19-13)^2} = \sqrt{85}$

5. Center: $\left(\frac{5}{6}, -\frac{1}{3}\right), r = \frac{5}{7}$

7. **(a)** $r = \sqrt{(0-(-2))^2 + (4-5)^2} = \sqrt{5}$

(b) $x^2 + (y-4)^2 = 5$

9. Center: (3, 0); major axis is vertical

11. **(a)** Solve $4x - 3y = -12$ for y. Substitute $\frac{4}{3}x + 4$ for y in the first equation:

$$16x^2 + 9\left(\frac{4}{3}x + 4\right)^2 = 144$$
$$16x^2 + 9\left(\frac{16}{9}x^2 + \frac{32}{3}x + 16\right) = 144$$
$$16x^2 + 16x^2 + 96x + 144 = 144$$
$$32x^2 + 96x = 0$$
$$32x(x+3) = 0$$

$32x = 0$ or $x + 3 = 0$
$x = 0$ or $x = -3$

Substituting $x = 0$ in $y = \frac{4}{3}x + 4$ gives

$y = \frac{4}{3}(0) + 4 = 4.$

Substituting $x = -3$ in $y = \frac{4}{3}x + 4$ gives

$y = \frac{4}{3}(-3) + 4 = -4 + 4 = 0.$

The points of intersection are: (0, 4) and (–3, 0).
The correct graph is i.

(b) Solve $4x - 3y = -12$ for y. Substitute $\frac{4}{3}x + 4$ for y in the first equation:

$$x^2 + 4\left(\frac{4}{3}x + 4\right)^2 = 4$$
$$x^2 + 4\left(\frac{16}{9}x^2 + \frac{32}{3}x + 16\right) = 4$$
$$x^2 + \frac{64}{9}x^2 + \frac{128}{3}x + 64 = 4$$
$$\frac{73}{9}x^2 + \frac{128}{3}x + 60 = 0$$
$$73x^2 + 384x + 540 = 0$$

Using the quadratic formula, there are no real roots. Thus, there are no points of intersection.
The correct graph is ii.

13. Adding the two equations gives:

$$50x^2 = 200$$
$$x^2 = 4$$
$$x = \pm 2$$

Substituting $x = 2$ in $25x^2 + 4y^2 = 100$ gives:

$$100 + 4y^2 = 100$$
$$4y^2 = 0$$
$$y = 0$$

Substituting $x = -2$ in $25x^2 + 4y^2 = 100$ gives:

$$100 + 4y^2 = 100$$
$$4y^2 = 0$$
$$y = 0$$

The points of intersection are: (2, 0) and (–2, 0).

15. The related equation $y = -\frac{1}{3}x + 1$ is a line with y-intercept (0, 1) and slope $-\frac{1}{3}$. Graph the related equation using a solid line since the symbol "≥" includes equality.

Test Point Above: (0, 2)

$y \geq -\frac{1}{3}x+1$

$2 \geq -\frac{1}{3}(0)+1$

$2 \geq 1$ True

Test Point Below: (0, 0)

$y \geq -\frac{1}{3}x+1$

$0 \geq -\frac{1}{3}(0)+1$

$0 \geq 1$ False

Since the inequality is true at the test point (0, 2), shade the region above the line.

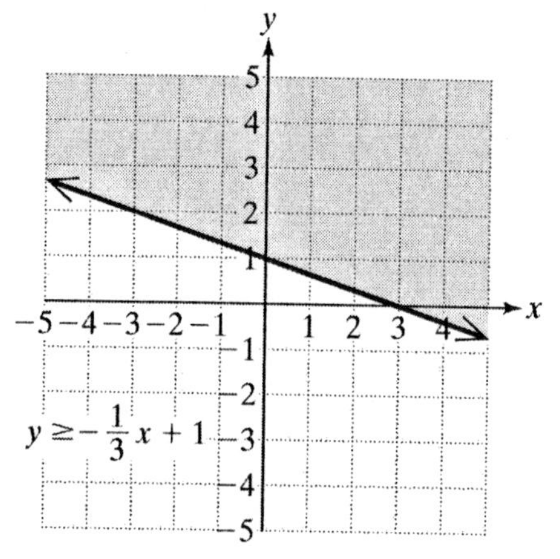

17. The solution to $y \leq \sqrt{x}$ is the set of points on or below the curve $y=\sqrt{x}$.

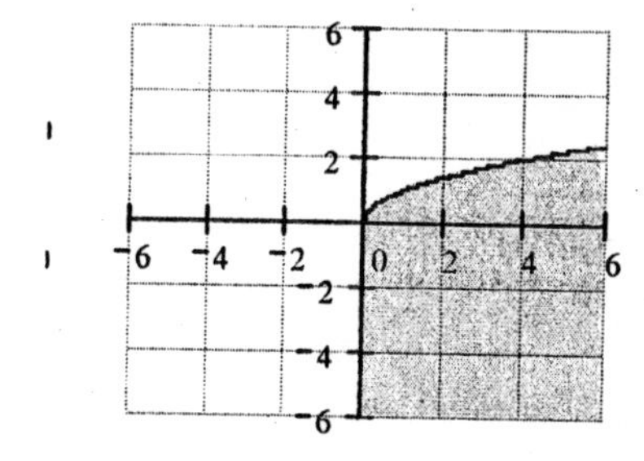

The solution to $y > x-2$ is the set of points above the line $y = x-2$.

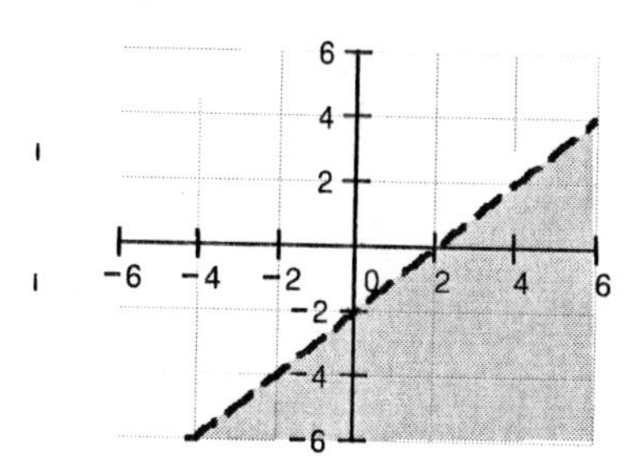

The solution to $x \geq 0$ is the set of points on and to the right of the line $x = 0$.

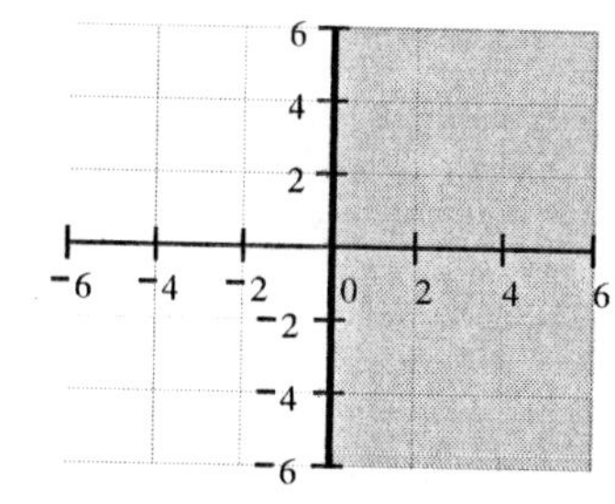

The solution to the system of inequalities is the intersection of the solution sets of the individual inequalities.

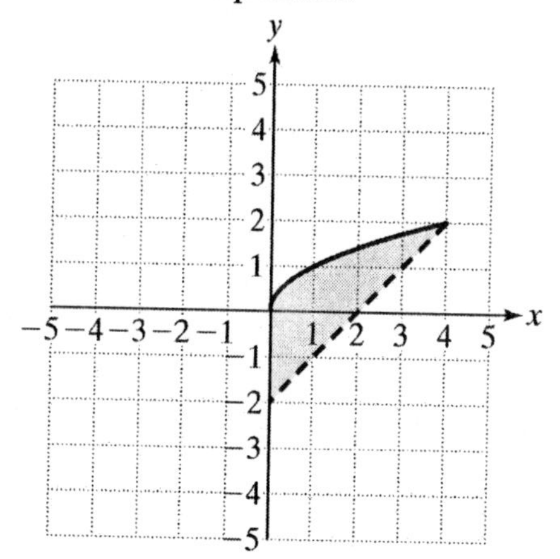

Cumulative Review Exercises, Chapters 1–13

1. $10y-5=2y-4+8y-1$

$10y-5=10y-5$

Identity. The solution is all real numbers.

3. Let x represent the first integer.
Let $2x - 5$ represent the second integer.

$$x(2x-5)=150$$
$$2x^2-5x=150$$
$$2x^2-5x-150=0$$
$$(2x+15)(x-10)=0$$
$$2x+15=0 \quad \text{or} \quad x-10=0$$
$$x=-\frac{15}{2} \quad \text{or} \quad x=10$$

$\left(-\frac{15}{2} \text{ is not an integer}\right)$ or $x = 10$

If $x = 10$, $2(10) - 5 = 15$
The two integers are 10 and 15.

5. **(a)** Using (0, 30) and (1, 40):

$$m=\frac{40-(30)}{1-0}=10$$

(b) Since the y-intercept is 30, $y = 10x + 30$.

(c) Let $x = 5$ (1990 is 5 years after 1985)
$y = 10(5) + 30 = \$80$ million

7. Multiply equation 1 by –2 to give the system:

$$\begin{cases} -2x-2y=2 \\ 2x-z=3 \\ y+2z=-1 \end{cases}$$

Add equations 1 and 2 to get $-2y - z = 5$.
The new system is:

$$\begin{cases} -2y-z=5 \\ y+2z=-1 \end{cases}$$

Now multiply the first row by 2 to eliminate z.

$$\begin{cases} -4y-2z=10 \\ y+2z=-1 \end{cases}$$

Add the equation to get $-3y = 9$.
$y = -3$
Substitute to find the remaining values.
$x + -3 = -1, x = 2;$
$2(2) - z = 3, z = 1$
The solution is (2, –3, 1).

9. $D=\begin{vmatrix} 4 & -2 \\ -3 & 5 \end{vmatrix}=20-6=14$

$$D_x=\begin{vmatrix} 7 & -2 \\ 0 & 5 \end{vmatrix}=35-0=35$$
$$D_y=\begin{vmatrix} 4 & 7 \\ -3 & 0 \end{vmatrix}=0-(-21)=21$$
$$x=\frac{D_x}{D}=\frac{35}{14}=\frac{5}{2}$$
$$y=\frac{D_y}{D}=\frac{21}{14}=\frac{3}{2}$$

The solution to the system is $\left(\frac{5}{2}, \frac{3}{2}\right)$.

11. $g(2) = 5;$
$g(8) = -1;$
$g(3) = 0;$
$g(-5) = 5$

13. **(a)** $(-2)^3+(-2)^2+(-2)+1=-5$

(b) $(x^3+x^2)+(x+1)$
$x^2(x+1)+(x+1)$
$(x+1)(x^2+1);$
$(-2+1)((-2)^2+1)$
$(-1)(5)$
-5

(c) They are the same.

15. **(a)** $24i+30i^2$
$24i+30(-1)$
$-30+24i$

(b)
$$\frac{3}{4-5i}\cdot\frac{4+5i}{4+5i}=\frac{12+15i}{16+20i-20i-25i^2}$$
$$=\frac{12+5i}{16+25}$$
$$=\frac{12+5i}{41}$$
$$=\frac{12}{41}+\frac{15}{41}i$$

17. $(5w+1)(25w^2-5w+1)=0$

$$5w=-1 \quad \text{or} \quad w=\frac{-(-5)\pm\sqrt{(-5)^2-4(25)(1)}}{2(25)}$$

$$w=-\frac{1}{5} \quad \text{or} \quad w=\frac{5\pm\sqrt{-75}}{50}$$

$$=\frac{5}{50}\pm\frac{5\sqrt{3}}{50}i$$

$$=\frac{1}{10}\pm\frac{\sqrt{3}}{10}i$$

19. $|x-9|<10$

$x-9<10$ and $x-9>-10$

$x<19$ and $x>-1$

Interval notation: $(-1, 19)$

21. $f(x)=(x^2+10x)-11$

$=(x^2+10x+25)-11-25$

$=(x+5)^2-36$

Vertex: $(-5, -36)$

23. The geometric figure would be a circle with equation: $x^2+(y-5)^2=16$.

25. Substitute $-x^2-4$ for y in the first equation:

$$x^2+(-x^2-4)^2=16$$
$$x^2+(x^4+8x^2+16)=16$$
$$x^4+9x^2+16=16$$
$$x^4+9x^2=0$$
$$x^2(x^2+9)=0$$

$x^2=0$ or $x^2+9=0$

$x=0$ or $x^2=-9$ (not real)

Substituting $x=0$ in $y=-x^2-4$ gives $y=-(0)^2-4=-4$.

The point of intersection is $(0, -4)$.

27. The solution to $y\ge\left(\frac{1}{2}\right)^x$ is the set of points on and above the curve $y=\left(\frac{1}{2}\right)^x$.

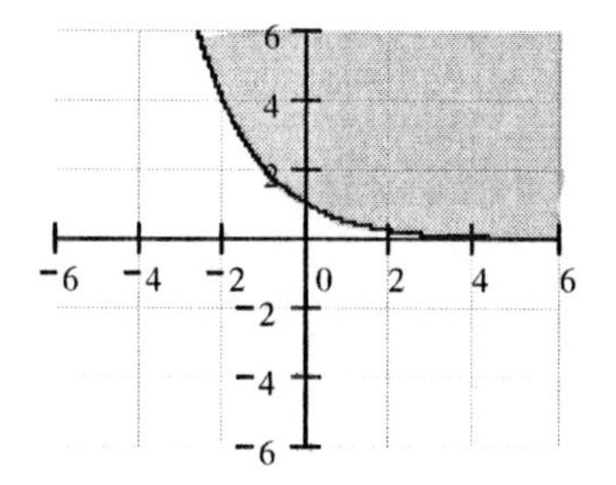

The solution to $x\le 0$ is the set of points on and to the left of the line $x=0$.

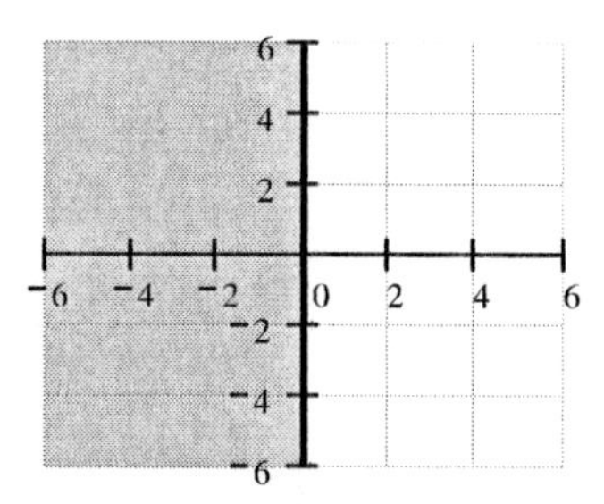

The solution to the system of inequalities is the intersection of the solution sets of the individual inequalities.

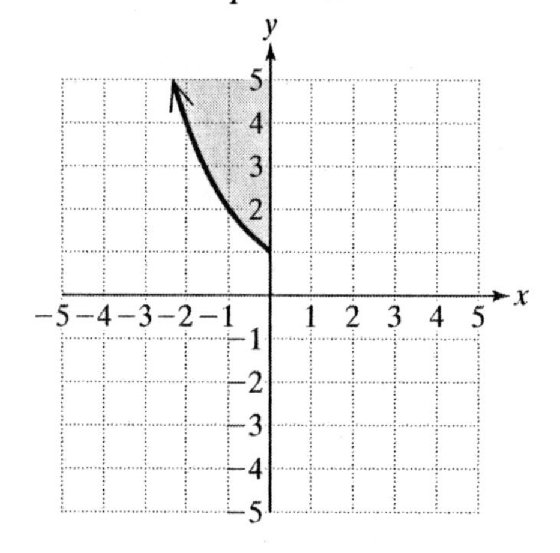

Chapter 14

Section 14.1 Practice Exercises

1. $a_1 = 3(1)+1 = 3+1 = 4$
$a_2 = 3(2)+1 = 6+1 = 7$
$a_3 = 3(3)+1 = 9+1 = 10$
$a_4 = 3(4)+1 = 12+1 = 13$

3. $a_1 = \sqrt{1+2} = \sqrt{3}$
$a_2 = \sqrt{2+2} = \sqrt{4} = 2$
$a_3 = \sqrt{3+2} = \sqrt{5}$
$a_4 = \sqrt{4+2} = \sqrt{6}$

5. $a_1 = \frac{3}{1} = 3$
$a_2 = \frac{3}{2}$
$a_3 = \frac{3}{3} = 1$
$a_4 = \frac{3}{4}$
$a_5 = \frac{3}{5}$

7. $a_1 = (-1)^1 \frac{1+1}{1+2} = -1\left(\frac{2}{3}\right) = -\frac{2}{3}$
$a_2 = (-1)^2 \frac{2+1}{2+2} = 1\left(\frac{3}{4}\right) = \frac{3}{4}$
$a_3 = (-1)^3 \frac{3+1}{3+2} = -1\left(\frac{4}{5}\right) = -\frac{4}{5}$
$a_4 = (-1)^4 \frac{4+1}{4+2} = 1\left(\frac{5}{6}\right) = \frac{5}{6}$

9. $a_1 = (-1)^{1+1}(1^2-1) = (-1)^2(1-1) = -1(0) = 0$
$a_2 = (-1)^{2+1}(2^2-1)$
$= (-1)^3(4-1)$
$= -1(3)$
$= -3$
$a_3 = (-1)^{3+1}(3^2-1) = (-1)^4(9-1) = 1(8) = 8$
$a_4 = (-1)^{4+1}(4^2-1)$
$= (-1)^5(16-1)$
$= -1(15)$
$= -15$

11. $a_2 = 3 - \frac{1}{2} = \frac{5}{2}$
$a_3 = 3 - \frac{1}{3} = \frac{8}{3}$
$a_4 = 3 - \frac{1}{4} = \frac{11}{4}$
$a_5 = 3 - \frac{1}{5} = \frac{14}{5}$

13. $a_1 = 1^2 - 1 = 1 - 1 = 0$
$a_2 = 2^2 - 2 = 4 - 2 = 2$
$a_3 = 3^2 - 3 = 9 - 3 = 6$
$a_4 = 4^2 - 4 = 16 - 4 = 12$

15. $a_1 = (-1)^1 3^1 = -1(3) = -3$
$a_2 = (-1)^2 3^2 = 1(9) = 9$
$a_3 = (-1)^3 3^3 = -1(27) = -27$
$a_4 = (-1)^4 3^4 = 1(81) = 81$

17. When n is odd, the term is negative. When n is even, the term is positive.

19. 50 + 0.02(500) = 50 + 10 = 60
50 + 0.02(500 – 60) = 50 + 8.80 = 58.80
50 + 0.02(440 – 58.80) = 50 + 7.62 = 57.62
$50 + 0.02(381.20 - 57.62) = 50 + 6.47$
$= 56.47$
$50 + 0.02(323.58 - 56.47) = 50 + 5.34$
$= 55.34$
$50 + 0.02(267.11 - 55.34) = 50 + 4.24$
$= 54.24$
\$60, \$58.80, \$57.62, \$56.47, \$55.34, \$54.24

21. 25000 = 25000
2(25000) = 50000
2(50000) = 100000
2(100000) = 200000
2(200000) = 400000
2(400000) = 800000
2(800000) = 1600000
25,000; 50,000; 100,000; 200,000; 400,000; 800,000; 1,600,000

23. For example: $a_n = 2n$

25. For example: $a_n = 2n - 1$

27. For example: $a_n = \frac{1}{n^2}$

29. For example: $a_n = (-1)^{n+1}$

31. For example: $a_n = (-1)^n 2^n$

33. For example: $a_n = \frac{3}{5^n}$

35. A sequence is an ordered list of terms. A series is the sum of the terms of a sequence.

37. $\sum_{i=1}^{4}(3i^2) = 3(1)^2 + 3(2)^2 + 3(3)^2 + 3(4)^2 = 3 + 12 + 27 + 48 = 90$

39. $\sum_{j=0}^{4}\left(\frac{1}{2}\right)^j = \left(\frac{1}{2}\right)^0 + \left(\frac{1}{2}\right)^1 + \left(\frac{1}{2}\right)^2 + \left(\frac{1}{2}\right)^3 + \left(\frac{1}{2}\right)^4 = 1 + \frac{1}{2} + \frac{1}{4} + \frac{1}{8} + \frac{1}{16} = \frac{31}{16}$

41. $\sum_{i=1}^{6} 5 = 5 + 5 + 5 + 5 + 5 + 5 = 30$

43.
$$\begin{aligned}\sum_{j=1}^{4}(-1)^j(5j) &= (-1)^1(5\cdot 1) + (-1)^2(5\cdot 2) + (-1)^3(5\cdot 3) + (-1)^4(5\cdot 4)\\ &= -1(5) + 1(10) - 1(15) + 1(20)\\ &= -5 + 10 - 15 + 20\\ &= 10\end{aligned}$$

45. $\sum_{i=1}^{4}\frac{i+1}{i} = \frac{1+1}{1} + \frac{2+1}{2} + \frac{3+1}{3} + \frac{4+1}{4} = 2 + \frac{3}{2} + \frac{4}{3} + \frac{5}{4} = \frac{73}{12}$

47. $\sum_{j=1}^{3}(j+1)(j+2) = (1+1)(1+2) + (2+1)(2+2) + (3+1)(3+2) = 2(3) + 3(4) + 4(5) = 6 + 12 + 20 = 38$

49.
$$\begin{aligned}\sum_{k=1}^{7}(-1)^k &= (-1)^1 + (-1)^2 + (-1)^3 + (-1)^4 + (-1)^5 + (-1)^6 + (-1)^7\\ &= -1 + 1 - 1 + 1 - 1 + 1 - 1\\ &= -1\end{aligned}$$

51. $\sum_{k=1}^{5} k^2 = 1^2 + 2^2 + 3^2 + 4^2 + 5^2 = 1 + 4 + 9 + 16 + 25 = 55$

53. $\sum_{n=1}^{6} n$

55. $\sum_{i=1}^{5} 4$

57. $\sum_{j=1}^{5} 4j$

59. $\sum_{k=1}^{4} (-1)^{k+1} \frac{1}{3^k}$

61. $\sum_{n=1}^{5} x^n$

63. $1 + 1\frac{1}{2} + 1\frac{1}{2} + 1\frac{1}{2} + 1\frac{1}{2} + 1\frac{1}{2} + 1\frac{1}{2} + 1\frac{1}{2} = 1 + \sum_{i=1}^{7} 1\frac{1}{2}$

The height after one week is $11\frac{1}{2}$ inches.

65. $\frac{1}{5}(10 + 15 + 12 + 18 + 22) = \frac{1}{5}(77) = 15.4$ g

67. $s = \sqrt{\frac{5(10^2 + 15^2 + 12^2 + 18^2 + 22^2) - (10 + 15 + 12 + 18 + 22)^2}{5(5-1)}}$

$s = \sqrt{\frac{5(100 + 225 + 144 + 324 + 484) - 77^2}{5(4)}}$

$s = \sqrt{\frac{5(1277) - 5929}{20}} = \sqrt{\frac{456}{20}} = 4.8$ g

69. $a_1 = -3$

$a_2 = a_1 + 5 = -3 + 5 = 2$

$a_3 = a_2 + 5 = 2 + 5 = 7$

$a_4 = a_3 + 5 = 7 + 5 = 12$

$a_5 = a_4 + 5 = 12 + 5 = 17$

$-3, 2, 7, 12, 17$

71. $a_1 = 5$
$a_2 = 4a_1 + 1 = 4(5) + 1 = 21$
$a_3 = 4a_2 + 1 = 4(21) + 1 = 84 + 1 = 85$
$a_4 = 4a_3 + 1 = 4(85) + 1 = 340 + 1 = 341$
$a_5 = 4a_4 + 1 = 4(341) + 1 = 1364 + 1 = 1365$
5, 21, 85, 341, 1365

73. 1, 1, 2, 3, 5, 8, 13, 21, 34, 55

75.
```
seq(8/n,n,1,5,1)
{8 4 2.66666666…
Ans▸Frac
 {8 4 8/3 2 8/5}
```

77.
```
seq((-1)^n*n²,n,
1,5,1)
{-1 4 -9 16 -25}
```

79.
```
sum(seq(8/n,n,1,
4,1)
       16.66666667
Ans▸Frac
              50/3
```

81.
```
sum(seq((-1)^n*n
²,n,1,100,1)
              5050
```

Section 14.2 Practice Exercises

1. A sequence is arithmetic if the difference between any term and the preceding term is constant.

3. $d = 8 - 2 = 6$

5. $d = 3 - 8 = -5$

7. $d = -11 - (-15) = -11 + 15 = 4$

9. $-3, -1, 1, 3, 5$

11. $0, \frac{1}{3}, \frac{2}{3}, 1, \frac{4}{3}$

13. $10, 4, -2, -8, -14$

15. $a_1 = 7,\ d = 12 - 7 = 5$
$a_n = 7 + (n-1)(5)$
$a_n = 7 + 5n - 5$
$a_n = 2 + 5n$

17. $a_1 = 1,\ d = -3 - 1 = -4$
$a_n = 1 + (n-1)(-4)$
$a_n = 1 - 4n + 4$
$a_n = 5 - 4n$

19. $a_1 = 1,\ d = \frac{4}{3} - 1 = \frac{1}{3}$
$a_n = 1 + (n-1)\left(\frac{1}{3}\right)$
$a_n = 1 + \frac{1}{3}n - \frac{1}{3}$
$a_n = \frac{2}{3} + \frac{1}{3}n$

21. $a_1 = 9,\ d = 6 - 9 = -3$
$a_n = 9 + (n-1)(-3)$
$a_n = 9 - 3n + 3$
$a_n = 12 - 3n$

23. $a_1 = -9,\ d = -1 - (-9) = -1 + 9 = 8$
$a_n = -9 + (n-1)(8)$
$a_n = -9 + 8n - 8$
$a_n = -17 + 8n$

25. $a_n = -3 + (n-1)(4)$
$a_n = -3 + 4n - 4$
$a_n = -7 + 4n$
$a_6 = -7 + 4(6) = -7 + 24 = 17$

27. $a_n = -1 + (n-1)(6)$
$a_n = -1 + 6n - 6$
$a_n = -7 + 6n$
$a_9 = -7 + 6(9) = -7 + 54 = 47$

29. $a_n = 1 + (n-1)(5)$
$a_n = 1 + 5n - 5$
$a_n = -4 + 5n$
$a_{10} = -4 + 5(10) = -4 + 50 = 46$

31. $a_n = 12 + (n-1)(-6)$
$a_n = 12 - 6n + 6$
$a_n = 18 - 6n$
$a_{11} = 18 - 6(11) = 18 - 66 = -48$

33. $a_1 = 8,\ d = 5,\ a_n = 98$
$a_n = a_1 + (n-1)d$
$98 = 8 + (n-1)(5)$
$90 = 5(n-1)$
$18 = n - 1$
$19 = n$

35. $a_1 = 1,\ d = 4,\ a_n = 85$
$a_n = a_1 + (n-1)d$
$85 = 1 + (n-1)(4)$
$84 = 4(n-1)$
$21 = n - 1$
$22 = n$

37. $a_1 = 2,\ d = \frac{1}{2},\ a_n = 13$
$a_n = a_1 + (n-1)d$
$13 = 2 + (n-1)\left(\frac{1}{2}\right)$
$11 = \frac{1}{2}(n-1)$
$22 = n - 1$
$n = 23$

39. $a_1 = \frac{13}{3},\ d = 2,\ a_n = \frac{73}{3}$
$a_n = a_1 + (n-1)d$
$\frac{73}{3} = \frac{13}{3} + (n-1)(2)$
$20 = 2(n-1)$
$10 = n - 1$
$11 = n$

41. $d = -11 - (-8) = -11 + 8 = -3$
2nd term $= -8 - (-3) = -8 + 3 = -5$
1st term $= -5 - (-3) = -5 + 3 = -2$
$a_1 = -2,\ a_2 = -5$

43. $a_1 = 3(1) + 2 = 3 + 2 = 5$
$a_{20} = 3(20) + 2 = 60 + 2 = 62$
$S_n = \frac{n}{2}(a_1 + a_n)$
$S_{20} = \frac{20}{2}(5 + 62) = 10(67) = 670$

45. $a_1 = 1 + 4 = 5$
$a_{20} = 20 + 4 = 24$
$S_n = \frac{n}{2}(a_1 + a_n)$
$S_{20} = \frac{20}{2}(5 + 24) = 10(29) = 290$

47. $a_1 = 4 - 1 = 3$
$a_{10} = 4 - 10 = -6$
$S_n = \frac{n}{2}(a_1 + a_n)$
$S_{10} = \frac{10}{2}(3 - 6) = 5(-3) = -15$

49. $a_1 = \frac{2}{3}(1) + 1 = \frac{5}{3}$
$a_{15} = \frac{2}{3}(15) + 1 = 11$
$S_n = \frac{n}{2}(a_1 + a_n)$
$S_{15} = \frac{15}{2}\left(\frac{5}{3} + 11\right) = \frac{15}{2}\left(\frac{38}{3}\right) = 95$

51. $a_1 = 4,\ a_n = 84,\ d = 4$
$a_n = a_1 + (n-1)d$
$84 = 4 + (n-1)(4)$
$80 = 4(n-1)$
$20 = n - 1$
$21 = n$
$S_n = \frac{n}{2}(a_1 + a_n)$
$S_{21} = \frac{21}{2}(4 + 84) = \frac{21}{2}(88) = 924$

53. $a_1 = 6,\ a_n = 34,\ d = 2$

$$a_n = a_1 + (n-1)d$$
$$34 = 6 + (n-1)(2)$$
$$28 = 2(n-1)$$
$$14 = n-1$$
$$15 = n$$
$$S_n = \frac{n}{2}(a_1 + a_n)$$
$$S_{15} = \frac{15}{2}(6+34) = \frac{15}{2}(40) = 300$$

55. $a_1 = -3,\ a_n = -39,\ d = -4$

$$a_n = a_1 + (n-1)d$$
$$-39 = -3 + (n-1)(-4)$$
$$-36 = -4(n-1)$$
$$9 = n-1$$
$$10 = n$$
$$S_n = \frac{n}{2}(a_1 + a_n)$$
$$S_{10} = \frac{10}{2}(-3 + (-39)) = 5(-42) = -210$$

57. $a_1 = 1,\ a_n = 100,\ n = 100$

$$S_n = \frac{n}{2}(a_1 + a_n)$$
$$S_{100} = \frac{100}{2}(1+100) = 50(101) = 5050$$

59. $a_1 = 1,\ a_n = 99,\ n = 50$

$$S_n = \frac{n}{2}(a_1 + a_n)$$
$$S_{50} = \frac{50}{2}(1+99) = 25(100) = 2500$$

61. $a_1 = 30,\ d = 2,\ n = 20$

$$a_n = a_1 + (n-1)d$$
$$a_n = 30 + (20-1)(2)$$
$$a_n = 30 + 19(2)$$
$$a_n = 68$$
$$S_n = \frac{n}{2}(a_1 + a_n)$$
$$S_{20} = \frac{20}{2}(30+68) = 10(98) = 980$$

980($15) = $14,700
There are 980 seats; total revenue is $14,700.

63. A sequence is geometric if the ratio between a term and the preceding term is constant.

65. $r = \frac{a_{n+1}}{a_n} = \frac{-1}{-2} = \frac{1}{2}$

67. $r = \frac{a_{n+1}}{a_n} = \frac{-12}{4} = -3$

69. $r = \frac{a_{n+1}}{a_n} = \frac{4}{1} = 4$

71. –4, 4, –4, 4, –4

73. $8,\ 2,\ \frac{1}{2},\ \frac{1}{8},\ \frac{1}{32}$

75. 2, –6, 18, –54, 162

77. $a_1 = 2,\ r = \frac{6}{2} = 3$

$$a_n = a_1 \cdot r^{n-1}$$
$$a_n = 2(3)^{n-1}$$

79. $a_1 = -6,\ r = \frac{12}{-6} = -2$

$$a_n = a_1 \cdot r^{n-1}$$
$$a_n = -6(-2)^{n-1}$$

81. $a_1 = \frac{16}{3},\ r = \frac{4}{\frac{16}{3}} = \frac{12}{16} = \frac{3}{4}$

$$a_n = a_1 \cdot r^{n-1}$$
$$a_n = \frac{16}{3}\left(\frac{3}{4}\right)^{n-1}$$

83. $a_n = -3\left(\frac{1}{2}\right)^{n-1}$

$$a_8 = -3\left(\frac{1}{2}\right)^{8-1}$$
$$= -3\left(\frac{1}{2}\right)^{7}$$
$$= -3\left(\frac{1}{128}\right)$$
$$= -\frac{3}{128}$$

85. $a_n = 6\left(-\frac{1}{3}\right)^{n-1}$

$$a_6 = 6\left(-\frac{1}{3}\right)^{6-1}$$
$$= 6\left(-\frac{1}{3}\right)^{5}$$
$$= 6\left(-\frac{1}{243}\right)$$
$$= -\frac{2}{81}$$

87. $a_n = 5(3)^{n-1}$

$$a_4 = 5(3)^{4-1} = 5(3)^3 = 5(27) = 135$$

89. $a_6 = \frac{5}{16}, r = -\frac{1}{2}$

$$a_n = a_1 \cdot r^{n-1}$$
$$\frac{5}{16} = a_1\left(-\frac{1}{2}\right)^{6-1}$$
$$\frac{5}{16} = a_1\left(-\frac{1}{2}\right)^{5}$$
$$\frac{5}{16} = a_1\left(-\frac{1}{32}\right)$$
$$\frac{5}{16}\left(-\frac{32}{1}\right) = a_1$$
$$-10 = a_1$$

91. $a_6 = 27,\ r = 3$

$$a_n = a_1 \cdot r^{n-1}$$
$$27 = a_1(3)^{6-1}$$
$$27 = a_1(3)^5$$
$$27 = a_1(243)$$
$$\frac{1}{9} = a_1$$

93. $a_2 = \frac{1}{3}, a_3 = \frac{1}{9}, r = \frac{\frac{1}{9}}{\frac{1}{3}} = \frac{1}{3},\ n = 3$

$$a_n = a_1 \cdot r^{n-1}$$
$$\frac{1}{9} = a_1\left(\frac{1}{3}\right)^{3-1}$$
$$\frac{1}{9} = a_1\left(\frac{1}{3}\right)^{2}$$
$$\frac{1}{9} = \frac{1}{9}a_1$$
$$1 = a_1$$

95. $a_1 = 10, r = \frac{2}{10} = \frac{1}{5},\ n = 5$

$$S_n = \frac{a_1(1-r^n)}{1-r}$$
$$S_5 = \frac{10\left(1-\left(\frac{1}{5}\right)^5\right)}{1-\frac{1}{5}}$$
$$= \frac{10\left(1-\frac{1}{3125}\right)}{\frac{4}{5}}$$
$$= 10\left(\frac{3124}{3125}\right)\left(\frac{5}{4}\right)$$
$$= \frac{1562}{125}$$

97. $a_1 = -2,\ r = \frac{1}{-2} = -\frac{1}{2},\ n = 5$

$$S_n = \frac{a_1(1-r^n)}{1-r}$$

$$S_5 = \frac{-2\left(1-\left(-\frac{1}{2}\right)^5\right)}{1+\frac{1}{2}} = \frac{-2\left(1+\frac{1}{32}\right)}{\frac{3}{2}} = -2\left(\frac{33}{32}\right)\left(\frac{2}{3}\right) = -\frac{33}{24} = -\frac{11}{8}$$

99. $a_1 = 12,\ r = \frac{16}{12} = \frac{4}{3},\ n = 5$

$$S_n = \frac{a_1(1-r^n)}{1-r}$$

$$S_5 = \frac{12\left(1-\left(\frac{4}{3}\right)^5\right)}{1-\frac{4}{3}} = \frac{12\left(1-\frac{1024}{243}\right)}{-\frac{1}{3}} = 12\left(\frac{-781}{243}\right)\left(-\frac{3}{1}\right) = \frac{3124}{27}$$

101. $a_1 = 1,\ r = \frac{\frac{2}{3}}{1} = \frac{2}{3},\ a_n = \frac{64}{729}$

$$a_n = a_1 \cdot r^{n-1}$$

$$\frac{64}{729} = 1\left(\frac{2}{3}\right)^{n-1}$$

$$\left(\frac{2}{3}\right)^6 = \left(\frac{2}{3}\right)^{n-1}$$

$$6 = n-1$$

$$7 = n$$

$$S_n = \frac{a_1(1-r^n)}{1-r}$$

$$S_7 = \frac{1\left(1-\left(\frac{2}{3}\right)^7\right)}{1-\frac{2}{3}} = \frac{1\left(1-\frac{128}{2187}\right)}{\frac{1}{3}} = \left(\frac{2059}{2187}\right)\left(\frac{3}{1}\right) = \frac{2059}{729}$$

103. $a_1 = -4,\ r = \frac{8}{-4} = -2,\ a_n = -256$

$$a_n = a_1 \cdot r^{n-1}$$

$$-256 = -4(-2)^{n-1}$$

$$64 = (-2)^{n-1}$$

$$(-2)^6 = (-2)^{n-1}$$

$$6 = n-1$$

$$7 = n$$

$$S_n = \frac{a_1(1-r^n)}{1-r}$$

$$S_7 = \frac{-4(1-(-2)^7)}{1-(-2)} = \frac{-4(1+128)}{3} = \frac{-4(129)}{3} = -172$$

105. (a) $a_n = 1000(1.05)^n$

$a_1 = 1000(1.05)^1 = 1000(1.05) = \$1050$

$a_2 = 1000(1.05)^2 = 1000(1.1025) = \$1102.50$

$a_3 = 1000(1.05)^3 = 1000(1.57625) = \$1157.63$

$a_4 = 1000(1.05)^4 = 1000(1.21551) = \$1215.51$

(b) $a_{10} = 1000(1.05)^{10}$
$= 1000(1.62889)$
$= \$1628.89$

$a_{20} = 1000(1.05)^{20}$
$= 1000(2.65330)$
$= \$2653.30$

$a_{40} = 1000(1.05)^{40}$
$= 1000(7.03999)$
$= \$7039.99$

107. $r = \frac{\frac{1}{6}}{1} = \frac{1}{6}$

$S = \frac{a_1}{1-r} = \frac{1}{1-\frac{1}{6}} = \frac{1}{\frac{5}{6}} = \frac{6}{5}$

109. $r = \frac{1}{-3} = -\frac{1}{3}$

$S = \frac{a_1}{1-r} = \frac{-3}{1+\frac{1}{3}} = \frac{-3}{\frac{4}{3}} = -\frac{9}{4}$

111. $r = \frac{-1}{\frac{2}{3}} = -\frac{3}{2}$

Sum does not exist because $|r| > 1$.

113. $r = \frac{200(0.75)}{200} = 0.75$

$S = \frac{a_1}{1-r} = \frac{200}{1-0.75} = \frac{200}{0.25} = \$800$ million

115. $r = \frac{2\left(\frac{3}{4}\right)^2(4)}{2\left(\frac{3}{4}\right)(4)} = \frac{3}{4}$

$S = \frac{a_1}{1-r} = \frac{2\left(\frac{3}{4}\right)(4)}{1-\frac{3}{4}} = \frac{6}{\frac{1}{4}} = 24$

24 + 4 = 28 feet

117. (a) $a_1 = \frac{7}{10}$

(b) $r = \frac{\frac{7}{100}}{\frac{7}{10}} = \frac{1}{10}$

(c) $S = \frac{a_1}{1-r} = \frac{\frac{7}{10}}{1-\frac{1}{10}} = \frac{\frac{7}{10}}{\frac{9}{10}} = \frac{7}{9}$

119. Geometric; $r = \frac{-\frac{3}{2}}{1} = -\frac{3}{2}$

121. Arithmetic; $d = \frac{1}{3} - \left(-\frac{1}{3}\right) = \frac{1}{3} + \frac{1}{3} = \frac{2}{3}$

123. Geometric; $r = \frac{6}{2} = 3$

125. Arithmetic; $d = 6 - 2 = 4$

127. Neither

129. Neither

131. Geometric; $r = \frac{1}{-1} = -1$

133. (a) $r = \frac{a_{n+1}}{a_1}$

$r = \frac{40000(1.04)}{40000} = 1.04$

$S_n = \frac{a_1(1-r^n)}{1-r}$

$S_{20} = \frac{40000(1-1.04^{20})}{1-1.04}$
$= \frac{40000(-1.191123)}{-0.04}$
$= \$1,191,123$

(b) $r = \frac{a_{n+1}}{a_1}$

$r = \frac{40000(1.045)}{40000} = 1.045$

$S_n = \frac{a_1(1-r^n)}{1-r}$

$S_{20} = \frac{40000(1-1.045^{20})}{1-1.045}$
$= \frac{40000(-1.411714)}{-0.045}$
$= \$1,254,857$

(c) $1,254,857 – $1,191,123 = $63,734

135.

```
seq(125000*1.04^
n,n,1,4,1)
{130000 135200 ...
```

137.

```
sum(seq(n,n,1,15
,1)
             120
```

139.

```
sum(seq(4*(1/2)^
(n-1),n,1,10,1)
       7.9921875
sum(seq(4*(1/2)^
(n-1),n,1,20,1)
     7.999992371
```

Section 14.3 Practice Exercises

1. $(x+y)^4 = x^4 + 4x^3y + 6x^2y^2 + 4xy^3 + y^4$

3. $(4+p)^3 = 4^3 + 3(4)^2p + 3(4)p^2 + p^3 = 64 + 48p + 12p^2 + p^3$

5. $(a^2+b)^6 = (a^2)^6 + 6(a^2)^5b + 15(a^2)^4b^2 + 20(a^2)^3b^3 + 15(a^2)^2b^4 + 6(a^2)b^5 + b^6$
$= a^{12} + 6a^{10}b + 15a^8b^2 + 20a^6b^3 + 15a^4b^4 + 6a^2b^5 + b^6$

7. $(p^2-w)^3 = (p^2+(-w))^3$
$= (p^2)^3 + 3(p^2)^2(-w) + 3(p^2)(-w)^2 + (-w)^3$
$= p^6 - 3p^4w + 3p^2w^2 - w^3$

9. The signs alternate on the terms of the expression $(a-b)^n$. The signs for the expression $(a+b)^n$ are all positive.

11. $3! = 3 \cdot 2 \cdot 1 = 6$

13. $1! = 1$

15. False

17. $9! = 9 \cdot (8 \cdot 7 \cdot 6 \cdot 5 \cdot 4 \cdot 3 \cdot 2 \cdot 1) = 9 \cdot 8!$

19. $\frac{8!}{4!} = \frac{8 \cdot 7 \cdot 6 \cdot 5 \cdot 4!}{4!} = 8 \cdot 7 \cdot 6 \cdot 5 = 1680$

21. $\frac{3!}{0!} = \frac{3 \cdot 2 \cdot 1}{1} = 6$

23. $\frac{8!}{3!5!} = \frac{8 \cdot 7 \cdot 6 \cdot 5!}{3 \cdot 2 \cdot 1 \cdot 5!} = \frac{8 \cdot 7 \cdot 6}{3 \cdot 2 \cdot 1} = \frac{336}{6} = 56$

25. $\frac{4!}{0!4!} = \frac{4!}{1 \cdot 4!} = 1$

27. $(s+t)^6 = \frac{6!}{6!0!}s^6 + \frac{6!}{5!1!}s^5t + \frac{6!}{4!2!}s^4t^2 + \frac{6!}{3!3!}s^3t^3 + \frac{6!}{2!4!}s^2t^4 + \frac{6!}{1!5!}s^1t^5 + \frac{6!}{0!6!}t^6$
$= s^6 + 6s^5t + 15s^4t^2 + 20s^3t^3 + 15s^2t^4 + 6st^5 + t^6$

29. $(b-3)^3 = (b+(-3))^3$
$= \frac{3!}{3!0!}b^3 + \frac{3!}{2!1!}b^2(-3) + \frac{3!}{1!2!}b(-3)^2 + \frac{3!}{0!3!}(-3)^3$
$= b^3 + 3b^2(-3) + 3b(9) + (-27)$
$= b^3 - 9b^2 + 27b - 27$

31. $(2x+y)^4 = \frac{4!}{4!0!}(2x)^4 + \frac{4!}{3!1!}(2x)^3y + \frac{4!}{2!2!}(2x)^2y^2 + \frac{4!}{1!3!}(2x)y^3 + \frac{4!}{0!4!}y^4$
$= 16x^4 + 4 \cdot 8x^3y + 6 \cdot 4x^2y^2 + 4 \cdot 2xy^3 + y^4$
$= 16x^4 + 32x^3y + 24x^2y^2 + 8xy^3 + y^4$

33. $(c^2-d)^7 = (c^2+(-d))^7$
$= \frac{7!}{7!0!}(c^2)^7 + \frac{7!}{6!1!}(c^2)^6(-d) + \frac{7!}{5!2!}(c^2)^5(-d)^2 + \frac{7!}{4!3!}(c^2)^4(-d)^3 + \frac{7!}{3!4!}(c^2)^3(-d)^4$
$+ \frac{7!}{2!5!}(c^2)^2(-d)^5 + \frac{7!}{1!6!}(c^2)(-d)^6 + \frac{7!}{0!7!}(-d)^7$
$= c^{14} - 7c^{12}d + 21c^{10}d^2 - 35c^8d^3 + 35c^6d^4 - 21c^4d^5 + 7c^2d^6 - d^7$

35. $\left(\frac{a}{2} - b\right)^5$
$= \left(\frac{a}{2} + (-b)\right)^5$
$= \frac{5!}{5!0!}\left(\frac{a}{2}\right)^5 + \frac{5!}{4!1!}\left(\frac{a}{2}\right)^4(-b) + \frac{5!}{3!2!}\left(\frac{a}{2}\right)^3(-b)^2 + \frac{5!}{2!3!}\left(\frac{a}{2}\right)^2(-b)^3 + \frac{5!}{1!4!}\left(\frac{a}{2}\right)(-b)^4 + \frac{5!}{0!5!}(-b)^5$
$= \frac{a^5}{32} + 5\left(\frac{a^4}{16}\right)(-b) + 10\left(\frac{a^3}{8}\right)(-b)^2 + 10\left(\frac{a^2}{4}\right)(-b)^3 + 5\left(\frac{a}{2}\right)(-b)^4 + (-b)^5$
$= \frac{1}{32}a^5 - \frac{5}{16}a^4b + \frac{5}{4}a^3b^2 - \frac{5}{2}a^2b^3 + \frac{5}{2}ab^4 - b^5$

37. $(m-n)^{11} = (m+(-n))^{11}$
$\frac{11!}{11!0!}m^{11} + \frac{11!}{10!1!}m^{10}(-n) + \frac{11!}{9!2!}m^9(-n)^2 = m^{11} - 11m^{10}n + 55m^9n^2$

39.

$$(u^2 - v)^{12} = (u^2 + (-v))^{12}$$

$$\frac{12!}{12!0!}(u^2)^{12} + \frac{12!}{11!1!}(u^2)^{11}(-v) + \frac{12!}{10!2!}(u^2)^{10}(-v)^2 = u^{24} - 12u^{22}v + 66u^{20}v^2$$

41. 9 terms

43. $\frac{11!}{6!5!}m^6(-n)^5 = -462m^6n^5$

45. $\frac{12!}{8!4!}(u^2)^8(-v)^4 = 495u^{16}v^4$

47. $\frac{10!}{0!10!}g^9 = g^9$

Section 14.4 Practice Exercises

1. $6! = 6 \cdot 5 \cdot 4 \cdot 3 \cdot 2 \cdot 1 = 720$

3. $0! = 1$

5. $_{10}P_3 = \frac{10!}{(10-3)!} = \frac{10!}{7!} = 10 \cdot 9 \cdot 8 = 720$

There are 720 ways in which 3 items can be selected from 10 items in a specified order.

7. $_{10}C_3 = \frac{10!}{(10-3)!3!} = \frac{10!}{7!3!} = \frac{10 \cdot 9 \cdot 8}{3 \cdot 2 \cdot 1} = 120$

There are 120 ways in which 3 items can be selected from 10 items in *no* specific order.

9.

$$\begin{aligned} _{12}P_9 &= \frac{12!}{(12-9)!} \\ &= \frac{12!}{3!} \\ &= 12 \cdot 11 \cdot 10 \cdot 9 \cdot 8 \cdot 7 \cdot 6 \cdot 5 \cdot 4 \\ &= 79{,}833{,}600 \end{aligned}$$

11. $_{12}C_9 = \frac{12!}{(12-9)!9!} = \frac{12!}{3!9!} = \frac{12 \cdot 11 \cdot 10}{3 \cdot 2 \cdot 1} = 220$

13. $_7P_1 = \frac{7!}{(7-1)!} = \frac{7!}{6!} = 7$

15. $_7C_1 = \frac{7!}{(7-1)!1!} = \frac{7!}{6!1!} = 7$

17.

$$\begin{aligned} _8P_8 &= \frac{8!}{(8-8)!} \\ &= \frac{8!}{0!} \\ &= 8 \cdot 7 \cdot 6 \cdot 5 \cdot 4 \cdot 3 \cdot 2 \cdot 1 \\ &= 40{,}320 \end{aligned}$$

19. $_8C_8 = \frac{8!}{(8-8)!8!} = \frac{8!}{0!8!} = 1$

21. **(a)** AB, BA, AC, CA, BC, CB

(b) AB, AC, BC

23. $6! = 6 \cdot 5 \cdot 4 \cdot 3 \cdot 2 \cdot 1 = 720$

25. $10^6 = 1{,}000{,}000$

27. $4 \cdot 3 \cdot 8 \cdot 6 = 576$

29. $_{10}P_3 = \frac{10!}{(10-3)!} = \frac{10!}{7!} = 10 \cdot 9 \cdot 8 = 720$

31.

$$\begin{aligned} _{11}C_6 &= \frac{11!}{(11-6)!6!} \\ &= \frac{11!}{5!6!} \\ &= \frac{11 \cdot 10 \cdot 9 \cdot 8 \cdot 7}{5 \cdot 4 \cdot 3 \cdot 2 \cdot 1} \\ &= 462 \end{aligned}$$

33.

$$\begin{aligned} _{10}C_5 &= \frac{10!}{(10-5)!5!} \\ &= \frac{10!}{5!5!} \\ &= \frac{10 \cdot 9 \cdot 8 \cdot 7 \cdot 6}{5 \cdot 4 \cdot 3 \cdot 2 \cdot 1} \\ &= 252 \end{aligned}$$

35. $_8P_2 = \frac{8!}{(8-2)!} = \frac{8!}{6!} = 8 \cdot 7 = 56$

37. $2 \cdot 25 \cdot 24 \cdot 23 = 27{,}600$

39. ${}_{10}C_4 = \dfrac{10!}{(10-4)!4!} = \dfrac{10!}{6!4!} = \dfrac{10 \cdot 9 \cdot 8 \cdot 7}{4 \cdot 3 \cdot 2 \cdot 1} = 210$

41. $3 \cdot 3 \cdot 2 \cdot 2 \cdot 1 \cdot 1 = 36$

43.
$$\begin{aligned} {}_9P_9 &= \frac{9!}{(9-9)!} \\ &= \frac{9!}{0!} \\ &= 9 \cdot 8 \cdot 7 \cdot 6 \cdot 5 \cdot 4 \cdot 3 \cdot 2 \cdot 1 \\ &= 362{,}880 \end{aligned}$$

45.
$$\begin{aligned} {}_{10}C_3 \cdot {}_8C_1 &= \frac{10!}{(10-3)!3!} \cdot \frac{8!}{(8-1)!1!} \\ &= \frac{10!}{7!3!} \cdot \frac{8!}{7!1!} \\ &= \frac{10 \cdot 9 \cdot 8}{3 \cdot 2 \cdot 1} \cdot 8 \\ &= 960 \end{aligned}$$

47. $2^3 = 8$

49.
$$\begin{aligned} &{}_{40}C_{12} \\ &= \frac{40!}{(40-12)!12!} \\ &= \frac{40!}{28! \cdot 12!} \\ &= \frac{40 \cdot 39 \cdot 38 \cdot 37 \cdot 36 \cdot 35 \cdot 34 \cdot 32 \cdot 31 \cdot 29}{12 \cdot 11 \cdot 10 \cdot 9 \cdot 8 \cdot 7 \cdot 6 \cdot 5 \cdot 4 \cdot 3 \cdot 2 \cdot 1} \\ &= 5{,}586{,}853{,}480 \end{aligned}$$

51. **(a)** ${}_{10}C_2 = \dfrac{10!}{(10-2)!2!} = \dfrac{10!}{8!\,2!} = \dfrac{10 \cdot 9}{2 \cdot 1} = 45$

(b) ${}_6C_2 = \dfrac{6!}{(6-2)!2!} = \dfrac{6!}{4!\,2!} = \dfrac{6 \cdot 5}{2 \cdot 1} = 15$

(c) ${}_4C_2 = \dfrac{4!}{(4-2)!2!} = \dfrac{4!}{2!\,2!} = \dfrac{4 \cdot 3}{2 \cdot 1} = 6$

(d)
$$\begin{aligned} {}_6C_1 \cdot {}_4C_1 &= \frac{6!}{(6-1)!1!} \cdot \frac{4!}{(4-1)!1!} \\ &= \frac{6!}{5!\,1!} \cdot \frac{4!}{3!\,1!} \\ &= 6 \cdot 4 \\ &= 24 \end{aligned}$$

53. $2^7 = 128$

55. **(a)** $9! = 9 \cdot 8 \cdot 7 \cdot 6 \cdot 5 \cdot 4 \cdot 3 \cdot 2 \cdot 1$
$= 362{,}880$

(b) $5 \cdot 4 \cdot 7! = 20 \cdot 5040 = 100{,}800$

(c) $5!4! = 120 \cdot 24 = 2{,}880$

57. **(a)** pot, pto, opt, otp, top, tpo

(b) tot, tto, ott

59. **(a)** $7^3 = 343$

(b) $7 \cdot 6 \cdot 5 = 210$

(c) $7 \cdot 7 \cdot 4 = 196$

61.
$$\begin{aligned} {}_{26}C_5 &= \frac{26!}{(26-5)!5!} \\ &= \frac{26!}{21!\,5!} \\ &= \frac{26 \cdot 25 \cdot 24 \cdot 23 \cdot 22}{5 \cdot 4 \cdot 3 \cdot 2 \cdot 1} \\ &= 65{,}780 \end{aligned}$$

63.
$$\begin{aligned} {}_{13}C_3 \cdot {}_{13}C_2 &= \frac{13!}{(13-3)!3!} \cdot \frac{13!}{(13-2)!2!} \\ &= \frac{13!}{10!\,3!} \cdot \frac{13!}{11!\,2!} \\ &= \frac{13 \cdot 12 \cdot 11}{3 \cdot 2 \cdot 1} \cdot \frac{13 \cdot 12}{2 \cdot 1} \\ &= 286 \cdot 78 \\ &= 22{,}308 \end{aligned}$$

Section 14.5 Practice Exercises

1. $_{10}C_4 = \frac{10!}{(10-4)!4!} = \frac{10!}{6!4!} = \frac{10 \cdot 9 \cdot 8 \cdot 7}{4 \cdot 3 \cdot 2 \cdot 1} = 210$

3. $5! = 5 \cdot 4 \cdot 3 \cdot 2 \cdot 1 = 120$

5. a, b, d, g, h

7. $\frac{3}{6} = \frac{1}{2}$

9. (a) $\frac{8}{59}$

(b) $\frac{15}{59}$

11. (a) $\frac{10}{80} = \frac{1}{8}$

(b) $\frac{60}{80} = \frac{3}{4}$

(c) $\frac{50}{80} = \frac{5}{8}$

(d) $\frac{22}{80} = \frac{11}{40}$

13. (a) $\frac{_3C_2}{_8C_2} = \frac{\frac{3!}{(3-2)!2!}}{\frac{8!}{(8-2)!2!}} = \frac{3}{28}$

(b) $\frac{_5C_2}{_8C_2} = \frac{\frac{5!}{(5-2)!2!}}{\frac{8!}{(8-2)!2!}} = \frac{10}{28} = \frac{5}{14}$

15. (a) $\frac{1}{_{39}C_5} = \frac{1}{\frac{39!}{(39-5)!5!}}$

$= \frac{1}{\frac{39 \cdot 38 \cdot 37 \cdot 36 \cdot 35}{5 \cdot 4 \cdot 3 \cdot 2 \cdot 1}}$

$= \frac{1}{575{,}757}$

(b) $\frac{575{,}756}{575{,}757}$

(c) The events of winning the grand prize and losing are not equally likely events.

17. $\frac{27}{130}$

19. $\frac{27}{130}$

21. $\frac{112}{130} = \frac{56}{65}$

23. $\frac{120}{300} = \frac{2}{5}$

25. $\frac{140}{300} = \frac{7}{15}$

27. $\frac{13}{52} = \frac{1}{4}$

29. $\frac{26}{52} = \frac{1}{2}$

31. $\frac{16}{52} = \frac{4}{13}$

33. $\frac{13}{52} = \frac{1}{4}$

35. $\frac{26}{52} = \frac{1}{2}$

37. $\frac{16}{52} = \frac{4}{13}$

39. (a) $\frac{120}{600} = \frac{1}{5}$

(b) $\frac{160}{600} = \frac{4}{15}$

(c) $\frac{580}{600} = \frac{29}{30}$

(d) $\frac{140}{480} = \frac{7}{24}$

(e) $\frac{20}{120} = \frac{1}{6}$

(f) $\frac{7}{24} \approx 0.292; \frac{1}{6} \approx 0.167$
Male firefighters are more likely to be promoted.

Chapter 14 Review Exercises

1. $a_1 = \left(\frac{3}{4}\right)^1 = \frac{3}{4}$
$a_2 = \left(\frac{3}{4}\right)^2 = \frac{9}{16}$
$a_3 = \left(\frac{3}{4}\right)^2 = \frac{27}{64}$

3. $d = 4 - 1 = 3$
$a_n = 1 + (n-1)(3)$

5. $\sum_{k=0}^{4} (-1)^k (2k) = (-1)^0(2\cdot 0) + (-1)^1(2\cdot 1) + (-1)^2(2\cdot 2) + (-1)^3(2\cdot 3) + (-1)^4(2\cdot 4)$
$= 0 - 2 + 4 - 6 + 8$
$= 4$

7. $\sum_{k=1}^{4} x^{3k}$

9. **(a)** $a_n = 100 + 10n$

(b) $a_1 = 100 + 10(1) = 110$
$a_2 = 100 + 10(2) = 120$
$a_3 = 100 + 10(3) = 130$
$a_4 = 100 + 10(4) = 140$
$a_5 = 100 + 10(5) = 150$

(c) $a_{27} = 100 + 10(27) = 370$
The cost of a speeding ticket is $370 for a driver traveling 27 mph over the speed limit.

11. $d = \frac{1}{6} - \frac{1}{2} = -\frac{1}{3}$

13. $d = 6 - 1 = 5$

$a_n = 1 + (n-1)(5)$

15. $a_n = 7 + (n-1)(-4)$

$a_{10} = 7 + (10-1)(-4) = 7 + (9)(-4) = 7 - 36 = -29$

17. $\sum_{i=1}^{50} (i-4)$

$a_1 = -3,\ a_{50} = 46,\ n = 50$

$S_{50} = \frac{50}{2}(-3+46) = 25(43) = 1075$

19. $r = \frac{-21}{24} = -\frac{1}{2}$

21. $r = \frac{6}{4} = \frac{3}{2}$

$a_n = 4\left(\frac{3}{2}\right)^{n-1}$

23. $\sum_{k=1}^{5} -4\left(\frac{3}{4}\right)^{k-1} = -4\left(\frac{3}{4}\right)^0 + (-4)\left(\frac{3}{4}\right)^1 + (-4)\left(\frac{3}{4}\right)^2 + (-4)\left(\frac{3}{4}\right)^3 + (-4)\left(\frac{3}{4}\right)^4$

$= -4 - 3 - \frac{9}{4} - \frac{27}{16} - \frac{81}{64}$

$= -\frac{781}{64}$

25. $r = \frac{\frac{1}{3}}{2} = \frac{1}{6}$

$S = \frac{2}{1 - \frac{1}{6}} = \frac{2}{\frac{5}{6}} = \frac{12}{5}$

27. $(x^2+4)^5 = \frac{5!}{5!\,0!}(x^2)^5 + \frac{5!}{4!\,1!}(x^2)^4(4) + \frac{5!}{3!\,2!}(x^2)^3(4)^2 + \frac{5!}{2!\,3!}(x^2)^2(4)^3 + \frac{5!}{1!\,4!}(x^2)(4)^4 + \frac{5!}{0!\,5!}(4)^5$

$= x^{10} + 20x^8 + 160x^6 + 640x^4 + 1280x^2 + 1024$

29. $\frac{7!}{3! \cdot 4!} = \frac{7 \cdot 6 \cdot 5 \cdot 4!}{3 \cdot 2 \cdot 1 \cdot 4!} = 35$

31. $(a+2b)^{11} = \frac{11!}{11!\,0!}(a)^{11} + \frac{11!}{10!\,1!}(a)^{10}(2b) + \frac{11!}{9!\,2!}(a)^9(2b)^2 = a^{11} + 22a^{10}b + 220a^9b^2$

33. $(a+2b)^6$

$\frac{6!}{3!3!}(a)^3(2b)^3 = 160a^3b^3$

35. $2^5 = 32$

37. ${}_{26}P_4 = \frac{26!}{(26-4)!} = 26 \cdot 25 \cdot 24 \cdot 23 = 358,800$

39. ${}_{8}C_3 \cdot {}_{10}C_1 = \frac{8!}{(8-3)!3!} \cdot \frac{10!}{(10-1)!1!}$

$= \frac{8 \cdot 7 \cdot 6}{3 \cdot 2 \cdot 1} \cdot 10$

$= 560$

41. **(a)** 18.6% = 0.186

(b) 16.2% + 12.7% = 28.9% = 0.289

(c) 25.6% = 0.256

Chapter 14 Test

1. $a_1 = -3\left(\frac{2}{3}\right)^0 = -3$

$a_2 = -3\left(\frac{2}{3}\right)^1 = -3\left(\frac{2}{3}\right) = -2$

$a_3 = -3\left(\frac{2}{3}\right)^2 = -3\left(\frac{4}{9}\right) = -\frac{4}{3}$

$a_4 = -3\left(\frac{2}{3}\right)^3 = -3\left(\frac{8}{27}\right) = -\frac{8}{9}$

Geometric

3. $\sum_{i=1}^{7} 5(-1)^{i-1} = 5-5+5-5+5-5+5 = 5$

5. $\sum_{i=1}^{7} \frac{3+i}{i}$

7. $a_1 = 7,\ a_{45} = 95,\ n = 45$

$S_{45} = \frac{45}{2}(7+95) = \frac{45}{2}(102) = 45(51) = 2295$

9. $a_5 = \frac{1}{3}(2)^5 = \frac{32}{3}$

11. $r = \frac{-8}{4} = -2$

$-2048 = 4(-2)^{n-1}$

$-512 = (-2)^{n-1}$

$(-2)^9 = (-2)^{n-1}$

$9 = n-1$

$10 = n$

$S_{10} = \frac{4(1-(-2)^{10})}{1-(-2)} = \frac{4(-1023)}{3} = -1364$

13. **(a)** 3(365) = $1,095

(b) 1095 + 1095(0.06) = $1,160.70

(c) $a_{29} = 1095(1.06)^{29}$ = $5,933.13

(d) $a_1 = 1095,\ r = 1.06,\ n = 30$

$S_{30} = \frac{1095(1-(1.06)^{30})}{1-1.06}$

$= \frac{1095(1-(1.06)^{30})}{-0.06}$

$= 86,568.71$

By saving money instead of buying cigarettes, the total savings plus interest amounts to $86,568.71.

15. $(a+b)^4 = a^4 + 4a^3b + 6a^2b^2 + 4ab^3 + b^4$

17. $\frac{7!}{5!\,2!}w^5(3z)^2 = \frac{42}{2}w^5(3z)^2 = 189w^5z^2$

19. ${}_{25}P_3 = \frac{25!}{(25-3)!} = 25 \cdot 24 \cdot 23 = 13,800$

21. ${}_{42}C_6 = \frac{42!}{(42-6)!6!}$

$= \frac{42 \cdot 41 \cdot 40 \cdot 39 \cdot 38 \cdot 37}{6 \cdot 5 \cdot 4 \cdot 3 \cdot 2 \cdot 1}$

$= 5,245,786$

23. **(a)** $\frac{20}{50} = \frac{2}{5}$

(b) $\frac{26}{50} = \frac{13}{25}$

(c) $\frac{36}{50} = \frac{18}{25}$

Cumulative Review Exercises, Chapters 1–14

1. $\frac{-2-3\sqrt{25-16}}{\sqrt{8^2+6^2}} = \frac{-2-3\sqrt{9}}{\sqrt{100}} = \frac{-2-9}{10} = -\frac{11}{10}$

3.
$$\begin{aligned} -8x+20-6x-2 &= -10+7x \\ -14x+18 &= -10+7x \\ -14x-7x &= -10-18 \\ -21x &= -28 \\ x &= \frac{-28}{-21} \\ x &= \frac{4}{3} \end{aligned}$$

5. Let x represent the measure of the first angle.
Let y represent the measure of the second angle.
$x + y = 90$
$x = 3y + 6$
Substitute $x = 3y + 6$ into the first equation and solve for y.
$$\begin{aligned} 3y+6+y &= 90 \\ 4y &= 84 \\ y &= 21 \end{aligned}$$
Substitute $y = 21$ into the second equation and solve for x.
$x = 3(21) + 6 = 63 + 6 = 69$
The angles are 69° and 21°.

7. x-intercept:
$0 = 3x$
$x = 0$
(0, 0)
y-intercept:
$y = 3(0)$
$y = 0$
(0, 0)

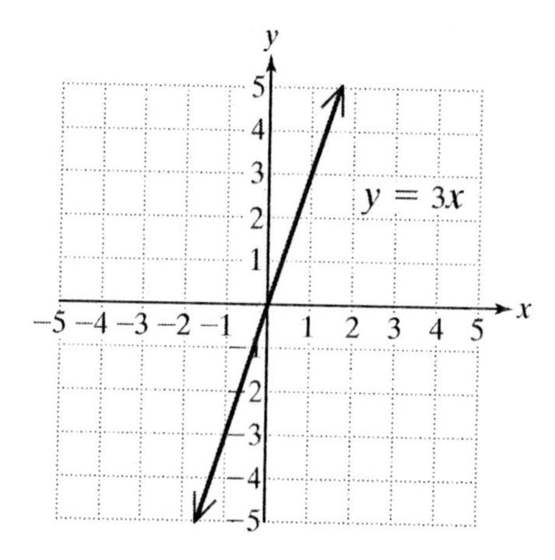

9. $m = \frac{1-3}{-3-2} = \frac{2}{5}$
$$\begin{aligned} y - y_1 &= m(x - x_1) \\ y - 3 &= \frac{2}{5}(x-2) \\ y - 3 &= \frac{2}{5}x - \frac{4}{5} \\ y &= \frac{2}{5}x + \frac{11}{5} \end{aligned}$$

11. A: $2x+3y-4z=17$
B: $x-2y+z=-5$
C: $3x+4y-3z=21$

Eliminate the x variable from equations A and B by multiplying equation B by −2.
$$\begin{aligned} 2x+3y-4z &= 17 \\ -2x+4y-2z &= 10 \\ \hline 7y-6z &= 27 \quad \text{Equation D} \end{aligned}$$

Eliminate the x variable from equations B and C by multiplying equation B by −3.
$$\begin{aligned} -3x+6y-3z &= 15 \\ 3x+4y-3z &= 21 \\ \hline 10y-6z &= 36 \quad \text{Equation E} \end{aligned}$$

Solve for y by eliminating the z variable from equations D and E by multiplying equation D by −1.
$$\begin{aligned} -7y+6z &= -27 \\ 10y-6z &= 36 \\ \hline 3y &= 9 \\ y &= 3 \end{aligned}$$

Substitute $y = 3$ into equation D and solve for z.

$$\begin{aligned} 7(3) - 6z &= 27 \\ 21 - 6z &= 27 \\ -6z &= 6 \\ z &= -1 \end{aligned}$$

Substitute $y = 3$ and $z = -1$ into equation B and solve for x.

$$\begin{aligned} x - 2(3) + (-1) &= -5 \\ x - 6 - 1 &= -5 \\ x &= 2 \end{aligned}$$

The solution to the system is (2, 3, –1).

13. $\left[\begin{array}{cc|c} 2 & -4 & -10 \\ -5 & 1 & 16 \end{array}\right]$

$\frac{1}{2}R_1 \Rightarrow R_1 \left[\begin{array}{cc|c} 1 & -2 & -5 \\ -5 & 1 & 16 \end{array}\right]$

$5R_1 + R_2 \Rightarrow \left[\begin{array}{cc|c} 1 & -2 & -5 \\ 0 & -9 & -9 \end{array}\right]$

$\frac{1}{9}R_2 \Rightarrow \left[\begin{array}{cc|c} 1 & -2 & -5 \\ 0 & 1 & 1 \end{array}\right]$

$2R_2 + R_1 \Rightarrow \left[\begin{array}{cc|c} 1 & 0 & -3 \\ 0 & 1 & 1 \end{array}\right]$

The solution to the system is (–3, 1).

15. $D = \begin{vmatrix} 2 & -3 & 0 \\ -1 & 4 & 1 \\ 0 & 1 & 3 \end{vmatrix}$

$$\begin{aligned} &= 2 \cdot \begin{vmatrix} 4 & 1 \\ 1 & 3 \end{vmatrix} + 1 \cdot \begin{vmatrix} -3 & 0 \\ 1 & 3 \end{vmatrix} + 0 \\ &= 2(12 - 1) + 1(-9 - 0) \\ &= 2(11) - 9 \\ &= 22 - 9 \\ &= 13 \end{aligned}$$

$D_y = \begin{vmatrix} 2 & 12 & 0 \\ -1 & 2 & 1 \\ 0 & 6 & 3 \end{vmatrix}$

$$\begin{aligned} &= 2 \cdot \begin{vmatrix} 2 & 1 \\ 6 & 3 \end{vmatrix} + 1 \cdot \begin{vmatrix} 12 & 0 \\ 6 & 3 \end{vmatrix} + 0 \\ &= 2(6 - 6) + 1(36 - 0) \\ &= 36 \end{aligned}$$

$y = \frac{D_y}{D} = \frac{36}{13}$

17.

$$\begin{aligned} w &= \frac{k}{f} \\ 300 &= \frac{k}{1200} \\ 360000 &= k \\ 450 &= \frac{360000}{f} \\ 450f &= 360000 \\ f &= 800 \end{aligned}$$

A 450 meter wave has a frequency of 800 kilohertz.

19. $\left(\frac{2.0 \times 3.2}{1.6}\right)\left(\frac{10^3 \times 10^{-7}}{10^{-9}}\right) = 4.0 \times 10^5$

21. $(x + 2y)^2 - 9 = x^2 + 4xy + 4y^2 - 9$

23. $\sqrt[4]{x^4} = |x|$

25. $8^{-2/3} = \left(\frac{1}{8}\right)^{2/3} = \left(\sqrt[3]{\frac{1}{8}}\right)^2 = \left(\frac{1}{2}\right)^2 = \frac{1}{4}$

27. $\sqrt[3]{8 \cdot 2 \cdot x^3 x^2 y^{12}} = 2xy^4 \sqrt[3]{2x^2}$

29.

$$\begin{aligned} \sqrt[3]{4x + 2} + 5 &= 3 \\ \sqrt[3]{4x + 2} &= -2 \\ 4x + 2 &= -8 \\ 4x &= -10 \\ x &= -\frac{5}{2} \end{aligned}$$

31.

$$\begin{aligned} \frac{-6}{3 - 7i} \cdot \frac{3 + 7i}{3 + 7i} &= \frac{-18 - 42i}{9 + 49} \\ &= \frac{-18}{58} - \frac{42i}{58} \\ &= -\frac{9}{29} - \frac{21}{29}i \end{aligned}$$

33.

$$\begin{aligned} 8x^6 + y^3 &= (2x^2)^3 + y^3 \\ &= (2x^2 + y)(4x^4 - 2x^2 y + y^2) \end{aligned}$$

35.
$$2x^2+12x=10$$
$$x^2+6x=5$$
$$x^2+6x+9=5+9$$
$$(x+3)^2=14$$
$$x+3=\pm\sqrt{14}$$
$$x=-3\pm\sqrt{14}$$

37.
$$x^4-x^2-12=0$$
$$(x^2-4)(x^2+3)=0$$
$$x^2-4=0 \quad\text{or}\quad x^2+3=0$$
$$x^2=4 \quad\text{or}\quad x^2=-3$$
$$x=\pm 2 \quad\text{or}\quad x=\pm i\sqrt{3}$$

39. $x^2-49\neq 0;\ x\neq 7, x\neq -7$
$(-\infty,-7)\cup(-7,7)\cup(7,\infty)$

41. Multiply each term by the common denominator x^2y.
$$\frac{x^2-y^2}{xy+x^2}=\frac{(x-y)(x+y)}{x(y+x)}=\frac{x-y}{x}$$

43. (a)
$$-3x+2\le 4 \quad\text{or}\quad 2x+3\le -5$$
$$-3x\le 2 \quad\text{or}\quad 2x\le -8$$
$$x\ge -\frac{2}{3} \quad\text{or}\quad x\le -4$$
$$(-\infty,-4]\cup\left[-\frac{2}{3},\infty\right)$$

(b)
$$-3x+2\le 4 \quad\text{or}\quad 2x+3\le -5$$
$$-3x\le 2 \quad\text{or}\quad 2x\le -8$$
$$x\ge -\frac{2}{3} \quad\text{or}\quad x\le -4$$
No solution

45. $|x-5|<7$
$$x-5<7 \quad\text{or}\quad x-5>-7$$
$$x<12 \quad\text{or}\quad x>-2$$
Interval notation: $(-2, 12)$

47.

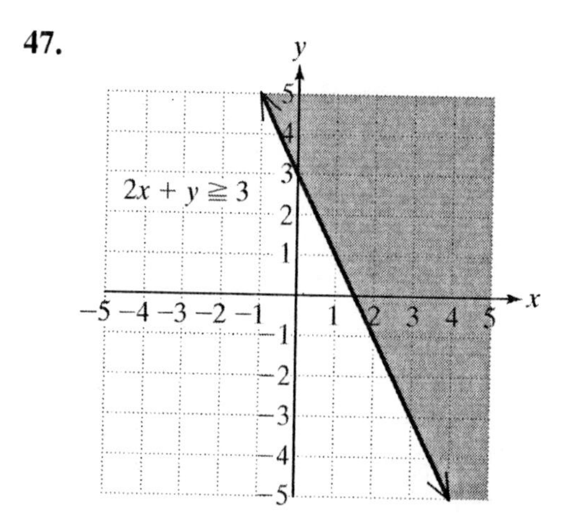

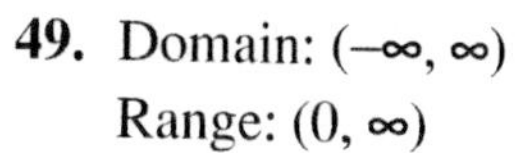

49. Domain: $(-\infty,\infty)$
Range: $(0,\infty)$

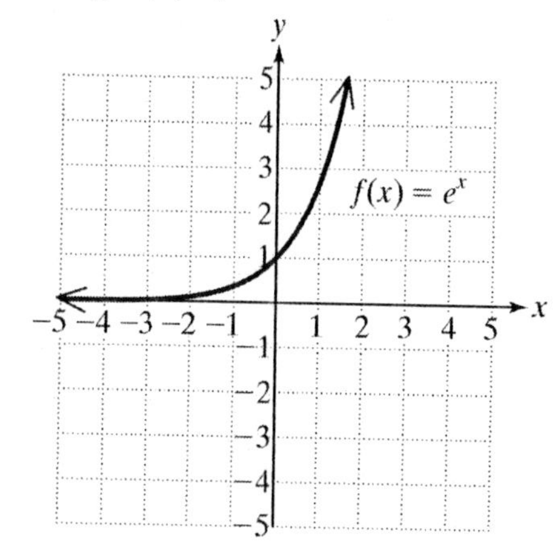

51.
$$\log_2 8=x$$
$$2^x=8$$
$$2^x=2^3$$
$$x=3$$

53.
$$3\ln x-\ln y+\frac{1}{2}\ln z=\ln x^3-\ln y+\ln\sqrt{z}$$
$$=\ln\left(\frac{x^3\sqrt{z}}{y}\right)$$

55.
$$\log((x+2)(x-1))=1$$
$$(x+2)(x-1)=10^1$$
$$x^2+x-2=10$$
$$x^2+x-12=0$$
$$(x+4)(x-3)=0$$
$x=-4$ or $x=3$
$x=3$ ($x=-4$ does not check)

57. $r = \frac{\frac{3}{2}}{\frac{3}{4}} = \frac{12}{6} = 2$

$a_n = \frac{3}{4}(2)^{n-1}$

Geometric

59. $a_1 = -5,\ d = 3,\ a_n = 88$

$88 = -5 + (n-1)(3)$

$93 = 3(n-1)$

$31 = n - 1$

$32 = n$

$S_{32} = \frac{32}{2}(-5 + 88) = 16(83) = 1328$

61. ${}_{41}C_6 = \frac{41!}{(41-6)!6!}$

$= \frac{41 \cdot 40 \cdot 39 \cdot 38 \cdot 37 \cdot 36}{6 \cdot 5 \cdot 4 \cdot 3 \cdot 2 \cdot 1}$

$= 4{,}496{,}388$

63. **(a)** $\frac{34}{74} = \frac{17}{37}$

(b) $\frac{46}{74} = \frac{23}{37}$

(c) $\frac{55}{74}$

(d) $\frac{41}{74}$

65. Center: (3, –4); Radius: $\sqrt{35}$

67.

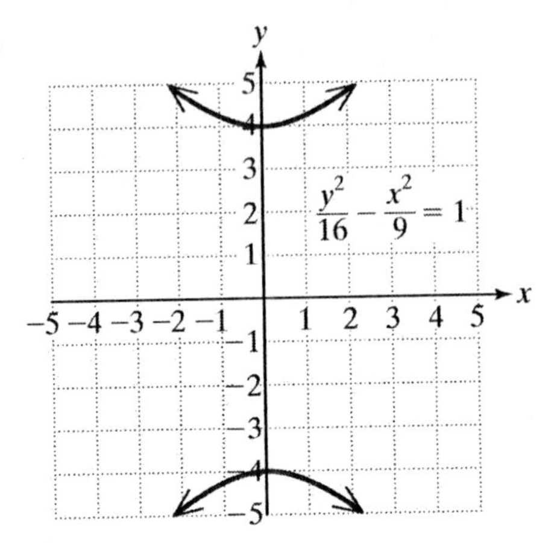

69. Substitute y for x^2 in the first equation:

$x^2 + y^2 = 20$

$(y) + y^2 = 20$

$y^2 + y - 20 = 0$

$(y+5)(y-4) = 0$

$y + 5 = 0$ or $y - 4 = 0$

$y = -5$ or $y = 4$

Substituting $y = -5$ in $y = x^2$ gives:

$-5 = x^2$

which has no real solutions.

Substituting $y = 4$ in $y = x^2$ gives:

$4 = x^2$

$\pm 2 = x$

The points of intersection are (2, 4) and (–2, 4).

Beginning Algebra Review

Review A Practice Exercises

1. $\left\{-6, -\frac{7}{8}, -0.\overline{3}, 0, \frac{4}{9}, 7, 10.4\right\}$

3. $\{-6, 0, 7\}$

5. $\{1, 8\}$

7. $\left\{-\pi, \sqrt{3}\right\}$

9. 5 is greater than −2.

11. −3 is greater than or equal to −4.

13. 10 is between 6 and 12.

15. 2 is not equal to −5.

17. $6 \cdot 3$

19. $20 \div 5$

21. $|-5|$

23. $2+|-8|$

25. $4 \cdot \frac{1}{3}$

27. $8 + (-10) = -2$

29. $51 - 32 = 51 + (-32) = 19$

31. $0 + 16 = 16$

33. $-13 - 20 = -13 + (-20) = -33$

35. $-2.1 + (-2.2) = -4.3$

37. $-19 - (-3) = -19 + 3 = -16$

39. $0 - 19 = 0 + (-19) = -19$

41. $5.2 - (-0.4) = 5.2 + 0.4 = 5.6$

43.
$$\begin{aligned} -\frac{3}{4} - \frac{3}{8} &= -\frac{3}{4} \cdot \frac{2}{3} - \frac{3}{8} \\ &= -\frac{6}{8} - \frac{3}{8} \\ &= -\frac{6}{8} + \left(-\frac{3}{8}\right) \\ &= -\frac{9}{8} \end{aligned}$$

45. $-\frac{7}{3} - \left(-\frac{1}{9}\right) = -\frac{7}{3} + \frac{1}{9} = -\frac{21}{9} + \frac{1}{9} = -\frac{20}{9}$

47. $3(-9) = -27$

49. $(-5)(-7) = 35$

51. $0 \cdot 32 = 0$

53. $(2.05)(-4.2) = -8.61$

55. $\left(-\frac{5}{9}\right) \cdot \left(-\frac{3}{2}\right) = \frac{5 \cdot 3}{9 \cdot 2} = \frac{5 \cdot 1}{3 \cdot 2} = \frac{5}{6}$

57. $-56 \div (-8) = 7$

59. $-78 \div 13 = -6$

61. $0 \div (-5) = 0$

63. $3.1 \div 0$ is undefined.

65. $\left(-\frac{7}{8}\right) \div \left(-\frac{7}{4}\right) = \frac{7}{8} \cdot \frac{4}{7} = \frac{7 \cdot 4}{8 \cdot 7} = \frac{1}{2}$

67. $|-14| - |-5| = 14 - 5 = 9$

69. $10 \div 5 \cdot 4 = 2 \cdot 4 = 8$

71. $6 + 42 \div 7 - 10 = 6 + 6 - 10 = 12 - 10 = 2$

73.
$$\begin{aligned} (8-5)^2 - (2+3)^2 &= 3^2 - 5^2 \\ &= 9 - 25 \\ &= 9 + (-25) \\ &= -16 \end{aligned}$$

75. $3^3 - 2 + 5(3-7) = 3^3 - 2 + 5(-4)$
$= 27 - 2 + 5(-4)$
$= 27 - 2 + (-20)$
$= 25 + (-20)$
$= 5$

77. $12 - \sqrt{25} + 4^2 \div 2 = 12 - 5 + 4^2 \div 2$
$= 12 - 5 + 16 \div 2$
$= 12 - 5 + 8$
$= 7 + 8$
$= 15$

79. $\dfrac{6+8-(-2)}{-40 \div 8 - 1} = \dfrac{6+8+2}{-5-1} = \dfrac{16}{-6} = -\dfrac{8}{3}$

81. $a^2 - b^2 = (3)^2 - (-5)^2 = 9 - 25 = -16$

83. $b + \sqrt{c} \cdot d = (-5) + \sqrt{9} \cdot (-2)$
$= -5 + 3 \cdot (-2)$
$= -5 + (-6)$
$= -11$

85. $A = \dfrac{1}{2}(b_1 + b_2)h$
$= \dfrac{1}{2}(28 + 18)20$
$= \dfrac{1}{2}(46)(20)$
$= 460 \text{ in}^2$

87. Associative property of multiplication

89. Distributive property of multiplication over addition

91. Commutative property of addition

93. 1; for example $1(3) = 3$

95. **(a)** reciprocal

(b) $\dfrac{1}{4}$

97. $5(a + 9) = 5 \cdot a + 5 \cdot 9 = 5a + 45$

99. $-7(y - z) = -7 \cdot y - (-7)z = -7y + 7z$

101. $\dfrac{1}{3}(6b + 9c - 15) = \dfrac{1}{3} \cdot 6b + \dfrac{1}{3} \cdot 9c - \dfrac{1}{3} \cdot 15$
$= 2b + 3c - 5$

103. $-(2x - 3y + 8) = (-1)(2x - 3y + 8)$
$= (-1)2x - (-1)3y + (-1)8$
$= -2x + 3y - 8$

105. $5a + 2(a + 7) = 5a + 2a + 14 = 7a + 14$

107. $8y - 3(y + 2) - 19 = 8y - 3y - 6 - 19$
$= 5y - 6 - 19$
$= 5y - 25$

109. $4w + 9u - (5w - 3u) = 4w + 9u - 5w + 3u$
$= (4w - 5w) + (9u + 3u)$
$= -w + 12u$

111. $2g - 4[2 - 3(g - 4h) - 2]$
$= 2g - 4[2 - 3g + 12h - 2]$
$= 2g - 4[-3g + 12h]$
$= 2g + 12g - 48h$
$= 14g - 48h$

Review B Practice Exercises

1. $4x = -84$
$\dfrac{4x}{4} = \dfrac{-84}{4}$
$x = -21$

3. $\dfrac{t}{5} = 12$
$5\left(\dfrac{t}{5}\right) = 5(12)$
$t = 60$

5. $6 + p = -14$
$-6 + 6 + p = -6 - 14$
$p = -20$

7. $7x + 12 = 33$
$7x + 12 - 12 = 33 - 12$
$7x = 21$
$x = 3$

9.
$$\begin{aligned}4(x+5)&=14\\4x+20&=14\\4x+20-20&=14-20\\4x&=-6\\x&=-\frac{6}{4}=-\frac{3}{2}\end{aligned}$$

11.
$$\begin{aligned}3b+25&=5b-21\\3b-5b+25&=5b-5b-21\\-2b+25&=-21\\-2b+25-25&=-21-25\\-2b&=-46\\b&=\frac{-46}{-2}=23\end{aligned}$$

13.
$$\begin{aligned}3w+2(w-7)&=7(w-1)\\3w+2w-14&=7w-7\\5w-14&=7w-7\\5w-7w-14&=7w-7w-7\\-2w-14&=-7\\-2w-14+14&=-7+14\\-2w&=7\\w&=-\frac{7}{2}\end{aligned}$$

15.
$$\begin{aligned}\frac{1}{5}y-\frac{3}{10}y&=y-\frac{2}{5}y+1\\10\left(\frac{1}{5}y-\frac{3}{10}y\right)&=10\left(y-\frac{2}{5}y+1\right)\\\frac{10}{1}\left(\frac{1}{5}y\right)-\frac{10}{1}\left(\frac{3}{10}y\right)&=\frac{10}{1}(y)-\frac{10}{1}\left(\frac{2}{5}y\right)+\frac{10}{1}(1)\\2y-3y&=10y-4y+10\\-y&=6y+10\\-7y&=10\\y&=-\frac{10}{7}\end{aligned}$$

17.
$$\frac{5}{6}(x+2)=\frac{1}{3}x-\frac{3}{2}$$
$$6\left[\frac{5}{6}(x+2)\right]=6\left(\frac{1}{3}x-\frac{3}{2}\right)$$
$$\frac{6}{1}\left(\frac{5}{6}\right)(x+2)=\frac{6}{1}\left(\frac{1}{3}x\right)-\frac{6}{1}\left(\frac{3}{2}\right)$$
$$5(x+2)=2x-9$$
$$5x+10=2x-9$$
$$3x+10=-9$$
$$3x=-19$$
$$x=-\frac{19}{3}$$

19.
$$0.2a+6=1.8a-2.8$$
$$0.2a-1.8a+6=1.8a-1.8a-2.8$$
$$-1.6a+6=-2.8$$
$$-1.6a+6-6=-2.8-6$$
$$-1.6a=-8.8$$
$$a=\frac{-8.8}{-1.6}=5.5$$

21.
$$0.72t-1.18=0.28(4-t)$$
$$0.72t-1.18=1.12-0.28t$$
$$0.72t+0.28t-1.18=1.12-0.28t+0.28t$$
$$t-1.18=1.12$$
$$t-1.18+1.18=1.12+1.18$$
$$t=2.3$$

23.
$$-4(x+1)-2x=-2(3x+2)-x$$
$$-4x-4-2x=-6x-4-x$$
$$-6x-4=-7x-4$$
$$-6x+7x-4=-4$$
$$x-4=-4$$
$$x=-4+4$$
$$x=0$$

25.
$$11-5(y+3)=-2[2y-3(1-y)]$$
$$11-5y-15=-2[2y-3+3y]$$
$$-5y-4=-2(5y-3)$$
$$-5y-4=-10y+6$$
$$-5y+10y-4=6$$
$$5y-4=6$$
$$5y=6+4$$
$$5y=10$$
$$y=2$$

27.
$$-8m-2[4-3(m+1)]=4(3-m)+4$$
$$-8m-2(4-3m-3)=12-4m+4$$
$$-8m-2(1-3m)=16-4m$$
$$-8m-2+6m=16-4m$$
$$-2m-2=16-4m$$
$$2m-2=16$$
$$2m=18$$
$$m=9$$

29.
$$4-6[2y-2(y+3)]=7-y$$
$$4-6(2y-2y-6)=7-y$$
$$4-6(-6)=7-y$$
$$4+36=7-y$$
$$40=7-y$$
$$33=-y$$
$$-33=y$$

31. Let x represent the unknown number.
$$5(8x)=20$$
$$40x=20$$
$$x=\frac{20}{40}=\frac{1}{2}$$

33. Let x represent the first integer.
$x+1$ and $x+2$ represent the next two integers.
$$x+(x+1)+(x+2)=-57$$
$$3x+3=-57$$
$$3x=-60$$
$$x=-20$$
$$x+1=-20+1=-19$$
$$x+2=-20+2=-18$$
$-20, -19, -18$

35. Let r represent the interest rate.
$$2500r=112.50$$
$$r=\frac{112.50}{2500}=0.045=4.5\%$$

37.

	4.2% Account	3.8% Account	Total
Principal	x	$3500 - x$	
Interest	$0.042x$	$0.038(3500 - x)$	141.40

$$\begin{aligned} 0.042x + 0.038(3500 - x) &= 141.40 \\ 0.042x + 133 - 0.038x &= 141.40 \\ 0.004x + 133 &= 141.40 \\ 0.004x &= 8.4 \\ x &= \frac{8.4}{0.004} = 2100 \end{aligned}$$

$3500 - x = 3500 - 2100 = 1400$

\$2100 in the 4.2% account, \$1400 in the 3.8% account.

39.

	Distance	Rate	Time
Car	$3x$	x	3
Truck	$3(x - 4)$	$x - 4$	3

$$\begin{aligned} 3x + 3(x - 4) &= 372 \\ 3x + 3x - 12 &= 372 \\ 6x - 12 &= 372 \\ 6x &= 384 \\ x &= 64 \end{aligned}$$

$x - 4 = 64 - 4 = 60$

Car: 64 mph; truck: 60 mph

41.

$$\begin{aligned} 3 - x &> 5 \\ -3 + 3 - x &> -3 + 5 \\ -x &> 2 \\ \frac{-x}{-1} &< \frac{2}{-1} \\ x &< -2 \end{aligned}$$

Interval notation: $(-\infty, -2)$

−7 −6 −5 −4 −3 −2 −1

43.
$$\begin{aligned} 4y+6 &\ge -18 \\ 4y+6-6 &\ge -18-6 \\ 4y &\ge -24 \\ \frac{4y}{4} &\ge \frac{-24}{4} \\ y &\ge -6 \end{aligned}$$

Interval notation: $[-6, \infty)$

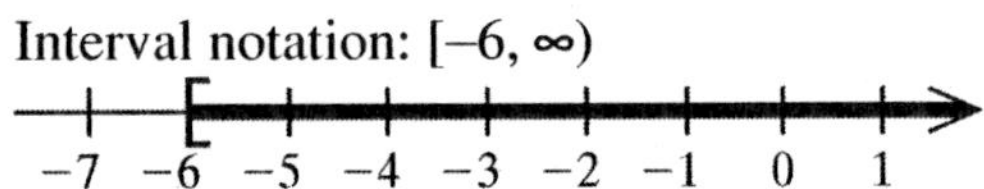

45.
$$\begin{aligned} 8m-15 &< 9m-13 \\ 8m-9m-15 &< 9m-9m-13 \\ -m-15 &< -13 \\ -m-15+15 &< -13+15 \\ -m &< 2 \\ \frac{-m}{-1} &> \frac{2}{-1} \\ m &> -2 \end{aligned}$$

Interval notation: $(-2, \infty)$

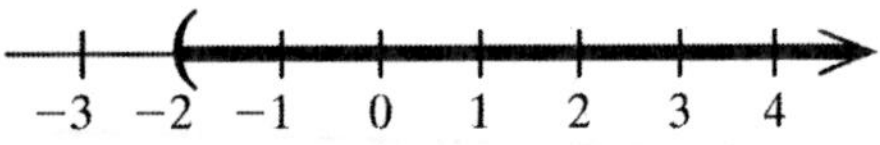

47.
$$\begin{aligned} 1-5(2t-7) &< 38t \\ 1-10t+35 &< 38t \\ 36-10t &< 38t \\ 36-10t+10t &< 38t+10t \\ 36 &< 48t \\ \frac{36}{48} &< \frac{48t}{48} \\ \frac{3}{4} &< t \end{aligned}$$

Interval notation: $\left(\frac{3}{4}, \infty\right)$

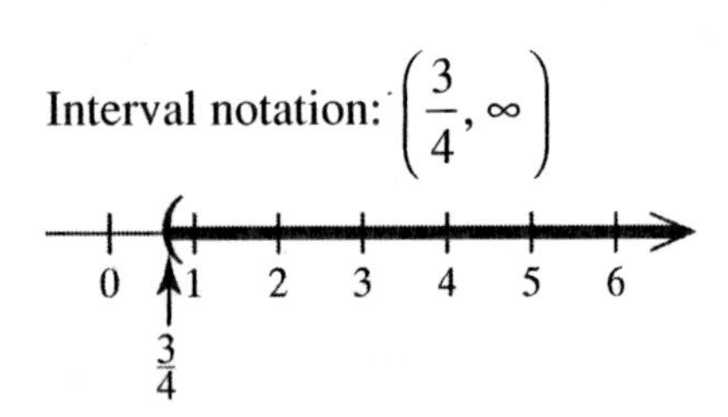

49.
$$\begin{aligned} -5 &< 4p+2 \le 6 \\ -5-2 &< 4p+2-2 \le 6-2 \\ -7 &< 4p \le 4 \\ -\frac{7}{4} &< \frac{4p}{4} \le \frac{4}{4} \\ -\frac{7}{4} &< p \le 1 \end{aligned}$$

Interval notation: $\left(-\frac{7}{4}, 1\right]$

−3 −2 −1 0 1 2

$-\frac{7}{4}$

51.
$$\begin{aligned} 2 &> -y+1 > -6 \\ 2-1 &> -y+1-1 > -6-1 \\ 1 &> -y > -7 \\ \frac{1}{-1} &< \frac{-y}{-1} < \frac{-7}{-1} \\ -1 &< y < 7 \end{aligned}$$

Interval notation: $(-1, 7)$

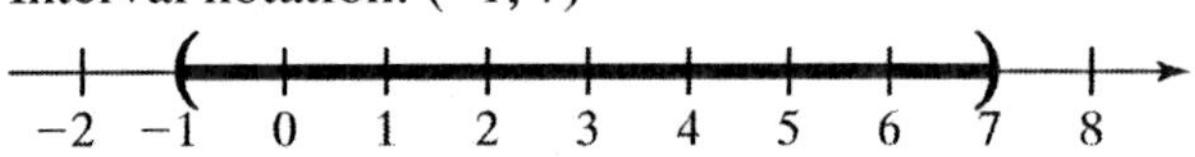

53.

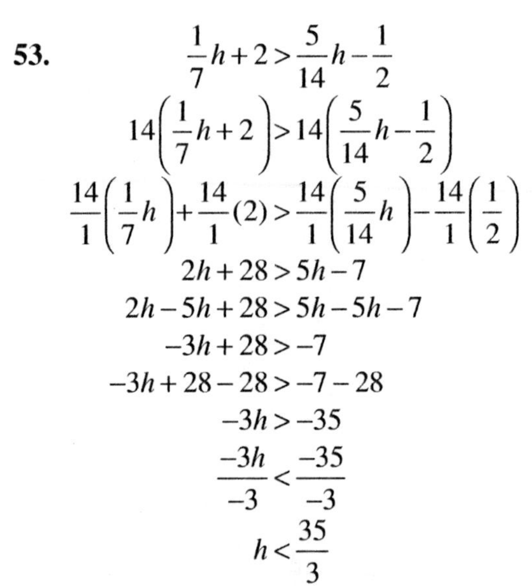
$$\begin{aligned} \frac{1}{7}h+2 &> \frac{5}{14}h-\frac{1}{2} \\ 14\left(\frac{1}{7}h+2\right) &> 14\left(\frac{5}{14}h-\frac{1}{2}\right) \\ \frac{14}{1}\left(\frac{1}{7}h\right)+\frac{14}{1}(2) &> \frac{14}{1}\left(\frac{5}{14}h\right)-\frac{14}{1}\left(\frac{1}{2}\right) \\ 2h+28 &> 5h-7 \\ 2h-5h+28 &> 5h-5h-7 \\ -3h+28 &> -7 \\ -3h+28-28 &> -7-28 \\ -3h &> -35 \\ \frac{-3h}{-3} &< \frac{-35}{-3} \\ h &< \frac{35}{3} \end{aligned}$$

Interval notation: $\left(-\infty, \frac{35}{3}\right)$

5 6 7 8 9 10 11 12 13

$\frac{35}{3}$

55. $0.4w + 3.1(w-1) \le -3.8$

$$0.4w + 3.1w - 3.1 \le -3.8$$
$$3.5w - 3.1 \le -3.8$$
$$3.5w - 3.1 + 3.1 \le -3.8 + 3.1$$
$$3.5w \le -0.7$$
$$\frac{3.5w}{3.5} \le \frac{-0.7}{3.5}$$
$$w \le -0.2$$

Interval notation: $(-\infty, -0.2]$

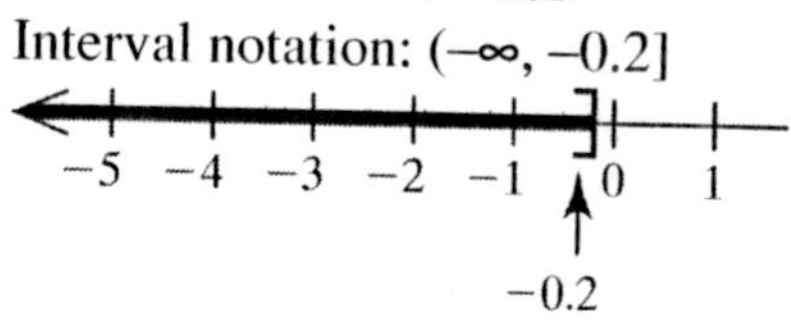

Review C Practice Exercises

1–7.

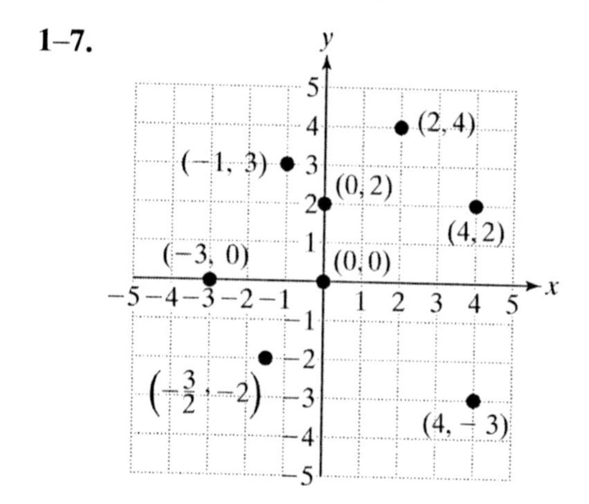

9. IV

11. III

13. I

15. II

17. $x + y = 4$

x	y
3	
	−1
0	

Substitute: $x = 3$

$$x + y = 4$$
$$3 + y = 4$$
$$y = 1$$

Substitute: $y = -1$

$$x + y = 4$$
$$x + (-1) = 4$$
$$x = 5$$

Substitute: $x = 0$

$$x + y = 4$$
$$0 + y = 4$$
$$y = 4$$

x	y
3	1
5	−1
0	4

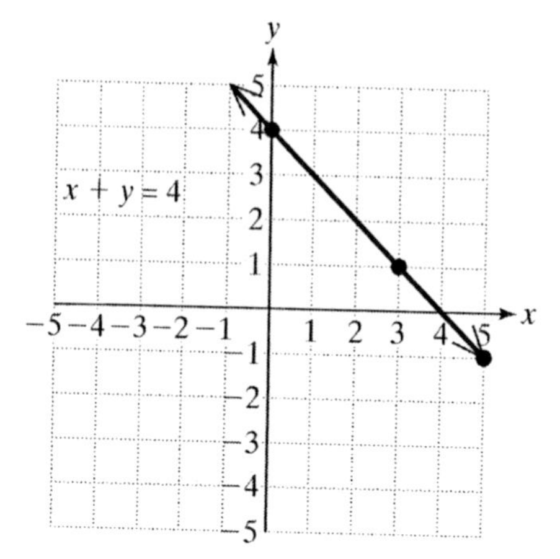

19.

x	y
0	
$\frac{2}{5}$	
−1	

$y = 5x$

$\rightarrow y = 5(0) \rightarrow$

$\rightarrow y = 5\left(\frac{2}{5}\right) \rightarrow$

$\rightarrow y = 5(-1) \rightarrow$

x	y
0	0
$\frac{2}{5}$	2
−1	−5

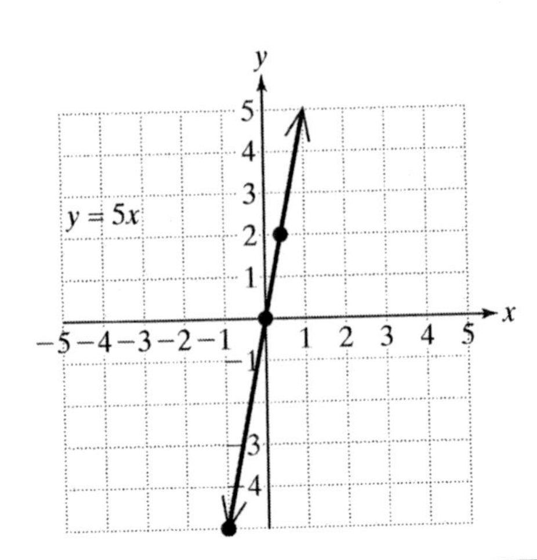

21.

x	y
3	
−3	
0	

$y = \frac{2}{3}x + 1$

→ $y = \frac{2}{3}(3) + 1$ →

→ $y = \frac{2}{3}(-3) + 1$ →

→ $y = \frac{2}{3}(0) + 1$ →

x	y
3	3
−3	−1
0	1

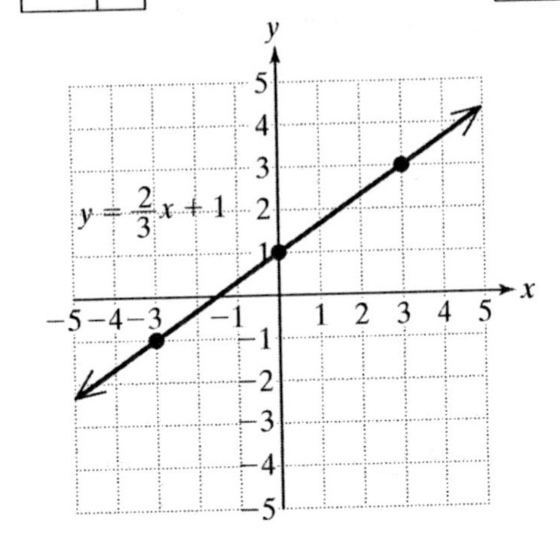

23. $3x - 2y = 18$
<u>x-intercept</u>
Substitute $y = 0$:
$$3x - 2(0) = 18$$
$$3x = 18$$
$$x = 6$$
$(6, 0)$
<u>y-intercept</u>
Substitute $x = 0$:
$$3(0) - 2y = 18$$
$$-2y = 18$$
$$y = -9$$
$(0, -9)$

25. $5x + y = -15$
<u>x-intercept</u>
Substitute $y = 0$:
$$5x + 0 = -15$$
$$x = -3$$
$(-3, 0)$
<u>y-intercept</u>
Substitute $x = 0$:
$$5(0) + y = -15$$
$$y = -15$$
$(0, -15)$

27. $\frac{1}{2}x + y = 2$
<u>x-intercept</u>
Substitute $y = 0$:
$$\frac{1}{2}x + 0 = 2$$
$$\frac{1}{2}x = 2$$
$$x = 4$$
$(4, 0)$
<u>y-intercept</u>
Substitute $x = 0$:
$$\frac{1}{2}(0) + y = 2$$
$$y = 2$$
$(0, 2)$

29. $-3x + 4y = 0$
<u>x-intercept</u>
Substitute $y = 0$:
$$-3x + 4(0) = 0$$
$$-3x = 0$$
$$x = 0$$
$(0, 0)$
<u>y-intercept</u>
Substitute $x = 0$:

$$-3(0)+4y=0$$
$$4y=0$$
$$y=0$$

(0, 0)

31. $y=-\frac{2}{3}x-3$

<u>x-intercept</u>

Substitute $y = 0$:

$$0=-\frac{2}{3}x-3$$
$$3=-\frac{2}{3}x$$
$$3\left(-\frac{3}{2}\right)=x$$
$$-\frac{9}{2}=x$$

$\left(-\frac{9}{2}, 0\right)$

<u>y-intercept</u>

Substitute $x = 0$:

$$y=-\frac{2}{3}(0)-3$$
$$y=-3$$

$(0, -3)$

33. $y = 6x - 5$

<u>x-intercept</u>

Substitute $y = 0$:

$$0=6x-5$$
$$5=6x$$
$$\frac{5}{6}=x$$

$\left(\frac{5}{6}, 0\right)$

<u>y-intercept</u>

Substitute $x = 0$:

$$y = 6(0) - 5$$
$$y = -5$$

$(0, -5)$

35. $2x=6$

$x=3$; vertical

37. $-3y+1=2$

$$-3y=1$$

$y=-\frac{1}{3}$; horizontal

39. $5x=0$

$x=0$; vertical

41. $4x-5y=6$

$$-5y=-4x+6$$
$$\frac{-5y}{-5}=\frac{-4x}{-5}+\frac{6}{-5}$$
$$y=\frac{4}{5}x-\frac{6}{5}$$

Slope is $\frac{4}{5}$; y-intercept $\left(0, -\frac{6}{5}\right)$

43. $-6x+2y=3$

$$2y=6x+3$$
$$\frac{2y}{2}=\frac{6x}{2}+\frac{3}{2}$$
$$y=3x+\frac{3}{2}$$

Slope is 3; y-intercept $\left(0, \frac{3}{2}\right)$

45. $5x-1=4$

$$5x=5$$
$$x=1$$

Undefined slope; no y-intercept

47. $x+4y=0$

$$4y=-x$$
$$\frac{4y}{4}=\frac{-x}{4}$$
$$y=-\frac{1}{4}x$$

Slope is $-\frac{1}{4}$; y-intercept (0, 0)

49. $y+4=8$

$$y=4$$

Slope is 0; y-intercept (0, 4)

51. Plot the y-intercept, $(0, -2)$. The slope is $\frac{1}{2}$.

From the y-intercept, move 1 unit up and 2 units right to find a second point.

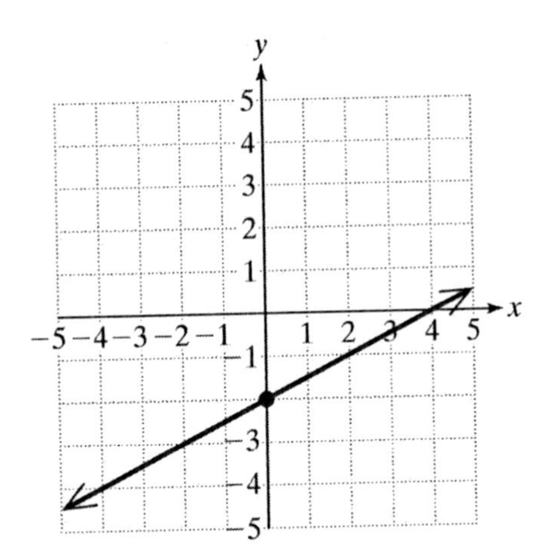

53. Plot the y-intercept, $(0, 3)$. The slope is $\frac{-3}{2}$.

From the y-intercept, move 3 units down and 2 units right to find a second point.

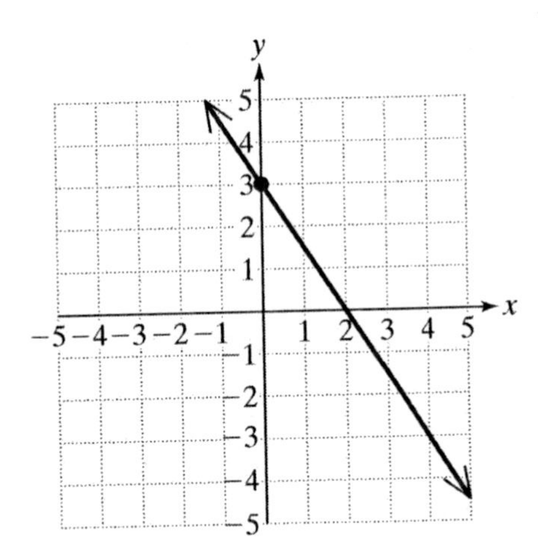

55. Plot the y-intercept, $(0, -2)$. The slope is $\frac{4}{1}$.

From the y-intercept, move 4 units up and 1 unit right to find a second point.

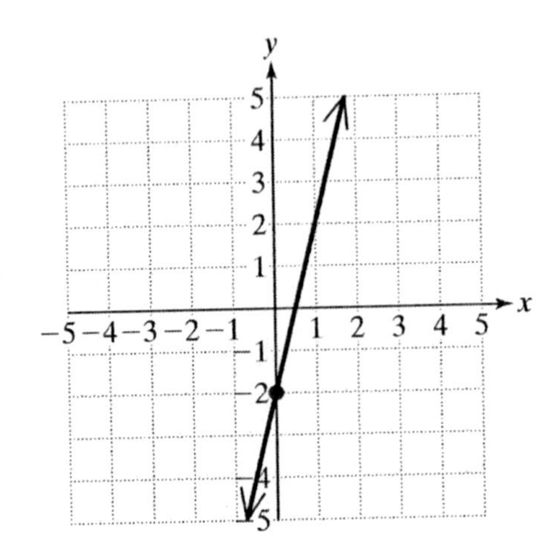

57. $2x - 3y = 12$

x-intercept: $2x - 3(0) = 12$

$$2x = 12$$
$$x = 6$$
$$(6, 0)$$

y-intercept: $2(0) - 3y = 12$

$$-3y = 12$$
$$y = -4$$
$$(0, -4)$$

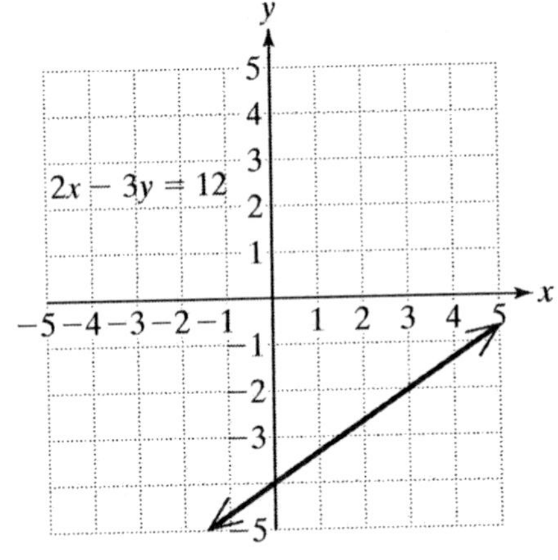

59. $x - 2y = -4$

x-intercept: $x - 2(0) = -4$

$$x = -4$$
$$(-4, 0)$$

y-intercept: $0 - 2y = -4$

$$-2y = -4$$
$$y = 2$$
$$(0, 2)$$

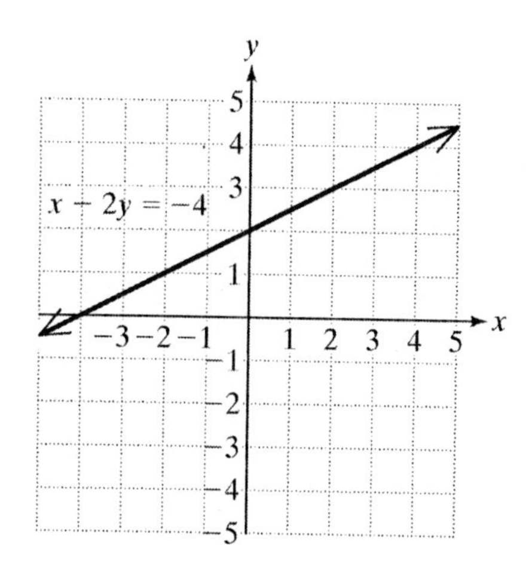

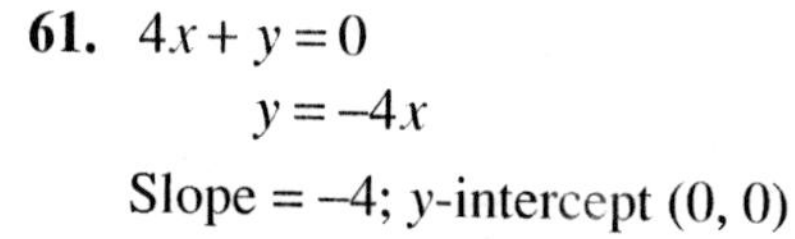

61. $4x + y = 0$

$y = -4x$

Slope $= -4$; y-intercept $(0, 0)$

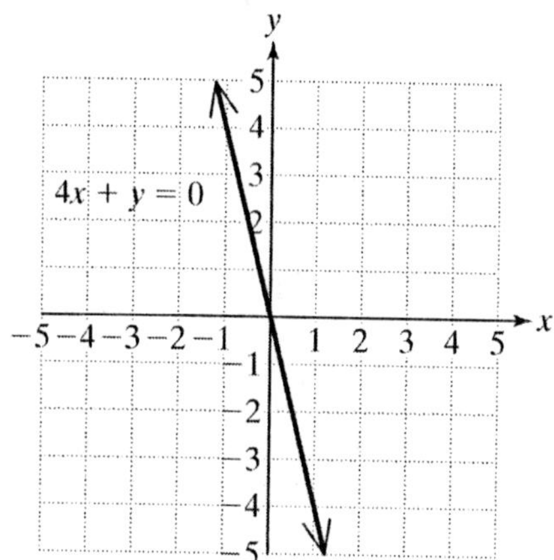

63. $y = -\frac{2}{3}x + 1$

Slope $= -\frac{2}{3}$; y-intercept $(0, 1)$

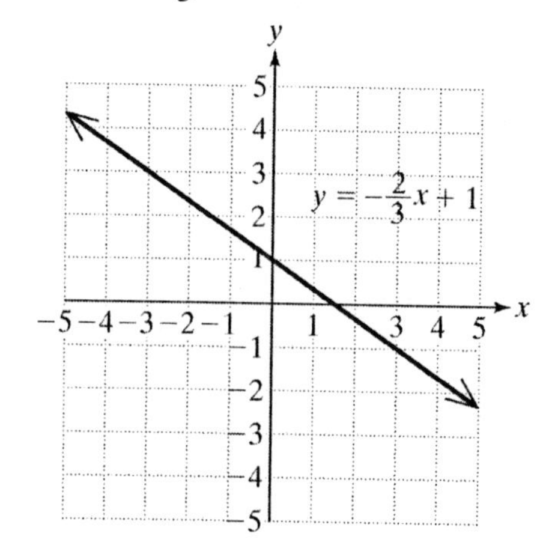

65. $-2x + 1 = 5$

$-2x = 4$

$x = -2$

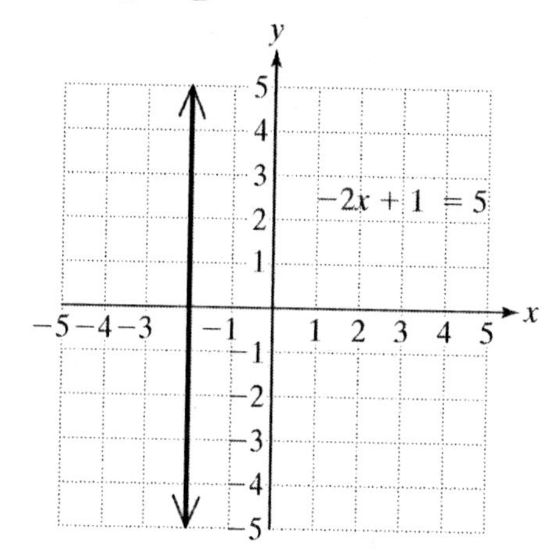

67. $y = 3x - 3$

Slope $= 3$; y-intercept $(0, -3)$

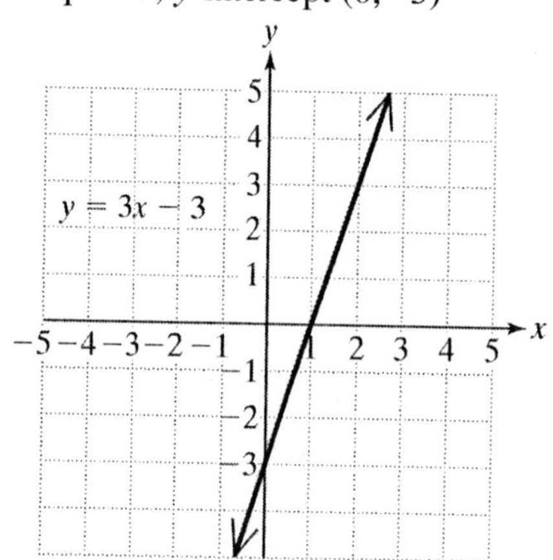

69. l_1: $-2x + y = 1$

$y = 2x + 1$

Slope $= 2$

l_2: $2x - y = 3$

$-y = -2x + 3$

$\frac{-y}{-1} = \frac{-2x}{-1} + \frac{3}{-1}$

$y = 2x - 3$

Slope $= 2$

The lines have the same slope. Therefore they are parallel.

71. l_1: $-2x+5y=5$

$$5y=2x+5$$
$$\frac{5y}{5}=\frac{2x}{5}+\frac{5}{5}$$
$$y=\frac{2}{5}x+1$$

Slope $=\frac{2}{5}$

l_2: $5x-2y=-2$

$$-2y=-5x-2$$
$$\frac{-2y}{-2}=\frac{-5x}{-2}-\frac{2}{-2}$$
$$y=\frac{5}{2}x+1$$

Slope $=\frac{5}{2}$

The lines are neither parallel nor perpendicular.

73. l_1: $x+y=7$

$$y=-x+7$$

Slope $=-1$

l_2: $x-y=-2$

$$-y=-x-2$$
$$\frac{-y}{-1}=\frac{-x}{-1}-\frac{2}{-1}$$
$$y=x+2$$

Slope $=1$

The slope of l_1 is the opposite of the reciprocal of the slope of l_2. Therefore, the lines are perpendicular.

75. l_1: $y=6$

l_2: $y=-2$

Both lines have slope 0. Therefore, they are parallel.

Review D Practice Exercises

1. $6^2 \cdot 6^5 = 6^{2+5} = 6^7$

multiplication of like bases

3. $\frac{y^{10}}{y^7} = y^{10-7} = y^3$

division of like bases

5. $\left(\frac{w}{4}\right)^4 = \frac{w^4}{4^4} = \frac{w^4}{256}$

power of a quotient

7. $(q^5)^6 = q^{5\cdot 6} = q^{30}$

power rule

9. $3^0 = 1$

11. $\left(\frac{1}{2}\right)^{-1} = \left(\frac{2}{1}\right)^1 = 2$

13. $(-5)^{-2} = \left(\frac{1}{-5}\right)^2 = \frac{1}{25}$

15. $6^{-1} + (-4)^0 = \frac{1}{6} + 1 = \frac{7}{6}$

17. $\frac{r^2 \cdot r^6}{r^{10}} = \frac{r^{2+6}}{r^{10}} = \frac{r^8}{r^{10}} = r^{8-10} = r^{-2} = \frac{1}{r^2}$

19. $\frac{b^3}{b^4 \cdot b^{-3}} = \frac{b^3}{b^{4-3}} = \frac{b^3}{b^1} = b^{3-1} = b^2$

21. $(3x^2)^3 = 3^3(x^2)^3 = 27x^{2\cdot 3} = 27x^6$

23. $\frac{6w^2z}{2z^{-1}} = \frac{3w^2z}{z^{-1}}$

$$= 3w^2z^{1-(-1)}$$
$$= 3w^2z^{1+1}$$
$$= 3w^2z^2$$

25. $\left(\dfrac{a^2b^{-1}}{c^3d^{-4}}\right)^{-2} = \left(\dfrac{c^3d^{-4}}{a^2b^{-1}}\right)^2$

$= \dfrac{(c^3)^2(d^{-4})^2}{(a^2)^2(b^{-1})^2}$

$= \dfrac{c^{3\cdot 2}d^{-4\cdot 2}}{a^{2\cdot 2}b^{-1\cdot 2}}$

$= \dfrac{c^6d^{-8}}{a^4b^{-2}}$

$= \dfrac{b^2c^6}{a^4d^8}$

27. $(5^0w^2z)\cdot(w^{-3}z^{-4}) = 1w^2w^{-3}z^1z^{-4}$

$= w^{2-3}z^{1-4}$

$= w^{-1}z^{-3}$

$= \dfrac{1}{wz^3}$

29. $3{,}050{,}000 = 3.05\times 10^6$

31. $0.0000251 = 2.51\times 10^{-5}$

33. $89{,}600{,}000 = 8.96\times 10^7$

35. $0.00039 = 3.9\times 10^{-4}$

37. $(2.4\times 10^1)(1.5\times 10^3) = (2.4)(1.5)\times 10^1 10^3$

$= 3.6\times 10^4$

39. $\dfrac{6.3\times 10^{14}}{3.0\times 10^{-2}} = \left(\dfrac{6.3}{3.0}\right)\times\dfrac{10^{14}}{10^{-2}}$

$= 2.1\times 10^{14-(-2)}$

$= 2.1\times 10^{16}$

41. $(3.0\times 10^{-5})(7.3\times 10^2) = (3.0)(7.3)\times 10^{-5}10^2$

$= 21.9\times 10^{-3}$

$= (2.19\times 10^1)\times 10^{-3}$

$= 2.19\times 10^{-2}$

43. $\dfrac{(4.0\times 10^3)(1.2\times 10^{-4})}{6.0\times 10^{10}} = \dfrac{(4.0)(1.2)\times 10^3 10^{-4}}{6.0\times 10^{10}}$

$= \dfrac{4.8\times 10^{-1}}{6.0\times 10^{10}}$

$= \left(\dfrac{4.8}{6.0}\right)\times\dfrac{10^{-1}}{10^{10}}$

$= 0.8\times 10^{-1-10}$

$= 0.8\times 10^{-11}$

$= (8.0\times 10^{-1})\times 10^{-11}$

$= 8.0\times 10^{-12}$

45. $(5p^3 + 2p - 3) + (8p^3 - 4p^2 + 14)$

$= 5p^3 + 2p - 3 + 8p^3 - 4p^2 + 14$

$= 5p^3 + 8p^3 - 4p^2 + 2p - 3 + 14$

$= 13p^3 - 4p^2 + 2p + 11$

47. $\left(10n^2 - \dfrac{3}{8}n + 2\right) - \left(11n^2 + \dfrac{5}{8}n - 6\right)$

$= 10n^2 - \dfrac{3}{8}n + 2 - 11n^2 - \dfrac{5}{8}n + 6$

$= 10n^2 - 11n^2 - \dfrac{3}{8}n - \dfrac{5}{8}n + 2 + 6$

$= -n^2 - n + 8$

49. $(-u^2v+6u^2-2uv^2)+(11u^2+7uv^2)-(9u^2v+u^2+3uv^2)$
$=-u^2v+6u^2-2uv^2+11u^2+7uv^2-9u^2v-u^2-3uv^2$
$=-u^2v-9u^2v+6u^2+11u^2-u^2-2uv^2+7uv^2-3uv^2$
$=-10u^2v+16u^2+2uv^2$

51. $4p(p^3-5p^2+2p+8)=4p(p^3)+4p(-5p^2)+4p(2p)+4p(8)=4p^4-20p^3+8p^2+32p$

53. $(3x+y)(-2x-5y)=3x(-2x)+3x(-5y)+y(-2x)+y(-5y)$
$=-6x^2-15xy-2xy-5y^2$
$=-6x^2-17xy-5y^2$

55. $(8b+4)(2b-3)=8b(2b)+8b(-3)+4(2b)+4(-3)=16b^2-24b+8b-12=16b^2-16b-12$

57. $(x-3y)(x^2-3xy+5y^2)=x(x^2)+x(-3xy)+x(5y^2)+(-3y)(x^2)+(-3y)(-3xy)+(-3y)(5y^2)$
$=x^3-3x^2y+5xy^2-3x^2y+9xy^2-15y^3$
$=x^3-3x^2y-3x^2y+5xy^2+9xy^2-15y^3$
$=x^3-6x^2y+14xy^2-15y^3$

59. $(3h-8)(3h+8)=(3h)^2-(8)^2=9h^2-64$

61. $(4x-5)^2=(4x)^2-2(4x)(5)+(5)^2=16x^2-40x+25$

63. $\left(\frac{1}{3}t^2-9\right)\left(\frac{1}{3}t^2+9\right)=\left(\frac{1}{3}t^2\right)^2-(9)^2=\frac{1}{9}t^4-81$

65. $(0.2x^2-3)^2=(0.2x^2)^2-2(0.2x^2)(3)+(3)^2=0.04x^4-1.2x^2+9$

67. $\frac{6a^3b^2-18a^2b^2+3ab^2-9ab}{3ab}=\frac{6a^3b^2}{3ab}-\frac{18a^2b^2}{3ab}+\frac{3ab^2}{3ab}-\frac{9ab}{3ab}=2a^2b-6ab+b-3$

69.
$$\begin{array}{r|l} & x^2-5x+11 \\ \hline x+3 & x^3-2x^2-4x+33 \\ & \underline{-(x^3+3x^2)} \\ & -5x^2-4x \\ & \underline{-(-5x^2-15x)} \\ & 11x+33 \\ & \underline{-(11x+33)} \\ & 0 \end{array}$$

$x^2-5x+11$

71.

$$
\begin{array}{r}
2p^2-7p+4 \\
p-5\overline{\big)\,2p^3-17p^2+39p-10} \\
\underline{-(2p^3-10p^2)} \\
-7p^2+39p \\
\underline{-(-7p^2+35p)} \\
4p-10 \\
\underline{-(4p-20)} \\
10
\end{array}
$$

$$2p^2-7p+4+\frac{10}{p-5}$$

73. $\dfrac{10m^5-16m^4+8m^3+8m^2-2m}{-2m}$

$$=\frac{10m^5}{-2m}-\frac{16m^4}{-2m}+\frac{8m^3}{-2m}+\frac{8m^2}{-2m}-\frac{2m}{-2m}$$
$$=-5m^4+8m^3-4m^2-4m+1$$

75.

$$
\begin{array}{r}
4a^2+6a+9 \\
2a-3\overline{\big)\,8a^3+0a^2+0a-9} \\
\underline{-(8a^3-12a^2)} \\
12a^2+0a \\
\underline{-(12a^2-18a)} \\
18a-9 \\
\underline{-(18a-27)} \\
18
\end{array}
$$

$$4a^2+6a+9+\frac{18}{2a-3}$$

Review E Practice Exercises

1. $a^2-b^2=(a+b)(a-b)$

3. $a^3+b^3=(a+b)(a^2-ab+b^2)$

5. **(a)** Difference of squares

(b) $t^2-100=t^2-10^2=(t-10)(t+10)$

7. **(a)** Sum of cubes

(b) $y^3+27=y^3+3^3=(y+3)(y^2-3y+9)$

9. **(a)** Trinomial (nonperfect square trinomial)

(b) $d^2+3d-28=(d+7)(d-4)$

11. **(a)** Trinomial (perfect square trinomial)

(b) $x^2-12x+36=x^2-2(x)(6)+(6)^2$
$=(x-6)^2$

13. **(a)** Four terms-grouping

(b) $2ax^2-5ax+2bx-5b$
$=ax(2x-5)+b(2x-5)$
$=(ax+b)(2x-5)$

15. **(a)** Trinomial (nonperfect square trinomial)

(b) $10y^2+3y-4=(2y-1)(5y+4)$

17. **(a)** Difference of squares

(b) $10p^2-640=10(p^2-64)$
$=10(p-8)(p+8)$

19. **(a)** Difference of cubes

(b) $z^4-64z=z(z^3-64)$
$=z(z-4)(z^2+4z+16)$

21. **(a)** Trinomial (nonperfect square trinomial)

(b) $b^3-4b^2-45b=b(b^2-4b-45)$
$=b(b-9)(b+5)$

23. **(a)** Trinomial (perfect square trinomial)

(b) $9w^2+24wx+16x^2$
$=(3w)^2+2(3w)(4x)+(4x)^2$
$=(3w+4x)^2$

25. **(a)** Four terms-grouping

(b) $60x^2-20x+30ax-10a$
$=10(6x^2-2x+3ax-a)$
$=10[2x(3x-1)+a(3x-1)]$
$=10(2x+a)(3x-1)$

27. **(a)** Four terms-grouping

(b) $x^3+4x^2-9x-36=x^2(x+4)-9(x+4)$
$=(x^2-9)(x+4)$
$=(x-3)(x+3)(x+4)$

29. **(a)** Difference of squares

(b) $w^4-16=(w^2-4)(w^2+4)$
$=(w-2)(w+2)(w^2+4)$

31. **(a)** Difference of cubes

(b) $t^6-8=(t^2)^3-2^3$
$=(t^2-2)(t^4+2t^2+4)$

33. **(a)** Trinomial (nonperfect square trinomial)

(b) $8p^2-22p+5=(4p-1)(2p-5)$

35. **(a)** Trinomial (perfect square trinomial)

(b) $36y^2-12y+1=(6y)^2-2(6y)(1)+(1)^2$
$=(6y-1)^2$

37. **(a)** Four terms-grouping

(b) $(x^2+4x+4)-y^2$
$=(x+2)^2-y^2$
$=(x+2-y)(x+2+y)$

39. **(a)** Sum of squares

(b) $2x^2+50=2(x^2+25)$

41. **(a)** Trinomial (nonperfect square trinomial)

(b) $12r^2s^2+7rs^2-10s^2$
$=s^2(12r^2+7r-10)$
$=s^2(4r+5)(3r-2)$

43. **(a)** Trinomial (nonperfect square trinomial)

(b) $x^2+8xy+33y^2=(x+3y)(x+11y)$

45. **(a)** Sum of cubes

(b) $m^6+n^3=(m^2)^3+n^3$
$=(m^2+n)(m^4-m^2n+n^2)$

47. **(a)** Trinomial (nonperfect square trinomial)

(b) $x^2(a+b)-x(a+b)-12(a+b)$
$=(x^2-x-12)(a+b)$
$=(x-4)(x+3)(a+b)$

49. **(a)** None of these

(b) $x^2-4x=x(x-4)$

51. $(2x+5)(x-3)=0$
$2x+5=0$ or $x-3=0$
$2x=-5$ or $x=3$
$x=-\frac{5}{2}$ or $x=3$

53. $x(x-5)=0$
$x=0$ or $x-5=0$
$x=0$ or $x=5$

55. $5(w+3)=0$
$5=0$ or $w+3=0$
no solution $w=-3$
The solution is $w=-3$.

57. $z^2-2z-8=0$
$(z-4)(z+2)=0$
$z-4=0$ or $z+2=0$
$z=4$ or $z=-2$

59. $6x^2-7x=5$
$6x^2-7x-5=0$
$(3x-5)(2x+1)=0$
$3x-5=0$ or $2x+1=0$
$3x=5$ or $2x=-1$
$x=\frac{5}{3}$ or $x=-\frac{1}{2}$

61.
$$2x(x-10)=-9x-12$$
$$2x^2-20x=-9x-12$$
$$2x^2-11x+12=0$$
$$(2x-3)(x-4)=0$$
$$2x-3=0 \quad \text{or} \quad x-4=0$$
$$2x=3 \quad \text{or} \quad x=4$$
$$x=\frac{3}{2} \quad \text{or} \quad x=4$$

63.
$$w^3-w^2-w+1=0$$
$$w^2(w-1)-1(w-1)=0$$
$$(w^2-1)(w-1)=0$$
$$(w-1)(w+1)(w-1)=0$$
$$w-1=0 \quad \text{or} \quad w+1=0 \quad \text{or} \quad w-1=0$$
$$w=1 \quad \text{or} \quad w=-1 \quad \text{or} \quad w=1$$
$$w=1 \quad \text{or} \quad w=-1$$

Review F Practice Exercises

1. $\frac{x+2}{x+4}$

The denominator is zero when

$x+4=0$

$x=-4$

The domain is $\{x \mid x \neq -4\}$.

3. $\frac{2x}{25x^2-9}$

The denominator is zero when
$$25x^2-9=0$$
$$(5x-3)(5x+3)=0$$
$$5x-3=0 \quad \text{or} \quad 5x+3=0$$
$$x=\frac{3}{5} \quad \text{or} \quad x=-\frac{3}{5}$$

The domain is $\left\{x \,\middle|\, x \neq \frac{3}{5} \text{ and } x \neq -\frac{3}{5}\right\}$.

5. $\frac{t-7}{8}$

The denominator is never zero.

The domain is $\{t \mid t \text{ is any real number}\}$.

7. $\frac{3x^4y^7}{12xy^8}=\frac{x^3}{4y}$

9. $\frac{t^2-4}{2t-4}=\frac{(t-2)(t+2)}{2(t-2)}=\frac{t+2}{2}$

11. $\frac{2y^2+5y-12}{2y^2+y-6}=\frac{(2y-3)(y+4)}{(2y-3)(y+2)}=\frac{y+4}{y+2}$

13. $\frac{2x}{15y^2}\cdot\frac{3y^5}{4x^2}=\frac{xy^5}{5\cdot 2y^2x^2}=\frac{y^3}{10x}$

15.
$$\frac{5y-15}{10y+40}\cdot\frac{2y^2+y-28}{2y^2-13y+21}$$
$$=\frac{5(y-3)}{10(y+4)}\cdot\frac{(2y-7)(y+4)}{(2y-7)(y-3)}$$
$$=\frac{1}{2}$$

17. $\frac{a^2b^3}{5c^2}\div\frac{ab}{15c^3}=\frac{a^2b^3}{5c^2}\cdot\frac{15c^3}{ab}=3ab^2c$

19.
$$\frac{p^2-36}{2p-4}\div\frac{2p+12}{p^2-2p}$$
$$=\frac{p^2-36}{2p-4}\cdot\frac{p^2-2p}{2p+12}$$
$$=\frac{(p-6)(p+6)}{2(p-2)}\cdot\frac{p(p-2)}{2(p+6)}$$
$$=\frac{p(p-6)}{4}$$

21.
$$\frac{t^2+7t}{3-t}\div\frac{t^2-49}{t^2-3t}=\frac{t^2+7t}{3-t}\cdot\frac{t^2-3t}{t^2-49}$$
$$=\frac{t(t+7)}{(-1)(t-3)}\cdot\frac{t(t-3)}{(t-7)(t+7)}$$
$$=-\frac{t^2}{t-7}$$

23.
$$\frac{1}{4x^3y^7}=\frac{1}{2^2x^3y^7}$$
$$\frac{1}{8xy^{10}}=\frac{1}{2^3xy^{10}}$$
$$\text{LCD}=2^3x^3y^{10}=8x^3y^{10}$$

25. $\dfrac{1}{x^2-x-12}=\dfrac{1}{(x-4)(x+3)}$

$\dfrac{1}{x^2-9}=\dfrac{1}{(x-3)(x+3)}$

$\text{LCD} = (x-3)(x+3)(x-4)$

27. $\dfrac{1}{2x}-\dfrac{5}{6x}=\dfrac{1\cdot 3}{2x\cdot 3}-\dfrac{5}{6x}=\dfrac{3}{6x}-\dfrac{5}{6x}=\dfrac{-2}{6x}=-\dfrac{1}{3x}$

29. $\dfrac{1}{2x^3y}+\dfrac{5}{4xy^2}=\dfrac{1\cdot 2y}{2x^3y\cdot 2y}+\dfrac{5\cdot x^2}{4xy^2\cdot x^2}=\dfrac{2y}{4x^3y^2}+\dfrac{5x^2}{4x^3y^2}=\dfrac{5x^2+2y}{4x^3y^2}$

31. $\dfrac{x^2}{x-7}-\dfrac{14x-49}{x-7}=\dfrac{x^2-(14x-49)}{x-7}=\dfrac{x^2-14x+49}{x-7}=\dfrac{(x-7)^2}{x-7}=x-7$

33. $\dfrac{7x}{4x-2}+\dfrac{5x}{2x-1}=\dfrac{7x}{2(2x-1)}+\dfrac{5x}{2x-1}=\dfrac{7x}{2(2x-1)}+\dfrac{5x\cdot 2}{(2x-1)\cdot 2}=\dfrac{7x+10x}{2(2x-1)}=\dfrac{17x}{2(2x-1)}$

35.
$$\begin{aligned}\frac{x}{x^2+x-12}-\frac{2}{x^2+3x-4}&=\frac{x}{(x+4)(x-3)}-\frac{2}{(x+4)(x-1)}\\&=\frac{x(x-1)}{(x+4)(x-3)(x-1)}-\frac{2(x-3)}{(x+4)(x-1)(x-3)}\\&=\frac{x(x-1)-2(x-3)}{(x+4)(x-3)(x-1)}\\&=\frac{x^2-x-2x+6}{(x+4)(x-3)(x-1)}\\&=\frac{x^2-3x+6}{(x+4)(x-3)(x-1)}\end{aligned}$$

37.
$$\begin{aligned}\frac{-3}{y-2}+\frac{2y+11}{y^2+y-6}-\frac{2}{y+3}&=\frac{-3}{y-2}+\frac{2y+11}{(y+3)(y-2)}-\frac{2}{y+3}\\&=\frac{-3(y+3)}{(y-2)(y+3)}+\frac{2y+11}{(y+3)(y-2)}-\frac{2(y-2)}{(y+3)(y-2)}\\&=\frac{-3(y+3)+(2y+11)-2(y-2)}{(y-2)(y+3)}\\&=\frac{-3y-9+2y+11-2y+4}{(y-2)(y+3)}\\&=\frac{-3y+6}{(y-2)(y+3)}\\&=\frac{-3(y-2)}{(y-2)(y+3)}\\&=\frac{-3}{y+3}\end{aligned}$$

39. $\frac{5}{x-3}-\frac{x+2}{x-3}=\frac{5-(x+2)}{x-3}=\frac{5-x-2}{x-3}=\frac{3-x}{x-3}=\frac{(-1)(x-3)}{x-3}=-1$

41. $\frac{\frac{3}{a}+\frac{4}{b}}{\frac{7}{a}-\frac{1}{b}}=\frac{ab\left(\frac{3}{a}+\frac{4}{b}\right)}{ab\left(\frac{7}{a}-\frac{1}{b}\right)}=\frac{ab\left(\frac{3}{a}\right)+ab\left(\frac{4}{b}\right)}{ab\left(\frac{7}{a}\right)-ab\left(\frac{1}{b}\right)}=\frac{3b+4a}{7b-a}$

43. $\frac{8x-\frac{1}{2x}}{1-\frac{1}{4x}}=\frac{4x\left(8x-\frac{1}{2x}\right)}{4x\left(1-\frac{1}{4x}\right)}=\frac{4x(8x)-4x\left(\frac{1}{2x}\right)}{4x(1)-4x\left(\frac{1}{4x}\right)}=\frac{32x^2-2}{4x-1}=\frac{2(16x^2-1)}{4x-1}=\frac{2(4x-1)(4x+1)}{4x-1}=2(4x+1)$

45. $\frac{\frac{u^2-v^2}{uv}}{\frac{u+v}{uv}}=\frac{uv\left(\frac{u^2-v^2}{uv}\right)}{uv\left(\frac{u+v}{uv}\right)}=\frac{u^2-v^2}{u+v}=\frac{(u-v)(u+v)}{u+v}=u-v$

47.
$$\frac{\frac{1}{y-5}}{\frac{2}{y+5}+\frac{1}{y^2-25}}=\frac{\frac{1}{y-5}}{\frac{2}{y+5}+\frac{1}{(y-5)(y+5)}}$$
$$=\frac{(y-5)(y+5)\left(\frac{1}{y-5}\right)}{(y-5)(y+5)\left[\frac{2}{y+5}+\frac{1}{(y-5)(y+5)}\right]}$$
$$=\frac{y+5}{(y-5)(y+5)\left(\frac{2}{y+5}\right)+(y-5)(y+5)\left[\frac{1}{(y-5)(y+5)}\right]}$$
$$=\frac{y+5}{2(y-5)+1}$$
$$=\frac{y+5}{2y-9}$$

49.
$$\frac{1}{8}-\frac{3}{4x}=\frac{1}{2x}$$
$$8x\left(\frac{1}{8}-\frac{3}{4x}\right)=8x\left(\frac{1}{2x}\right)$$
$$8x\left(\frac{1}{8}\right)-8x\left(\frac{3}{4x}\right)=8x\left(\frac{1}{2x}\right)$$
$$x-6=4$$
$$x=10$$

Check: $\frac{1}{8}-\frac{3}{4(10)}=\frac{1}{2(10)}$

$$\frac{1}{8}-\frac{3}{40}=\frac{1}{20}$$
$$\frac{1}{20}=\frac{1}{20}\quad\text{True}$$

51.
$$\frac{5}{x-3}=\frac{2}{x-3}$$
$$(x-3)\left(\frac{5}{x-3}\right)=(x-3)\left(\frac{2}{x-3}\right)$$
$$5=2$$
No solution

53.
$$1+\frac{6}{x}=-\frac{8}{x^2}$$
$$x^2\left(1+\frac{6}{x}\right)=x^2\left(-\frac{8}{x^2}\right)$$
$$x^2(1)+x^2\left(\frac{6}{x}\right)=x^2\left(-\frac{8}{x^2}\right)$$
$$x^2+6x=-8$$
$$x^2+6x+8=0$$
$$(x+4)(x+2)=0$$
$$x+4=0 \quad \text{or} \quad x+2=0$$
$$x=-4 \quad \text{or} \quad x=-2$$
Check $x=3$: $1=\frac{8}{(3)}-\frac{15}{(3)^2}$
$$1=\frac{8}{3}-\frac{15}{9}$$
$1=1$ True

Check $x=5$: $1=\frac{8}{(5)}-\frac{15}{(5)^2}$
$$1=\frac{8}{5}-\frac{15}{25}$$
$1=1$ True

55.
$$\frac{4x+11}{x^2-4}-\frac{5}{x+2}=\frac{2x}{x^2-4}$$
$$(x-2)(x+2)\left[\frac{4x+11}{(x-2)(x+2)}-\frac{5}{x+2}\right]=(x-2)(x+2)\left[\frac{2x}{(x-2)(x+2)}\right]$$
$$(x-2)(x+2)\left[\frac{4x+11}{(x-2)(x+2)}\right]-(x-2)(x+2)\left(\frac{5}{x+2}\right)=(x-2)(x+2)\left[\frac{2x}{(x-2)(x+2)}\right]$$
$$4x+11-5(x-2)=2x$$
$$4x+11-5x+10=2x$$
$$-x+21=2x$$
$$-3x+21=0$$
$$-3x=-21$$
$$x=7$$

Check: $\frac{4(7)+11}{(7)^2-4}-\frac{5}{(7)+2}=\frac{2(7)}{(7)^2-4}$

$$\frac{39}{45}-\frac{5}{9}=\frac{14}{45}$$

$$\frac{14}{45}=\frac{14}{45} \quad \text{True}$$

Notes

Notes

Notes

Notes

Notes

Notes

Notes

Notes

Notes

Notes